Solid State Ionics—2002

MATERIALS RESEARCH SOCIETY
SYMPOSIUM PROCEEDINGS VOLUME 756

Solid State Ionics—2002

Symposium held December 2–5, 2002, Boston, Massachusetts, U.S.A.

EDITORS (SYMPOSIUM EE):

Philippe Knauth
Université de Provence
Marseille, France

Jean-Marie Tarascon
Université de Picardie
Amiens, France

Enrico Traversa
Università di Roma "Tor Vergata"
Rome, Italy

Harry L. Tuller
Massachusetts Institute of Technology
Cambridge, Massachusetts, U.S.A.

ORGANIZERS (SYMPOSIUM FF):

Hubert A. Gasteiger
GM Global R&D
Honeoye Falls, New York, U.S.A.

Larry Pederson
Pacific Northwest Laboratory
Richland, Washington, U.S.A.

Brant A. Peppley
Royal Military College of Canada
Kingston, Ontario, Canada

Levi T. Thompson
University of Michigan
Ann Arbor, Michigan, U.S.A.

Thomas Zawodzinski
Los Alamos National Laboratory
Los Alamos, New Mexico, U.S.A.

Materials Research Society
Warrendale, Pennsylvania

CAMBRIDGE
UNIVERSITY PRESS

32 Avenue of the Americas, New York NY 10013-2473, USA

Cambridge University Press is part of the University of Cambridge.

It furthers the University's mission by disseminating knowledge in the pursuit of education, learning and research at the highest international levels of excellence.

www.cambridge.org
Information on this title: www.cambridge.org/9781558996939

CODEN: MRSPDH

Copyright 2003 by Materials Research Society.

ISBN 978-1-558-99693-9 Hardback

CONTENTS

OXIDE ELECTROCERAMICS FOR
SEPARATION MEMBRANES AND
GAS SENSORS

*Invited Paper

CATHODE MATERIALS FOR
LITHIUM BATTERIES

*Invited Paper

PROTON EXCHANGE MEMBRANE
FUEL CELLS

*Invited Paper

*Invited Paper

SOLID OXIDE FUEL CELLS

*Invited Paper

PREFACE

This proceedings volume includes invited and contributed papers presented at Symposium EE, "Solid State Ionics," and Symposium FF, "Materials for Fuel Cells and Fuel Processors," held December 2–5 at the 2002 MRS Fall Meeting in Boston, Massachusetts. It gives an account of current developments in Solid State Ionics and highlights advances in the domains of energy storage and conversion and environmental monitoring. The largest clusters of papers are devoted to lithium batteries and fuel cell technology. This reflects the dominating relevance of these applications for fostering the modern telecommunications and information society (energy storage systems for laptops and portable telephones) and the development of environmentally friendly electric transportation and energy transformation systems, replacing traditional combustion technology. One can securely predict that fundamental and applied research in Solid State Ionics will continue to grow at a fast pace in the coming years.

Part I of the proceedings is devoted to "Theory/Inorganic Ion Conductors." The invited paper by K. Funke treats the ion dynamics in disordered solid electrolytes in the framework of the MIGRATION concept. The frequency dependence of the ionic conductivity, but also, for the first time, of the dielectric permittivity, can now be theoretically described. Several papers demonstrate the great potential of simulation techniques (Monte-Carlo, ab-initio...) for achieving an improved understanding of the mechanisms of ionic motion in solids. The new domain of "nano-ionics," i.e. ionic conduction in nanostructured materials, is also addressed. A. Tschöpe shows in an invited paper how space charge theory provides, for the first time, a quantitative description of the grain-size dependent electrical conductivity of CeO_2. The invited paper by T.O. Mason and coworkers gives a theoretical analysis of impedance spectra of nanocrystalline electroceramics and discusses the limits of the classical brick layer model. Three papers are devoted to ion conduction properties in the $(La,Li)TiO_3$ perovskite system. Furthermore, tunnel compounds, cation conducting composites with mesoporous Al_2O_3, an unusual cation conduction mechanism in $LiNaSO_4$ and a new metastable fluoride ion conductor, $CaSn_2F_6$, are presented.

Part II discusses "Oxide Electroceramics for Separation Membranes and Gas Sensors." The bismuth-base materials, especially the BiMeVOx system, are discussed in the invited paper by R.-N. Vannier et al. Various materials for oxygen permeation membranes are considered in contributed papers, including La-Sr-Co-Fe oxides, Ba-In oxides and oxide composites. The invited paper by E. Di Bartolomeo describes potentiometric NOx sensors based on stabilized zirconia and various oxides as auxiliary electrode. Contributed papers describe NOx sensors with other metal oxides, and oxides that could be used in resistive oxygen sensors, especially doped $SrTiO_3$ and pure and doped CeO_2. A new type of resonant device, based on the high temperature piezoelectric langasite, transduces a change of the oxygen stoichiometry of a thin-film oxide into a change of the resonance frequency. This is the high temperature analogue of the well-known quartz microbalance. Finally, zeolites are investigated as new sensor materials, the preparation of TiO_2 (anatase) nanoceramics and the phase diagram of the In_2O_3-WO_3 system, useful for electrochromics, are presented.

Part III presents "Cathode Materials for Lithium Batteries." Conventional wisdom was that only certain intercalation compounds can insert and de-insert lithium ions and can be used as anode and cathode materials. Recent experiments show, however, that simple binary compounds, hitherto neglected in the search for better cathode and anode materials, can reversibly exchange lithium ions if the particle size is sufficiently small. This is a major discovery, implying that a large number of compounds have to be checked again from this point of view. In his invited paper, G. Amatucci et al. investigate such a compound, FeF_3, as new high performance cathode material. Several contributed papers are devoted to the "classical" cathode materials—manganese dioxide, manganese spinel and the $Li(Co,Ni)O_2$ system—describing structural and chemical disorder and advanced preparation techniques. Other papers discuss transition metal phosphates as well as vanadium-oxide and molybdenum-oxide based cathode materials and characterization of the electrochemical cycling process by sophisticated structural techniques, such as EXAFS and XANES or Scanning Probe Microscopy.

Part IV presents "Anode Materials for Lithium Batteries and Polymer Electrolytes." A wide range of anode materials is described, ranging from nitrides, over silicon to metals, such as tin, and mesoporous silicates. Ball-milling is presented as an alternative way for anode preparation. In the context of lithium batteries and fuel cells, polymer-based nanostructured and nanocomposite solid electrolytes are an important subject. The use of functionalized oxide nanoparticles for the preparation of polymer nanocomposites for solid electrolytes and biomedical applications is described in several papers. Ionic conduction mechanisms in PEO and PEO-clay composites are investigated by Nuclear Magnetic Resonance and Neutron Scattering.

Part V treats "Proton Exchange Membrane Fuel Cells." A particular important aspect that is addressed concerns the catalytic properties of electrode materials, given the importance of *in situ* reforming for this kind of device. The invited paper by Ball and Thompsett reports the development of CO-tolerant Pt-based electrode materials. Contributed papers present strategies for CO removal, other electrode materials and electrocatalysts, particularly based on composites and nanoporous carbon. Optimization of a carbon composite bipolar plate is discussed in the invited paper by Besmann and coworkers. Other topics include hydrogen separation, mechanistic investigations by x-ray absorption spectroscopy and the development of a microbial fuel cell.

Part VI is devoted to "Solid Oxide Fuel Cells" (SOFC). The reports presented at the symposium suggest that the breakthrough to commercial use is not far away, given especially impressive improvements in the mechanical properties of solid oxide membranes. The invited paper by H.-D. Wiemhöfer gives an overview on advanced electrical characterization for oxygen ion conductors, including ceria and lanthanum gallates. Contributed papers treat the preparation and properties of apatites as new solid oxygen ion conductors, Gd-doped CeO_2, Aurivillius phases, and the doped LSM system. Several articles are devoted to the classical solid electrolyte for SOFC—Y-stabilized ZrO_2 in thin-film or bulk form—and discuss the influence of interfaces on the ion conduction properties. The possibility of spurious substrate effects on thin-film conductivity is pointed out. Various other aspects of SOFC developments are also addressed, such as lanthanum cobaltite electrode materials, multilayer anodes, braze development and corrosion problems of metal interconnects. Given that the commercial application is not far away, the last paper presents a cost model for SOFC.

We acknowledge the contributions of all of the authors of this proceedings volume, which provide an excellent overview of the current status of "Solid State Ionics—2002." Financial support of the symposium by Centre National de la Recherche Scientifique, Robert Bosch AG (Stuttgart, Germany) and SETNAG (Marseille, France), is greatly appreciated.

Philippe Knauth
Jean-Marie Tarascon
Enrico Traversa
Harry L. Tuller

February 2003

MATERIALS RESEARCH SOCIETY SYMPOSIUM PROCEEDINGS

MATERIALS RESEARCH SOCIETY SYMPOSIUM PROCEEDINGS

Prior Materials Research Society Symposium Proceedings available by contacting Materials Research Society

Theory/Inorganic
Ion Conductors

Mat. Res. Soc. Symp. Proc. Vol. 756 © 2003 Materials Research Society

THE *MIGRATION* CONCEPT FOR IONIC MOTION
IN MATERIALS WITH DISORDERED STRUCTURES

K. FUNKE and R.D. BANHATTI
University of Münster, Institute of Physical Chemistry and Sonderforschungsbereich 458,
Schlossplatz 4, D – 48149 Münster, Germany
E-mail: K.Funke@uni-muenster.de and banhatt@uni-muenster.de

ABSTRACT

The dynamics of the mobile ions in materials with disordered structures are a challengingly complicated many-particle process. In this paper, we consider characteristic frequency-dependent conductivities and permittivities of such materials and show that they can be well reproduced within the framework of the MIGRATION concept. The meaning of the acronym is MIsmatch Generated Relaxation for the Accommodation and Transport of IONs. In the MIGRATION concept, we attempt to grasp the essence of the ion dynamics in a simple set of rules which convey a physical picture of the most relevant elementary processes. The rules are expressed in terms of three coupled rate equations which then form the basis for deriving frequency-dependent model conductivities and permittivities.

I INTRODUCTION

Ion conducting materials with disordered structures comprise glasses as well as structurally disordered crystalline electrolytes. The frequency-dependent electric and dielectric properties of such materials are determined by the hopping dynamics of the mobile ions. Therefore, valuable information on the elementary hopping processes and, in particular, on correlations between hops are obtainable from a study of the frequency-dependent conductivity [1], $\sigma(\omega)$, and the (relative) frequency-dependent permittivity, $\varepsilon(\omega)$. The two functions are the constituent parts of the complex conductivity, $\hat{\sigma}(\omega) = \sigma(\omega) + i\omega\varepsilon_0\varepsilon(\omega)$. Here, ω denotes the angular frequency, while ε_0 is the permittivity of free space.

In the following, we consider data only below microwave frequencies, i.e., in a frequency range where $\sigma(\omega)$ has to be attributed entirely to the hopping motion of the ions, while vibrational contributions to $\sigma(\omega)$ may still be neglected. In this frequency range, vibrations and fast ionic polarizations as well as even faster (electronic) processes contribute a constant value, $\varepsilon(\infty)$, to $\varepsilon(\omega)$, while the remaining part, $\varepsilon(\omega) - \varepsilon(\infty)$, results from the hopping motion. Sometimes, we will use the notation $\hat{\sigma}_{HOP}(\omega) = \hat{\sigma}(\omega) - i\omega\varepsilon_0\varepsilon(\infty)$.

A surprising experimental observation is made when conductivity spectra, $\sigma(\omega)$, of different fast ion conducting materials, i.e. of structurally disordered crystals and glassy

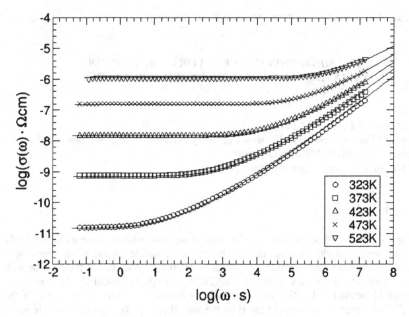

Fig. 1a: Conductivity isotherms of $0.2Na_2O \cdot 0.8GeO_2$ glass. Fits are made using the CMR model (Eq.(9)).

electrolytes, are compared with each other. In fact, many of them are found to be virtually identical in shape, not only for a given material at different temperatures, but even for different materials. This is most clearly seen when a *scaled* conductivity is plotted versus a *scaled* angular frequency. This property of scaling is known as the time-temperature superposition principle [2-12]. In our notation, $\sigma_s(\omega_s)$, the scaled conductivity is meant to be $\sigma_s(\omega) = \sigma(\omega)/\sigma(0)$, and the scaled angular frequency is $\omega_s = \omega/\omega_0$. Here, ω_0 marks the onset of the conductivity dispersion on the ω scale. A quantitative definition will be given later, see Eq. (12).

Figures 1a and 1b are, respectively, plots of unscaled and scaled ionic conductivities of a particular glassy ion conductor, namely, a sodium germanate glass of composition $0.2Na_2O \cdot 0.8GeO_2$ [8,13]. Virtually the same scaled curve, $\sigma_s(\omega_s)$, as in Fig. 1b has also been obtained for many other glassy and crystalline fast ion conductors [2-12]. Therefore, this "master curve" must be considered representative.

Recently, a model concept has been introduced which is able to explain experimental data such as those of Fig. 1 on the basis of two simple coupled rate equations. The model rate equations are meant to grasp the essence of the rules that determine the dynamics of the mobile ions. This model, called the concept of mismatch and relaxation (CMR) [14], will be outlined in section III. In fact, the solid lines in Figs. 1a and 1b, which reproduce the experimental conductivities very well, have been derived from the CMR.

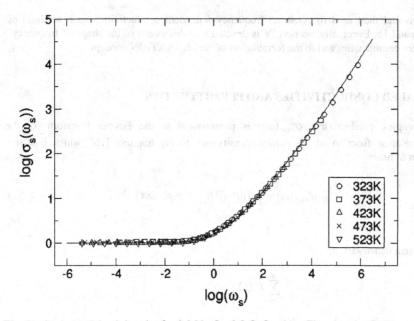

Fig. 1b: Scaled conductivity plot for 0.2 Na$_2$O · 0.8 GeO$_2$ glass. Fit using the CMR model with exponent K=2 (see Eq.(10)).

The main purpose of the present contribution is, however, to extend our model treatment to include also the frequency dependence of the permittivity, $\varepsilon(\omega)$. As will be shown in sections II and IV, it is useful to introduce and discuss a scaled version of this function, $\varepsilon_s(\omega_s)$, in analogy to $\sigma_s(\omega_s)$. In section IV, we will further see that the extreme long-time/low-frequency behavior inherent in the CMR does not match the low-frequency behavior of $\varepsilon(\omega)$ and, in particular, does not reproduce its dc plateau, $\varepsilon(0)$. Therefore, we have to reformulate our model. We do so by suggesting a subtle change which has a clear physical meaning and becomes significant only at long times. This change hardly affects the results obtained for $\sigma(\omega)$ and $\sigma_s(\omega_s)$, but yields the proper long-time behavior determining the shape of $\varepsilon(\omega)$ at low frequencies.

At the advanced stage to be presented in section IV, our model is called the MIGRATION concept, the acronym standing for MIsmatch Generated Relaxation for the Accommodation and Transport of IONs. In the MIGRATION concept, the dynamics of the mobile ions are described in terms of three coupled rate equations (instead of two in the CMR). Experimental spectra, $\sigma(\omega)$ and $\varepsilon(\omega)$, as well as the scaled functions, $\sigma_s(\omega_s)$ and $\varepsilon_s(\omega_s)$, are well reproduced by the MIGRATION concept.

This paper is organized as follows. In section II, the time-dependent correlation factor, $W(t)$ [15], is introduced in order to provide an effective means to express both unscaled and scaled conductivities and permittivities. In section III, $W(t)$ is obtained from the CMR rate equations and then used to construct frequency-dependent conductivities such as those of Figs. 1a and 1b. Eventually, section IV is devoted to a discussion of the shape of frequency-dependent permittivities and to the formulation of the MIGRATION concept.

II SCALED CONDUCTIVITIES AND PERMITTIVITIES

The complex conductivity, $\hat{\sigma}_{HOP}(\omega)$, is proportional to the Fourier transform of the autocorrelation function of the current density due to the hopping [16], which is a real function of time:

$$\hat{\sigma}_{HOP}(\omega) \propto \int_0^\infty \langle \vec{i}(0) \cdot \vec{i}(t) \rangle_{HOP} \cdot \exp(-i\omega t) \cdot dt . \qquad (1)$$

The current density itself,

$$\vec{i}(t) = \frac{q}{V} \sum_{i=1}^N \vec{v}_i(t) , \qquad (2)$$

is proportional to the ionic charge, q, inversely proportional to the volume of the sample, V, and proportional to the sum of the velocities of the mobile ions, $\vec{v}_i(t)$. Therefore, its autocorrelation function contains a large number of cross terms, $\langle \vec{v}_i(0) \cdot \vec{v}_j(t) \rangle$, with $i \neq j$. In our treatment we, however, assume that their role is negligible, which implies that the Haven ratio, i.e., the factor in the Nernst-Einstein relation, is close to unity [17]. In this case, $\hat{\sigma}_{HOP}(\omega)$ will be proportional to the Fourier transform of the velocity autocorrelation function, $\langle \vec{v}(0) \cdot \vec{v}(t) \rangle_{HOP}$. In order to derive realistic conductivity spectra, we have to envisage a shape of this function as schematically outlined in Fig. 2.

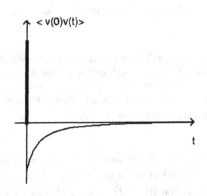

Fig.2: Schematic representation of the velocity autocorrelation function.

The intense narrow component at very short times corresponds to the velocity autocorrelation during one single hop, while the decaying negative component at longer times results from a decaying probability for a correlated "backward" hop to occur after the "initial forward" hop. The time-dependent correlation factor, $W(t)$, which is now introduced, is the normalized integral of $\langle \vec{v}(0) \cdot \vec{v}(t) \rangle_{HOP}$, with $W(0) = 1$. At long times, $W(t)$ tends to $W(\infty)$. This value, $W(\infty)$, is the fraction of "successful" hops. While all hops contribute to the ionic conductivity at sufficiently high frequencies, i.e., in the high-frequency plateau, $\sigma = \sigma(\infty)$, only the successful ones contribute to the conductivity at low frequencies, i.e., in the low-frequency plateau, $\sigma = \sigma(0)$. Note that the high-frequency plateau is not included in Figs. 1a and 1b.

The relationship between $\hat{\sigma}_{HOP}(\omega)$ and $W(t)$ is then

$$\frac{\hat{\sigma}_{HOP}(\omega)}{\sigma(\infty)} = 1 + \int_0^\infty \dot{W}(t) \cdot \exp(-i\omega t) \cdot dt, \tag{3}$$

where the dot denotes a differentiation with respect to time.

In view of experimental data such as those of Fig. 1, it is advantageous to formulate the ratio, $\hat{\sigma}_{HOP}(\omega)/\sigma(0)$, which we call $\hat{\sigma}_s(\omega)$, in terms of $W(t)$ or, rather, in terms of the scaled correlation factor, $W_s(t) = W(t)/W(\infty)$:

$$\hat{\sigma}_s(\omega) = W_s(0) + \int_0^\infty \dot{W}_s(t) \cdot \exp(-i\omega t) \cdot dt. \tag{4}$$

After replacing $\dot{W}_s(t)$ with $d(W_s(t) - 1)/dt$, integration by parts yields

$$\hat{\sigma}_s(\omega) = 1 + i\omega \int_0^\infty (W_s(t) - 1) \cdot \exp(-i\omega t) \cdot dt. \tag{5}$$

Introducing scaled times and angular frequencies via $t_s = t/t_0$ and $\omega_s = \omega/\omega_0$, with t_0 and $\omega_0 = 1/t_0$ to be defined in the next section, we find for the real part of Eq. (5):

$$\sigma_s(\omega_s) = 1 + \omega_s \int_0^\infty (W_s(t_s) - 1) \cdot \sin(\omega_s t_s) \cdot dt_s. \tag{6}$$

The experimental observation of scaling, cf. Fig. 1b, means that variations of $W_s(t_s)$ with temperature or with the material under consideration have only negligible influence on $\sigma_s(\omega_s)$. This is a remarkable statement.

In a second step, we consider a scaled permittivity,

7

$$\varepsilon_S(\omega_S) = \text{Im}\,\hat{\sigma}_S(\omega_S)/\omega_S = \frac{\varepsilon_0 \omega_0}{\sigma(0)} \cdot (\varepsilon(\omega_S) - \varepsilon(\infty)). \tag{7}$$

From the imaginary part of Eq. (5), $\varepsilon_S(\omega_S)$ is found to be

$$\varepsilon_S(\omega_S) = \int_0^\infty (W_S(t_S) - 1) \cdot \cos(\omega_S t_S) \cdot dt_S . \tag{8}$$

In a formal sense, Eq. (8) is similar to Eq. (6). The question, however, of whether or not $\varepsilon_S(\omega_S)$ will also display the property of scaling, is not easy to answer *a priori* from Eq. (8), since the temperature dependence of $W_S(t_S)$ at short scaled times will have much more effect on the cosine transform of Eq. (8) than on the sine transform of Eq. (6).

In this section, the problem of reproducing experimental (unscaled and scaled) conductivity and permittivity spectra has been reduced to the problem of finding a realistic scaled correlation factor, $W_S(t_S)$.

III THE CMR AND SCALED CONDUCTIVITY

The solid line included in Fig. 1b, which reproduces the scaled conductivity spectrum very well, has been calculated from Eqs. (9) and (6). Equation (9) is a rate equation for $W_S(t_S)$:

$$-\frac{dW_S(t_S)}{dt_S} = (W_S(t_S) \cdot \ln W_S(t_S))^2 . \tag{9}$$

It results from the CMR rate equations [14],

$$-\dot{g}(t) = A \cdot g^2(t) \cdot W(t) \tag{10}$$

$$-\dot{W}(t) = -B \cdot \dot{g}(t) \cdot W(t) , \tag{11}$$

and from the definition of t_S, see Eq. (13) below.

In Eqs. (10) and (11), $g(t)$ is a normalized mismatch function, see [14,15], while A and B are temperature-dependent constants. Suppose an ion (the "central" ion) performs a hop from site X to site Y at time $t = 0$. It thereby creates mismatch between its own position and the position where it would be optimally relaxed with respect to the arrangement of its neighbors.

8

In Eqs. (10) and (11) we then consider a later time, $t > 0$, and assume that the ion is now (still or again) at site Y, which is the case with probability $W(t)$. In this situation, we observe the system reducing its mismatch. This is possible either on the single-particle route, with the ion hopping back to X, or on the many-particle route, with the mobile neighbors rearranging. Equation (11) claims that the rates on these routes, $-\dot{W}(t)/W(t)$ and $-g(t)$, respectively, are proportional to each other at all times. The function $g(t)$ decays from $g(0) = 1$ to $g(\infty) = 0$, while $W(t)$ decays from $W(0) = 1$ to $W(\infty) = \exp(-B)$. Equation (11) implies that $1 + (\ln W(t))/B$ and $g(t)$ are identical at all times.

The other equation, Eq. (10), describes the decay of $g(t)$ on the many-particle route. Here it is useful to regard $g(t)$ as a normalized electric dipole moment. Its dipole field influences the hopping motion of the other mobile ions in the neighborhood, facilitating hops in their respective preferred directions. This has two consequences, which are both included in Eq. (10). In the first place, $g(t)$ will decay with time and, secondly, the dipole field becomes increasingly shielded. The rate of reduction of g at time t is proportional to the convolution of the driving force, which is g itself, and the velocity autocorrelation function of the neighboring ions. The latter function is assumed to be the same as for the "central" ion. Apart from a normalization factor, the product $g(t) \cdot W(t)$ turns out to be an excellent approximation for this convolution. The progressive shielding is taken into account by a time-dependent effective number, $N(t)$, of neighboring mobile ions which at time t still experience the dipole field. Then Eq. (10) results from the empirical, yet plausible assumption that $N(t)$ is itself proportional to $g(t)$.

Equation (9) is obtained from Eqs. (10) and (11) in two steps. In the first step, $g(t)$ is eliminated. The resulting function $W(t)$ allows us to reproduce unscaled conductivity spectra such as those of Fig. 1a via Eq. (3). In the second step, we define an onset angular frequency of the conductivity dispersion, ω_0, by

$$\omega_0 = \frac{A}{B} \cdot \exp(-B). \tag{12}$$

The inverse of the onset angular frequency, $t_0 = 1/\omega_0$, may be regarded as the time at which the hopping motion of the ions becomes random. The scaled time used in Eq. (9) is then

$$t_s = \frac{t}{t_0} = t \cdot \omega_0. \tag{13}$$

Although examples have been found where the exponent K in Eq. (10) differs from two, cf. Refs. [14,18], these are not considered in this paper. Rather, we focus on those numerous glassy and crystalline electrolytes for which the exponent K is two.

The remarkable result of this section consists in having found a unique function $W_s(t_s)$ and, therefore, a unique function $\sigma_s(\omega_s)$, which is indeed in good agreement with the experimental data of many ion-conducting materials, cf. the characteristic example of Fig. 1.

In order to transform the spectra of Fig. 1a into the scaled spectrum of Fig. 1b, the ω axis has been replaced by an ω / ω_0 axis, with $\omega_0 = (A/B) \cdot \exp(-B)$. It is interesting to note that this scaling is equivalent to the so-called Summerfeld scaling [19] where, apart from a constant, the product, $T \cdot \sigma(0)$, is used for normalizing the ω scale. The reason for the equivalence is as follows. Comparing model conductivity isotherms and experimental ones [14], we always find $A \propto \sigma(\infty)$ and, as a consequence, $A \cdot \exp(-B) \propto \sigma(0)$. Since both $\sigma(\infty)$ and $\sigma(0)$ are Arrhenius activated, the same holds true for $\sigma(0)/\sigma(\infty) = \exp(-B)$, which implies $1/B \propto T$. Therefore, forming $\omega_0 = (A/B) \cdot \exp(-B)$, we now realize that $\omega_0 \propto T \cdot \sigma(0)$ must hold, which means that our model scaling and the Summerfield scaling are equivalent. In either case, the basic assumption is that the only effect of temperature is to speed up or slow down the ionic hopping processes, while the underlying mechanism remains unchanged.

IV THE *MIGRATION* CONCEPT AND SCALED PERMITTIVITY

In Fig. 3a, we present frequency-dependent permittivities, $\varepsilon(\omega)$, of the glassy electrolyte $0.2Na_2O \cdot 0.8GeO_2$ at different temperatures. These correspond to the conductivities of Fig. 1a. Likewise, Fig. 3b is a plot of the scaled permittivity of this system, $\varepsilon_S(\omega_S)$, as constructed from the experimental data of Fig. 3a via Eq. (7). Figure 3b thus corresponds to the scaled conductivity of Fig. 1b. However, Fig. 3b should not be taken as "representative" for all materials (see below).

At this stage, two statements can be made. In the first place we note that, in the scaled representation of Fig. 3b, the permittivity data do, indeed, superimpose. Secondly, we see that $\varepsilon_S(\omega_S)$ has a temperature-independent, finite low-frequency value, $\varepsilon_S(0)$.

The solid lines included in Figs. 3a and 3b have been obtained from the MIGRATION concept, which will be presented in this section.

The necessity to replace the CMR with a more appropriate treatment, i.e. with the MIGRATION concept, arises from the following observation. Trying to explain the data of Figs. 3a and 3b on the basis of the CMR equations, Eqs. (10) and (11), we encounter a problem. In the extreme long-time limit, when $W(t)$ approaches $W(\infty)$ asymptotically, we find from these equations that both $-\dot{g}$ and $-\dot{W}$ vary with time as $1/t$. Inserting the proportionality, $-\dot{W}_S \propto 1/t$, into Eq. (4) and considering the imaginary part of this equation, we find that $\varepsilon(\omega)$, in the limit of $\omega \to 0$, diverges logarithmically, in disagreement with experimental results, which clearly prove that $\varepsilon(0)$ is finite. Therefore, Eqs. (10), (11) cannot be correct in the long-time limit.

The puzzle can be solved by a more careful and physically more realistic choice of the number function, $N(t)$, in the limit of long times. As outlined in the previous section, Eq. (10) is based on the idea that

$$-\frac{\dot{g}(t)}{g(t)} \propto W(t) \cdot N(t). \tag{14}$$

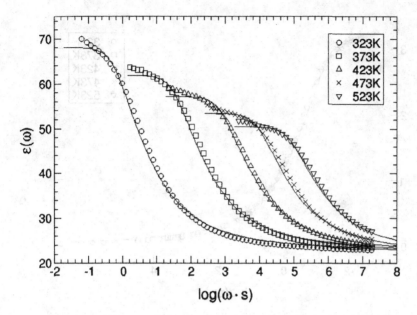

Fig. 3a: Permittivity spectra of $0.2Na_2O \cdot 0.8GeO_2$ glass at different temperatures. Note that the low frequency data of $\varepsilon(\omega)$ are affected by electrode polarization effects. At low frequencies, the permittivity due to the hopping has to be constant. Fits have been made using the treatment of the MIGRATION concept with $N(\infty) = 0.075$ (see Eq. (21)).

Therefore, $-\dot{g}(t)/g(t) \propto N(t)$ holds true at very long times, when $W(t) \cong W(\infty)$. For $\varepsilon(0)$ to be finite, we expect $g(t)$ to decay exponentially at long times. Therefore, $N(t)$ should tend to a positive value, $N(\infty) > 0$, contradicting the assumption that $N(t)$ is proportional to $g(t)$. Physically, $N(\infty) > 0$ appears very reasonable, since the nearest neighbors of the "central" ion will always experience the unshielded dipole field. Accordingly, $N(t) \propto g(t)$ is now replaced by

$$(N(t) - N(\infty)) \propto g(t). \tag{15}$$

The scaled rate equations, Eqs. (17-19), which define the MIGRATION concept, are free of any temperature dependent parameters, if we choose

$$(N(t) - N(\infty)) = B \cdot g(t) = g_s(t). \tag{16}$$

Here, $g_s(t) = \ln W_s(t)$ denotes a suitably scaled version of the mismatch function.

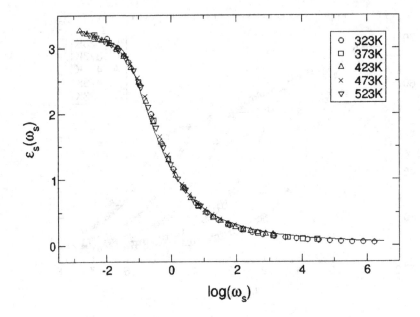

Fig. 3b: Scaled permittivity of $0.2Na_2O \cdot 0.8GeO_2$ glass as defined by Eq. (7). Solid curve obtained from the MIGRATION concept with $N(\infty) = 0.075$.

Equations (10) and (11) are now replaced by the following three rate equations:

$$-\frac{\dot{W}_s(t_s)}{W_s(t_s)} = -\dot{g}_s(t_s) \tag{17}$$

$$-\frac{\dot{g}_s(t_s)}{g_s(t_s)} = W_s(t_s) \cdot N(t_s) \tag{18}$$

$$-\frac{\dot{N}(t_s)}{N(t_s)} = W_s(t_s) \cdot (N(t_s) - N(\infty)). \tag{19}$$

Here, the dots denote differentiation with respect to scaled time. Note that Eq. (18) implies

$$-\frac{d(N(t_s) - N(\infty))/dt_s}{N(t_s) - N(\infty)} = W_s(t_s) \cdot N(t_s) = -\frac{\dot{g}_s(t_s)}{g_s(t_s)} \tag{20}$$

and is hence consistent with Eq. (16).

Combining Eqs. (16), (17) and (18) results in

$$-\frac{dW_S}{dt_S} = W_S^2 \cdot \ln W_S \cdot (\ln W_S + N(\infty)),$$ (21)

which replaces Eq. (9). The arguments outlined above, including the three rate equations and, eventually, Eq. (21) for finding the scaled time-dependent correlation factor, constitute the MIGRATION concept. For constructing scaled functions, $\sigma_S(\omega_S)$ and $\varepsilon_S(\omega_S)$, we again use Eqs. (6) and (8), now inserting $W_S(t_S)$ from Eq. (21), with $N(\infty)$ being independent of temperature.

As it turns out, the choice of $N(\infty)$ has hardly any effect on $\sigma(\omega)$ and $\sigma_S(\omega_S)$. Indeed, virtually the same conductivity spectra are obtained, when $N(\infty)$ is varied or is even zero as in the CMR. On the other hand, $\varepsilon(0)$ and $\varepsilon_S(0)$ now become finite with $\varepsilon_S(0)$ determined by $N(\infty) > 0$. Sodium germanate glasses with different compositions display different values of $\varepsilon_S(0)$. Accordingly, different values of $N(\infty)$ are required in order to reproduce the permittivity spectra of these glasses. In each glass, $\varepsilon_S(0)$ is independent of temperature. This is reflected by the temperature independence of $N(\infty)$.

Eventually, we would like to emphasize that the MIGRATION concept enables us to model electric and dielectric properties of solid electrolytes in any particular representation. This includes not only the complex conductivity, but also the complex modulus and complex impedance representations.

ACKNOWLEDGEMENTS:

It is a pleasure to thank M.D. Ingram and A. Heuer for many fruitful discussions. M.D. Ingram also helped with a critical reading of the manuscript. We would like to thank B. Roling and C. Martiny for the use of their impedance data on sodium germanate. S. Brunklaus (née S. Brueckner) is gratefully acknowledged for providing Fig. 2. Financial support from the Fonds der Chemischen Industrie is hereby acknowledged. Sonderforschungsbereich 458 is funded by the German Science Foundation, DFG.

REFERENCES:

[1] K. Funke and C. Cramer, *Curr. Opin. Solid State Mat. Sci.*, **2**, 483 (1997).
[2] H.E. Taylor, Trans. *Faraday Soc.*, **52**, 873 (1956).
[3] J.O. Isard, *J. Non-Cryst. Solids*, **4**, 357 (1970).
[4] J.R. Dyre, *J.Appl. Phys.*, **64**, 2456 (1988).
[5] H. Kahnt, *Ber. Bunsen-Ges. Phys. Chem.*, **95**, 1021 (1991).
[6] B. Roling, K. Funke, A. Happe and M.D. Ingram, *Phys. Rev. Lett.*, **78**, 2160 (1997).

[7] D.L. Sidebottom, *Phys. Rev. Lett.*, **82**, 3653 (1999).

[8] B. Roling, C. Martiny and K. Funke, *J.Non-Cryst. Solids*, **249**, 201 (1999)

[9] B. Roling, C. Martiny and S. Brueckner, *Phys. Rev. B*, **63**, 214203 (2000).

[10] K. Funke, S. Brueckner, C. Cramer and D. Wilmer, *J.Non-Cryst. Solids*, **307-310**, 921 (2002).

[11] C. Cramer, S. Brueckner, Y. Gao, K. Funke, R. Belin, G. Taillades and A. Pradel, *J.Non-Cryst. Solids*, **307-310**, 905 (2002).

[12] M. Cutroni, A. Mandanici, P. Mustarelli, C. Tomasi and M. Federico, *J. Non-Cryst. Solids*, **307-310**, 963 (2002).

[13] C. Martiny, Diplom Thesis, University of Muenster (1998).

[14] K. Funke, R.D. Banhatti, S. Brueckner, C. Cramer, C. Krieger, A. Mandanici, C. Martiny and I. Ross, *Phys. Chem. Chem. Phys.*, **4**, 3155 (2002).

[15] K. Funke, *Prog. Solid State Chem.*, **22**, 111 (1993).

[16] R. Kubo, *J.Phys. Soc. of Japan*, **12**, 570 (1957).

[17] M.P. Thomas and N.L. Peterson, *Solid State Ionics*, **14**, 297 (1984).

[18] C. Cramer, S. Brueckner, Y. Gao and K. Funke, *Phys. Chem. Chem. Phys.*, **4**, 3214 (2002).

[19] S. Summerfeld, *Philos. Mag. B*, **52**, 9 (1985).

Mat. Res. Soc. Symp. Proc. Vol. 756 © 2003 Materials Research Society   EE1.2

Bond valence analysis of ion transport in reverse Monte Carlo models of mixed alkali glasses

Stefan Adams[1] and Jan Swenson[2]
[1] GZG, Abt. Kristallographie, Universität Göttingen,
D-37077 Göttingen (Germany)
[2] Department of Applied Physics, Chalmers University of Technology,
S-412 96 Göteborg, Sweden

ABSTRACT

An analysis of RMC structure models of ion conducting glasses in terms of our bond softness sensitive bond-valence method enables us to identify the conduction pathways for a mobile ion as regions of sufficiently low valence mismatch. The strong correlation between the volume fraction F of the "infinite pathway cluster" and the transport properties yields a prediction of both the absolute value and activation energy of the dc ionic conductivities directly from the structural models. Separate correlations for various types of mobile cations can be unified by employing the square root of the cation mass as a scaling factor. From the application of this procedure to RMC models of mixed alkali glasses, the mixed alkali effect, i.e. the extreme drop of the ionic conductivity when a fraction of the mobile ions is substituted by another type of mobile ions may be attributed mainly to the blocking of conduction pathways by unlike cations. The high efficiency of the blocking can be explained by the reduced fractal dimension of the pathways on the length scale of individual ion transport steps.

INTRODUCTION

Solid electrolytes presently attract considerable scientific interest because of their potential applications in electrochemical devices. While investigations mostly focus on the optimization of conductivities in superionic phases, it is as important to understand, why the high selective ionic mobility through the nearly frozen matrix in a solid electrolyte declines dramatically when a fraction of the mobile ions is substituted by another type of mobile ions. As this well-known but hardly understood conductivity drop is most pronounced for solid electrolytes containing alkali ions, it is generally referred to as the mixed alkali effect (MAE) [1-3]. The here studied MAE in glasses is connected to large non-linear changes in those properties that are directly linked to ionic transport (ionic conductivity, ionic diffusion, dielectric relaxation etc.) while macroscopic properties such as density, elastic moduli, refractive index etc. exhibit only small deviations from the normal gradual variations with composition.

One of the main obstacles for an understanding of ionic conduction in glasses in general and particularly of the MAE is the lack of detailed knowledge on the local structure and thus on the conduction pathways for the mobile ions. As the structural information from diffraction data of amorphous solids is not sufficiently detailed to grant for unique structure solutions, RMC fits represent the only viable technique to convert the experimental data into local structure models. From these RMC structure models transport pathways for the mobile ions are identified by a modified bond valence approach and characterized by means of statistical methods. In recent years we have developed this approach and applied it successfully to a large variety of glass systems containing a single type of mobile ion M (M=Ag^+[4,5] or alkali [6,7]). Here we report on the extension of the application range to the ion transport pathways in the prototype MAE system $Li_xRb_{1-x}PO_3$ as a function of the composition.

BOND SOFTNESS SENSITIVE BOND VALENCES

The length R_{A-X} of a bond between a "cation" A and an "anion" X can be related to its bond valence $s_{A-X}=exp[(R_0-R_{A-X})/b]$, with tabulated empirical bond valence (BV) parameters R_0 and b [8-10]. The simplicity of the method and the availability of BV parameters for many atom pairs [9-12] stimulated the extensive use of BV calculations in crystal chemistry, e.g. to judge the plausibility of structure determinations. BV investigations can be used to investigate structure-conductivity correlations in both crystalline [14,15] and amorphous [4-7] solid electrolytes, if accessible sites for mobile ions A^+ in local structure models of the solid electrolytes are identified with sites where the BV sum $V(A) = \Sigma_x \, s_{A-X}$ approaches the formal valence $V_{ideal}(A)$. Pathways between equilibrium sites along which the BV sum deviation $|V(A)-V_{ideal}(A)|$ remains as low as possible then should represent probable low-energy pathways for the ion transport. Literature BV parameter determinations mostly postulate that the BV sum of a central cation should be fully determined by interactions to its first coordination shell. Since the small range of bond lengths to nearest neighbors impedes an independent refinement of R_0 and b from reference structures, Brown and Altermatt [16] suggested to regard b = 0.37Å as a universal constant. This facilitates the determination of BV parameters for a wide variety of atom pairs, but as it predetermines the shape of the interatomic potential for a given charge and coordination type irrespective of the polarizabilities of the interacting species, correct valence sums can only be expected at equilibrium sites. BV calculations at non-equilibrium sites (e.g. interstitial sites in a conduction pathway) require more realistic estimates of the potential shape by an adjustment of b to the "softness" of the specific bond. Moreover an application of the nearest neighbor convention would generally lead to unphysical sudden variations of the BV sum along pathways connecting different coordination shells. As a comparison of ionic conductors with different mobile ions requires BV parameters with consistent boundary conditions, we determined the *softBV* parameter set [17, 18] that systematically accounts for the "bond softness" of the respective atom pair, exploiting an empirical correlation between b and the difference of the Parr & Pearson's "absolute softnesses" σ [19]

$$\frac{1}{\sigma} = \frac{1}{2}\left(\frac{\partial \mu}{\partial N}\right)_v \approx \frac{IE - EA}{2}, \qquad (1)$$

where μ represents the electronic chemical potential (-μ is equivalent to Mulliken's absolute electronegativity χ [20]), N the number of electrons, v the potential due to the nucleus and external influences, EA the electron affinity and IE the ionization enthalpy. Absolute cation softnesses can be defined alike using $EA(A^{z+}) = IE(A^{(z-1)+})$ [21], whereas the softness of the respective neutral atom is commonly used as a crude approximation for the anion softness. In harmony with the *hard and soft acids and bases* (HSAB) concept it may be supposed that the bond softness is related to the difference between the softnesses of the interacting particles.

STRUCTURE-CONDUCTIVITY RELATIONS IN GLASSES

The generation of local structure models by the reverse Monte Carlo (RMC) method made it possible to apply the BV analysis to ion conducting glasses. RMC fits [22-24] essentially yield "snapshot"-type structure model from experimental data and thus unequivocally refer to the temperature and pressure of the sample in the diffraction experiment (provided that the RMC model is large enough to permit a replacement of time-averaging by space-averaging to account for thermal vibrations). Three dimensional BV sum maps for the RMC models of ion conducting glasses containing ca. 4000 atoms were calculated by dividing the model into ca. 4 million volume elements and calculating the valence sum for a hypothetical cation of the mobile species A^+ at the center of each volume element. If the valence sum mismatch $|\Delta V(A)|$ is below a certain threshold or $\Delta V(A)$ changes its sign across the volume element, the whole volume element is classified as *"accessible"* for the mobile ion. A comparatively high threshold value of 0.2 valence units (v.u.) has been chosen here to reduce the statistical uncertainty in the number of accessible sites for the mixed alkali systems. Clusters of adjacent "accessible" sites constitute local transport pathways for the mobile ions. Due to the non-statistical distribution of accessible volume (and of the mobile cations) an accessible volume fraction of a few percent is sufficient to ensure the existence of an infinite pathway cluster. The *dc* conductivity should be determined by the ionic motion within this infinite pathway cluster. For ion conducting glasses with a single mobile cation we had observed [4,5] that the cube root of the pathway volume fraction F for a given mismatch threshold is linearly related to both the absolute conductivity and the activation energy of the conduction process. Common linear correlations for all investigated glass systems are obtained if the pathway volume fraction and the absolute conductivities are scaled by the square root of M (cf. Fig. 1). The empirical constants A_i, B_i depend on the chosen $|\Delta V|$ threshold.

$$A_1 \cdot \ln\left(\sigma T \sqrt{M}\right) + B_1 = \left(F\sqrt{M}\right)^{1/3} = A_2\left(-\frac{E_A}{k_B T}\right) + B_2 \qquad (2)$$

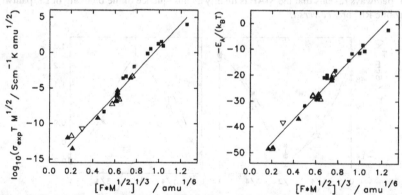

Figure 1: Correlations between the scaled cube root of the pathway volume fraction F (for $|\Delta V| = 0.2$ v.u.) and the absolute value σ_{exp} (l.h.s.) or the activation energy E_A (r.h.s.) of the experimental conductivity. Symbols indicate the type of mobile ion ($\blacktriangle$: Li$^+$; $\triangle$: Na$^+$; $\triangledown$: Rb$^+$; $\blacksquare$: Ag$^+$). Both pathway volume fractions and absolute conductivities are scaled by the square root of the cation mass M (in atomic mass units *amu*).

The BV method focuses on interactions of the mobile cation with counterions in the immobile matrix, while interactions between mobile and immobile cations are considered by exclusion radii only (for details see Ref. [7]) and interactions among mobile cations are completely neglected. Fortunately, the concentration of mobile cations within the "accessible" volume hardly changes over the range of systems under study, so that the extent of interactions among mobile ions should vary only marginally with F and cause a nearly constant offset in the structure conductivity correlation (For mixed alkali glasses the sum of the concentrations of both mobile species approaches the value found for glasses with a single mobile cation).

APPLICATION TO MIXED ALKALI GLASSES

As the conductivity drop in mixed alkali glasses tends to increase with the size difference between the alkali ion types, we have chosen the glass system $Li_xRb_{1-x}PO_3$ (x=0, 0.25, 0.5, 0.75, and 1) for our investigations [25]. RMC structure models were derived from wide-angle neutron and x-ray diffraction data taken at the diffractometers LAD (Rutherford Appleton Laboratory) and GILDA (ESRF, Grenoble). See Refs. [26,27] for details on the experiments and modeling. The structure models show no major structural alterations to which the breakdown of the conductivity might be attributed. In line with further experimental [28,29] and simulation [30-32] studies it may be presumed that the alkali ions retain their local environment and the two types of alkali ions are randomly mixed [27, 3, 34]. Theoretical models of the MAE (see e.g. [35-38]) are mostly based on unverified assumptions, such as a selective hopping mechanism, or a crucial role of Coulomb interactions among the mobile ions or of site relaxation. In contrast to that, the application of the BV analysis to quantitative structure models straightforwardly produces a nearly quantitative prediction of the MAE. This is demonstrated in Fig. 2, where the experimental composition dependence of E_A and σ_{dc} for the $Li_xRb_{1-x}PO_3$ system is compared to the predictions from the linear relations displayed in Fig. 1. A closer inspection of the distinct Li^+ and Rb^+ pathways reveals that the MAE is mainly a consequence of the blocking of Li^+ pathways by immobile Rb^+ or *vice versa*.

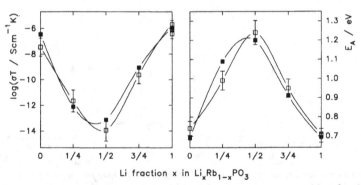

Figure 2: Activation energy E_A and *dc* conductivity σ versus composition for the glass system $Li_xRb_{1-x}PO_3$ at 300 K. Open squares are experimental data points [40], and filled squares correspond to the values predicted from the pathway volume fractions F of the structural models. Solid lines are guide to the eye.

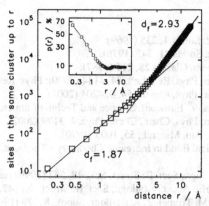

Figure 3: Log-log plot of the number of sites in the infinite Li⁺ pathway cluster of LiPO₃ *vs.* distance from a starting point in the pathway (averaged over all cluster sites, $\Delta V = 0.2$). The slopes represent the fractal dimension d_f of the pathway. The insert displays the probability *p(r)* for a site at a distance *r* from the starting point to be a part of the same pathway cluster.

The high effectiveness of the blocking can be rationalized by the reduced fractal dimension d_f of the infinite pathway cluster as illustrated in Fig. 3. d_f (the slope of the log-log plot) depends only slightly on the mismatch threshold ΔV and is clearly < 2 for the length scales of thermal vibrations and individual transport steps, while it approaches 3 for longer length scales. Thus, it may be suggested that the MAE is a natural consequence whenever the mobile ions have different sizes and / or different "bond softnesses" so that type A cations are immobile if they are located in type B pathways and *vice versa*.

The bond valence approach moreover qualitatively harmonizes to various other properties of the MAE. The blocking of energetically favorable jump destinations for a cation naturally explains the increase of the activation energy. As the long range connectivity of the pathways becomes irrelevant for high frequency experiments, our approach also predicts a considerably less pronounced MAE as experimentally observed. For increasing temperatures the generally higher volume fraction of accessible sites and the thereby higher fractal dimension of the pathways should tend to reduce the MAE again in agreement with experimental findings.

The combination of RMC structure modeling of mixed alkali phosphate glasses with the bond valence analysis makes it possible to predict the MAE from the local structure models of the glasses. The two types of alkali ions in a mixed alkali glass have distinctly different low dimensional pathways, which leads to an highly effective blocking of A ions even by a low number of B ions and *vice versa*. This interruption of conduction pathways is the main reason for the experimentally observed MAE. Finally, it might be noted that a slight depression of the conductivity would occur if the blocking effect was ignored. Even if the volume occupied by Rb⁺ is counted as free for mobile Li, the pathway volume in Li₀.₅Rb₀.₅PO₃ corresponds to a conductivity that is about one order of magnitude lower than that in LiPO₃, because the mixed alkali structure is less well adapted to Li⁺ cations.

ACKNOWLEDGMENTS
We thank C. Karlsson for providing us with conductivity data on the Li$_x$Rb$_{1-x}$PO₃ glass system.

19

REFERENCES

1. J. O. Isard, J. Non-Cryst. Solids 1, 235 (1969).
2. D. E. Day, J. Non-Cryst. Solids 21, 343 (1976).
3. M. D. Ingram, Phys. Chem. Glasses 28, 215 (1987).
4. St. Adams & J. Swenson, Phys. Rev. Lett. 84, 4144 (2000); Phys. Rev. B 63, 054201 (2000).
5. J. Swenson & St. Adams; Phys. Rev. B. 64, 024204 (2001).
6. J. Swenson & St. Adams; 9th Euroconf. Science and Techn. of Ionics, Rhodos, Sept. 2002.
7. St. Adams & J. Swenson; Phys. Chem. Chem. Phys. 4, 3179 (2002).
8. G. Donnay, R. Allmann, Am. Mineral., 55, 1003 (1970).
9. I. D. Brown, The Chemical Bond in Inorganic Chemistry - The bond valence model, Oxford University Press (2002).
10. I.D. Brown, Acta crystallogr., Sect. B: Struct. Sci., 48, 553 (19992) and 53, 381 (1997).
11. N.E. Brese & M. O' Keeffe Acta crystallogr., Sect. B: Struct. Sci., 47, 192 (1991).
12. S.F. Radaev, L. Fink & M. Trömel, Z. Kristallogr. Suppl., 8, 628 (1994).
13. I.D. Brown, J. Appl. Crystallogr. 29, 479 (1996).
14. St. Adams, J. Maier, Solid State Ionics, 105, 67 (1998).
15. St. Adams, Solid State Ionics, 136/137, 1351 (2000).
16. I.D. Brown & D. Altermatt, Acta crystallogr., Sect. B: Struct. Sci., 41, 244 (1985).
17. St. Adams, Acta crystallogr., Sect. B: Struct. Sci., 57, 278 (2001).
18. St. Adams, softBV parameter tables; http://kristall.uni-mki.gwdg.de/softBV/index.html
19. R.G. Parr & R.G. Pearson, J. Am. Chem. Soc., 105, 1503 (1983).
20. R.G. Pearson, J. Am. Chem. Soc., 107, 6801 (1985) and Inorg. Chem., 27, 734 (1988).
21. R.S. Mulliken, J. Chem. Phys., 3, 573 (1935).
22. R.L. McGreevy & L. Pusztai, Molec. Simul., 1, 359 (1988).
23. R.L. McGreevy, Ann. Rev. Mat. Sci., 22, 217 (1992).
24. R.L. McGreevy, Nucl. Inst. Meth. In Phys. Res. A, 354, 1 (1995).
25. J. Swenson & Stefan Adams , submitted to Phys. Rev. Lett.
26. J. Swenson, A. Matic, A. Brodin, L. Börjesson & W. S. Howells, Phys. Rev. B 58, 11331 (1998).
27. J. Swenson et al., Phys. Rev. B 63, 132202 (2001).
28. G. P. Jr. Rouse, P. J. Miller & W. M. Risen, J. Non-Cryst. Solids 28, 193 (1978).
29. A. C. Hannon, B. Vessal & J. M. Parker, J. Non-Cryst. Solids 150, 97 (1992).
30. T. Uchino, T. Sakka, Y. Ogata & M. A. Iwasaki, J. Non-Cryst. Solids 146, 26 (1992).
31. J. Habasaki, I. Okada & Y. Hiwatari, J. Non-Cryst. Solids 208, 181 (1996).
32. J. Swenson, L. Börjesson & W.S. Howells, J. Phys: Cond. Matter, 1999, 11, 9275.
33. B. Gee & H. Eckert, J. Phys. Chem. 100, 3705 (1996).
34. F. Ali, A. V. Chadwick, G. N. Greaves, M. C. Jermy, K. L. Ngai & M. E. Smith; Solid State NMR 5, 133 (1995).
35. P. Maass, A. Bunde & M. D. Ingram, Phys. Rev. Lett. 68, 3064 (1992); A. Bunde, M. D. Ingram & P. Maass, J. Non-Cryst. Solids 172-174, 1222 (1994).
36. P. Maass, J. Non-Cryst. Solids 255, 35 (1999).
37. G. N. Greaves & K. L. Ngai, Phys. Rev. B 52, 6358 (1995).
38. R. Kirchheim, J. Non-Cryst. Solids 272, 85 (2000).
39. G. N. Greaves, J. Non-Cryst. Solids 71, 203 (1985).
40. C. Karlsson, A. Mandanici, A. Matic, J. Swenson & L. Börjesson, submitted to Phys. Rev. B.

Mat. Res. Soc. Symp. Proc. Vol. 756 © 2003 Materials Research Society　　　　　　　　　　　　　　　　EE1.3

Influence of doping on the electrochemical properties of anatase

Marina V. Koudriachova and Simon W. de Leeuw
Computational Physics, Dept. of Multiscale Physics, Delft University of Technology,
Lorentzweg 1, 2628 CJ Delft, the Netherlands

The effect of substitution on the intercalation properties of anatase-structured titania has been investigated in first principles calculations. Ti^{4+}-ions were substituted by Zr^{4+}, Al^{3+} and Sc^{3+} respectively and O^{2-}-ions by N^{3-}. For each compound the open circuit voltage profile (OCV) was calculated and compared to anatase. Lithium intercalation proceeds as in pure anatase through a phase separation into a Li-rich and a Li-poor phase in all cases examined here. The Li-content of the phases depends on the nature of the dopant and its concentration. Substitution by N^{3-}-ions does not lead to lower potentials, whereas doping with trivalent Sc^{3+}- and Al^{3+}- ions decreases the intercalation voltage. Substitution by tetravalent Zr^{4+}-ions within the range of solubility does not significantly affect the OCV of anatase. A correlation is observed between the predicted equilibrium voltage and the participation of the Ti^{4+}-ions in accommodating the donated electron density upon lithiation.

I.　　Introduction

Anatase-structured titania TiO_2 is a technologically important material. It has applications in solar cells, electrochromic displays and hydrogen sensors[1-5]. The low potential for Li-insertion (1.8 eV Li^+/Li) and the remarkable stability of the electrochemical performance upon lithiation make anatase a promising candidate as anode material. For this application a lower intercalation potential is desirable.

A promising way to influence the properties of electrode materials is doping with heteroatoms. A considerable amount of research has been carried out to improve electrochemical properties of a number of important cathode materials. It has been shown that suitable doping allows an increase of the equilibrium voltage, suppresses undesirable structural effects and improves the kinetics of Li-insertion/extraction[6-10].

In the traditional viewpoint the redox potential of the transition-metal (TM) ion, which changes valence upon Li-insertion, determines the intercalation voltage. Substitution by another TM-ion adds new features to the open circuit voltage (OCV), which correspond to a reduction of the guest TM ion[6]. The oxidation state of a TM ion in the solid is strongly influenced by structural factors. If the substitution stabilizes the host structure upon charge/discharge a higher intercalation voltage is observed[11].

Recently a different viewpoint on the reduction reaction has been put forward. Consideration of charge densities obtained from ab initio calculations has led to the suggestion that the potential for Li intercalation in late transition metal oxides is mainly determined by the participation of O-ions in electron exchange[12,13]. If this is the case substitution of TM-framework ions by electrochemically non-active elements with a fixed valence (sp-elements) will force electron density to the O-sites and thus lead to higher intercalation voltages. Similarly substitution of the framework ions on the anion sublattice should have major consequences for the capacity for Li-intercalation. In agreement with this idea higher voltages have been observed for $LiCoO_2$ doped with sp-elements (Al, Ga, Mg)[9,13] and lower voltages were predicted from first principles calculations for the (hypothetical) layered compounds $LiCoS_2$ and $LiCoSe_2$[12]. However both, higher and lower voltages have been reported for Al-doped $LiNiO_2$[8,9].

In order to investigate how the intercalation potential of anatase TiO_2 is affected by doping an ab initio study has been undertaken. First principle calculations provide accurate energetics of Li-insertion and are a useful tool in understanding and predicting the electrochemical properties of electrode materials. We have considered substitution of Ti^{4+}-ions by tetravalent Zr^{4+}- and trivalent Al^{3+}- and Sc^{3+}- ions on the cation sublattice and replacement of O^{2-}-ions by N^{3-} on the anion sublattice. The choice of Zr^{4+}- represents a traditional point of view on the reduction reaction. Zr^{4+}- is isovalent to Ti^{4+}- and has a lower redox potential. N^{3-}- and Al^{3+}-ions were chosen to investigate the role of anions and sp-elements. Since N^{3-}-ions are more electronegative than O^{2-}-ions doping should lead to lower intercalation potential. If the oxygen participation in the electron exchange is important, as has been argued for $LiCoO_2$, the intercalation potential for Al^{3+}-doped titania is expected to be higher than for the pure compound. Trivalent Sc^{3+}-ions might stabilize the anatase structure upon lithiation, since $LiScO_2$ adopts a structure similar to anatase.

Zr^{4+}-, Al^{3+}- and N^{3-}-ions are soluble in anatase. N^{3-}-ions can be incorporated into the lattice from an atmosphere of NH_3[14,15], Zr^{4+}- and Al^{3+}-ions from corresponding oxides, ZrO_2 and Al_2O_3[16,17]. The maximum solubility of Al_2O_3 is reported to be 22 wt%[17], that of Zr^{4+}-ions is about 0.075 Zr/Ti[16]. Data on the solubility of N^{3-}- and Sc^{3+}-ions are not available.

II   Details of simulations

All calculations were performed using the CASTEP code[18,19] within the pseudo-potential plane-wave formalism, spin polarized density functional theory in the generalized gradient approximation[20] and ultrasoft pseudopotentials[21]. The k-space sampling was performed on a regular grid with a spacing of $0.1 Å^{-1}$. A plane wave cutoff energy of 380 eV was found to converge the total energy to 0.01 eV per formula unit. The computational procedure has been validated for a number of parent and mixed oxides as well as for the $Li_{0.5}TiO_2$- orthorhombic phase. Preliminary calculations showed that the dopant ions prefer to enter the anatase lattice substitutionally.

The calculations were performed for a model structure corresponding to a high dopant concentration, 25% for the cation and 12.5% for the anion sublattice. This corresponds to a replacement of either a Ti^{4+}- or an O^{2-} ion in the primitive unit cell of anatase (Ti_4O_8). These concentrations may be too high to be achieved experimentally. By confining ourselves to a primitive unit cell of the anatase structure the computing time was reduced considerably without changing the trends in intercalation behavior. The supercells used in calculations contain 4, 2 or 1 primitive units. Substitution by Zr^{4+}-ions was performed also in a super cell composed of 4 primitive unit cells of anatase ($ZrTi_{15}O_{32}$). This corresponds to Zr/Ti=0.0625, close to the solubility limit observed experimentally.

For each dopant the lowest energy configurations were sought in the full range of the Li-insertion concentration x ($0<x<1$). The equilibrium voltage profile was computed through numerical differentiation of the total energy with respect to metallic Li and a bulk structure (Zr^{4+}-doped anatase or anatase doped with a trivalent cation or anion and a charge compensating ion).

III.   Results and Discussion
Structure

Anatase TiO_2 adopts a tetragonal structure, space group *I41/amd*, with cell parameters a=b=3.78 Å, c=9.51 Å[22] (calculated values a=b=3.80 Å, c=9.62 Å). Lithium insertion proceeds through the occupancy of octahedral sites. There are four octahedral sites in a unit cell of anatase. Occupation of two octahedral sites results in the formation of the $Li_{0.5}TiO_2$ phase. Due to a Jahn-Teller like distortion this compound adopts an orthorhombic structure[23]. At normal conditions Li-insertion proceeds through a two-phase equilibrium between a Li-poor and a Li-rich ($Li_{0.5}TiO_2$) phase[24-25]. At elevated temperatures higher Li-concentrations can be achieved. It has been shown that intercalation at x>0.5 is diffusion limited[25]. The lowest energy configurations of Li_xTiO_2 have been discussed extensively in a previous communication[25]. In Zr^{4+}-substituted anatase the ordering is very similar to that in the pure TiO_2 structure. Li^+-ions tend to avoid the sites near the zirconium ions, since their capability for accommodating the charge density donated by lithium is smaller than titanium ions. At low dopant rates and low and intermediate Li-concentrations the intercalation reaction should not be affected much by Zr^{4+}-doping. Substitution of tetravalent Ti^{4+}-ions by trivalent Al^{3+}- or Sc^{3+}-ions or divalent O^{2-} ions by N^{3-} requires charge compensation. At x=0.25 the lowest energy configurations correspond to Li^+-ions occupying octahedral positions in the immediate neighborhood of the dopant in the *ab*-planes. At higher concentrations (up to x=0.75) the neighboring sites in the *c*-direction are gradually filled. The lowest energy configuration for $Li_{0.75}Al_{0.25}TiO_2$ is shown in fig. 1.

Fig. 1: Lowest energy configuration of $Li_{0.75}Al_{0.25}TiO_2$. Al-ions are shown in black,
O-ions – in white, Ti-ions in dark gray, Li ions in light gray.

Equilibrium voltage profile

The calculated OCV profiles for doped anatase structures are shown in fig.2-3. The characteristic shape of a van der Waals loop similar to that predicted for pure anatase[25] indicates that doping does not change the two-phase nature of the insertion reaction. However, doping does change the composition of the Li-rich and Li-poor phases. For Zr^{4+}- and Sc^{3+}-doped anatase the stoichiometry of the Li-rich phase is similar to that for pure anatase, whereas for Al^{3+}- and N^{3-}-doped anatase it is closer to x~0.7 and x~0.6 respectively. This reflects the distribution of the electron density from the charge compensating ions. Another consequence of the charge compensation is a significant loss of capacity upon doping with trivalent elements and the composition of the Li-poor phase at the doping concentrations considered is close to x=0.3.

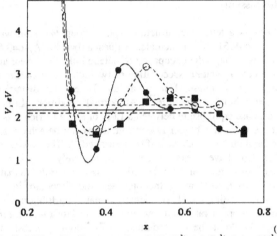

Fig. 2: OCV profiles for anatase doped with Al^{3+}- (■), Sc^{3+}- (●) and N^{3-} (○) ions.
A straight line shows the equilibrium voltage and concentrations.

Doping by Zr^{4+}-ions within solubility limit (Zr/Ti=0.0626) does not influence the equilibrium voltage (2.22 eV against 2.20 eV for pure anatase[25]). If higher doping concentrations could be achieved the potential for Li-insertion below x=0.5 would be considerably lower (1.96eV, fig.3). Our calculations show that substitution at these high concentrations is still energetically favorable. Substitution by Al^{3+}- and Sc^{3+}-ions lead to a lower equilibrium voltage (2.09eV and 2.16 eV respectively). Doping with N^{3-}- ions increases the potential for Li insertion (2.28 eV). This result is unexpected if the charge transfer to oxygen ions in the insertion reaction plays a major role.

Charge distribution

A Mulliken population analysis was carried out to examine the distribution of the electron density donated to the host material upon Li-insertion. The calculations were performed on a primitive unit cell of anatase (Ti_4O_8) containing a single substitutional ion. For later reference, in pure anatase the titanium ions carry a charge of 1.36e resulting in a charge of -0.68e for oxygen. Doping as well as charge compensation leads to a redistribution of the charge on the anion and cation sublattices.

Consider first substitution by ions with a lower electronegativity, i.e. Zr^{4+}, Sc^{3+} for Ti^{4+} or N^{3-} for O^{2-}. These ions are charge deficient and the neighboring Ti^{4+}-ions are slightly reduced (Zr: by -0.03e, Sc: by -0.06e and N: by -0.07e). The ionic charge of the dopant ion recovers upon donation of the charge from the Li^+-ions. Adding charge-compensating ions decreases the charge on N^{3-} ions from -0.62e to -0.83e. This value is similar to the ionic charge of nitrogen in $TiN_{0.61}$ (-0.79e), a compound with a structure belonging to the same space group as anatase. Zr^{4+}-ions in Ti^{4+}-sites have an ionic charge of 1.68e as compared to 1.56e in zirconia. When 2 Li are added (x=0.5), a considerable amount of extra charge is transferred to the Zr^{+4}-ions resulting in an ionic charge of 1.54e, similar to that in pure zirconia. Consequently the charge transfer to Ti^{+4}-ions is reduced (68% of the Li-charge compared to 80% in pure anatase). In the same way a smaller amount (78%) of extra charge is the transferred to Ti^{4+}-ions upon lithiation of scandium-doped anatase $LiScTi_3O_8$. The distribution of

24

the added electron density upon lithiation over the Ti^{4+}- and O^{2-} sublattices remains similar to that in pure anatase. In the case of nitrogen-doped anatase more electron density is transferred to the cation sublattice reflecting the lower electronegativy of nitrogen compared to oxygen.

Aluminum is more electronegative than titanium. Its ionic charge in Al-substituted TiO_2 is computed to be 1.64e, which is close to its charge in pure Al_2O_3. Ti^{4+}-ions in positions neighboring the Al^{3+}-ion loose 0.08e, which is recovered upon charge compensation. Upon further lithiation the Al^{3+}-ions remain inert and hardly participate in the accommodation of the added electron density. The bulk of the added charge density (75%) is donated to the Ti^{4+}-ions.

The total amount of charge donated to the host lattice by charge compensating ions strongly depends on the nature of the dopant (Al: -0.9e, Sc:-0.78e and N:-0.86e) and is correlated to its electronegativity. Though the amount of charge donated by the intercalating Li-ions also depends on the nature of the dopant, it is closer to its value in pure anatase: -0.86e and -0.80e at x=0.5 and 0.75 respectively. In the Li-rich phase of Sc^{3+}- and Zr^{4+}-doped anatase at x=0.5 each Li donates on average -0.82e, in N^{3-} doped anatase -0.88e and -0.76e in Al^{3+}-doped anatase at x=0.75.

The data above demonstrate that the majority of the charge donated upon intercalation of lithium resides on the Ti^{4+}-ions. Fig. 4 shows that the equilibrium voltage is correlated with the participation of the Ti^{4+}-ions in accommodating the extra electron density from intercalated ions. Though doping has a major effect on the charge of the anions no correlation has been observed between the induced charges and the equilibrium voltage. In agreement with the results from Ceder et. al.[13] Al^{3+} substitution forces electron density to move to oxygen ions coordinating the Al^{3+}-ions, which receive -0.16e (4) and -0.10 (2). When a charge compensating Li^+-ion is introduced more charge density is added to these oxygen ions, resulting in a charge of -0.95e. A similar although less pronounced effect is predicted for substitution with Zr^{4+}- and Sc^{3+}-ions. For instance, upon substitution with Zr^{4+}-ions the surrounding O^{2-} ions receive respectively -0.1e (4) and -0.03e (2). Additionally the charge transfer to the anion sublattice in nitrogen-doped anatase is predicted to be smaller than in pure anatase, yet the estimated intercalation voltage is higher. Hence, in contrast to what has been observed for the late transition metal oxides[13] it is the change of the oxidation state of the titanium ions and not the transfer to the oxygen ions which controls the capacity for Li intercalation in substituted anatase TiO_2.

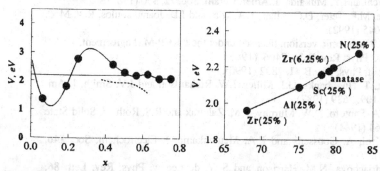

Fig. 3: OCV profiles for anatase doped with Zr^{4+}-ions. The dashed line corresponds to a doping concentration of 25% Zr^{4+}. A straight line shows the equilibrium voltage and concentrations.

Fig.4: Equilibrium voltage vs. percentage participation of Ti^{4+}-ions in accommodation of charge donated by Li-ions.

IV. Conclusions.

First principle calculations of Li-insertion into anatase TiO_2 doped with Al^{3+}-, Sc^{3+}-, N^{3-}- and Zr^{4+}-ions shows that the Ti^{4+}-ions play a major role in the accommodation of the electron density donated by Li. Increased participation of the Ti^{4+}-ions on doping with N^{3-} is accompanied by a higher equilibrium voltage. In case of substitution with Zr^{4+}-, Al^{3+} or N^{3-} ions a smaller amount of charge is transferred to the Ti^{4+}-ions and a lower potential for L intercalation is predicted. Replacement of Ti^{4+}- by Zr^{4+}-ions within the solubility limit does not lead to a lower potential due to the small dopant concentration.

References

1. H.O.Finklea, in *Semiconductor Electrodes, Studies in Physical and Theoretical Chemistry*, Elsevier, Amsterdam, Vol. **55**, 43, 1988.
2. D. O'Hare, *Inorganic Intercalation Compounds*, in *Inorganic Materials*, D.W. Bruce and D. O'Hare (Eds), p171, John Wiley & Sons Ltd, UK, 1996.
3. S.Huang, L. Kavan, A. Kay and M. Grätzel, J. Electrochem. Soc. **141**, 142 (1995).
4. L. Kavan, M. Grätzel, S.E. Gilbert, C. Klemenz and H.J. Scheel, J. Am. Chem. Soc. **118**, 6716 (1996).
5. T. Ohzuku, Z. Takehara and S. Yoshizawa, Electrochim. Acta **27**, 1263 (1982).
6. Y.P. Yu, E. Rham and R.Holze, Electrochim. Acta **47**, 4391 (2002).
7. C. Delmas, M. Menetrier, L. Croguennec, I. Saadounne, A. Rougier, C. Pouillerie, G Prado, M. Grüne and L. Fournes, Electrochim. Acta **45**, 243 (1999).
8. T. Ohzuku, K. Nakura and T. Aioli, Electrochim. Acta **45**, 151 (1999).
9. C. Julien, G.A. Nazri and A. Rougier, Solid State Ionics **135**, 121 (2000).
10. C. Julien, M.A. Camacho-Lopez, M. Lemal and S. Ziolkiewicz, Mat. Sci. Eng. **B95**, 6 (2002).
11. J. Kim and K. Amine, Electrochem. Comm. **3**, 52 (2001).
12. G. Ceder, M.K. Aydinol and A.F. Kohan, Comp. Mater. Sci. **8**, 161 (1997).
13. G. Ceder, Y.-M. Chang, D.R. Sadoway, M.K. Aydinol, Y.I. Jang and B. Huang, Nature **392**, 694 (1998).
14. C. Blaauw, H.M. Naguib, A. Ahmed, S.W. Ahmed, J.L. Whitton and J.R. Leslie, Mater. Res. Bull. **18**, 87 (1983).
15. K. Fukushima and I. Yamada, Jap. J. Appl. Phys. **35**, 5790 (1996).
16. J. Yang and J.M.F. Ferreira, Mater. Res. Bull. **33**, 389 (1998).
17. O. Yamaguchi and Y. Mukaida, J. Amer. Ceram. Soc. **72**, 330 (1989).
18. M.C. Payne, M.P. Teter, D.C. Allan, T.A.Arias and J.D. Joannapoulos, Rev. Mod. Phys. **64**, 1045 (1992).
19. CASTEP 3.9 Academic version, licensed under the UKCP-MSI agreement, 1999.
20. J.P. Perdew, Phys. Rev. B **34**, 7406 (1986).
21. D. Vanderbilt, Phys. Rev.B **41**, 7892 (1990).
22. J.K. Burdet, T. Hughbanks, G.J. Miller, J.W. Richardson and J.V. Smith, J. Am. Chem. Soc **109**, 3639 (1987).
23. R.J. Cava, A. Santoro, D.W. Murphy, S.M. Zahurak and R.S. Roth, J. Solid State Chem. **83**, 64 (1984).
24. R.van de Krol, A. Goossens and E.A. Meulenkamp, J. Electrochem. Soc. **146**, 3150 (1999).
25. M.V. Koudriachova, N.M. Harrison and S.W. de Leeuw, Phys. Rev. Lett. **86**, 1275 (2001).

Mat. Res. Soc. Symp. Proc. Vol. 756 © 2003 Materials Research Society

Space Charge Layers in Polycrystalline Cerium Oxide

Andreas Tschöpe
Universität des Saarlandes, Technische Physik, Gebäude 43B, 66041 Saarbrücken, Germany

ABSTRACT

The effect of space charge layers in polycrystalline cerium oxide was analyzed by comparing experimental results of grain size-dependent electrical conductivity with theoretical models. Modeling included the calculation of space charge segregation of acceptor ions and of the effective electrical conductivity of polycrystalline cerium oxide in both the macroscopic and mesoscopic range of grain sizes. It is shown that an L^{-3} power law for the electronic conductivity in the nm-regime is characteristic for the equilibrium space charge model and different from the scaling behavior of alternative models. The origin of space charge potential was investigated by numerical calculation of the electrical potential in a two-phase model. It was found, that a positive excess charge at grain boundaries of cerium oxide is caused by an enhanced oxygen deficiency at the grain boundary core. The influence of acceptor ion doping in the dilute limit and of non-equilibrium distribution of acceptor ions on electrical conductivity was also studied.

INTRODUCTION

Cerium oxide is a mixed ionic/electronic conductor (MIEC). Both, electronic and ionic charge carriers are generated in thermodynamic equilibrium during partial reduction of ceria to nonstoichiometric CeO_{2-x}. The ratio of electronic to ionic conductivity can be manipulated by the addition of lower-valent cations. These acceptor ions participate in the charge balance with the result, that electronic conductivity decreases and ionic conductivity increases when the concentration of acceptor ions is raised. Defect chemistry of the bulk phase, which combines the law of mass action for defect equilibria with the requirement of local charge neutrality allows to calculate the concentration of charge carriers and conductivity for a given temperature, oxygen partial pressure and acceptor concentration [1-3].

In the recent past, the effect of grain size on ionic and electronic conductivity in cerium oxide has been investigated, as it may provide additional options for tailoring materials properties. The major differences found in nanocrystalline cerium oxide as compared to microcrystalline counterparts were (i) a predominantly electronic rather than ionic conductivity, (ii) an enhanced absolute value of electronic conductivity, and (iii) a reduced effective activation energy for electronic conductivity [3-9]. The conclusion, that electronic conductivity was dominating in the investigated nanocrystalline samples, was first based on the observation of a distinct oxygen partial pressure dependence, and was later confirmed by measurements of thermopower [5,10] and measurements of conductivity using ion-blocking electrodes [9]. An enhanced oxygen deficiency at grain boundaries has been suggested as origin of the large electronic conductivity and the reduced activation energy for σ_{el} in nanocrystalline ceria. In this case, redistribution of mobile charge carriers between the grain boundary core and the bulk crystallites in equilibrium is expected and should lead to the formation of space charge layers [11]. It was shown, that experimental results could be consistently explained by space charge models for polycrystalline cerium oxide [9,12]. Nevertheless, several alternative models and also different variants of the space charge model are in discussion. Hence, it seems necessary to (i) find a criterion, that allows to better evaluate the consistency of possible models, and (ii) to investigate, which parameters in the suggested models are relevant and which may be negligible. A detailed analysis of grain size dependence of physical properties and of the origin of space charge potential could be a first step in this direction. As an example, the grain size dependence of thermopower in polycrystalline cerium oxide exhibited good agreement with the equilibrium space charge model [10]. Similar agreement with grain size-dependent electronic conductivity in the nanometer regime was also reported, but the macroscopic space charge model did not allow a direct comparison with measured data of nanocrystalline materials [13].

The first objective of the present study was to model electrical conductivity of polycrystalline cerium oxide in the mesoscopic (nanometer) regime for direct comparison with experimental data. Numerical modelling is necessary, because an analytical description of interpenetrating space charge layers in a 3-dimensional crystal of a size, that is comparable to the screening length, is not available. Numerical methods, that have been developed for the simulation of semiconductor devices will be employed to calculate charge transport in three dimensions [14,15].

The second objective is to investigate the scaling of σ_{el} with grain size for different models and their comparison with experimental data. Finally, numerical modeling of the electrochemical equilibrium in the space charge layer along grain boundaries, using a simple 'second phase'-approach to introduce the defect chemistry in the grain boundaries will be presented [16]. The origin of the space charge layer, the effect of acceptor ions and the implications of non-equilibrium distribution were also investigated by numerical modeling.

THEORY

When atomic disorder in an ionic solid in thermodynamic equilibrium leads to the formation of two types of defects, A' and B$^\bullet$, a potential difference $\Delta\Phi_{surf}=(g_A-g_B)/2e_0$ is established between the bulk and the surface in order to compensate for the difference in the standard free enthalpies of defect formation, g_A and g_B [17]. This argument has been applied by Kliewer and Koehler to calculate the space charge potential at the surface of NaCl [18], and was also transferred to grain boundaries and dislocations. One drawback of this approach is, that the standard free enthalpies of defect formation for the individual species A' and B$^\bullet$ are generally unknown. In the bulk phase, the defect concentrations are completely determined by the law of mass action

$$[A'] \cdot [B^\bullet] = \exp\left\{ -\frac{g_{A'} + g_{B^\bullet}}{k_B T} \right\} \tag{1}$$

of the underlying defect equilibrium and the requirement of charge neutrality. Hence, the bulk phase can be chosen as reference point and the space charge layers described by the Gouy-

Chapman theory [19]. This second approach has been successfully used in the space charge model of solid ionic materials [11] and was applied to cerium oxide [12]. Of course, the two models are interchangeable and the individual standard free enthalpies of defect formation are completely determined by the materials constants in the law of mass action and the space charge potential. As shown in Appendix A, the standard free enthalpies of formation for oxygen vacancies and electrons in cerium oxide can be formally calculated for a given temperature, oxygen partial pressure and space charge potential.

The electrical potential $\Phi(\mathbf{r})$ in thermodynamic equilibrium can be obtained by solving the Poisson-Boltzmann equation (PBE),

$$\nabla^2\Phi(\mathbf{r}) = -\frac{e_0}{\varepsilon\varepsilon_0}\cdot\left[\exp\left\{-\frac{g_{B'}+e_0\Phi(\mathbf{r})}{k_B T}\right\}-\exp\left\{-\frac{g_{A'}-e_0\Phi(\mathbf{r})}{k_B T}\right\}\right]. \tag{2}$$

Analytical solutions of this differential equation exist for the 1-dimensional problem with symmetric electrolytes, i.e. when the absolute values of the defect charges are equal [18]. The analytical solutions are still good approximations for non-symmetric electrolytes, when the space charge potential is large so that the charge density is dominated by the accumulation of one type of point defects. In the macroscopic size range, where the grain size L is much larger than the screening length λ_D, the space charge layers around a crystallite contribute independently to the effective conductivity. Therefore, an individual crystallite (the building unit of a polycrystalline material) could be separated into a bulk core and surrounding boundary layers. The concentration profiles within the boundary layers were described by the analytical solutions of the BPE, and the effective conductivity was obtained by combining the various components into the equivalent circuit of the brick-layer model [11,12].

If the grain size is of the same order of magnitude as the screening length, so that the space charge layers overlap, then the brick-layer model is no longer appropriate and the 3-dimensional Poisson-Boltzmann equation must be solved numerically. Furthermore, the non-equilibrium process of charge transport as result of an externally applied voltage must also be calculated numerically. Numerical methods for the solution of such problems have been developed for the simulation of semiconductor devices. With the help of a mathematical transformation, it is possible to split the solution of the non-equilibrium process into two parts. In the first part, the equilibrium distribution of electrical potential and charge carrier density is calculated by by solving the PBE. Fig. 1 is a plot of the potential field in the (011)-plane across a cubic crystallite before an external voltage is applied. For a grain size $L\approx1.7\cdot\lambda_D$, the potential difference between the boundary and the core was only 50% of the bulk value.

In the second step, the density of ionic and electronic current after application of an external voltage is calculated by solving the drift diffusion equations,

$$\mathbf{j}_{A'} = \mu_{A'}Nk_B T\exp\left\{\frac{e_0\Phi}{k_B T}\right\}\mathrm{grad}(u), \tag{3a}$$

$$\mathbf{j}_{B'} = -\mu_{B'}Nk_B T\exp\left\{\frac{-e_0\Phi}{k_B T}\right\}\mathrm{grad}(v), \tag{3b}$$

$$\mathrm{div}(\mathbf{j}_{A'}) = \mathrm{div}(\mathbf{j}_{B'}) = 0. \tag{4}$$

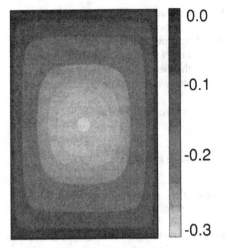

0.0

-0.1

-0.2

-0.3

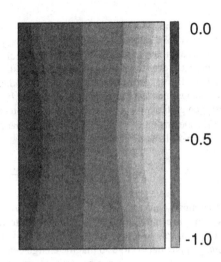

0.0

-0.5

-1.0

Figure 1. Electrical Potential Φ in the (011)-plane of a cubic crystallite with size $L \approx 1.7\lambda_D$ in equilibrium.

Figure 2. Distribution of the Slotboom variable u in the (011)-plane .

In these equations, the Slotboom variables

$$u = \exp\left\{\frac{-e_0\varphi}{k_B T}\right\} \quad \text{and} \quad v = \exp\left\{\frac{e_0\varphi}{k_B T}\right\} \tag{5}$$

are introduced, which represent the deviation of the charge carrier densities from equilibrium in terms of an electrical potential φ, that is caused by the external voltage and superimposed to the equilibrium potential [14]. The distribution of u in the (011)-plane of a cubic crystallite with a potential gradient along the x-axis is shown in fig.2. Integration of the current density then leads to the electrical conductivity. Further details of this model and the numerical methods will be published elsewhere.

RESULTS AND DISCUSSION

Grain size-dependent electrical conductivity

The results, obtained for the grain size dependence of electrical conductivity in the mesoscopic range can be combined with the earlier results of the macroscopic space charge model for larger

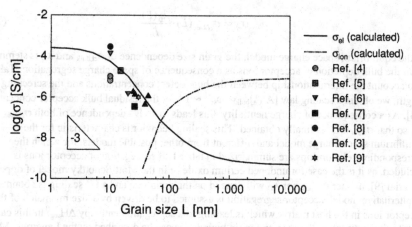

Figure 3. Comparison of experimental and theoretical results on electrical conductivity of cerium oxide at 500°C as function of grain size. Measured values were collected from various publication and the theoretical data were obtained from the space charge model with a space charge potential of 0.6 V.

grain sizes [12,13] and compared with experimental data, that were compiled from several publications, fig.3. The calculations were performed with different values for the space charge potential and the best agreement was obtained for a value $\Delta\Phi=0.6\pm0.05$V. While ionic conductivity is found to dominate at large grain sizes, a transition to predominantly electronic conductivity below 100 nm was obtained. An interesting result is the steep increase of electronic conductivity in the mesoscopic transition range between the macroscopic grain sizes and the flat-band limit at very small grain sizes. This steep increase could be described by a power-law dependence $\sigma_{el} \propto L^{-3}$ which was found to be characteristic for the equilibrium space charge model as shown in the following section.

<u>**Scaling law analysis**</u>

For large effects, i.e. when the space charge potential is sufficiently large ($|\Delta\Phi|\geq0.3$V) and the concentration of acceptor ions is not too high (≤1000 ppm), the total electronic conductivity is given as

$$\sigma_{el} \approx F(\Phi) \cdot \sigma_{el,bulk} \cdot \varphi_{sc},\tag{6}$$

where φ_{sc} is the volume fraction of space charge layers in the crystallites and F is an enhancement factor, that depends on the space charge potential. Since the volume fraction is proportional to λ_D/L, we obtain a relationship

$$\sigma_{el} \propto \sigma_{el,bulk}(L) \cdot \frac{\lambda_D(L)}{L}. \tag{7}$$

In the equilibrium space charge model, the grain size dependence of $\sigma_{el,bulk}$ and λ_D is stemming from the bulk depletion of acceptor ions as a consequence of space charge segregation. Taking into account the interrelationship between the bulk defect concentrations and the screening length, we obtain a scaling law $[A'_{Ce}]_{bulk} \propto \varphi_{sc}^{-1} \propto L^2$ for the residual bulk acceptor concentration [13]. As a consequence of charge neutrality, this leads to an L^{-1}-dependence of both $\sigma_{el,bulk}$ and λ_D, so that $\sigma_{el,tot} \propto L^{-3}$ is finally obtained. This scaling behavior is characteristic for the equilibrium space charge model and different from other possible models, for which the corresponding relationships are summarized in table 1. If segregation of acceptor ions is excluded, as it is the case for undoped cerium oxide or in the Mott-Schottky model of doped material [9], neither $\sigma_{el,bulk}$ nor λ_D will depend on the grain size and a L^{-1} scaling is obtained. In an alternative model, acceptor segregation is assumed to be driven by a size mismatch of the acceptor ions in the host matrix which is leading to a segregation enthalpy ΔH_{seg}. In this case, the distribution between bulk phase and grain boundaries can be described by the Langmuir-McLean adsorption isotherm [20],

$$\frac{\Theta}{1-\Theta} \approx [A'_{Ce}]_{bulk} \exp\left\{-\frac{\Delta H_{seg}}{k_B T}\right\}. \tag{8}$$

The total acceptor concentration is given as the sum of bulk concentration and the grain boundary excess, which is proportional to the grain boundary area per unit volume $A_{gb}/V = 2/L$ [21]. For large effects, the residual bulk acceptor concentration is then proportional to L, finally leading to an L^{-2} scaling law for the total electronic conductivity.

Table I. Scaling law analysis for various models

Proportionality	$[A'_{Ce}]_{bulk}$	$\sigma_{el,bulk}$	λ_D	$\sigma_{el,tot}$
Equilibrium space charge model	L^2	L^{-1}	L^{-1}	L^{-3}
No segregation: Undoped CeO$_2$, Mott-Schottky	Const.	Const.	Const.	L^{-1}
Langmuir-McLean type segregation	L	$L^{-1/2}$	$L^{-1/2}$	L^{-2}

It should be noticed that core models, in which the enhanced electronic conductivity is explained by an increased electron density and/or electron mobility in the grain boundary core, also lead to an L^{-1} scaling, since these effects would be proportional to the grain boundary area per unit volume. The different scaling laws may help to evaluate concurring models. Based on the experimental data that are currently available, the space charge segregation model seems to be appropriate. However, further measurements on a more consistent series of samples will be necessary in order to evaluate the various models.

Origin of space charge potential

Positive space charge layers have been reported in acceptor-doped alkaline earth titanates and were attributed to donor states in the grain boundaries [22]. In order to identify the nature and investigate the impact of possible donor states in cerium oxide quantitatively, a simple two-phase model, in which the grain boundary is treated as a thin layer of a second phase [16] between two cerium oxide crystals has been used. In order to reduce the computational expenses, numerical simulations were performed on a 1-dimensional problem. A tri-layer crystal is composed of three layers of cerium oxide bulk phase separated by layers of a second phase, representing the grain boundaries. The equilibrium distribution of electrical potential perpendicular to the interfaces was calculated by solving the Poisson-Boltzmann equation with different values for the free enthalpies of defect formation in the bulk phase and grain boundary phase. For the bulk phase, values for g_{Vo} and g_e were calculated as described above and in App.A. For the grain boundary, it was assumed that the enthalpies of defect formation differ from the bulk values by a certain amount, i.e. $g_{Vo,gb}=g_{Vo,bulk}+\Delta g_{Vo}$. An example of the calculated electrical potential across the tri-layer is shown in fig. 4. The potential difference between the surfaces on the left and right side and the bulk phase are determined by the absolute values of g_{Vo} and g_e. In the numerical calculation, the width of the grain boundary phase, the values for g_{Vo}, g_e, and Δg, and the width L of the inner crystal layer were varied and the space charge potential at grain boundaries was determined.

The calculated grain boundary space charge potentials are shown in fig. 5 as function of Δg. The first important results was, that the obtained space charge potentials were not very sensitive to the width of the grain boundary layer L_B. The solid lines in fig. 5 were obtained for a width of 5 Å and the dashed lines above and below correspond to 7 Å and 3 Å, respectively. The second

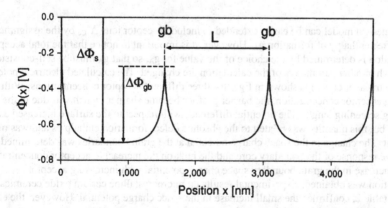

Figure 4. Electrical potential across a tri-layer crystal with two surfaces and two grain boundaries, calculated by numerical solution of the Poisson-Boltzmann equation.

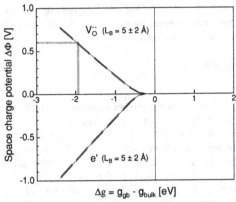

Figure 5. Grain boundary space charge potential in undoped cerium oxide at 500°C as function of Δh for oxygen vacancies and electrons.

result was, that the space charge potential only depends on the difference Δg, but not the absolute values of the individual free enthalpies of defect formation. Hence, the arbitrary choice of the surface potential was not significant. And the third result was, that a positive space charge potential, which would be in agreement with the experimental data, could only be obtained, when the formation enthalpy for oxygen vacancies was reduced in the grain boundary core as compared to the bulk phase. This implies, that the grain boundary is more reduced than the bulk. The quantitative comparison revealed, that a space charge potential of 0.6 V corresponds to a reduction in the enthalpy of vacancy formation by almost 2 eV.

Effect of acceptor doping

The numerical model can be easily extended to include acceptor ions A'_{Ce} by the assignment of a formal free enthalpy of formation g_A. However, it is important to notice that the total acceptor concentration is determined by the choice of the value for g_A, so that g_A must be self-consistently adjusted when other parameters of the calculation are changed. The calculated electrical potential within the tri-layer crystal is shown in fig.6 for three different acceptor concentrations. With increasing acceptor concentration, the barrier profiles became sharper, which was due to the decreasing screening length. The potential difference with respect to the surface increased as expected, but this quantity was related to the absolute defect formation enthalpies and was not significant. The change in the space charge potential at the grain boundaries was determined by the relative response of the boundary core and the bulk on the increasing acceptor concentration. A small increase in the grain boundary space charge potential with increasing acceptor concentration was obtained. Experimental results on microcrystalline cerium oxide ceramics, shown in table 2, confirmed the small increase in the space charge potential. However, the effect was rather small if compared to the experimental error.

Table II. Experimental results for the grain boundary space charge potential in microcrystalline cerium oxide as function of total acceptor concentration (determined by calibrated X-ray fluorescence analysis).

$[A'_{Ce}]$	$\Delta\Phi_{gb}$ [V]
$7 \cdot 10^{-4}$	0.66 ± 0.05
$6 \cdot 10^{-3}$	0.70 ± 0.05
$1.2 \cdot 10^{-2}$	0.72 ± 0.05

So far, thermodynamic equilibrium was assumed. While oxygen vacancies and electrons are mobile and may therefore be distributed in equilibrium, this is not possible for the acceptor ions at temperatures as low as 500°C. This leads to the question of the effect of non-equilibrium distribution of acceptor ions on the grain size dependence of electrical conductivity. There are different possible scenarios for non-equilibrium distribution. Since all materials that have been investigated were sintered to ceramics at higher temperatures before the measurements, it is plausible to assume, that the acceptor ions may distribute in equilibrium at the high sintering temperature, but are then frozen in their position during cooling. The effect of this quenching can also be modeled numerically. In the calculation protocol, the equilibrium concentration profiles were calculated for a temperature of 1000°C. Then, the acceptor profiles were frozen while electrons and vacancies were allowed to equilibrate during cooling to 500°C. From the resulting concentration profiles, the electrical conductivity parallel to the grain boundary layers were

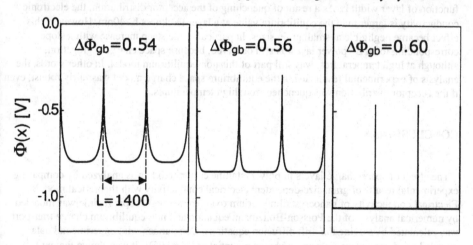

Figure 6. Profiles of electrical potential across a tri-layer crystal for increasing acceptor concentration of $1 \cdot 10^{-8}$, $1 \cdot 10^{-6}$ and $1 \cdot 10^{-4}$. T=500°C, Δg_{V_O}=-1.8eV.

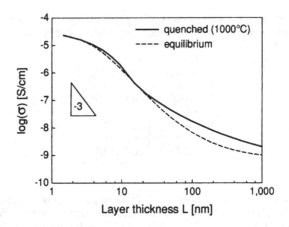

Figure 7. Electronic sheet conductivity parallel to grain boundaries as function of layer thickness L for equilibrium distribution of acceptor ions at 500°C and non-equilibrium acceptor profiles, quenched from 1000°C.

calculated by integration over the cross section of the internal crystal. Since space charge layers parallel to the direction of current provide the major contribution to electronic conductivity, this calculated sheet conductivity is also representative for a 3-dimensional crystal. In fig.7, the resulting sheet conductivity of the quenched state and of the equilibrium state are plotted as function of layer width L. As a result of quenching of the acceptor distribution, the electronic conductivity is larger than the equilibrium value at a layer thickness $L \geq 40$nm. However, this effect became negligible at small grain sizes. In particular, the steep increase with a slope corresponding to an L^{-3} power law was still observed, because space charge segregation, although at high temperatures, was still part of this non-equilibrium model. In other words, the analysis of experimental results using the equilibrium space charge model was fairly robust, even if the acceptor distribution was quenched from high temperatures.

CONCLUSIONS

The effect of space charge layers in polycrystalline cerium oxide was analyzed by comparing experimental results of grain size-dependent electrical conductivity with theoretical models. Electrical conductivity of nanocrystalline cerium oxide in the mesoscopic regime was modeled by numerical analysis of the Poisson-Boltzmann equation and non-equilibrium charge transport was calculated by solving the drift diffusion equations. Comparison with experimental data revealed good agreement for a space charge potential of 0.6 ± 0.05V. It was shown that an L^{-3} power law for the electronic conductivity could be derived for the equilibrium space charge model, while other scaling behavior were obtained for alternative models. The origin of space charge potential was investigated by numerical calculation of the electrical potential in a two-phase model. It was found, that the positive space charge potential at grain boundaries of cerium oxide is caused by a reduced enthalpy of formation of oxygen vacancies at the grain boundary core, with a value about 2 eV lower than in the bulk. Doping with acceptor ions was found to result in a small increase in the space charge potential, in agreement with experimental data. Quenching of high-temperature equilibrium distribution of acceptor ions to lower temperatures was found have a negligible effect on electrical conductivity of nanocrystalline materials.

ACKNOWLEDGMENTS

The author thanks C. Bäuerle for the numerical analysis of the 3-dimensional problem, S. Kilassonia for experimental assistance and R. Birringer for valuable discussions. Financial support by the German National Science Foundation (Deutsche Forschungsgemeinschaft, Cooperative Research Center 277, 'Interface controlled materials') and the Fonds der Chemischen Industrie is gratefully acknowledged.

APPENDIX A

As noticed by Frenkel [17] and later Kliewer and Koehler [18], a space charge layer should exist at the surface of an ionic material in order to compensate the difference in standard enthalpies of formation for the various defects in a particular material. In pure cerium oxide, oxygen vacancies and electrons are the significant charge carriers, and their concentrations are determined by the oxygen exchange equilibrium

$$O_O^x \Leftrightarrow V_O^{\cdot\cdot} + 2e' + \frac{1}{2}O_2 \qquad (A1)$$

with the law of mass action

$$n^2 \frac{[V_O^{\cdot\cdot}]}{[O_O^x]} = K_0 \exp\left\{-\frac{h_{red}}{k_B T}\right\} p_{O_2}^{-1/2} = \exp\left\{-\frac{2g_{e'}^0 + g_{V_O^{\cdot\cdot}}^0}{k_B T}\right\}, \qquad (A2)$$

and the charge neutrality equation

$$2[O_O^x] \exp\left\{-\frac{g_{V_O^{\cdot\cdot}}^0 + 2e_0 \Phi_{bulk}}{k_B T}\right\} = \exp\left\{-\frac{g_{e'}^0 - e_0 \Phi_{bulk}}{k_B T}\right\}. \qquad (A3)$$

Please notice that all concentrations are expressed as molar fraction per CeO_2 formula unit so that $[O_O^x]=2$ (this factor of 2 is usually omitted and included in the prefactor K_0). As a further deviation from the usual notation of the law of mass action, both K_0 and $p_{O_2}^{1/2}$ are tranfered into the argument of the exponential function in eqn. A2 with the result that the free enthalpies of defect formation become dependent on temperature and oxygen partial pressure. With the known quantities for $K_0 = 3.73 \cdot 10^6$ atm$^{-1/2}$ and $h_{red} = 4.67$ eV [1], the two relationships allow to calculate the individual enthalpies of defect formation g^0 of both electrons and oxygen vacancies for a given temperature T, oxygen partial pressure p_{O2} and bulk potential $\Delta\Phi_{bulk}$.

REFERENCES

1. H.L. Tuller, A.S. Nowick, *J. Electrochem. Soc.* **126**, 209 (1979).
2. H.L. Tuller in *Nonstoichiometric Oxides*, edited by T.O. Sørensen, (Academic Press, New York, 1981) p.271.
3. A. Tschöpe, E. Sommer, R. Birringer, *Solid State Ionics* **139**, 255 (2001).
4. Y.-M. Chiang, E.B. Lavik, I. Kosacki, H.L. Tuller, and J.Y. Ying, *J. Electroceramics* **1**, 7 (1997).
5. J.H. Hwang, T.O. Mason, *Z. Phys. Chem.* **207**, 21 (1998).
6. I. Koacki, T. Suzuki, H.U. Anderson, in *Solid State Ionic Devices*, edited by E.D. Wachsman, M.-L. Liu, J.R. Akridge, N. Yamazoe (Electrochemical Society Proc. **99-13**, Pennington, 1999), p.190.
7. E. Sommer, *diploma thesis*, Universität des Saarlandes, 1997.
8. A. Tschöpe, J.Y. Ying, H.L. Tuller, *Sensor & Actuators* **B31**, 111 (1992).
9. S. Kim, J. Maier, *J. Electrochem. Soc.* **149**, J73-J83 (2002).
10. A. Tschöpe, S. Kilassonia, B. Zapp, R. Birringer, *Solid State Ionics* **149**, 261 (2002).
11. J. Maier, *Prog. Solid State Chem.* **23**, 171 (1995).
12. A. Tschöpe, *Solid State Ionics* **139**, 267 (2001).
13. A. Tschöpe, R.B. Birringer, *J. Electroceramics* **7**, 169 (2001).
14. P.A. Markovich, "Semiconductor Equations" (Springer Verlag, Wien, 1990).
15. C. Bäuerle, *diploma thesis*, Universität des Saarlandes, 2001.
16. J. Jamnik, J. Maier, S. Pejovnik, *Solid State Ionics* **75**, 51 (1995).
17. J. Frenkel, "Kinetic Theory of Liquids" (Oxford Univ. Press, New York, 1946).
18. K.L. Kliewer, J.S. Koehler, *Phys. Rev.* **140**, 1226 (1965).
19. D.F. Evans, H. Wennerström, "The Colloidal Domain" (VCH Publ., New York, 1994) p.110.
20. D. McLean, "Grain boundaries in Metals" (Oxford Univ. Press, 1957).
21. R.T. DeHoff in *Applied Metallography*, edited by G.F. Van der Voort (Van Nostrand, New York, 1986) p. 89.
22. R. Waser, R. Hagenbeck, *Acta mater.* **48**, 797 (2000).

Mat. Res. Soc. Symp. Proc. Vol. 756 © 2003 Materials Research Society   EE4.6

Impedance/Dielectric Spectroscopy of Electroceramics in the Nanograin Regime

N. J. Kidner, B. J. Ingram, Z. J. Homrighaus, T. O. Mason, and E. J. Garboczi[1]
Department of Materials Science and Engineering and Materials Research Center, Northwestern
University, Evanston, IL 60208, U.S.A.
[1]Materials and Construction Research, Building and Fire Research Laboratory, National Institute
of Standards and Technology, Gaithersburg, MD 20899, U. S. A.

ABSTRACT

In the microcrystalline regime, the behavior of grain boundary-controlled electroceramics is
well described by the "brick layer model" (BLM). In the nanocrystalline regime, however, grain
boundary layers can represent a significant volume fraction of the overall microstructure and
simple layer models are no longer valid. This work describes the development of a pixel-based
finite-difference approach to treat a "nested cube model" (NCM), which more accurately
calculates the current distribution in polycrystalline ceramics when grain core and grain
boundary dimensions become comparable. Furthermore, the NCM approaches layer model
behavior as the volume fraction of grain cores approaches unity (thin boundary layers) and it
matches standard effective medium treatments as the volume fraction of grain cores approaches
zero. Therefore, the NCM can model electroceramic behavior at all grain sizes, from nanoscale
to microscale. It can also be modified to handle multi-layer grain boundaries and property
gradient effects (e.g., due to space charge regions).

INTRODUCTION

There are a number of existing and proposed applications of electroceramics in
nanocrystalline form, including batteries, fuel cells, gas separation membranes, solar cells, etc.
[1] Nanoceramics are utilized as chemical catalysts and as chemical sensors. Their
microcrystalline counterparts are often used as active electrical devices (e.g., varistors and
thermistors). In certain cases, like the latter, grain boundaries are necessary to impart the
required electro-active or thermo-active responses. In other cases, grain boundaries act as
undesirable barriers limiting transport (e.g., in ionic conductors). In still others, boundaries
between dissimilar ceramics can impart enhanced ion transport due to high mobility space charge
regions (e.g., in "dispersed ionic conductors") [2,3]. Given the high surface-to-volume ratios in
nanoceramics, grain boundaries can be expected to exert greater influence over
electrical/dielectric properties than in conventional microcrystalline ceramics.

There are several problems with existing grain boundary layer models (see below) insofar as
describing the electrical/dielectric response of nanoceramics is concerned. First, in the nanograin
regime, boundary layers such as space-charge regions or local oxidation layers can represent a
significant volume fraction of the overall microstructure (see Figure 1). Conventional layer
models, such as the "brick layer model" (BLM), are hardly adequate for such a situation.
Second, as pointed out by Maier [4], there can be differential transport coefficients parallel vs.
perpendicular to the grain boundaries. Finally, space charge regions represent spatially varying
electrical properties, which are not consistent with the simple property step functions assumed in
most layer models.

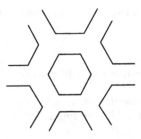

Figure 1. Schematic of nanostructure with a significant grain boundary layer

The BLM was first conceived 25 years ago by Beekmans and Heyne [5], although Burggraaf and co-workers are credited with coining the "brick layer" name [6,7]. The microstructural picture is represented by Figure 2. The simplest form of the BLM, which we refer to as the series-BLM (S-BLM), ignores the side-wall contributions (on the left) and considers only the serial connections of grain cores and capping grain boundary layers (on the right). The corresponding equivalent circuit is shown in Figure 3a, where the open box represents the equivalent circuit ($R_{gc}C_{gc}$) of the grain cores and the shaded box represents the equivalent circuit ($R_{gb}C_{gb}$) of the grain boundaries. Using the notation of Boukamp [8], this series combination of two (RC) parallel circuits can be represented as ($R_{gb}C_{gb}$)($R_{gc}C_{gc}$). This model is quite appropriate for thin, continuous, and highly resistive second phase films such as siliceous layers in low purity, microcrystalline ionic conductors [9].

The major deficiency of the S-BLM, ignoring side-wall contributions, was addressed by Näfe [10], who developed a series/parallel BLM (SP-BLM) by connecting the central grain core/grain boundary serial path of Figure 2 (on the right) in parallel with the side-wall grain boundary path (on the left). The corresponding equivalent circuit is shown in Figure 3b. We recently applied the SP-BLM to the analysis of nanoceramic impedance/dielectric spectra [11].

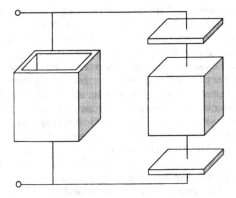

Figure 2. The brick layer model with series/parallel connectivity.

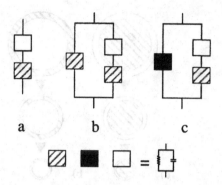

a   b   c

Figure 3. Equivalent circuit representations for a) the series brick layer model (S-BLM), b) the series/parallel BLM (SP-BLM), and c) the SP-BLM with different electrical properties parallel vs. perpendicular to the grain boundary.

A modified form, which we refer to as the SP'-BLM, was developed by Maier and coworkers [3,12] to allow for differential electrical conductivity perpendicular vs. parallel to the grain boundary. The SP'-BLM equivalent circuit model is shown in Figure 3c.

A limitation common to both SP-BLM and SP'-BLM models is that current flow is restricted to either the central core (series) path or the outer (parallel) path, which is clearly not the case in an actual nanostructure (see Figure 1). Bonanos et al. [9] commented that the SP-BLM "is valid at high or low conductivity ratios…" but had "reservations about the use of this model over the entire σ_{gc}/σ_{gb} range, since it is not clear how, when $\sigma_{gb}\sim\sigma_{gc}$, this assumption that the current flows via two separate mechanisms can be tenable." Based upon comparison with the Maxwell-Wagner/Hashin-Shtrikman effective medium theory, which sets the absolute upper and lower limits of conductivity for isotropic two-phase composites, McLachlan et al. [13] concluded that "…where it (the SP-BLM) lies outside the MW-HS limits (which it does at most intermediate grain core volume fractions) it is fundamentally wrong."

Effective medium theory (EMT) has also been applied to the complex impedance/ dielectric response of electroceramics. EMT models obviate the limitations of the layer models by taking into account real current distributions in heterogeneous media. They therefore provide important benchmarks against which to compare microstructurally-based models.

As early as 1914, Wagner [14] showed that Maxwell's equation for DC conductivity [15] also worked for the complex conductivity. A MW medium can be visualized as built up from a space-filling array of coated spheres, as in Figure 4, with each sphere surrounded by a mixture of the two components having the mean or effective property value of the medium. As pointed out by McLachlan et al. [13], the MW model is equivalent to the Hashin-Shtrikman upper and lower bounds for conductivity of an isotropic two-phase mixture [16] and the well known Clausius-Mossoti equation for dielectrics. In the limit that the volume fraction of the continuous (matrix) phase becomes small (thin boundary layers), we showed that the impedance/dielectric response becomes indistinguishable from the brick layer models [13]. Thin coatings, whether insulating

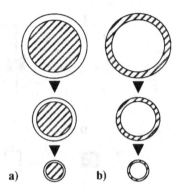

Figure 4. The basic building blocks for the Maxwell-Wagner/Hashin Shtrikman effective media, with either a) conductive coatings or b) insulating coating.

or conductive relative to the cores, behave identically regardless of the grain morphology (i.e., spheres vs. cubic "bricks"). Finite element analyses on "real" 2-D microstructures agreed well with BLM predictions unless grain shape became highly distorted or a bimodal distribution of grain sizes was present [17-19]. This means that simplified morphologies, whether spherical (i.e., the various EMT models) or cubic (e.g., the nested-cube model below), stand a very good chance of accurately describing the impedance/dielectric response of nanostructures with equiaxed, mono-sized grains.

The range of minority phase volume fractions over which the MW-HS model is believed to be valid is $0 \leq \phi \leq 0.3$ [9]. We are interested in developing a model capable of traversing the entire range of grain core volume fraction from 0 (nanoscale) to 1 (microscale), assuming nanometer scale boundary layers, e.g., consistent with space charge layers in electroceramics. One model pertinent to the present work is that of Zuzovsky and Brenner [20]. The Zuzovsky-Brenner Model (ZBM) consists of a cubic array of second phase spherical particles suspended in a continuous matrix phase, the unit cell of which is shown in Figure 5a. This model is perhaps the

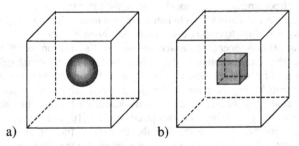

Figure 5. Unit cells of the a) Zuzovsky/Brenner model, with spherical second phase particles (grain cores) on a simple cubic lattice and b) the nested cube model, with cubic second phase particles (grain cores) on a simple cubic lattice.

most representative of the situation in nanoceramics with relatively thick grain boundary layers; grain core morphology will most likely *not* maintain overall grain shape due to smearing/rounding of boundary layers (e.g., space charge regions, see figure 1). The problem with the ZBM is that a percolation threshold (grain core-to-grain core) is reached at a grain core volume fraction of 0.52.

The present work reports the development of an analogous nested-cube model or NCM, the unit cell of which is shown in Figure 5b. The NCM has no percolation threshold, and is capable of describing impedance/dielectric behavior over the entire range of grain core fractions, from small values (where it matches ZBM behavior and also agrees with MW-HS model results) to large values (where it matches the brick layer model results). As we will show, the NCM also has potential for describing multi-layer grain boundary structures and property gradients at grain boundaries (e.g., in space charge regions).

EXPERIMENTAL DETAILS

The nested cube model is not tractable analytically. A FORTRAN-77 finite-difference numerical program, **ac3d.f**, was therefore modified to carry out pixel-based computer calculations at finite frequencies. This program, developed at NIST, can be accessed at http://ciks.cbt.nist.gov/monograph/, Chapter 2, along with a manual in HTML format [21]. The program was designed to compute the electrical properties of random materials whose microstructure can be represented by a 3-D digital image. It can also be used to simulate non-random, but analytically intractable geometries, as in the present work. A system size of between 20^3 and 80^3 pixels was employed to represent the 3-D structure of the NC model in Fig. 5b. Depending on the grain size, pixels are either grain core or grain boundary (for a single grain boundary layer). In the computation process, a finite-difference node is set up in the middle of each pixel. As part of the computation, bonds are assigned between each pair of nodes reflecting the (RC) values assigned to each pixel. A conjugate gradient method is then used to solve Laplace's equation at each frequency to give the complex conductivity of the microstructure. Real and imaginary conductivities are then converted to impedance and modulus quantities via standard equations.

To generate the periodic simple cubic lattice of the NCM, it is necessary to add a shell of imaginary states around the main system to maintain the periodic boundary conditions. For a given grain core volume fraction, the system size is varied to assess the effect of spatial resolution. A plot of conductivity vs. $1/N$ (where N is the number of pixels) is extrapolated to give the conductivity at $1/N \rightarrow 0$. Computing time restricted system size to below 100^3 pixels.

The NCM was compared to EMT models (MW-HS) and the ZBM at small-to-intermediate grain core volume fractions, and to the S-BLM and SP-BLM models at intermediate-to-large grain core volume fractions. Unlike the NCM, analytical equations exist for each of these models, which could be expressed in terms of complex conductivities, $\sigma^* = \sigma_r + i\sigma_i$, involving both real ($\sigma_r$) and imaginary ($\sigma_i$) components. Standard equations were employed in each case to convert to impedance and modulus formats. We also considered Bode plots (log-log plots of real and imaginary impedance or capacitance vs. frequency).

RESULTS AND DISCUSSION

First to be considered is the case of large grain core volume fraction, i.e., thin grain boundary layers. Figure 6 shows a Nyquist (impedance or Z-plane) plot of NCM results vs. the two brick layer models. The ZBM is not valid at this volume fraction of grain cores (0.927), since this is above the percolation threshold of the ZBM (0.52). The ratio of grain boundary-to-grain core conductivity (σ_{gb}/σ_{gc}) was set at 0.1 and the dielectric constants were the same ($\varepsilon_{gb}=\varepsilon_{gc}$). The NCM is in good agreement with both the S-BLM and the SP-BLM, as also seen in modulus and Bode plots (not shown). This is to be expected, since the grain boundary layers are quite thin at this value of grain core volume fraction; D/d, the ratio of grain core dimension to grain boundary thickness is approximately 39. The NCM picture at large grain core volume fraction closely resembles that of the boundary layer models, especially the SP-BLM.

At the other end of the relative size spectrum, Z-plane results for a small grain core volume fraction (0.162) are shown in Figure 7. The ratio of grain core-to-grain boundary dimension is now D/d~1.20. For the calculations, the ratios of grain boundary-to-grain core properties were set at σ_{gb}/σ_{gc}=0.10 and $\varepsilon_{gb}/\varepsilon_{gc}$=10, respectively. The NCM results are seen to approach the MW-HS predictions, which is important, since realistic models must agree with EMT predictions at small volume fractions. This agreement was also observed in Modulus and Bode plots (not shown). There is also reasonable agreement between the NCM results and ZBM predictions. Some differences can be anticipated based on the difference in grain core morphologies (see Figure 5). If we consider the dilute limit ($\phi<0.1$), the conductivity of a composite (σ) relative to that of the matrix (σ_m) should vary with the volume fraction of highly conductive second phase particles according to [22]:

$$(\sigma/\sigma_m) = 1 + [\sigma]_\infty \, \phi \tag{1}$$

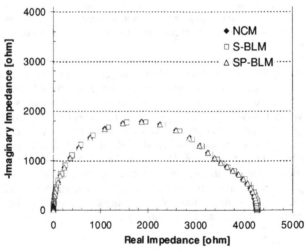

Figure 6. Simulated impedance response for various models assuming $\sigma_{gb}/\sigma_{gc} = 0.1$, $\varepsilon_{gb}=\varepsilon_{gc}$, and a volume fraction of 0.927 for grain cores. See text for code to models.

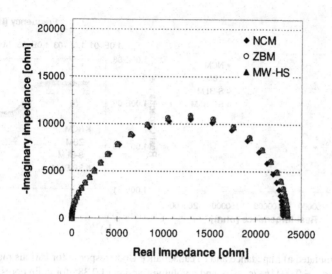

Figure 7. Simulated impedance response for various models assuming $\sigma_{gb}/\sigma_{gc} = 0.1$, $\varepsilon_{gb}/\varepsilon_{gc}=10$, and a volume fraction of 0.162 for grain cores. See text for code to models.

where $[\sigma]_\infty$ is the "intrinsic conductivity" of the conducting particles, each shape having a characteristic value. It has been established that the intrinsic conductivity of a conductive cube is 3.4 whereas that of a conductive sphere is 3.0 [22]. Therefore, it is not surprising to see some differences, but otherwise close similarity between the two models. In the intrinsic range ($\phi<0.1$) we found that both the ZBM and NCM agreed well with the MW-HS model.

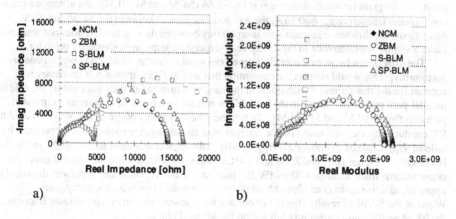

a)   b)

Figure 8. Simulated a) impedance and b) modulus response for various models assuming σ_{gb}/σ_{gc} = 0.1, $\varepsilon_{gb}/\varepsilon_{gc}$ =100, and a volume fraction of 0.385 for grain cores. See text for code to models.

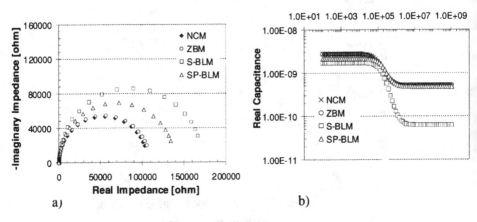

Figure 9. Simulated a) impedance and b) capacitance Bode response for various models assuming $\sigma_{gb}/\sigma_{gc} = 0.001$, $\varepsilon_{gb}/\varepsilon_{gc}=10$, and a volume fraction of 0.385 for grain cores. See text for code to models

At intermediate values of grain core volume fraction, significant differences between the NCM (or ZBM) and the brick layer models emerge. Figures 8a and b show Z-plane and M-plane plots, respectively, for a conductivity ratio of $\sigma_{gb}/\sigma_{gc}=0.1$ and a dielectric constant ratio of $\varepsilon_{gb}/\varepsilon_{gc}$ =100 at a grain core volume fraction of 0.385. Similarly, Figures 9a and b show Z-plane and C-Bode plots, respectively, for the same grain core volume fraction (0.385), but with a conductivity ratio of $\sigma_{gb}/\sigma_{gc}=0.001$ and a dielectric constant ratio of $\varepsilon_{gb}/\varepsilon_{gc}$ =10. In both cases the ratio of grain core-to-grain boundary dimension is D/d~2.66 (NCM and SP-BLM). As at smaller grain core volume fraction (e.g., $\phi=0.162$), there is good agreement between the NCM and ZBM predictions; the difference in grain core morphology between the two models does not seem to make a significant difference in their frequency-dependent impedance/dielectric behavior.

The differences between the brick layer model results and the NCM/ZBM predictions are noteworthy. This would support the contention that an EMT-like approach is necessary to account for the true current distributions in the nanostructure, rather than discounting the role of parallel-path grain boundaries (in the S-BLM) or restricting current flow in series vs. parallel paths (in the SP-BLM and SP'-BLM). To test the general validity of the NCM, we calculated the DC conductivity vs. grain core volume fraction over its entire range (0<ϕ<1) for a conductivity ratio of σ_{gb}/σ_{gc} =0.01 or conversely σ_{gb}/σ_{gc} =10. This is plotted against the SP-BLM and MW-HS models in Figure 10. (The ZBM and S-BLM were not plotted, since they do not cover the entire volume fraction range.) The MW-HS lines are definitive, since they represent the absolute upper (conductive matrix) and lower bounds (resistive matrix) for isotropic composites. Whereas the SP-BLM results clearly fall outside the allowed range at certain volume fractions, the NCM predictions consistently fall within the allowed range.

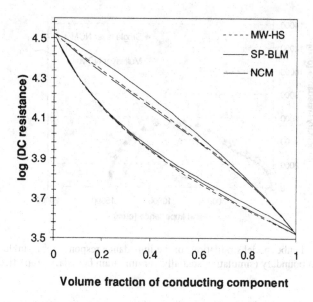

Figure 10. DC resistivity bounds for various models assuming $\sigma_{gb}/\sigma_{gc} = 0.1$ for the upper bound and $\sigma_{gb}/\sigma_{gc} = 10$ for the lower bound. See text for code to models.

The C-Bode plot differences in Figure 9b are also of note. This dual-plateau behavior is characteristic of (RC)(RC) equivalent circuit interpretation (as in Figure 3a). We have shown that the high frequency plateau is given by [23]:

$$C_{re}(hiv) = C_2C_1/(C_1+C_2) \tag{2A}$$

and the low frequency plateau is given by:

$$C_{re}(lov) = (R_2^2C_2+R_1^2C_1)/(R_1+R_2)^2 \tag{2B}$$

Only in the case of $C_2 >> C_1$ is Cre(hiv)≈C_1, and with the additional constraint that $R_2 >> R_1$ is Cre(lov)≈C_2. This corresponds to the classical BLM instance where the second (R_2C_2) component corresponds to very thin (therefore high capacitance) grain boundaries, usually with a high resistance compared to the grain cores. It follows, however, that if the grain boundaries are neither thin (as in Figure 1) nor much more resistive than the grain cores, the high frequency C-Bode plateau will be a combination of the two component capacitances (Eq. 2A) whereas the low frequency plateau with be a still more complex combination of all four parameters (Eq. 2B). Unfortunately, many impedance/dielectric spectroscopy practitioners are prone to derive the dielectric constant for a given microstructural element directly from what they interpret to be the corresponding C-Bode plateau. This is highly suspect in the nanoscale regime, where the C-

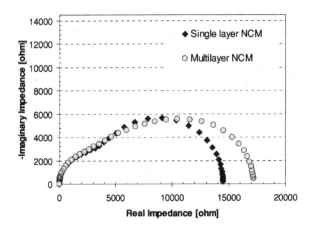

Figure 11. Nested cube model simulations of the impedance response for a single layer vs. a multi-layer grain boundary (simulating spatially varying grain boundary properties). See text for code to models.

Bode plateaus are bound to be complex functions of the four constituent electrical parameters (σ_1, σ_2, ε_1, ε_2) and the volume fraction of grain cores. The NCM should be much more reliable insofar as fitting impedance/dielectric spectra in the nanoregime in terms of local component electrical properties and volume fractions.

The nested cube approach can be modified to account for multi-layer grain boundaries and/or spatial gradients of electrical properties, e.g., associated with space charge regions. Figure 11 shows preliminary Z-plane simulations for a grain boundary with three layers. Here we have subdivided a grain boundary layer representing 61.5% of the overall grain volume whose values are $\sigma_{gb}/\sigma_{gc}=0.1$ and $\varepsilon_{gb}/\varepsilon_{gc}=100$ into three layers of equal width (1/3 each) totaling 61.5% of the microstructure, and whose resistivities vary from 150% of the average in the near-grain boundary layer(s) to 100% of the average in the middle layer to 50% of the average in the near-grain core layer. Similarly, the local dielectric constant has been varied in the same manner from layer to layer. This structure is hardly representative of the variation in properties in a space charge region. Nevertheless, it offers some insight into the effect of local property variations on the resulting impedance response. There is noticeable arc-depression for the multilayer NCM as compared to the single layer NCM in Figure 11. Similar arc-depression has been reported for nanoceramics in the literature [23]. Further work in this area is warranted.

CONCLUSIONS

A novel "nested cube model" (NCM) has been developed to describe the frequency-dependent behavior of electroceramics in the nanocrystalline regime. The NCM is capable of describing behavior over the entire range of grain sizes, from nanocrystalline (where grain core volume fractions are small) to microcrystalline (where grain boundary thicknesses are small). It was shown that the NCM agrees with effective medium theory (Maxwell-Wagner/Hashin-Shtrikman model) at small grain core volume fractions, with the Zuzovsky-Brenner model (grain core spheres on a simple cubic lattice) at intermediate grain core volume fractions, and with the brick layer models at large grain core volume fractions, as expected. In the intermediate volume fraction regime, the NCM is a more accurate model to describe the current distributions between grain cores and grain boundaries. Such a model is necessary to accurately deconvolute local electrical properties (conductivity, dielectric constant) from impedance/dielectric spectra. The NCM can be modified to account for multi-layers and/or spatial property gradients at grain boundaries.

ACKNOWLEDGMENTS

This work was supported in part by the U.S. Department of Energy under grant no. DE-FG02-84ER45097 and in part by the National Science Foundation under grant no. DMR-0076097 through the Materials Research Science and Engineering Center program.

REFERENCES

1. Y.–M. Chiang, , J. Electroceram. **1** (3), 205 (1997).
2. N.J. Dudney, Ann. Rev. Mater. Sci. **19**, 103 (1989).
3. J. Maier, Prog. Solid State Chem. **23**, 171 (1995).
4. J. Maier, Ber. Busenges. Phys. Chem. **90**, 26 (1986).
5. N.M. Beekmans and L. Heyne, Electrochim. Acta **21**, 303 (1976).
6. T. van Dijk and A.J. Burggraaf, Phys. Stat. Solidi A **63**, 229 (1981).
7. M.J. Verkerk, B.J. Middlehuis, and A.J. Burggraaf, Solid State Ionics **6**, 159 (1982).
8. B.A. Boukamp, "Equivalent Circuit (EQUIVCRT.PAS)," University of Twente, The Netherlands (1990).
9. N. Bonanos, B.C.H. Steele, E.P. Butler, W.B. Johnson, W.L. Worrell, D.D. Macdonald, and M.C.H. McKubre, in *Impedance Spectroscopy: Emphasizing Solid Materials and Systems*, edited by J.R. Macdonald, (Wiley and Sons, New York, 1987) pp. 191-205 .
10. H. Näfe, Solid State Ionics **13**, 255 (1984).
11. J.–H. Hwang, D.S. McLachlan, and T.O. Mason, J. Electroceram. **3** (1), 7 (1999).
12. S. Kim and J. Maier, J. Electrochem. Soc., preprint.
13. D.S. McLachlan, J.–H. Hwang, and T.O. Mason, J. Electroceram. **5** (1), 37 (2000).
14. K.M. Wagner, in *Arkiv Electrotechnik*, edited by H. Schering, (Springer-Verlag, Berlin, 1914).
15. J.C. Maxwell, *A Treatise on Electricity and Magnetism*, 2nd ed. (Clarendon Press, Oxford, 1881).
16. Z. Hashin and S. Shtrikman, J. Appl. Phys. **33**, 3125 (1962).
17. J. Fleig and J. Maier, J. Electrochem. Soc. **145**, 2081 (1998).
18. J. Fleig and J. Maier, J. Eur. Ceram. Soc. **19**, 693 (1999).
19. R. Hagenbeck and R. Waser, Ber. Busenges. Phys. Chem. **101**, 1238 (1997).
20. M. Zuzovsky and H. Brenner, J. Appl. Math. Phys. **28**, 979 (1977).

21. E.J. Garboczi, NIST Internal Report **6269** (1998). Also available at http://ciks.cbt.nist.gov/monograph/, Chap. 2.
22. J.F. Douglas and E.J. Garboczi, in Adv. Chem. Phys., Vol. XCI, edited by I. Prigogine and S.A. Rice, (Wiley & Sons, 1995), p. 85.
23. T.O. Mason, J.–H. Hwang, N. Mansourian-Hadavi, G.B. Gonzalez, B.J. Ingram, and Z.J. Homrighaus, in Nanocrystalline Metals and Oxides: Selected Properties and Applications, edited by P. Kauth and J. Schoonman, (Kluwer Academic Publishers, Norwell, MA, 2002), p. 111.

Mat. Res. Soc. Symp. Proc. Vol. 756 © 2003 Materials Research Society

Li mobility in $(Li,Na)_yLa_{0.66-y/3}TiO_3$ perovskites $(0.09<y\leq0.5)$. A model system for the percolation theory.

J. Sanz[1], A. Rivera[1,2], C. León[2], J. Santamaría[2], A. Várez[3], O. V'yunov[4], A.G. Belous[4],

[1] Instituto Ciencia de Materiales de Madrid - CSIC, 28049 Cantoblanco, Spain.
[2] GFMC, Dpto. Física Aplicada III, Universidad Complutense, 28040 Madrid, Spain.
[3] Dpto. Ciencia de Materiales. Universidad Carlos III de Madrid. 28911 Leganés, Spain.
[4] Solid State Chem., Inst. Inorg. Chem., Ukrainian Academy Sciences, 252680, Ukraine.

ABSTRACT

The dependence of ionic transport on structure and composition of perovskites $Li_yLa_{0.66-y/3}TiO_3$ ($0.09 \leq y \leq 0.5$) and $Li_{0.5-x}Na_xLa_{0.5}TiO_3$ ($0 \leq x \leq 0.5$) has been analyzed by means of Neutron Diffraction, NMR and Impedance Spectroscopy. In the first series, ion conductivity displays a non-Arrhenius behavior, decreasing activation energy at increasing temperatures. Li mobility is accompanied by a considerable increase of Li thermal factor deduced from ND data. In the second series, local lithium mobility decrease monotonously with the sodium content at room temperature; however, long-range dc conductivity decreases sharply at x = 0.2 more than six orders of magnitude. This decrease on dc conductivity is discussed in terms of a three-dimensional percolation theory of vacant A-sites. In this model, the number of vacant sites is controlled by the amount of Na and La of the perovskite.

INTRODUCTION

Interest in ionic conducting solids is increasing in the last years because of their potential application as solid electrolytes in batteries, fuel cells and other electrochemial devices[1]. $Li_yLa_{0.66-y/3}TiO_3$ perovskites, with $0.09<y<0.5$, are among the best Li ion conductors, showing a dc conductivity of 10^{-3} S/cm at room temperature[2,3]. Since the discovery of its outstanding electrical properties, several groups have investigated structural features of these perovskites; however, reasons that enhances Li mobility have not been completely identified.

X-ray diffraction patterns of LLTO perovskites have been interpreted assuming a doubled perovskite along the c-axis ($a_p, a_p, 2a_p$ unit cells); however the symmetry changes along the series, from orthorhombic (S.G. Pmmm) to tetragonal (S.G. P4/mmm), when the Li content increases[3-5]. In samples quenched from 1350 °C, unit cell becomes cubic with parameters a_p, a_p, a_p (S.G. Pm3m) [4,6]. These changes of symmetry are related to the cation vacancies ordering. In Li-poor samples, cation vacancies are disposed in alternating planes along the c-axis; however, in Li-rich samples, vacancies become disordered. For samples quenched from high temperature the highest disordering is achieved[6].

The octahedral tilting was analyzed with neutron diffraction. The structural analysis of the Li-poor $Li_{0.18}La_{0.61}TiO_3$ perovskite, confirmed that vacant A-sites are disposed in alternated planes along the c-axis. Moreover, an octahedral tilting along the **b** axis produced the observed $2a_p \times 2a_p \times 2a_p$ superstructure[7-9]. Neutron diffraction pattern of the quenched Li-rich member $Li_{0.5}La_{0.5}TiO_3$ was described with the rhombohedral unit cell $\sqrt{2}a, \sqrt{2}a, 2\sqrt{3}a$ (S.G. R-3c). In this sample, octahedral tilting was produced along the three axes and vacancies were fully disordered[6]. Substitution of Li by Na does not affect unit cell dimensions of rhombohedral perovskites. In these samples, Na ions occupy A sites of the perovskite while Li ions are located at the center of the square planar windows connecting contiguous A sites[10].

In order to analyze structural factors that affect transport properties in lanthanum titanates perovskites, we have studied solid solutions $Li_yLa_{0.66-y/3}TiO_3$ ($0.09 \leq y \leq 0.5$) and $Li_{0.5-x}Na_xLa_{0.5}TiO_3$ ($0 \leq x \leq 0.5$) (LLTO and LNLTO series). This study has been carried out by neutron diffraction (ND), Nuclear Magnetic Resonance (NMR) and Impedance Spectroscopy (IS). In particular, short and long-range mobilities were analyzed with NMR and IS techniques.

EXPERIMENTAL

Samples were obtained from stoichiometric amounts of dried Li_2CO_3 (Merck), Na_2CO_3 (Merck), La_2O_3 (Aldrich 99.99%) and TiO_2 (Aldrich 99 %) by solid state reaction, following the procedure used in previous works[4,10]. The mixture was grounded together in an agate mortar and heated at 800 ºC for 4 hours in order to eliminate CO_2. The reground products were cold-pressed at 150 MPa and heated at 1150 ºC for 12 hours. Finally, LNLTO and LLTO samples were pressed and heated at 1200 ºC and 1350 ºC (6 h) respectively.

^{7}Li and ^{23}Na NMR spectra were obtained at room temperature in a MSL-400 Bruker spectrometer working at 155.45 MHz and 104.8 MHz respectively. Spectra were taken after irradiation of the sample with a $\pi/2$ pulse (3 μs). Spin-spin relaxation times (T_2) were deduced from the linewidth of the spectra. Spin-lattice relaxation times (T_1) were measured between 170 K and 500 K by using the classical (π-τ-$\pi/2$) sequence in a SXP 4/100 Bruker spectrometer, working at 10.6 MHz.

Sintered cylindrical pellets 5 mm in diameter and 1 mm thick, with evaporated silver electrodes, were used for electrical measurements. Impedance Spectroscopy measurements were conducted between 125 K and 330 K, using precision LCR meters HP4284A and HP4285A in the frequency range 20 Hz - 30 MHz. These measurements were carried out under N_2 flow to ensure an inert atmosphere.

RESULTS AND DISCUSSION

Li mobility in $Li_yLa_{0.66-y/3}TiO_3$ series

Figure 1 shows the plot of dc conductivity versus reciprocal temperature for orthorhombic and tetragonal perovskites, $Li_yLa_{0.66-y/3}TiO_3$ ($y = 0.18$ and 0.5). Both samples present a similar behavior, displaying activation energies that progressively decrease from 0.4 to 0.26 eV, when temperature increases. In analyzed series, dc conductivity increases with the Li content, reaching a *plateau* for $y > 0.25$[4].

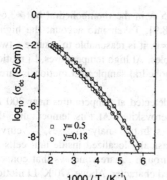
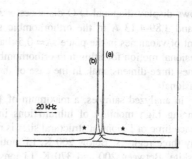

Figure 1.- Arrhenius plot of dc-conductivity of LLTO samples with y= 0.18 and 0.5

Figure 2.- R.T. ^{7}Li NMR spectra of LLTO samples with **(a)** y= 0.18 and **(b)** y=0.5.

^{7}Li (I=3/2) NMR spectra of $Li_{0.18}La_{0.61}TiO_3$ (a) and $Li_{0.5}La_{0.5}TiO_3$ (b) are given in figure 2. The spectrum of the Li-poor sample, x = 0.18, is formed by the central and two satellite transitions. The analysis of Li NMR spectra showed the presence of two lithium species with different mobility that exchange their positions to give a mean quadrupolar constant, $C_Q \sim 60$ kHz, at room temperature[7]. In Li-rich samples (y = 0.5), only a single component with very low C_Q was detected, indicating that mobile species predominate in this perovskite[11]. ND experiments of this samples, showed that Li ions were four-fold coordinated to oxygens in a planar square configuration occupying the center of the square windows that connect contiguous A-sites of the perovskite[6].

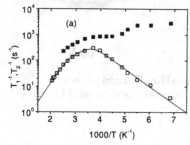
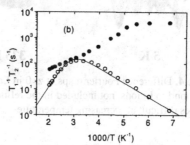

Figure 3.- Temperature dependence of T_1^{-1} (open symbols) and T_2^{-1} (solid symbols) for samples with y=0.18 **(a)** and y=0.5 **(b)**.

In figure 3, ^{7}Li spin-lattice (T_1^{-1}) and spin-spin (T_2^{-1}) relaxation rates of $Li_{0.18}La_{0.61}TiO_3$ (a) and $Li_{0.5}La_{0.5}TiO_3$ (b) perovskites are plotted versus reciprocal temperature. In both samples, Li mobility increases considerably above 180K, producing a considerable decrease of the linewidth (T_2^{-1}) with temperature. In the case of orthorhombic samples, the decrease of T_2^{-1} values is not monotonous, observing a *plateau* between 200-300 K. This behavior has been ascribed to the existence for lithium of a two-dimensional motion in ordered samples[4,7,12]. At this point, it is interesting to compare square windows that connect contiguous A sites of the perovskites. ND data showed that these windows display diagonal O-O distances ≈ 3.65-4.15 Å in the rhombohedral but ≈3.64-3.88 and 3.89-4.13 Å in the orthorhombic sample[8,9]. Taken into account the higher amount of vacancies in the plane $z/c= 0.5$ than in $z/c=0$, it is reasonable to assume a two-dimensional motion for lithium in orthorhombic samples. At high temperatures, Li motion becomes three-dimensional. In the case of the rhombohedral samples, Li motion is three-dimensional.

In analyzed samples, a maximum of $1/T_1$ is detected at temperatures near 300 K, evidencing high mobility of lithium ions in two perovskites. At this temperature, the residence time of Li ions at structural sites is near 10^{-8} s. In two analyzed cases, T_1^{-1} curves display three regimes. Below 200 K, Li motion is basically localized inside unit cells of perovskite. Between 200 and 370 K, Li ions pass through square windows that connect contiguous A-sites, preserving the Li motion a localized character. Above 370 K, Li motion becomes extended and a long range ionic transport is achieved[11,12]. In $Li_{0.5}La_{0.5}TiO_3$ sample, the last two regimes are poorly resolved.

Localization of lithium at unit cell faces of the perovskite, deduced from difference Fourier maps in $La_{0.5}Li_{0.5}TiO_3$, decreases considerably at increasing temperatures. Above 573 K, localization of Li becomes difficult with diffraction techniques (see Fig. 4). Taken into account that conductivity is similar in all analyzed $Li_yLa_{0.66-y/3}TiO_3$ perovskites, we can conclude that the octahedral tilting and distribution of vacancies do not limit seriously Li diffusion in LLTO perovskites.

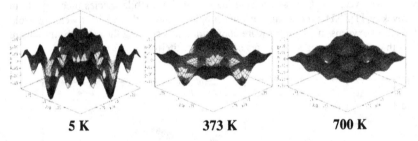

5 K   **373 K**   **700 K**

Figure 4. Difference Fourier maps at $z=0$ in $La_{0.5}Li_{0.5}TiO_3$. Detected minima correspond to Li ions, not included in the refinement model. Localization of Li becomes difficult at increasing temperatures.

Li mobility in $Li_{0.5-x}Na_xLa_{0.5}TiO_3$

^{7}Li MAS-NMR spectra of $Li_{0.5-x}Na_xLa_{0.5}TiO_3$ perovskites showed that the central line is more intense than that would correspond to satellite transitions (modulated by equally spaced lateral bands), indicating the presence of mobile species whose concentration decreases in a monotonous way with the sodium content (Figure 5a). ^{7}Li MAS NMR spectra of less mobile species were fitted with the same parameters than in LLTO series ($C_Q= 60$ kHz, $\eta= 0.5$), suggesting that Li^+ ions occupy similar positions in both series. In Na-rich samples, most Li displays very low mobility. However, conductivity σ_{dc} values in LNLTO series showed an abrupt decrease at $x = 0.31$ that cannot be explained

in terms of the number of charge carriers, n_c (Figure 5b). This fact evidences the critical effect of Na on the long-range ionic transport in LNLTO perovskites.

In a recent study of the $Li_{0.5-x}Na_xLa_{0.5}TiO_3$ ($0 \leq x \leq 0.5$) series, it was shown that all samples display a rhombohedral R-3c symmetry, with unit cell parameters that do not change appreciably with the composition[10]. The octahedral tilting produces Ti-O-Ti angles near 167º and an oblique distortion of square windows that connect contiguous A sites of the perovskite, impeding Na ions to pass through windows. If Li ions were located at the A sites, the amount of vacant A-sites and conductivity should be very low in this series. However, dc conductivity of Li-rich samples (x<0.2) is very high at room temperature, suggesting that A-sites associated with Li are not occupied and Li pass through square windows.

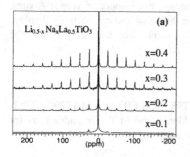 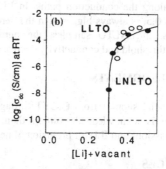

Figure 5.- (a) ^7Li MAS-NMR spectra of $Li_{0.5-x}Na_xLa_{0.5}TiO_3$ perovskites with increasing amount of sodium **(b)** σ_{dc} vs. Li plus vacant sites in LLTO and LNLTO.

Taken into account that Na occupy preferentially A-sites, we have considered a percolation model in which A sites associated with Li ions are vacant and participate in ionic diffusion, but La^{3+} and Na^+ ions occupy A sites blocking conduction pathways of the perovskite[10]. In this model, vacant distribution is random and dc conductivity is given by the expression $\sigma_{dc}=K(n-n_p)^2$, where n is the number of vacant A-sites and n_p is the percolation threshold for a three-dimensional cubic network ($n_p \approx 0.31$). In agreement with these predictions, ion conductivity is very high when the sodium content is lower than 0.2, but very low above this value. At x = 0.2, the amount of vacancies is close to the percolation threshold, and conductivity decreases drastically (see Figure 5b)[10,13].

Based on structural and spectroscopic results, the amount of vacant A-sites could correspond to the sum of nominal vacancies plus Li content. The plot of dc-conductivity vs Li + vacancies is given in the Figure 4b. Conductivity values of LLTO and LNLTO series agree with those deduced from the percolation model, confirming that sites A associated with lithium ions are vacant and participate to Li conduction. According to this fact, the amount of vacancies in LLTO samples (0.3-0.5) is higher than the percolation threshold and conductivity of all these samples is important. In the case of LNLTO series, the amount of vacancies decrease below the percolation threshold and dc-conductivity decrease dramatically.

CONCLUSIONS

Cation mobility has been studied in crystalline $Li_yLa_{0.66-y/3}TiO_3$ ($0.09 \leq y \leq 0.5$) and $Li_{0.5-x}Na_xLa_{0.5}TiO_3$ ($0 \leq x \leq 0.5$) perovskites by means of NMR and Impedance spectroscopies. In the first series, modifications produced in bottlenecks by tilting of octahedra do not affect seriously Li diffusion. In these perovskites, Li occupies unit cell faces of the perovskite, increasing the number of vacant A-sites that participate in Li motion. In orthorhombic perovskites, distribution of vacancies favors the two-dimensional motion of Li.

In $Li_{0.5-x}Na_xLa_{0.5}TiO_3$ perovskites, Na and La ions are located at A sites, reducing the amount of vacant A sites that participate in ion motion. According to this fact, σ_{dc} conductivity display a sharp decrease at $x = 0.2$ that has been ascribed to a percolation of Li motion along the conduction paths. In LLTO samples, the amount of vacant A-sites, $0.33 \leq n_V \leq 0.5$, is always bigger than the percolation threshold and conductivity displays high values. In LNLTO samples, the number of vacant sites decreases below the percolation threshold and conductivity decreases drastically.

ACKNOWLEDGMENTS

Financial support from CICYT through MAT98 1053-C03 and MAT2001 3713-C03 projects is acknowledged. Authors thank J.A. Alonso and M.T. Fernández-Díaz for helpful discussions and ILL for provision of neutron beam time.

REFERENCES

1. K. M. Colbow, J. R. Dahn, R. R. Hearing, *J. Power Sources* **26**, 397 (1989).
2. A. G. Belous, G. N. Novitskaya, S. V. Polyanetskaya, Yu. I. Gornikov, Izv. Akad. Nauk SSSR, *Neorg. Mater.* **23**, 470 (1987).
3. Y. Inaguma, C. Liquan, M. Itoh, T. Nakamura, T. Uchida, H. Ikuta, M. Wakihara, *Solid State Commun.* **86**, 689 (1993).
4. J. Ibarra, A. Várez, C. León, J. Santamaría, L.M. Torres-Martínez, J. Sanz, *Solid State Ionics* **134**, 219 (2000).
5. J. L. Fourquet, H. Duroy, M.P. Crosnier-Lopez, *J. Solid Stat. Chem.* **127**, 283 (1996).
6. J. A. Alonso, J. Sanz, J. Santamaría, C. León, A. Várez, M. T. Fernández-Díaz, *Angew. Chem. Int. Ed.* **39**, 619 (2000).
7. M. A. París, J. Sanz, C. León, J. Santamaría, J. Ibarra, A. Várez, *Chem. Mat.* **12**, 1694 (2000).
8. J. Sanz, J. A. Alonso, A. Varez, M. T. Fernández-Díaz, *J. Chem. Soc. Dalton Trans.* 1406 (2002).
9. Y. Inaguma, T. Katsumata, M. Itoh, Y. Morii, *J. Solid State Chem.* **166**, 67 (2002).
10. A. Rivera, C. León, J. Santamaría, A. Varez, O. V'yunov, A. G. Belous, J. A. Alonso, J. Sanz, *Chem. Mat.* (2002) (in press).
11. C. León, J. Santamaría, M. A. París, J. Sanz, J. Ibarra, L. M. Torres, *Phys. Rev. B* **56**, 5302, (1997).
12. A. Rivera, C. León, J. Santamaría, A. Varez, M. A. París, J. Sanz, *J. Non-Crystalline Solids* **307-310**, 992 (2002).
13. Y. Inaguma, M. Itoh, *Solid State Ionics* **86-88**, 257 (1996).

Mat. Res. Soc. Symp. Proc. Vol. 756 © 2003 Materials Research Society EE2.2

Preparation of (La, Li)TiO₃ Dense Ceramics using Sol-Gel and Ion-Exchange Process

Seiichi Suda, Hiroyuki Ishii and Kiyoshi Kanamura

Graduate School of Engineering, Tokyo Metropolitan University,
1-1 Minami-Ohsawa, Hachioji, Tokyo 192-0397, Japan.

ABSTRACT

Lithium ionic conductor, (La, Li)TiO₃, has synthesized with La/Li-TiO₂ amorphous spheres that were obtained by sol-gel and ion-exchange method, and succeeding La^{3+}/Li^+ partial ion exchange. In this work, La^{3+}/Li^+ ion exchange conditions were mainly investigated in order to obtain dense (La, Li)TiO₃ ceramics that have highly ionic conductivities. La^{3+}/Li^+ ion exchange behavior was changed with ion-exchange solutions, and the Li/Ti ratio was increased with an increase in ethanol/water ratio in the solvent used for La^{3+}/Li^+ partial ion exchange. The use of an adequate ethanol/water ratio resulted in La/Li-TiO₂ amorphous spheres with the composition of La/Li/Ti=0.54/0.34/1.00, and sintering of the spheres at 1200°C for 5 h in air led to dense (La, Li)TiO₃ ceramics which exhibit the conductivity of 4.0×10^{-3} S cm^{-1} at 25°C.

INTRODUCTION

Perovskite-type lanthanum lithium titanate, (La, Li)TiO₃, has highly lithium ionic conductivity, which exhibits more than 10^{-3} S cm^{-1} at room temperature [1-3]. This ceramic can be applicable to solid electrolyte of all-solid-state rechargeable lithium batteries or gas sensing devices such as NO$_x$ and SO$_x$ because of high ionic conduction. The conductivity of (La, Li)TiO₃ depends on A-site occupancy of lithium ions, lanthanum and vacancy, which is formed by introducing lanthanum ions into the sites, because lithium ions pass through A-sites of the perovskite structure. Therefore, precise control of the compositions is indispensable to obtain (La, Li)TiO₃ ceramics that exhibit high lithium ionic conductivity at room temperatures. Both experimental results for (La, Li)TiO₃ ceramics synthesized by solid-state reactions and simulation results applied by three-dimensional percolation theory have revealed the relationship between the compositions and lithium ionic conductivity for (La, Li)TiO₃ and the highest bulk conductivity, which was reported to be 1.5x10^{-3} S cm^{-1} at 23°C, was provided at the lithium composition of $3x$=0.34 in the expression of $La_{2/3-x}Li_{3x}\square_{1/3-2x}TiO_3$ ($\square$: vacancy) [4-6].

Application of lithium ion conductors, such as (La, Li)TiO₃ ceramics, to solid electrolytes of rechargeable lithium ionic batteries requires high lithium ionic conductivity not only in bulk but also at grain boundaries. The sintering of (La, Li)TiO₃ ceramics generally requires to be at temperatures higher than 1350°C, and the conductivities at grain boundaries exhibit extremely lower than that in bulk. We have synthesized (La, Li)TiO₃ ceramics using amorphous spheres in order to increase density of the ceramics and conductivity derived from grain boundaries. (La, Li)TiO₃ ceramics was thus obtained by La/Li-TiO₂ amorphous spheres (LLT particles). LLT particles were synthesized by partial ion-exchange between lanthanum ions and lithium in Li-TiO₂ amorphous spheres (LT particles), and LT particles were synthesized by sol-gel and ion-exchange method.

We previously found that ion exchange between alkali ions and protons of OH groups that were formed by hydrolysis of metal alkoxides occur under a high alkaline condition, and the ion exchange leads to bonding similar to M-O-A (M: Ti, Si or Zr, A: K, Na, Li) [7, 8]. The sol-gel method incorporated with the ion exchange is referred to as sol-gel and ion-exchange method. This sol-gel and ion-exchange method leads to gels dispersing alkali ions homogeneously [9-11]. Lithium dispersing amorphous titanate spheres, LT particles, were synthesized by promoting sol-gel and ion-exchange method in emulsion. Schematic scheme for preparation of LLT particles and (La, Li)TiO$_3$ ceramics using sol-gel and ion-exchange method is shown in figure 1. Density and conductivities of (La, Li)TiO$_3$ ceramics derived from LLT particles were investigated to discuss the improvement of conductivities at grain boundaries.

EXPERIMENT

LT particles were synthesized in emulsion as follows [12]. Titanium tetra-*iso*-propoxide (TTIP) was mixed with *n*-octanol, and stirred at 40°C for 30 min in dry N$_2$. Acetonitrile and hydroxypropyl cellulose, which was used as a dispersing agent [13, 14], were added into the TTIP solution, and the mixed solution was stirred for 30min. Concentrated lithium hydroxide solution, *n*-octanol, and *n*-butanol were added into the mixed solution, and stirred for 60 min. This addition of the LiOH solution promotes not only hydrolysis and Ti-O-Ti condensation but also the ion exchange between lithium ions and protons of Ti-OH during the condensation almost simultaneously in the emulsion. Obtained LT particles were separated from the solutions using a centrifuge, washed with ethanol, and dried with a freeze-drier.

LLT particles were prepared by the partial ion exchange between lanthanum ion and lithium in LT particles. La(CH$_3$COO)$_3$ was dissolved in ethanol/water mixed solution, and the ethanol/water molar ratio was set at the range of 5-40. LT particles were put in the mixed solution, and stirred at 40°C for 24 h. Morphologies of LLT and LT particles were examined with a scanning electron microscope (SEM), and compositions of the particles were estimated by inductively coupled plasma arc emission spectrometry (ICP).

Decomposition of organic compounds or water from LLT particles was examined by thermogravimetry (TG). LLT particles were calcined at 700°C for 5 h in air, and (La, Li)TiO$_3$

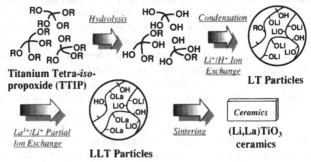

Figure 1. Schematic scheme of preparation for LLT particles and (La, Li)TiO$_3$ ceramics by sol-gel and ion-exchange method, and succeeding La^{3+}/Li$^+$ partial ion exchange.

ceramics were obtained by sintering calcined particles at 1200°C for 12 h in air. Crystal phases of obtained ceramics were identified by powder X-ray diffraction (XRD), and ionic conductivity was measured by ac two-probe method. Impedance of the ceramics was measured at the frequency range from 5 Hz to 13 MHz. The ionic conductivities derived from bulk and grain boundaries of the ceramics were separated using complex admittance plots.

RESULTS

LT particles were synthesized with various [LiOH]/[TTIP] batch compositions. Figure 2 shows the relationship between the batch compositions and Li/Ti molar ratio of LT particles measured by ICP. Li/Ti ratio of LT particles was increased with an increase in batch composition of [LiOH]/[TTIP], and LT particles with Li/Ti ratios less than Li/Ti=1.0 were obtained using the solutions with various batch compositions. Figure 3 shows SEM photograph of LT particles synthesized with the solution at the batch composition of [LiOH]/[TTIP]=1.4. The composition of the particles was Li/Ti=0.8. This showed LT particles spherical in shape but bimodal distribution of particle size. The size of LT particles was estimated to be 1.0 μm and 3.0 μm, and the bimodal distribution would be caused by insufficient conditions to form emulsion in which sol-gel and ion exchange occurred homogeneously. LT particles of 3.0 μm were obtained by allowing sol-gel reactions in the emulsion, and other LT particles smaller than 1.0 μm would be obtained by nucleation and particle growth out of the emulsion.

LLT particles were then synthesized using LT particles that have a composition of Li/Ti=0.8. The La(CH₃COO)₃ concentration of the solution that was used for La³⁺/Li⁺ ion exchange was adjusted at 4.0 mM. LLT particles were first synthesized using the solutions with various ratios of ethanol/water. Figure 4(a) shows the relationship between ethanol/water ratio and Li/Ti or La/Ti ratio of LLT particles. La/Ti ratio of LLT particles was independent of ethanol/water ratio of the ion-exchange solution, but Li/Ti ratio was increased with an increase in ethanol/water ratio. Li/Ti and La/Ti ratios were also investigated by changing concentrations of La(CH₃COO)₃ at the ethanol/water ratio of 11, 14 or 19. The relationship between La(CH₃COO)₃ concentration in the ion-exchange solutions and compositions of LLT particles was then investigated to prepare LLT particles with various compositions. The La/Ti ratio of LLT particles were increased

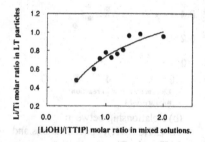

[LiOH]/[TTIP] molar ratio in mixed solutions.

Figure 2. The relationship between Li/Ti ratio in LT particles and [LiOH]/[TTIP] batch composition.

Figure 3. SEM photograph of LT particles with a composition of Li/Ti=0.8.

with an increase in $La(CH_3CO)_3$ concentration of the ion-exchange solution, and the Li/Ti ratio were decreased simultaneously(Figure 4(b)). The ion exchange between La^{3+} and Li^+ would successfully occur under this condition, but only Li/Ti ratio was changed by changing ethanol/water ratio as shown in figure 4. Dielectric constant of the ion-exchange solution decreased with an increase in ethanol/water ratio, and effect of ethanol/water ratio on La/Ti of LLT particles scarcely appeared, whereas Li/Ti much depended on the ethanol/water ratio. The decrease in Li/Ti ratio on the La^{3+}/Li^+ ion exchange was caused not only by the ion exchange but by the dissolution of Li^+ ions into the ion-exchange solution, and the amount of dissolution would be depressed by using ion-exchange solutions at relatively higher ethanol/water ratios. LLT particles with various La/Ti and Li/Ti ratios were thus obtained by changing three parameters; Li/Ti ratio of LT particles controlled using [LiOH]/[TTIP] batch composition, Li/Ti and La/Ti ratios of LLT particles using La^{3+}/Li^+ ion-exchange solutions with various $La(CH_3COO)_3$ concentrations, and Li/Ti ratio of LLT particles using the solutions with various ethanol/water ratios. Figure 5 shows the composition of LLT particles obtained using LT particles with the composition of Li/Ti=0.8 in this experiment.

(La, Li)TiO_3 ceramics were then prepared with various LLT particles. The temperature at which LLT particles were calcined was investigated by TG and XRD. The decomposition of organic compounds occurred at temperatures lower than 440°C, but small amount of La_2CO_5 would arise at 477°C (Figure 5). The calcining temperature for LLT particles was fixed at 700°C because La_2CO_5 decomposed at 690°C. After calcined at 700°C for 5 h in air, LLT particles were sintered at 1200°C for 12 h in air. Figure 6 shows SEM photographs for the ceramics derived from LLT particles with the compositions of La/Li/Ti=0.54/0.34/1.00(LLT-a) and 0.38/0.32/1.00(LLT-b). Sintering at 1200°C for 12 h resulted in dense (La, Li)TiO_3 ceramics, and the ceramic derived from LLT-b has a slightly high density than that from LLT-a. Crystal phases of the ceramics were identified by XRD. The ceramic derived from LLT-a had a single phase of the perovskite of (La, Li)TiO_3, whereas the ceramic from LLT-b has phases mixed with the perovskite and rulile of TiO_2. LLT-a had a similar composition to ideal structure

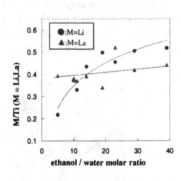

(a) Relationship between ethanol/water ratio and Li/Ti or La/Ti ratio in LLT particles. $La(CH_3COOH)_3$ concentration was adjusted to be 4mM.

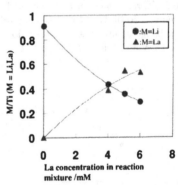

(b) Relationship between $La(CH_3COOH)_3$ concentrations and Li/Ti or La/Ti ratios in LLT particles. Ethanol/water ratio

Figure 4. Relationship between ethanol/water ratio and Li/Ti aor La/Ti ratio in LLT particles. $La(CH_3COO)_3$ concentration was adjusted to be 4mM.

of the perovskite, that is La$_{2/3-x}$Li$_{3x}$□$_{1/3-2x}$TiO$_3$ (□ : vacancy) , but LLT-b had too small concentration of lanthanum to construct the perovskite structure and this insufficient composition led to form titanium oxide. The presence of TiO$_2$ at grain boundaries may lead to increase density of the ceramics. Lithium ionic conductivities for both the ceramics were investigated using complex admittance plots. Bulk conductivity for LLT-a derived ceramic showed linear Arrehenius line and 4.01x10^{-3} S cm^{-1} at 25°C, whereas the straight line of Arrehenius plot for LLT-b derived ceramic bent at 250-300°C, and the conductivity at 25°C was estimated to be 1.35x10^{-3} S cm^{-1}. Activation energy for LLT-b at higher temperature region (250-350°C) would have comparable value to that for LLT-a, but that at lower temperature region (<250°C) was extremely larger than that for LLT-a. The formation of rutile phase would cause large activation energy for LLT-b at lower temperature region. (La, Li)TiO$_3$ ceramics that exhibited high bulk conductivities were obtained using LLT particles, but the conductivities at grain boundary were estimated to be about 10^{-5} S cm^{-1} at 25°C, which were lower than those in bulk (Figure 7). Application of (La, Li)TiO$_3$ ceramics to micro-devices can relieve the influence of low conductivity at grain boundaries slightly because small number of grain boundaries included in a micro-device, but the improvement of conductivity at grain boundaries is essential to make the best use of the high bulk conductivity of (La, Li)TiO$_3$ ceramics. Control of particle size or size distribution of LT or LLT particles was under investigation in order to enhance sinterability of ceramics derived from the amorphous spheres.

CONCLUSIONS

(La, Li)TiO$_3$ ceramics was prepared with amorphous spheres. LT particles were first synthesized by sol-gel and ion-exchange method. LT particles exhibited spherical in shape but bimodal distribution of particle size. The composition of LT particles was controlled at the Li/Ti ratio less than 1.0. LLT particles were then synthesized with LT particles by La^{3+}/Li^{+}

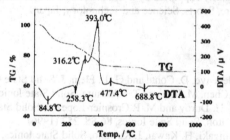

Figure 5. TG-DTA curves of LLT particles

Figure 6. SEM photographs for ceramics derived from LLT particles with the compositions of La/Li/Ti=0.54/0.34/1.00(a) and 0.38/0.32/1.00(b).

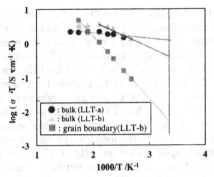

Figure 7 Arrhenius plots of (La, Li)TiO₃ ceramics derived from LLT particles.

partial ion exchange. La/Ti ratio was independent of ethanol/water ratio of the ion-exchange solutions whereas Li/Ti ratio much depended on the ethanol/water ratio. The control of both Li/Ti ratio of LT particles and La/Ti and Li/Ti ratios of LLT particles resulted in LLT particles with the composition of La/Li/Ti=0.54/0.34/1.00, which are similar to an adequate composition for the (La, Li)TiO₃ ceramics that have a maximum bulk conductivity at room temperature. The sintering of LLT particles at 1200°C for 5 h in air led to dense (La, Li)TiO₃ ceramics which exhibit a high lithium ionic conductivity of 4.0×10^{-3} S cm^{-1} at 25°C.

ACKNOWLEDGMENTS

This work was partly supported by a Grant-in-Aid for Scientific Research from the Ministry of Education, Culture, Sports, Science and Technology, Japan.

REFERENCES

1. L. Latie, G. Villeneuve, D. Conte and G. L. Flem, J. Solid State Chem., **51**, 293 (1984).
2. Y. Inaguma, L. Chen, M. Itoh and T. nakamura, Solid State Ionics, **70/71**, 196 (1994).
3. J. L. Foourquet, H. Duroy and M. P. Crosnier-Lopez, J. Solid State Chem., **127**, 283 (1996).
4. Y. Inaguma, M. Itoh, Solid State Ionics, **86-88**, 257 (1996).
5. Y. Harada, T. Ishigaki, H. Kawai, J. Kuwano, Solid State Ionics, **108**, 407 (1998).
6. T.Katsumata, Y. Inaguma, M. Itoh and K. Kawanura, Chem. Mater., **14**, 3930-3936 (2002).
7. S. Suda, M. Iwaida, K. Yamashita and T. Umegaki, J. Non-Cryst. Soldis, **176**, 26 (1994).
8. S. Suda, M. Iwaida, K. Yamashita and T. Umegaki, J. Non-Cryst. Solids, **197**, 65 (1996).
9. S. Suda, T. Tashiro and T. Umegaki, J. Non-Cryst. Solids, **255**, 178 (1999).
10. M. Matsumoto, S. Suda, K. Yamashita and T. Umegaki, Trans. Mater. Res. Soc. Japan, **25**, 257 (2000).
11. S. Suda, T. Koiwa, K. Kanamura and T. Umegaki, Key Eng. Mater., **206-213**, 127 (2002).
12. T. Ogihara, T. Yanagawa, N. Ogata and K. Yoshida, J. Ceram. Soc. Japan, **101**, 315 (1993).
13. J. H. Jean and T. A. Ring, Am. Ceram. Soc. Bull., **65**, 1574 (1986).
14. J. H. Jean and T. A. Ring, Colloids Surf., **29**, 273 (1988).

Mat. Res. Soc. Symp. Proc. Vol. 756 © 2003 Materials Research Society   EE3.8

Discussion about non-Arrhenius behavior of high Li-ion conductor, (La,Li)TiO$_3$

Tetsuhiro Katsumata, Yoshiyuki Inaguma, Satoshi Baba[1], Ko-ichi Hiraki[1] and Toshihiro Takahashi[1]
Department of Chemistry, Faculty of Science, Gakushuin University
1-5-1 Mejiro, Toshimaku, 171-8588 Japan
[1]Department of Physics, Faculty of Science, Gakushuin University
1-5-1 Mejiro, Toshimaku, 171-8588 Japan

ABSTRACT

We investigated the dielectric properties of La$_{0.53}$Na$_{0.41-x}$Li$_x$TiO$_3$ ($x<0.26$) and ^{7}Li NMR for La$_{0.53}$Na$_{0.34}$Li$_{0.17}$TiO$_3$ ($x=0.17$). As results, relaxation process were observed at 40 K and 225 K for La$_{0.53}$Na$_{0.34}$Li$_{0.17}$TiO$_3$ ($x=0.17$). The activation energy of the dielectric relaxation at 40 K is in accordance with the that obtained by NMR measurement. On the other hand, the activation energy of the relaxation at 225 K accords with that of the Li ion conduction in the low temperature region for (La,Li)TiO$_3$. These results indicate that different transport mechanisms intrinsically exist in (La,Li)TiO$_3$ and one of reasons for the non-Arrhenius behavior of (La,Li)TiO$_3$ is that the transport mechanism mainly related to the dc conductivity varies with the temperature.

INTRODUCTION

The A-site deficient perovskite-type oxide, La$_{2/3-x}$Li$_{3x}$TiO$_3$ shows high Li ion conductivity. When the concentrations of Li ion and vacancy were optimized, i.e. La$_{0.55}$Li$_{0.35}$TiO$_3$ the Li ion conductivity of the bulk part attained 10^{-3}S/cm at the room temperature which is one of the highest value in solid Li ion conductors[1,2]. These compounds, however, have interesting features except for its high ionic conductivity. One of them is the non-Arrhenius temperature dependence of the dc ionic conductivity. As shown in Fig.1, the temperature dependence of the dc conductivity has a curvature and the activation energy of the low temperature region is higher than that of the high temperature region in (La,Li)TiO$_3$. Furthermore, the activation energy in the high temperature region is equal to that obtained by the NMR measurement[3-5]. While many models have been proposed in order to explain the mechanism of this temperature dependence, there is still no general agreement on interpretation of the experimental data.

On the other hand, the dc ionic conductivity of (La,Li)TiO$_3$ was influenced by the site percolation and must be diminished when the number of objects formed the conduction path is less than percolation limit. The ionic conductivity of xLa$_{0.55}$Li$_{0.35}$TiO$_3$-(1-x)La$_{0.5}$Na$_{0.5}$TiO$_3$ decreased with a decrease in the sum of concentrations of Li ion and vacancy, 0.45x and became

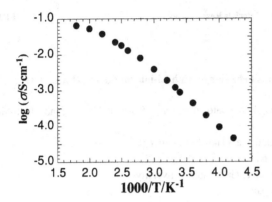

Figure 1. Arrhenius plot of the dc ionic conductivity for $La_{0.55}Li_{0.35}TiO_3$.

an immeasurably small value with approaching $0.45x$ to the percolation limit of the primitive cell, 0.3117[6]. Therefore, a perovskite-type compound, $La_{0.53}Na_{0.41-x}Li_xTiO_3$ ($x<0.26$), in which sum of the concentrations of the Li ion and the vacancy is less than percolation limit, can be considered as a dielectric material and the transport mechanism of Li ion may be able to discuss by investigating the dielectric properties of the $La_{0.53}Na_{0.41-x}Li_xTiO_3$ ($x<0.26$).

In this study, we discuss the transport mechanism, especially the non-Arrhenius temperature dependence of $(La,Li)TiO_3$ by elucidating the dielectric properties for $La_{0.53}Na_{0.41-x}Li_xTiO_3$ ($x<0.26$). In addition, ^7Li NMR measurement was carried out for $La_{0.53}Na_{0.34}Li_{0.17}TiO_3$ ($x=0.17$) and the activation energy obtained by the NMR measurement compares with that obtained by the dielectric measurement.

EXPERIMENTAL

Samples were synthesized by conventional solid state reaction technique. Starting materials were La_2O_3(4N), Li_2CO_3(3N), Na_2CO_3(3N) and TiO_2(3N). The La content of the La_2O_3 was determined by titration analysis. The mixture of these starting materials was calcined at 1073 K for 8 h and 1323 K for 24 h. The calcined powder was ground and pressed into pellets with a diameter 7 mm and a thickness 1.5 mm. These pellets were wrapped by a Pt foil and sintered at 1723 K for 3 h ($x=0.00$ and $x=0.04$) or 1523 K for 3 h ($x=0.08$, $x=0.11$ and $x=0.17$). The phase identification was carried out by powder X-ray diffraction method (Rigaku RINT 2100). The metal content of the synthesized samples were analyzed by induced couple plasma (ICP) spectroscopy. Dielectric constant and dielectric loss were measured by an Agilent technology 4284A precision LCR meter in the temperature range from 8 K to room temperature at frequency

f=1 kHz, 10 kHz, 100 kHz, 200 kHz and 500 kHz.

Measurements of ^{7}Li NMR spectra and spinlattice relaxation rate, $1/T_1$, for La$_{0.53}$Na$_{0.34}$Li$_{0.17}$TiO$_3$ (x=0.17) were done by a phase coherent hand made NMR spectrometer at resonance frequency of 51.4 MHz and 88.1 MHz in the temperature ranges 124-300 K and 10-240 K, respectively. The NMR spectra were obtained by the Fast Fourier Transformation of spine echo signal after $\pi/2$-τ'-π pulse sequence[7]. The $1/T_1$ data obtained from a recovery of nuclear magnetization after saturation comb pulses.

RESULTS AND DISCUSSION

According to the analysis of La$_{0.53}$Na$_{0.41-x}$Li$_x$TiO$_3$ (x<0.26), the sum of the concentrations of the Li ions and the vacancies does not exceed the percolation limit of the primitive cell for these compounds, which means that these compounds can not be considered as an ionic conductor, but a dielectric material. In fact, the dc conductivity of La$_{0.53}$Na$_{0.34}$Li$_{0.17}$TiO$_3$ (x=0.17) at room temperature was below the measurement limit in this study, 10^{-8} S/cm.

Figure 2 shows the temperature dependence of the dielectric constant and the dielectric loss at frequency f=10 kHz for La$_{0.53}$Na$_{0.41-x}$Li$_x$TiO$_3$[8]. For La$_{0.53}$Na$_{0.41}$TiO$_3$ (x=0.00), the dielectric constant monotonously increases with decreasing the temperature in the temperature range 50 K<T<300 K and is almost independent on the temperature below 50 K. The temperature dependence of the dielectric constant was fitted with the Barrett-type formula[9], thus this can be

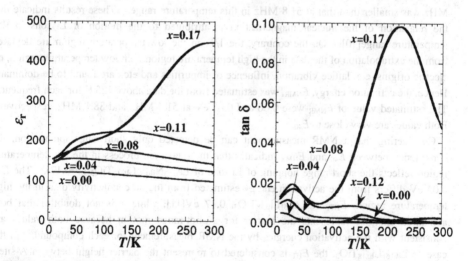

Figure 2. Temperature dependence of the electric permitivity and the dielectric loss for La$_{0.53}$Na$_{0.41-x}$Li$_x$TiO$_3$ (x<0.26).

found to be a quantum paraelectric compound. For other compounds, same anomalies appear by the substitution of Li for Na. For $x=0.04$ the dielectric constant increases with decreasing temperature similar to that of $x=0.00$ in the temperature $T \geq 50$ K. However, below 50 K the dielectric constant is not constant, but decreases with the temperature. For $x=0.08$ the broad and asymmetric peak is observed in the vicinity of 70 K. For $x=0.11$ the peak was observed more clearly and for $x=0.17$ apparent two peaks are observed in the vicinity of 50 K and 290 K. Since these anomalies appear more apparently with substitution of Li ion, they must be caused by the motion of Li ions.

The anomalies were also observed in the temperature dependence of the dielectric loss. For $La_{0.53}Na_{0.41}TiO_3$ ($x=0.00$), the peak was observed at 160 K. While the reason of this peak has not been clear, it is predicted to be due to the vacancy introduced in the A-site because such peak was not observed for $La_{0.5}Na_{0.5}TiO_3$ which is the quantum paraelectric. For other compounds, the peaks are observed in the vicinity of 40 K for $x=0.04$, 0.08 0.12 and 0.17. In addition, for $x=0.17$ the peak also was observed in the vicinity of 225 K. These results indicate that two different relaxation processes attributed to the motion of Li ions exist and these processes have the different activation energy. The peaks in the vicinity of 40 K and 225 K for $La_{0.53}Na_{0.34}Li_{0.17}TiO_3$ ($x=0.17$) show the frequency dependence. The activation energy estimated from the frequency dependence of the peaks in the vicinity of 40 K and 225 K, named E_{40} and E_{225}, were 0.05 eV and 0.37 eV, respectively.

Figure 3 shows the temperature dependence of the spinlattice relaxation rate, $1/T_1$, obtained at 51.8 MHz and 88.1 MHz. As shown in this figure, the slope of the temperature dependence of $1/T_1$ at 88.1 MHz is almost same as that at 51.8 MHz above 125 K, while the $1/T_1$ values at 88.1 MHz was smaller than that at 51.8 MHz in this temperature range. These results indicate that the relaxation of the nuclear magnetization is attributed to the motion of Li ions in this temperature range[10]. On the contrary, the $1/T_1$ in the low temperature region are deviated from the extrapolation of the data in the high temperature region. In low temperature region, no electric origins, e.g. lattice vibration, influence of impurities and etc., are found to be dominant. Hence, the activation energy, E_{NMR}, was estimated from the data above 125 K for each frequency. The estimated value of E_{NMR} were 0.06 and 0.05 eV at 51.4 MHz and 88.1 MHz, respectively. Both values are very close to E_{40}.

Considering that a NMR measurement can be detected only short-range ion motion, the consistency between E_{40} and E_{NMR} indicates that the relaxation process in the low temperature region reflects the short-range hopping of Li ions of $La_{0.53}Na_{0.34}Li_{0.17}TiO_3$ ($x=0.17$). The E_{40}, 0.05 eV differs from the activation energy estimated from the dc conductivity data in the high temperature region, E_{HT}, of $La_{0.61}Li_{0.18}TiO_3$, 0.17 eV[11], while it is not doubtful that both activation energies represents the barrier for a shot-range motion because both values are consistent with the activation energies by the NMR measurement for each compound. In the case of $La_{0.61}Li_{0.18}TiO_3$, the E_{HT} is considered to represent the barrier height between A-sites. However, for perovskite-type Li ion conductors the existence of Li ions at an off-center position in A-site was pointed out[4,5]. Therefore, two possibilities, i.e. barrier heights between A-sites

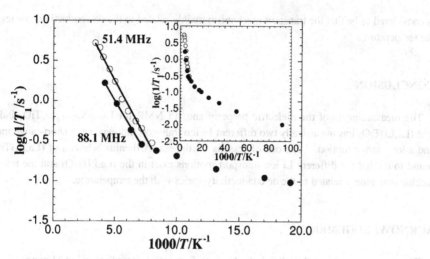

Figure 3. Arrhenius plot of spinlattice relaxation rate, $1T_1$ for $La_{0.53}Na_{0.34}Li_{0.17}TiO_3$ ($x=0.17$). The open circles and closed circles show the data measured at 51.4 MHz and 88.1 MHz, respectively.

and off-center positions in A-site, have to be took into consideration, when what is represented by this activation energy, 0.05 eV is discussed. While we could not unfortunately determine it in this study, the discrepancy between E_{40} and E_{HT} of $La_{0.61}Li_{0.18}TiO_3$ may be attributed to the change of the mechanism of the short-range motion, i.e. from hoping between off-center positions to hopping between A-sites.

On the other hands, the relaxation process in the high temperature region is found to reflect the long-range motion of Li ions, because this relaxation process has not been detected by the NMR measurement. As above mentioned, $La_{0.53}Na_{0.34}Li_{0.17}TiO_3$ ($x=0.17$) did not show the dc ionic conduction. However, this long-range motion is considered to occur in clusters of Li ion and vacancy. For $La_{0.53}Na_{0.34}Li_{0.17}TiO_3$ ($x=0.17$), it is highly possible that Li ions and vacancies distributes in A-sites with a cluster. This is supported by the experimental results, that the relaxation in the high temperature region is observed only for the compound in which the certain amount Li ions was introduced into A-site. Therefore, this long-range motion occurs in clusters, even though no dc ionic conductivity was observed.

An interesting point is that the E_{225} is in agreement with the activation energy of the dc conductivity in the low temperature region, E_{LT}, for (La,Li)TiO$_3$, e.g. E_{LT} of $La_{0.63}Li_{0.10}TiO_3$ is 0.36 eV[12] which is very close to E_{225}. This result indicates that the E_{LT} of (La,Li)TiO$_3$ represents the only long-range motion for these transport motions. Accordingly, the activation energies of dc ionic conductivity in the high and low temperature region are related to the each different transport motions and the one of reasons for the non-Arrhenius behavior of (La,Li)TiO$_3$

is considered to be that the transport mechanism mainly related to the dc conductivity varies with the temperature.

CONCLUSION

The measurements of the dielectric property and ^{7}Li NMR for $La_{0.53}Na_{0.41-x}Li_xTiO_3$ indicate that $(La,Li)TiO_3$ has intrinsically two different Li ion transport motions, i.e. a short-range motion and a long-range motion. The one of reasons for the non-Arrhenius behavior of $(La,Li)TiO_3$ is found to be that the different Li ion transport motions exist in the $(La,Li)TiO_3$ and the transport mechanism mainly related to the dc conductivity varies with the temperature.

ACKNOWLEDGEMENT

This work was supported by "High-Technology Research Center Project" of Ministry of Education, Science, Sports and Culture.

REFERENCES

1   A. G. Belous, G. N. Novitiskaya, S. V. Polyanetskaya, I. Gornikov, *Izv. Akad. Nauk. SSSR, Neorg. Mater.* **23**, 470 (1987).
2   Y. Inaguma, L. Chen, M. Itoh, T. Nakamura, T. Uchida, M. Ikuta, M. Wakihara, *Solid State Commun.* **86**, 689 (1993).
3   C. Léon, M. L. Lucía, J. Santamaría, M. A. París, J. Sanz, and A. Várez, *Phys. Rev. B* **54**, 184 (1996)
4   J. Emery, J. Y. Buzare, O. Bohnke, J. L. Fourquet, *Solid State Ionics* **99**, 41 (1997).
5   M. A. París, J. Sanz, C. Léon, J. Santamaría, J. Ibarra and A. Várez, *Chem. Mater.* **12**, 1964 (2000).
6   Y. Inaguma, M. Itoh, *Solid State Ionics*, **86-88**, 257 (1996).
7   C. P. Slichter *"Principles of Magnetic Resonance"* (Springer 3rd Ed).
8   T. Katsumata, Y. Inaguma, *Solid State Ionics*, In print.
9   J. H. Barrett, *Phys Rev.* **86**, 118 (1952)
10   A. Abragam *"Principles of Nuclear Magnetism"* (Oxford Science Publications).
11   C. Léon, A. Rivera, A. Várez, J. Sanz, J. Santamaría, and K. L. Ngai, *Phys. Rev. letter* **86**, 1279 (2001).
12   Y. Inaguma, L. Chen, M. Itoh, T. Nakamura, *Solid State Ionics* **70/71**, 196 (1994).

Mat. Res. Soc. Symp. Proc. Vol. 756 © 2003 Materials Research Society

Phase Stability of Beta-gallia Rutile Intergrowths: $(Ga,In)_4(Sn,Ti)_{n-4}O_{2n-2}$

Malin Charoenwongsa and Doreen D. Edwards
School of Ceramic Engineering and Material Science, Alfred University,
Alfred, NY, 14802, U.S.A.

ABSTRACT

Beta-gallia-rutile (BGR) intergrowths possess one-dimensional tunnels that are suitable hosts for small-to-medium cations, thereby making them potential candidates for ion conductors, ion separators, battery electrodes, and chemical sensors. The BGR intergrowths are a series of homologous compounds expressed generically as $Ga_4M_{n-4}O_{2n-2}$, where n is an integer; and M is a tetravalent cation that forms a rutile-type oxide. In an attempt to identify materials with high tunnel densities and higher contents of a reducible M^{4+} cation, we are mapping the compositional stability regions of intergrowths expressed as $Ga_{4-4x}In_{4x}Sn_{(n-4)(1-y)}Ti_{(n-4)y}O_{2n-2}$ where n = 6, 7 and 9 and $0.15 < x < 0.30$ and $0 < y < 1.0$. Polycrystalline samples were prepared by solid-state reaction at 1250 – 1400 °C and characterized by X-ray diffraction. Factors that affect phase stability are discussed.

INTRODUCTION

Beta-gallia-rutile (BGR) intergrowths and their derivatives are potential candidates for ion conductors, ion separators, battery electrodes, and chemical sensors. The BGR intergrowths possess one-dimensional tunnels, approximately 2.5Å in diameter, making them suitable hosts for small-to-medium size cations such as Li^+ and Na^+. BGR intergrowths form when beta-gallia, Ga_2O_3, and oxides possessing the rutile structure, MO_2, react at high temperatures to form a series of homologous compounds expressed as $Ga_4M_{n-4}O_{2n-2}$. In the Ga_2O_3-TiO_2 system, several BGR intergrowths with $9 > n > 51$ (odd) have been reported. In the Ga_2O_3-SnO_2 and Ga_2O_3-GeO_2 systems, BGR intergrowths with n = 5 have been reported.[1-6]

Tunnel density increases with decreasing n, thus BGR intergrowths with lower n values are expected to be have higher ion storage capacity. Nevertheless, the known lower-n intergrowths, such as Ga_4SnO_8 and Ga_4GeO_8, do not possess an easily reducible cation that can participate in ion insertion reactions. An underlying hypothesis of this work is that Ti-containing BGR intergrowths with low n values can be stabilized through cation substitution of the parent BGR structure.

In previous work, one of the authors[2] demonstrated that indium substitution for gallium in $Ga_4Sn_{n-4}O_{2n-2}$ resulted in the stabilization of the n = 6, 7 – 17 (odd) intergrowths. The stabilization of new intergrowth phases was attributed to a decrease in the lattice mismatch between the beta-gallia and rutile components of the intergrowth structure. In the current work, we are mapping the compositional stability ranges of BGR intergrowths expressed as $Ga_{4-4x}In_{4x}Sn_{(n-4)(1-y)}Ti_{(n-4)y}O_{2n-2}$, n = 6, 7 and 9 and $0.15 < x < 0.30$ and $0 < y < 1.0$. Samples were prepared by solid-state reaction at 1250 – 1400 °C. Phase composition and lattice parameters of the equilibrated samples were analyzed using X-ray diffraction.

EXPERIMENTAL DETAILS

Several samples were prepared by high-temperature solid-state reaction according to the chemical formula $Ga_{4-4x}In_{4x}Sn_{(n-4)(1-y)}Ti_{(n-4)y}O_{2n-2}$ as summarized in Table I. Throughout this paper, sample compositions are indicated by three values, corresponding to n, x, and y in the chemical formula.

Table I. Summary of Starting Compositions of Samples Prepared as $Ga_{4-4x}In_{4x}Sn_{(n-4)(1-y)}Ti_{(n-4)y}O_{2n-2}$

n	x = [In] / ([In] + [Ga])	y = [Ti] / ([Ti] + [Sn])
6	0.10, 0.15, 0.17, 0.19, 0.25	0.0, 0.1, 0.25, 0.3, 0.5, 0.75, 1.0
7	0.10, 0.17, 0.19, 0.25	0.0, 0.05, 0.06, 0.1, 0.133, 0.15, 0.2, 0.33
9	0.10, 0.15, 0.17, 0.19, 0.25, 0.30	0.0, 0.04, 0.08, 0.12, 0.2, 0.3

Commercial oxide powders (>99.9% purity on a cation basis, Aldrich Chemical Co.) were heated overnight at 700 °C to removed adsorbed water and stored in a desiccator prior to use. Appropriate amounts of starting oxides were moistened with acetone and mechanically mixed in an agate mortar. The mixed oxide powders were uniaxially pressed into 12.5 cm diameter pellets at 150 MPa and then heated in air for 7-20 days at 1150-1400 °C, depending on the time needed to reach equilibrium. Every 2-3 days, samples were reground and put back to the furnace, to hasten the reaction. All samples, sandwiched between two sacrificial pellets of the same composition, were placed in high-purity alumina crucibles. After firing, samples were dry quenched in air. The samples were reground, repressed, and reheated until the equilibrium was achieved as determined by X-ray diffraction.

X-ray diffraction analysis was conducted on a Philips XRG3100 diffractometer using Cu $K\alpha$ radiation (40kV, 20mA). Powder samples were mixed with lithium fluoride powder, which was used as an internal standard, and applied to a zero background sample holder (single-crystal quartz) lightly coated with petroleum jelly. Data were collected over 10-70° 2θ in 0.05° steps, counting 1 second per step. Lattice parameters were determined by the least-squares method.

RESULTS and DISCUSSION

All samples listed in Table I were white after firing at 1250 – 1400 °C. Samples were identified either as phase pure or multiphase based on XRD patterns of n= 6, 7 and 9 intergrowths reported previously.[2, 3] We observed that samples heated at 1400 °C reached equilibrium faster than those heated at 1250 and 1300 °C. For example, samples prepared with n = 9 and varied titanium content were phase pure after being heated at 1300 °C for 15 days, while those heated at 1400 °C were phase pure after being heated for 5 to 7 days. There was no evidence of significant vaporization of any component, as determined by weight loss measurements. The rate of reaction was also found to be highly dependent on composition and the purity of the starting materials. For example, samples prepared with n = 9 typically achieved equilibrium within 5 to 7 days, whereas samples prepared with n = 6 and n = 7 were slower to equilibrate. Samples prepared with

99.99+ % Ga_2O_3 and 99.9+ % SnO_2 formed intergrowth phases more readily than those prepared with higher purity starting materials. (The 99.9+ % SnO_2, containing >100 ppm of Fe and Sb as well as >10 ppm of Ca, As, Al, Pb, Cu, Bi, S, and Si, was the primary source of impurity cations. All other starting materials contained < 10 ppm of any reported impurity.) We were unable to obtain phase-pure material for some compositions regardless of the heating time.

Typical X-ray diffraction patterns for nominally phase-pure samples prepared with n = 6, n = 7, and = 9 are showed in Figure 1. Although the diffraction patterns show several common reflections, the phases are discernable from one another due to a number of characteristic low-angle peaks. The lattice parameters of the phase-pure intergrowths prepared with y = 0 were in good agreement with those reported previously for the Ga_2O_3-In_2O_3-SnO_2 system.[3]

Figure 2 summarizes the compositional stability ranges for n = 6, n = 7, and n = 9 intergrowths in the Ga_2O_3-In_2O_3-SnO_2-TiO_2 system at 1400 °C. In these diagrams, the indium content in the beta-gallia component is plotted as the horizontal axes and the titanium content in the rutile component is plotted on the vertical axes. Solid diamonds indicate compositions that resulted in nominally phase-pure materials, whereas open diamonds indicate compositions that resulted in mixed phase assemblages. The diagonal lines in the graphs indicate the degree of lattice mismatch between the beta-gallia and rutile components, defined as $[(b_{\beta\text{-gallia}} - c_{rutile})/b_{\beta\text{-gallia}}] \times 100$.

As shown in Figure 2a, only two phase-pure n = 6 samples were obtained at 1400 °C. All of the other samples prepared as n = 6 contained various mixtures of n = 5, n = 6, n = 7, and n = 9 intergrowths. Our results are in disagreement with previous reports, which indicate that the compositional stability of $Ga_{4-4x}In_{4x}Sn_2O_{10}$ at 1250 °C ranges from x = 0.19 to x = 0.30.[3]

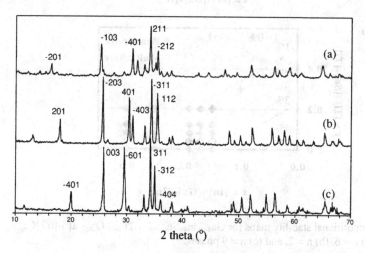

Fig. 1. X-ray diffraction patterns of (a) $Ga_3InSn_{1.4}Ti_{0.6}O_{10}$ (n = 6),
(b) $Ga_{3.24}In_{0.76}Sn_{2.6}Ti_{0.4}O_{12}$ (n = 7), and (c) $Ga_{3.24}In_{0.76}Sn_{4.6}Ti_{0.4}O_{12}$ (n = 9).

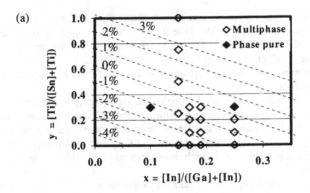

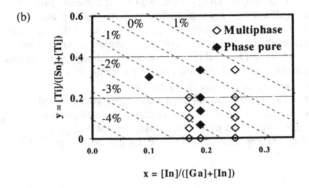

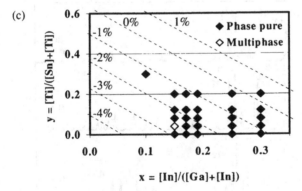

Fig. 2. Compositional stability maps for $Ga_{4-4x}In_{4x}Sn_{(n-4)(1-y)}Ti_{(n-4)y}O_{2n-2}$ at 1400 °C: (a) n = 6, (b) n = 7, and (c) n = 9 phases.

Several phase-pure n = 7 samples were obtained when samples were heated at 1400 °C for 10 days, as shown in Figure 2b. Samples fired 1250 and 1300 °C showed mixed phases regardless of heating time. Again, our results are in disagreement with previous reports, which indicate that n = 7 intergrowths form readily at 1250 °C and that the compositional stability of $Ga_{4-4x}In_{4x}Sn_3O_{12}$ at 1250 °C ranges from x = 0.17 to x = 0.30.[3]

Most of the samples prepared with n = 9 were phase pure, as shown in Figure 2c. Lattice parameter measurements of samples prepared with n = 9 and y = 0 indicated that substitution of Ga^{3+} (r = 0.62 Å) with In^{3+} (r = 0.80 Å) results in an increase in the lattice parameters a, b, and c and a decrease in the lattice parameter β, which is consistent with earlier reports.[3] Figure 3 shows that substitution of Sn^{4+} (r = 0.69 Å) with Ti^{4+} (r = 0.605 Å) results in a decrease in the unit cell volume. The effective ionic radii we used were taken from Shannon.[7] The data in Figures 2c and 3 indicate that the solubility limit of Ti in $Ga_{4-4x}In_{4x}Sn_{5(1-y)}Ti_{5y}O_{16}$ is greater than y = 0.2.

CONCLUSIONS

The phase stability of beta-gallia-rutile intergrowths in the Ga_2O_3-In_2O_3-SnO_2-TiO_2 system was investigated. Samples with a range of compositions, expressed as $Ga_{4-4x}In_{4x}Sn_{(n-4)(1-y)}Ti_{(n-4)y}O_{2n-2}$, n = 6, 7 and 9, 0.15 <x < 0.30, and 0 < y < 1.0, were prepared by solid-state reaction at 1250 °C – 1400 °C and characterized using x-ray diffraction. While nominally phase-pure materials were obtained for each of the three intergrowths investigated, the compositional stability ranges for the n = 6 and n = 7 intergrowths were smaller than that of the n = 9 intergrowths and could not be correlated to lattice matching between the beta-gallia and rutile component of the intergrowth

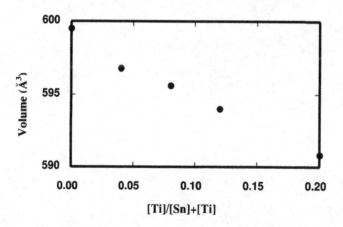

Fig. 3. Unit cell volume of $Ga_{4-4x}In_{4x}Sn_{5(1-y)}Ti_{5y}O_{16}$ as a function of [Ti]/[Sn]+[Ti].

structure. The findings of this work are in disagreement with previous reports regarding the phase stability of intergrowths in the Ga_2O_3-In_2O_3-SnO_2 system. The reaction kinetics (and perhaps even the phase stability) of the intergrowths are highly dependent on minor impurities in the starting materials leading us to speculate that the disagreement of this work with previous work may result from the differences in traces impurity levels.

ACKNOWLEDGEMENTS

This work was supported by the National Science Foundation (DMR-0093690). Graduate stipend for Malin Charoenwongsa was provided by the Royal Thai Government.

REFERENCES

1 L. A. Bursill and G. G. Stone, J. Solid State Chem. **38**, 149-157 (1981).

2. D. D. Edwards and T. O. Mason, J. Am. Ceram. Soc. **81**, [12] 3285-92 (1998)

3. D. D. Edwards, T.O. Mason, W. Sinkler, L. D. Marks, K. R. Poeppelmeier, Z. Hu, and J. D. Jorgensen, J. Solid State Chem. **150**, 294-304 (2000).

4. M. B. Varfolomeev, A. S. Mironova, and N. T. K. Nein, Russ. J. of Inorg. Chem. **20**, 3140-3141 (1975).

5. S. Kamiya and J. D. Tilley, J. Solid State Chem. **22**, 205-216 (1977).

6. A. V. Kahn, D. Michel, and M. P. Yorba, J. Solid State Chem. **65**, 377-382 (1986).

7. R.D. Shannon, *Acta Cryst*. **A32**, 751 (1976).

Mat. Res. Soc. Symp. Proc. Vol. 756 © 2003 Materials Research Society

Enhanced Ionic Conduction Observed for Ordered-Mesoporous Alumina-Ionic Conductor Composites

Hideki Maekawa,[1,2] Ryo Tanaka,[1] and Tsutomu Yamamura[1]
[1]Tohoku Univ, Dept of Metallurgy, Sendai, Japan
[2]PRESTO, Japan Science and Technology Corporation, Japan.

ABSTRACT

Ordered-mesoporous Al_2O_3 was synthesized by the sol-gel method using neutral surfactants as templates. The pore size can be controlled over the range of 2.8~12.5 nm by using different surfactant copolymers and by different synthetic conditions. By utilizing cyclohexane as a co-solvent, mesoporous Al_2O_3 having relatively mono-dispersed particle size was obtained. Composites composed of the synthesized mesoporous Al_2O_3 and the lithium ion conductor (LiI) was prepared. The dc electrical conductivity of $50LiI \cdot 50$(mesoporous Al_2O_3) was 2.6×10^{-4} S cm^{-1} at room temperature, which is more than 100 times higher than that of pure LiI. The pore size dependence of the conductivity of LiI-mesoporous Al_2O_3 composite was examined. A systematic dependence of conductivity upon pore size was observed, in which the conductivity increased with decreasing the pore size.

INTRODUCTION

A dispersion of insoluble dielectric oxide particles such as Al_2O_3 in certain ionic conductors is known to increase total electrical conductivity [1-3]. This effect was observed for the first time on the LiI-Al_2O_3 system [1]. The mechanism of the conductivity enhancement was suggested to be the increase of cationic defect concentration at a space-charge region near the insulator-ionic conductor interfaces [1-3]. The width of the space charge is characterized by the Debye length which is in the range of several 10 nm for certain ionic conductors [4]. In the case where the distance between two interfaces is comparable to the Debye length, a further conductivity enhancement due to a "nano-size" effect was predicted, as a result of an overlapping space-charge layer [5]. The purpose of the present investigation is to prepare the LiI-mesoporous Al_2O_3 composites with various pore sizes and to investigate the channel size dependence on the conductivity.

EXPERIMENTAL DETAILS

Preparation and Characterization of the Mesoporous Al_2O_3

Powdered mesoporous Al_2O_3 was prepared by the sol-gel method utilizing neutral surfactants as templates [6-8]. Surfactants used were Span80, Span85 and Triton X114 (Aldrich). Surfactants, $Al^s(OBt)_3$ (Kanto Chemicals), and $LaCl_3 \cdot 7H_2O$ (Nacalai Tesque) were dissolved in the solvent (cycohexane, Wako Chemicals) and stirred for 30 minutes. To this solution, water was added dropwise and the solution was let stirred for 15 h at 318 K. The resulting sol was filtered and dried at 373 K. Mesoporous-Al_2O_3 was obtained by calcination of the sol at 773 K for 6 h. Pore size, pore volume and BET surface area were obtained by N_2 adsorption and desorption isotherms at 77 K by using Micromeritics ASAP2010. X-ray powder diffraction

Table 1. Synthetic conditions and physical properties of the mesoporous Al_2O_3.

Sample	Surfactants	Composition* $x : y : z$	BET Surface Area $(m^2 g-1)$	Pore Size (nm)	Pore Volume $(cm^3 g^{-1})$
A	Span 80	2.70 : 0.30 : 0.08	360	3.0	0.30
B	Span 85	1.35 : 0.15 : 0.10	315	4.2	0.32
C	Triton X-114	2.70 : 0.30 : 0.15	315	7.0	0.63
D	Triton X-114	0.68 : 0.15 : 0.08	400	9.0	0.96
E	Triton X-114	1.35 : 0.15 : 0.08	390	12.5	1.20

*The composition denotes molar fractions of x;cyclohexane, y;$Al^s(OBt)_3$, and z;surfactant per 1 mole of H_2O. The molar fraction of $LaCl_3 \cdot 7H_2O$ was fixed at 4×10^{-3} mol per 1 mole of H_2O.

patterns of pores were obtained using Cu K_α radiation (RINT2000, Rigaku). Particle size and shape were observed by FE-SEM (S-4300E, Hitachi). Table 1 summarizes the synthetic conditions of the synthesized mesoporous-Al_2O_3.

Preparation and Characterization of the Mesoporous Al_2O_3-LiI composite

Mesoporous Al_2O_3 with various pore sizes were employed for the fabrication of a composite with LiI. Mesoporous Al_2O_3 and LiI were mixed with a mortar and pressed into a pellet under the globebox atmosphere (< 1 ppm v H_2O). The pellet was sealed in the Pyrex glass under vacuum and the ampoule was heated at 773 K for 15 h. An ac electrical conductivity of the composite was measured by using Hewlett Packard 4194A impedance meter. The frequency range measured was 100 ~ 10 MHz and the temperature range was 297 K~343 K.

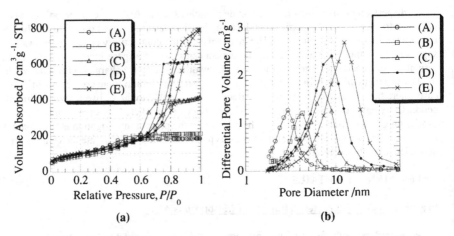

(a)   (b)

Figure 1. (a); N_2 absorption and desorption isotherm plots for mesoporous Al_2O_3 of samples (A)-(E) as described in Table 1. (b); Pore size distribution obtained from the adsorption branch of N_2 adsorption isotherm (Figure 1(a)) for mesoporous Al_2O_3 of samples (A)-(E).

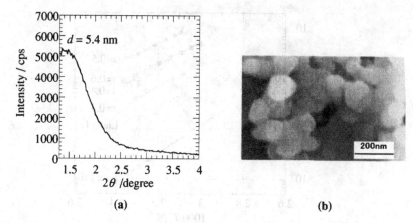

Figure 2. (a); XRD pattern and (b); SEM image of sample (A) described in Table 1.

The dc conductivity was obtained from the diameter of a single arc observed on the Cole-Cole plot of the ac impedance spectra.

RESULTS

Mesoporous Al$_2$O$_3$

The N$_2$ adsorption and desorption isotherms of the synthesized mesoporous Al$_2$O$_3$ (Figure 1(a)) were Type IV, with a small hysteresis that indicates a minor deviation of the overall pore structure from a regular array of cylinders. The pore-diameter-distribution curve shown in Figure 1(b) was obtained from the adsorption branch of the N$_2$ adsorption isotherm and calculated by the BJH method [9]. A sharply peaked distribution was observed reflecting a uniform mesopore size. Pore size can be controlled in the range of 3~12.5 nm by changing surfactant species and the composition of the sol-gel solution. The pore size, the BET surface areas and the pore volume have been summarized in Table 1.

A low angle X-ray diffraction (XRD) pattern and a SEM image of sample (A) are shown in Figure 2. A single broad peak from ordered mesopore was observed at 5.4 nm (Figure 2 (a)). Assuming a (100) reflection from the hexagonal array of pores, the wall thickness was estimated as 3.2 nm by comparing the hexagonal unit cell length ($2d_{100}/3^{-1/2}$) and the pore size (3.0 nm). The SEM image (Figure 2(b)) shows the particle size of mesoporous Al$_2$O$_3$ is relatively mono-dispersed and their size was in the range of several hundred nanometers.

Mesoporous Al$_2$O$_3$-LiI composite

Figure 3 shows the logarithm of the dc conductivity *vs.* 1/T for the x(mesoporous Al$_2$O$_3$ (pore size = 4nm))-(1-x)LiI together with the literature value of Al$_2$O$_3$ particle dispersed composite [1]. The conductivity was increased with increasing the mesoporous Al$_2$O$_3$ contents up to x = 0.5. The maximum conductivity was 2.6×10^{-4} S cm^{-1} at 298K, which is more than 2 orders of

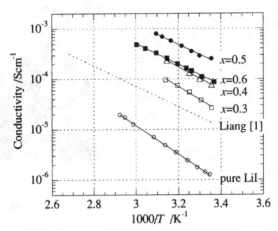

Figure 3. Temperature dependences of the dc electrical conductivity for x(mesoporous Al_2O_3 (pore size = 4nm))-(1-x)LiI composites.

magnitude higher than that of LiI, and 10 times higher than that of alumina particle dispersed composite [1]. Although slight decrease with increasing mesoporous Al_2O_3 content was observed, the activation energy of the dc electrical conductivity showed no significant compositional dependence. This behavior suggests an increase of a cation carrier concentration with increasing mesoporous Al_2O_3 concentration. The decrease of conductivity at x=0.6 can be explained by the decrease of percolation pathways.

Figure 4 shows the pore size dependence of the dc conductivity of mesoporous Al_2O_3-LiI composites. All composites show a maximum conductivity at a composition near x=0.5.

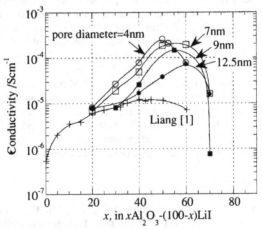

Figure 4. Compositional and mesopore size dependences of the dc conductivity at 297 K for x(mesoporous Al_2O)-(1-x)LiI composites.

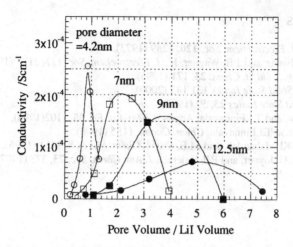

Figure 5. Relationship between the dc electrical conductivity and the ratio of (pore volume)/(LiI volume) for x(mesoporous Al_2O_3)-$(1-x)$LiI composites.

The composition of the maximum conductivity shifted to LiI side with decreasing pore size. The conductivity increased with decreasing pore size.

DISCUSSIONS

In Figure 5 the horizontal axis of Figure 4 is replaced by the (pore volume)/(LiI volume) ratio calculated from the LiI density and the N_2 sorption pore volume. The maximum conductivity is observed at different (pore volume)/(LiI volume) ratio for different pore diameters. The composite with 4.2 nm pore shows a maximum near (pore volume)/(LiI volume)=1, whereas other composites show a maximum at a larger (pore volume)/(LiI volume) ratio. This suggests different sized pores are differently filled with LiI, if one expects the maximum conductivity at the maximum volume of LiI loaded in the mesopore.

Under this assumption, 4.2 nm pore is filled with LiI, whereas other sized pores are partially filled with LiI. Estimated adsorption layers of LiI are approximately 2.1, 1.0, 0.8, and 0.6 nm for 4.2, 7, 9, and 12.5 nm mesopores, respectively. A further control of the LiI layer thickness inside the mesopore will be possible by fabricating the composite under different temperature and pressure conditions.

ACKNOWLEDGMENTS

This work was supported under PRESTO of Japan Science and Technology Corporation

REFERENCES

1. C.C. Liang, *J. Electrochem. Soc.* **120**, 1289 (1973).
2. S. Pack, B. Owens, and J.B. Wagner, Jr., *J. Electrochem. Soc.* **127**, 2177 (1980).
3. J. Maier, *Prog. Solid St. Chem.* **23**, 171 (1995).
4. H. L. Tuller, *Solid State Ionics* **131** 143 (2000).
5. J. Maier, *Solid State Ionics* **23**, 59 (1987).
6. S.A. Bagshaw, and T.J.Pinnavaia, *Angew. Chem. Int. Ed.* **35**, 1102(1996).
7. W.Z. Zhang, and T.J.Pinnavaia, *Chem. Comm.* 1185 (1998).
8. F. Vaudry, S. Khodabandeh, and M.E. Davis, *Chem. Mater.* **8**, 1451 (1996).
9. E.P. Barret, L.G. Joyner, and P.P. Halenda, *J. Am. Chem. Soc.* **73**, 373 (1987).

Mat. Res. Soc. Symp. Proc. Vol. 756 © 2003 Materials Research Society   EE2.4

Unusual Fast Cation Conduction in the High-Temperature Phase of Lithium Sodium Sulfate

H. Feldmann, R.E. Lechner[1], and D. Wilmer
Münster University, Institute of Physical Chemistry and Sonderforschungsbereich 458,
Schlossplatz 4/7, 48149 Münster, Germany
[1]Hahn-Meitner-Institut, Glienicker Str. 100, 14109 Berlin, Germany

ABSTRACT

Lithium sodium sulfate ($LiNaSO_4$) belongs to a group of simple inorganic salts exhibiting fast-cation conducting high-temperature phases with rotationally disordered anions. The analysis of a combination of quasielastic neutron scattering and high-frequency (10 MHz to 60 GHz) conductivity measurements in the high-temperature phase of $LiNaSO_4$ reveals an unusual cation conduction mechanism: the Haven ratio, $H_R = D^*/D_\sigma$, turns out to be considerably larger than one. This behavior, to our knowledge detected for the first time in a typical fast ion conductor, can be traced back to a charge correlation factor clearly smaller than unity, indicating that charge transport is less effective than tracer transport in this material.

INTRODUCTION

A number of crystalline materials have high-temperature phases which exhibit both high cation conductivity and rotational anion disorder. These are known as fast ion conducting plastic phases.

The mechanisms of ion conduction in the plastic phases of compounds like Li_2SO_4, $LiNaSO_4$, $LiAgSO_4$, or Na_3PO_4 have been discussed for a long time. It has been suggested that the transport of cations is greatly enhanced by an intimate coupling to the reorientational motion of the complex anions. There is an ongoing debate between the proponents of this so-called "paddle-wheel" mechanism [1] and those who favor an explanation in terms of a percolation-type mechanism [2].

In order to determine the relevance of the dynamic coupling in plastic-phase fast-ion conductors, it is highly desirable to examine the dynamics of cations and anions in entirely dynamic experiments.

In this paper, we give a short report on experiments on $LiNaSO_4$ using quasielastic neutron scattering and high frequency conductivity measurements.

The room temperature β-phase of $LiNaSO_4$ is stable up to $T_{tr} = 788$ K. Its space group symmetry is $P31c$ with $Z = 6$. The structure is built by SO_4 and LiO_4 tetrahedra. Upon transition into the α phase, the cation conductivity rises by more than two orders of magnitude to about $0.07\,\Omega^{-1}\,cm^{-1}$. In α-$LiNaSO_4$, the rotationally disordered anions form a BCC arrangement with $a = 5.75$ Å (829 K). Here, the sodium ions are located in octahedral sites $(0, \frac{1}{2}, \frac{1}{2})$, while the smaller lithium ions prefer a tetrahedral environment $(\frac{1}{4}, 0, \frac{1}{2})$.

EXPERIMENTAL

Sample preparation

Powder samples of $LiNaSO_4$ were prepared by fusing Li_2SO_4 and Na_2SO_4 at 973 K in a platinum crucible. Isotope-enriched $^7Li_2SO_4$ was used for the neutron scattering samples in order to reduce neutron absorption by ^6Li. After cooling to room temperature the crystalline samples were ground to a powder and kept in an evacuated drying chamber at 473 K for 72 hours. Sample identification as LiNaSO$_4$ was performed by X-ray diffraction. Samples were kept dry during all stages of handling and measurement.

Conductivity measurements at radio and microwave frequencies

The frequency-dependent electrical conductivity above about 10 MHz is derived from the transmission and reflection behavior of a sample with respect to an incoming electromagnetic wave. The conductivity at the respective frequency is determined with the help of the continuity conditions for the electromagnetic field at the sample/air phase boundaries. From 10 MHz to 60 GHz, coaxial and rectangular waveguide systems were used. For the measurements, a short waveguide section was filled with a pressed powder sample of LiNaSO$_4$. The sample waveguide section was then connected to the test ports of a vector network analyser (Anritsu MS 4623B and 37397C for coaxial and rectangular waveguide measurements, respectively), which records the amplitudes and phases of the transmitted and reflected waves relative to the incoming wave as a function of frequency.

Quasielastic neutron scattering

Time-of-flight experiments on ^{7}LiNaSO$_4$ were performed at the cold-neutron spectrometer NEAT [3], located at the Berlin Neutron Scattering Center (Hahn-Meitner-Institut, Germany). A powder sample was kept in a closed ceramic sample holder made of Alsint (99.7 % Al$_2$O$_3$). Spectra were taken at four temperatures in the α phase with an incident neutron wavelength of 5.1 Å, resulting in an elastic energy resolution of about 100 μeV. The data were normalized and corrected for self absorption, detector efficiency and detailed balance using the standard program packages for data analysis available at the spectrometers. Q ranges exhibiting Bragg peaks have been excluded from further analysis.

RESULTS AND DISCUSSION

High-frequency conductivity

Fig. 1 (lhs) shows the frequency-dependent conductivity of LiNaSO$_4$ in the α phase, measured at four different temperatures. At low frequencies, the conductivity remains constant up to about 300 MHz. Here the measured conductivity corresponds to the dc conductivity, σ_{dc}. The values, shown in an Arrhenius-representation in Fig. 1 (rhs), follow straight lines, resulting in activation energies of (0.59 ± 0.03) eV and (1.34 ± 0.08) eV in the α and β phases, respectively. At frequencies beyond about 300 MHz, however, the conductivity increases markedly with frequency. Such a dispersion is not unusual in solid electrolytes [4]; it is surprising, however, to find clear dispersion at such high conductivity levels. In RbAg$_4$I$_5$, e.g., the prototypal fast silver ion conductor, the conductivity dispersion is hardly detectable at room temperature where the material has a dc conductivity of about $0.3\,\Omega^{-1}\,cm^{-1}$. An explanation for the conductivity dispersion will be given later, see below.

Quasielastic neutron scattering

The first step of quasielastic neutron scattering interpretation is a phenomenological data analysis, i.e., fitting one or more Lorentzian lines, broadened by convolution with the energy resolution function, to the spectra measured at different scattering angles. In the case of LiNaSO$_4$, the data

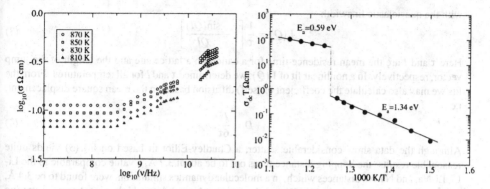

Figure 1: Frequency-dependent conductivity of LiNaSO$_4$ in the BCC high-temperature phase (left). Arrhenius representation of the dc conductivities (right).

obtained on NEAT immediately proved the necessity to use two Lorentzian lines to represent different components of the quasielastic spectra. In Fig. 2, we show the linewidths of the *narrower* component of a free fit using two Lorentzian lines as a function of the wave vector transfer Q. In our case of moderate energy transfers, Q is to a good approximation given by

$$Q_{elastic} = \frac{4\pi}{\lambda} \sin \frac{\theta}{2},$$
(1)

where λ and θ are the neutron wavelength and the scattering angle, respectively. Evidently, the

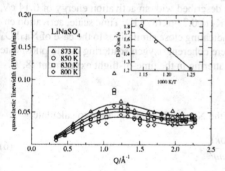

Figure 2: Q-dependent quasielastic linewidths (smaller component) of LiNaSO$_4$. Time-of-flight results from NEAT. Elastic resolution was about 100 μeV. Solid lines are fit results based on the Chudley-Elliott model. Insert: Arrhenius plot of the diffusion coefficients as defined in Eq. (4)

Q-dependence of the quasielastic linewidth shows the characteristic traits of jump diffusion. The Chudley-Elliott model [5] predicts for the incoherent dynamic structure factor (powder average)

$$S_{inc}(Q,\omega) = \frac{1}{\pi} \cdot \frac{\Gamma(Q)}{\Gamma^2(Q) + \omega^2}$$
(2)

83

with the quasielastic linewidth

$$\Gamma(Q) = \frac{1}{\tau}\left[1 - \frac{\sin(Ql)}{Ql}\right].$$ (3)

Here τ and l are the mean residence time for a cation on a lattice site and the length of the jump vector, respectively. In a nonlinear fit of $\Gamma(Q)$, we determined τ and l for all temperatures. From the fits we may also calculate the coefficient of (self) diffusion based on the mean square displacement, i.e.,

$$D = \frac{l^2}{6\tau}.$$ (4)

Although the data show considerable scatter, a Chudley-Elliot fit based on Eq. (3) yields quite reasonable results: the hopping distance turns out to be about 3.7 Å, a value comparable to the Li-Li, Li-Na, and Na-Na distances which, in a molecular dynamics simulation, were found to be 3.4 Å, 3.6 Å and 3.8 Å, respectively [6]. Due to the fact that Na exhibits a higher incoherent scattering cross section, it is no surprise that our experiments yield a hopping distance which is closer to the distances involving sodium ions. Using Eq. (4) to determine coefficients of self-diffusion with subsequent linear regression yields

$$D(T) = 1.2 \cdot 10^{-3}\,\mathrm{cm^2/s} \cdot \exp(-0.32\,\mathrm{eV}/k_B T).$$ (5)

The insert of Fig. 2 contains the diffusion coefficients along with the regression output. The results attain typical liquid-like values which emphasizes the extraordinary cation mobility in these materials. The range of diffusion coefficients agrees well with the results of tracer measurements [7].

Using the parameters of the Chudley-Elliott fit to fix the linewidth of the narrower quasielastic components now allows for a better characterization of the second (broader) quasielastic component. Its linewidth hardly depends on Q, indicating that we are observing a *localized* motion here. Its temperature dependence can approximately be described with an activation energy of 0.14 eV, the characteristic times ranging from 0.94 ps (800 K) to 0.79 ps (873 K). Time scale, activation energy and the lack of a Q dependence are features indicating close similarities to the case of Na_3PO_4 and solid solutions $xNa_2SO_4 \cdot (1-x)Na_3PO_4$ where coherent oxygen scattering due to phosphate reorientation was the only cause of quasielastic scattering in the time-of-flight experiment [8, 9].

Interpretation in terms of diffusivities

For comparison with the self-diffusivities, we use the Nernst-Einstein equation to calculate

$$D_\sigma = \frac{\sigma kT}{nq^2},$$ (6)

where n, q and k are the number density of the mobile ions, their charge and Boltzmann's constant, respectively. While q is easily identified as the elementary charge e, the number density needs some discussion. Since the tracer diffusivities of Li^+ and Na^+ are nearly identical [7], we may assume all cations as mobile and thus calculate n on the basis of the total number of cations, i.e., $n = 2.1 \cdot 10^{28}\,\mathrm{m^3}$. Fig. 3 gives an overview of all diffusivities in an Arrhenius representation. It is evident that our neutron scattering results agree well with the tracer diffusivities D^* of both Li^+ and Na^+. The values of D_σ, however, when calculated from the dc conductivies σ_{dc}, are lower by nearly an order of magnitude. We obtain a Haven ratio,

$$H_R = D^*/D_\sigma$$ (7)

84

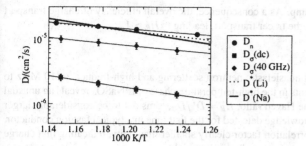

Figure 3: Arrhenius representation of different diffusion coefficients determined in LiNaSO₄. Tracer diffusion coefficients D^* taken from [7]. Values of D_σ (Eq. (6)) calculated using σ_{dc} (squares) and $\sigma(40\,\mathrm{GHz})$.

of about ten. This is surprising, since in most solid fast ion conductors, typical correlation effects result in $H_R < 1$ [10, 11]. Even if we assume that only one type of cations contributes to σ_{dc}, thereby reducing n and increasing D_σ, H_R still has a value of about five.

A discussion of correlation effects should be based on the correlation factors f and f_I for D^* and D_σ, respectively, which enter the Haven ratio such that

$$H_R = f/f_I. \tag{8}$$

f is a measure of the efficiency of cation jumps contributing to D^*. $f < 1$ signals a preference for correlated backward jumps; $f > 1$ means that several jumps occur preferentially in the same direction, while $f = 1$ indicates no correlation between jumps of an individual ion at all. Since a forward correlation ($f > 1$) is only conceivable in, e.g. channel-like systems, we may exclude this possibility here. Rather, $H_r > 1$ must be due to $f_I < 1$. f_I describes correlation between *charges*, including cross-correlations with other ions. If $f_I < 1$ were caused by a backward correlation of the same ion alone, this correlation would appear in the tracer correlation as well, and these contributions would cancel in Eq. (8). Only a correlated backward motion of *other* ions can explain the observed behavior. A cation that leaves its site induces correlated jumps of neighboring ions in the opposite direction. While the original cation jump still contributes to the self-diffusivity, the backward-correlated jumps of surrounding ions reduce the efficiency of the charge transport.

We cannot, however, assume that the correlated motion of neighboring ions starts immediately. Rather, we expect that, at short times after the initial jump, the neighboring ions have not yet reacted on the new situation. Since the short time ion behavior is monitored in the high-frequency conductivity, we should expect a *decreasing* effect of the backward correlation of other ions, i.e., an *increasing* conductivity with *increasing* frequency. This is, indeed, observed experimentally, cf. Fig. 1. The conductivity dispersion is thus explained by the time-dependent onset of co-operative backward jumps. If we substitute σ_{dc} by $\sigma(40\,\mathrm{GHz})$ in Eq. (6), we obtain values of D_σ which are much closer to D^*, resulting in $H_R \approx 2$, see Fig. 3.

The unusual cation transport behavior in LiNaSO₄ may be discussed in terms of a possible cation-anion interaction in this material: a jump of an individual cation from one site to another may be supported by the rotational motion of an anion which, at the same time, makes it less likely for the ion to jump back to its original site. Relaxation by backward jumps of the same ion is thus less likely. Rather, the relaxation is performed by jumps of neighboring ions. Owing to the potential gradient created by the initial jump, the neighboring ions are more likely to move in a direction opposite to the initial jump. As a consequence, the overall efficiency of charge transport is lower than the the efficiency of the tracer transport, leading to $H_R > 1$.

CONCLUSION

The analysis of a combination of quasielastic neutron scattering and high-frequency (10 MHz to 60 GHz) conductivity measurements in the high-temperature phase of $LiNaSO_4$ reveals an unusual cation conduction mechanism: the Haven ratio, $H_R = D^*/D_\sigma$, turns out to be considerably larger than one. This behavior, to our knowledge detected for the first time in a typical fast ion conductor, can be traced back to a charge correlation factor clearly smaller than unity, indicating that charge transport is less effective than tracer transport in this material.

Upon leaving their sites, cations in $LiNaSO_4$ (both Li and Na are mobile) obviously lead neighboring cations to jump in the opposite direction, thereby reducing the efficiency of the initial jump for the charge transport. Since one cannot expect an immediate response of the neighboring ions, we may predict that at short times after the initial hop, jumps of neighboring cations have not yet occurred. This would lead to an increase of D_σ and to a tendendy of H_R to approach unity with increasing frequency, which is, indeed, experimentally observed.

The unusual cation transport behavior can be discussed on the basis of a dynamic interaction between cations and anions. A cation that jumps from one site to another may be supported by the rotational motion of an anion which, at the same time, makes it less likely for the ion to jump back to its original site. Instead, the relaxation of the increased potential energy has to be performed by jumps of neighboring ions. As these jumps are more likely to occur in a direction opposite to the initial jump, the overall efficiency of charge transport is lower than the tracer transport, leading to $H_R > 1$.

ACKNOWLEDGEMENTS

A. Lundén and L. Nilsson provided considerable amounts of isotope-enriched $^7Li_2SO_4$. We have profited by stimulating discussions with K. Funke, and A. Putnis.

Financial support by the Deutsche Forschungsgemeinschaft in the framework of the Sonder-forschungsbereich 458 and by the Fonds der Chemischen Industrie is gratefully acknowledged.

REFERENCES

1. N.H. Andersen, P.W.S.K. Bandaranayake, M.A. Careem, M.A.K.L. Dissanayake, C.N. Wijayasekera, R. Kaber, A. Lundén, B.-E. Mellander, L. Nilsson, J.O. Thomas, *Solid State Ionics* **57**, 203 (1992).
2. E.A. Secco, *Solid State Ionics* **60**, 233 (1993).
3. R. E. Lechner, R. Melzer, J. Fitter, *Physica B* **226**, 86 (1996).
4. K. Funke, D. Wilmer, *Radiation Effects and Defects in Solids* **155**, 387 (2001).
5. C. T. Chudley, R. J. Elliott, *Proc. Phys. Soc.* **77**, 353 (1961).
6. M. Ferrario, M.L. Klein, I.R. McDonald, *Mol. Phys.* **86**, 923 (1995).
7. R. Tärneberg, A. Lundén, *Solid State Ionics* **90**, 209 (1996).
8. D. Wilmer, K. Funke, M. Witschas, R. D. Banhatti, M. Jansen, G. Korus, J. Fitter, R. E. Lechner, *Physica B* **266**, 60 (1999).
9. D. Wilmer, H. Feldmann, R.E. Lechner, *Phys. Chem. Chem. Phys.* **4**, 3260 (2002).
10. G. E. Murch, *Solid State Ionics* **7**, 177 (1982).
11. J.O. Isard, *J. Non-Cryst. Solids* **246**, 19 (1999).

Mat. Res. Soc. Symp. Proc. Vol. 756 © 2003 Materials Research Society

IONIC CONDUCTIVITY OF THE NEW FLUORIDE-ION CONDUCTOR CaSn₂F₆

Michael F. BELL, Georges DÉNÉS[1] and Zhimeng ZHU
Laboratory of Solid State Chemistry and Mössbauer spectroscopy, Laboratories for Inorganic Materials, Department of Chemistry and Biochemistry, Concordia University, Montréal, Québec, Canada

ABSTRACT

Metastable CaSn₂F₆ has been prepared for the first time and characterized. It is a well crystalline material that leaches SnF₂ in water to give the microcrystalline fluorite-type Ca₁₋ₓSnₓF₂ solid solution. In both materials, tin(II) is covalently bonded to fluorine, and thus carries a stereoactive non-bonding electronic pair. The electrical conductivity of CaSn₂F₆ was measured by the complex impedance method. The CaSn₂F₆ material was found to be a mixed conductor ($\tau_i = 0.50$), with a F⁻ conductivity a little below that of α-SnF₂. On heating to 250°C, it decomposes irreversibly to give SnF₂ and probably amorphous CaF₂ (undetected).

INTRODUCTION

The most efficient fluoride-ion conductors contain divalent tin and/or have a structure derived from that of the CaF₂ fluorite-type structure. The conductivity of tetrameric α-SnF₂ is higher than that of β-PbF₂, the best fluoride-ion conductor among the fluorite type, however, at the α → γ transition, the conductivity of α-SnF₂ decreases below that of β-PbF₂ [1]. The MSnF₄ materials (M = Ba and Pb) have a conductivity three orders of magnitude higher than that of the corresponding MF₂, with PbSnF₄, being the very best [2 & 3].

The first preparation of MF₂/SnF₂ materials was carried out by Donaldson and Senior, who prepared PbSnF₄, MSn₂F₆ and MSn₄F₁₀ (M = Sr, Ba and Pb) [4]. However, these authors were unable to obtain well defined materials in the CaF₂/SnF₂ system, and precipitation reactions between calcium nitrate and tin(II) fluoride in aqueous solution were reported to be not reproducible and give semi-amorphous materials with a variable stoichiometry. No reaction was obtained by one of us (GD) when CaF₂ and SnF₂ were heated together [5]. The preparation, structures and phase transitions of new materials in the MF₂/SnF₂ systems (M = Sr, Ba and Pb) was later studied in details by Dénès et al [5]. It was found that many of the phases obtained have a structure related to the fluorite-type, some with M/Sn order, while others are partially or fully disordered [6 & 7]. It was also found that PbSnF₄ undergoes many phase transitions , making it one of the most complex material known [8].

In the present work, we report the first preparation of crystalline CaSn₂F₆, its characterization, and the study of its electrical properties. It was found to have a conductivity slightly lower than that of α-SnF₂, and higher than that of γ-SnF₂, with a transport number for ions, $\tau_i = 0.50$, and to decompose irreversibly when heated up to 270°C.

---

[1] *To whom all correspondence should be addressed: gdenes@vax2.concordia.ca*

EXPERIMENTAL PROCEDURES

$CaSn_2F_6$ was obtained by precipitation when a 1.5M solution of $Ca(NO_3)_2.2H_2O$ (Fisher, 99%) was added on stirring to a 1.5M solution of SnF_2 (Ozark Mahoning, 99%). The precipitate was filtered, washed with a minimum of cold water, and allowed to dry in air at ambient temperature. Chemical analysis was performed on samples dissolved in HCl, by *Atomic Absorption Spectrometry* (AAS) for Ca and Sn, and by use of an Orion fluoride-ion electrode for F on solutions buffered with TISAB II (Total Ionic Strength Adjustment Buffer). The presence of nitrate ions was checked by the *Ultraviolet Spectrophotometric Screening Method*, using a Fischer Scientific spectrophotometer model 1001. The bulk density was measured by the Archimedean method of displacement in CCl_4. X-ray powder diffraction and ^{119}Sn Mössbauer spectroscopy were carried out at ambient temperature on instruments that have already been described [8]. Complex impedance measurements were performed on pressed powders, using two copper disks as ionically blocking electrodes, a Solartron 1174 *Frequency Response Analyzer* in the frequency range 100Hz to 100kHZ, from ambient temperature, up to 300°C. The electronic conductivity was determined by the Hebb-Wagner *polarization method*, i.e. by measuring the I.V current on pressed powders sandwiched between a positively polarized ionically blocking copper disk and a negatively polarized pellet of pressed powdered tin metal. A Princeton Applied Research Polarographic Analyzer, model 364, was used, with a constant potential scan rate of 5mV/sec.

RESULTS AND DISCUSSION

1. Material preparation and characterization

$CaSn_2F_6$ was obtained by precipitation for a low calcium nitrate molar ratio X in the reaction mixture, where $X = n_{Ca}/(n_{Ca}+n_{Sn}) < 0.23$, and n_{Ca} and n_{Sn} are the number of moles of calcium nitrate and tin(II) fluoride, respectively. In contrast with the early work of Donaldson and Senior [4], the reaction was found to provide reproducible results, and chemical analysis showed it is consistent with $CaSn_2F_6$ (mol% obs: Ca: 11.3, Sn: 20.8, F: 67.9, calc: Ca: 11.1, Sn: 22.2, F: 66.7), provided X<0.23 and the amount of washing water was kept to a minimum. The bulk density was measured to be equal to $4.20g.cm^{-3}$. This is 2.7% lower than the weighed average densities of CaF_2 and SnF_2, and shows that, at the atomic level, the packing of Ca, Sn and F in $CaSn_2F_6$ is less efficient than in CaF_2 and in SnF_2. A lower packing means larger interionic/interatomic distances and therefore weaker interactions since coulombic interactions decrease with increasing distances. It results that the lattice energy of $CaSn_2F_6$ is lower (i.e. less negative) than the weighed average of those of CaF_2 and of SnF_2. Thus, $CaSn_2F_6$ is less stable than the mixture of CaF_2 and SnF_2, and it is in a metastable state, and this is confirmed in the last part of this work. Its formation is probably due to the medium conditions at the reaction site where precipitation takes place, and its metastability can be accounted for by the insufficient thermal energy of room temperature for decomposing $CaSn_2F_6$ and reforming more stable CaF_2 and SnF_2.

X-ray powder diffraction shows a diffraction pattern made of very narrow lines (fig. 1c) characteristic of a highly crystalline material that bears no resemblance to that of CaF_2 or SnF_2, or to the $Ca_{1-x}Sn_xF_2$ solid solution (figs. 1a & 1b). It is also totally unrelated to the diffraction patterns of $SrSn_2F_6$ and $BaSn_2F_6$ [4]. Computer Search-Match did not identify it to any known phase in the JCPDS database, and since no report of the existence of $CaSn_2F_6$ was found in the

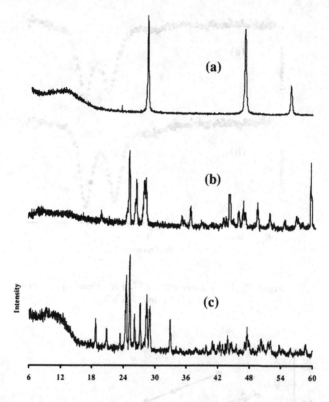

Figure 1: X-ray diffraction patterns of: (a) CaF₂, (b) α-SnF₂, and (c) CaSn₂F₆
(the low angle hump is due to the plexiglass sample holder)

literature, it was assumed to be a new compound. Tin-119 Mössbauer spectroscopy at ambient temperature (fig. 2) gives a large tin(II) quadrupole doublet ($\delta = 3.18$mm.s^{-1}, $\Delta = 1.90$mm.s^{-1}) characteristic of covalently bonded divalent tin, that has a hybridized stereoactive non-bonded electron pair (lone pair) [9].

2. Electrical properties

The electrical conductivity of $CaSn_2F_6$ is somewhat similar to that of SnF_2, and it is much lower than that of fast-ion conductors $PbSnF_4$ and $BaSnF_4$ (fig. 3). Its activation energy is similar to that of α-SnF_2 above 80°C (0.53eV), and this is quite higher than for the highest performing F⁻ conductors ($PbSnF_4$: 0.39eV, $BaSnF_4$: 0.32eV). In addition, while the transport number for fluoride ions is very high for SnF_2 and $MSnF_4$ ($\tau_i > 0.99$), it was found to be only 0.50 for $CaSn_2F_6$, making it a mixed conductor. Its conductivity being lower than that of the highest performance $MSnF_4$, and its activation energy being higher, are undoubtedly a reflection of its different crystal structure. The $MSnF_4$ structure is related to that of the best MF_2 fluoride ion conductors, the fluorite-type. Although the vacant F_x cubes present in the fluorite-type MF_2

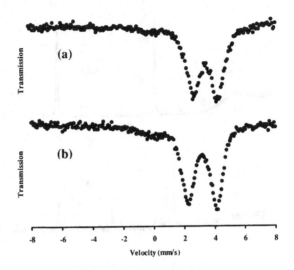

Figure 2: Ambient temperature ^{119}Sn Mössbauer spectrum of:
(a) α-SnF$_2$, and (b) CaSn$_2$F$_6$

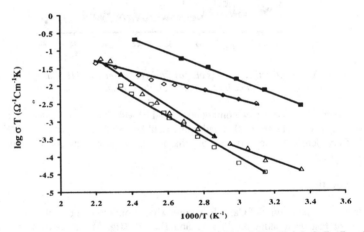

Figure 3: Ionic conductivity of: (a) ■ PbSnF$_4$ (E$_a$ = 0.39), (b) ◊ BaSnF$_4$ (E$_a$ = 0.30), (c) Δ =
α-SnF$_2$ (E$_a$ = 0.40, 0.69 and (d) □ CaSn$_2$F$_6$ (E$_a$ = 0.64)

conductors, assumed to be used to form interstitial fluoride-ion sites, are held responsible for their high fluoride ion mobility, no such empty cubes exist in MSnF$_4$ [10]. The exceptionally high conductivity of the latter has been attributed to the partial population of the spacing between adjacent tin layers, where the tin(II) lone pairs are located [11]. Since the crystal structure of

$CaSn_2F_6$ is not known at this stage, one cannot explain how the fluoride ions can move over long distances in this material, however, it is much less efficient than in $MSnF_4$. In addition, although the mechanism of electron motion is not known, one can say that it is not due to the tin non-bonded electron pair, since the lone pair was shown by Mössbauer spectroscopy to be stereoactive (fig. 2), therefore locked in a hybrid orbital belonging to a single tin atom. This contrasts with the case of non-stereoactive lone pairs, found in many tin(II) containing bromides and iodides, like in $CsSnBr_3$ [12]. In such cases, the lone pair being on the unhybridized 5s orbital of tin, can be easily partially transferred to the conduction band of the solid. Such solids are brightly colored materials. This is not the case of $CaSn_2F_6$, which is white. The participation of electrons to the conduction mechanism in white tin(II) solids with a stereoactive lone pair was also observed in poorly conducting tin(II) chloride fluorides, $SnClF$ and Sn_2ClF_3 [13].

3. Decomposition at high temperature

When $CaSn_2F_6$ is heated, the activation energy drops to nearly zero at ca. 170°C (fig. 4a). When heated up to 270°C, and then cooled, below ca. 250°C, the conductivity decreases rapidly (fig. 4b) and the curve seems to extrapolate to the conductivity of γ-SnF_2 (fig. 4c), that was measured earlier [1]. X-ray diffraction of the sample after cooling to ambient temperature shows a complete change of set of peaks (fig. 5). No $CaSn_2F_6$ was left in the sample after heating, and only the peaks of α-SnF_2 were present. Therefore, above 270°C, all the $CaSn_2F_6$ had decomposed to liquid SnF_2 (Mp = 215°C) and amorphous CaF_2 (not detected by diffraction). The conductivity and X-ray diffraction observations are in agreement with the know set of phase transitions of SnF_2 [14]. Decomposition of $CaSn_2F_6$ occurred above the melting point of SnF_2 (215°C). Solidification gives γ-SnF_2, which undergoes a displacive second order paraelastic to ferroelectric phase transition to β-SnF_2 at 66°C on cooling [14]. Below 150°C, γ-SnF_2, and β-SnF_2 below 66°C, are in a metastable state, and may change to stable α-SnF_2, at any time, usually quite fast [14]. In the present case, X-ray diffraction shows it happened (fig.5).

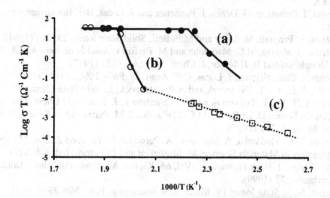

Figure 4: *Ionic conductivity of $CaSn_2F_6$ through the decomposition point: (a) heating curve, (b) cooling curve, (c) = γ-SnF_2*

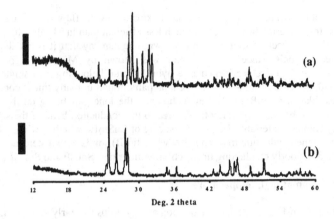

Figure 5: X-ray diffraction patterns of a $CaSn_2F_6$ sample: (a) as prepared
(only $CaSn_2F_6$ peaks are observed), and (b) after the conductivity measurements

ACKNOWLEDGEMENT

The Natural Science and Engineering Research Council of Canada and Concordia University are acknowledged for supporting this work. Dr. A. Muntasar, Concordia University, is gratefully acknowledged for helping with the graphics of this manuscript.

REFERENCES

1. D. Ansel, J. Debuigne, G. Dénès, J. Pannetier and J. Lucas, Ber. Bunsenges. Phys. Chem. **82**, 376 (1978).
2. G. Dénès, T. Birchall, M. Sayer and M.F. Bell, Solid State Ionics **13**, 213 (1984).
3. G. Dénès, G. Milova, M.C. Madamba and M. Perfiliev, Solid State Ionic **86-88**, 77 (1996).
4. J. D. Donaldson and B. J. Senior, J. Chem. Soc. (A), 1821 (1967).
5. G. Dénès, J. Pannetier, and J. Lucas, C.R. Acad. Sc. Paris **280C**, 831 (1975).
6. G. Dénès, Y.H. Yu, T. Tyliszczak and A.P. Hitchcock, J. Solid State Chem. **91**, 1 (1991).
7. G. Dénès, Y.H. Yu, T. Tyliszczak and A.P. Hitchcock, J. Solid State Chem. **104**, 239 (1993).
8. A. Collin, G. Dénès, D. Le Roux, M.C. Madamba, J. M. Parris and A. Salaün, Intern. J. Inor. Mater. **1**, 289 (1999).
9. G. Dénès, M.C. Madamba, A. Muntasar, A. Peroutka, K. Tam and Z. Zhu, Mössbauer Spectroscopy in Materials Science, M. Miglierini and D. Petridis (eds.), NATO Science Series, 3. High Technology, Vol. **66**, Kluwer Academic Publishers, Dordretch (Netherlands), 25 (1999).
10. G. Dénès, Solid State Ionics IV, Mater. Res. Soc. Symp. Proc. **369**, 295 (1995).
11. R. Kanno, K. Ohno, H. Izumi, Y. Kawamoto, T. Kamiyama, H. Asano and E. Izumi, Solid State Ionics **70/71**, 253 (1994).
12. J. D. Donaldson and J. Silver, J. Solid State Chem. **18**, 117 (1976).
13. P. Claudy, J. M. Letoffe, S. Vilminot, W. Granier, Z. Al Ozaibi and L. Cot, J. Fluorine Chem. **18**, 203 (1981).
14. G. Dénès, Mat. Res. Bull. **15**, 807 (1980)

Oxide Electroceramics for Separation Membranes and Gas Sensors

Mat. Res. Soc. Symp. Proc. Vol. 756 © 2003 Materials Research Society

Oxide Ion Transport in Bismuth-Based Materials

Rose-Noëlle Vannier, Edouard Capoen, Caroline Pirovano, César Steil, Guy Nowogrocki, Gaëtan Mairesse
Laboratoire de Cristallochimie et Physicochimie du Solide, CNRS UMR 8012, ENSCL, Université des Sciences et Technologies de Lille, B.P. 108, 59652 Villeneuve d'Ascq Cedex, France

Richard J. Chater, Stephen J. Skinner, John A. Kilner
Centre for Ion Conducting Membranes (CICM), Department of Materials, Imperial College London, Prince Consort Road, London SW7 2BP, UK

ABSTRACT

When used as ceramic membranes for the electrically driven separation of oxygen from air, BIMEVOX materials allow the production of high oxygen fluxes at moderate temperature, 300-600°C. However, $^{18}O/^{16}O$ Isotope Exchange Depth Profile Technique revealed low kinetics of oxygen transfer at the surface of these ceramics when studied under equilibrium. These kinetics were considerably enhanced when a current was applied. The same membranes were characterized under working conditions using X-ray synchrotron and neutron radiations. Their dynamical transformation under bias was confirmed and explained by a slight reduction of the BIMEVOX electrolyte under working conditions.

INTRODUCTION

Bismuth-based materials exhibit attractive oxide ion conductivity. Among these, the BIMEVOX compounds are considered as the best oxide ion conductors at moderate temperatures, 300-600°C. They are derived from the parent compound $Bi_4V_2O_{11}$ by partial substitution for vanadium with a metal. A wide range of elements (Cu, Co, Ni, Ta, Nb, Sb...) are able to substitute for vanadium and this allows the stabilization at room temperature of the highly conductive γ form of the parent compound [1]. BICUVOX.10, for instance, is obtained by partial substitution for vanadium with 10% of copper. Because of their high performances, these materials could be used as membrane for electrically driven Ceramic Oxygen Generators. The principle of such a device is shown in figure 1. It is very similar to the Solid Oxide Fuel Cell (SOFC) and relies on the ability of oxide ions to migrate through a ceramic material under an electric field. In a first step, oxygen molecules are dissociated into oxide ions at the cathode according to the reaction:

$$O_2 + 4e^- \rightarrow 2O^{2-}$$

These oxide ions migrate, under the influence of the electric field, to the anode where they recombine into oxygen molecules according to the reverse reaction. This process allows for the production of controlled amounts of very high purity (>99.99%) oxygen, which can be delivered under pressure without the use of any mechanical device.

Two steps govern the oxygen transport is such membranes i) the oxygen exchange at the surface of the ceramic, ii) the oxygen diffusion through the ceramic. The limiting step in the whole process is often the gas-solid transfer at the surface of the material and, to help this transfer, electrode materials are usually added at the surface of the membrane.

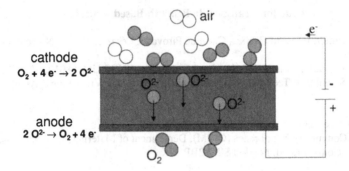

Figure 1. The principle of a Ceramic Oxygen Generator. Oxygen molecules are dissociated into oxide ions at the cathode according to the reaction $O_2 + 4e^- \rightarrow 2O^{2-}$. Then, these oxide ions, under the electric field, migrate to the anode where they recombine into oxygen molecule according to the reverse reaction.

High oxygen fluxes were obtained with membranes composed of a BIMEVOX electrolyte that were simply co-sintered between two gold grids [2]. Several compositions (BICUVOX, BICOVOX, BIZNVOX) were studied at temperatures between 430 and 600°C. Excellent performances with close to 100% efficiency and current densities up to 1A/cm^2 were observed. No additional electrode materials was added, except for the gold grids, which were primarily acting as a current collector.

One technique to characterize the oxygen transfer in such membranes is the $^{18}O/^{16}O$ Isotope Exchange Depth Profile Technique (IEDP) based on Secondary Ion Mass Spectroscopy (SIMS). These materials were investigated using the IEDP-SIMS technique under equilibrium and under an electrical bias. From these measurements, and combining with in situ X-ray and neutron diffraction experiments, performed under working conditions the mechanisms involved in the oxygen transfer were explained.

EXPERIMENTAL DETAILS

Cylindrical pellets (10mm diameter and 3-5mm thick), with a relative density higher than 95%, were prepared from attrition-milled powder of BIBIVOX.02 (a composition close to that of the parent compound $Bi_4V_2O_{11}$), BICUVOX.10 and BICOVOX.10 compositions. To perform the $^{18}O/^{16}O$ Isotope Exchange experiments, the surfaces were polished down to 0.25μm to obtain a mirror like surface. Oxygen exchanges were carried out during 12 hours at about 700°C and oxygen partial pressure of 200mbar. The pellets were first annealed in research grade $^{16}O_2$ (99.996%) under the same conditions as the subsequent isotopic exchange to ensure that the samples were in chemical equilibrium with their surroundings and to remove any damage caused by the polishing. These experiments were performed under dry atmosphere with the moisture content directly monitored by sampling of the surrounding gas. In each case the moisture content was found to be a few ppm. After the isotopic exchange anneal, the oxygen penetration was measured using secondary ion mass spectroscopy (SIMS, Atomika 6500) in both depth profile and line scan modes [3].

A second set of isotopic exchange experiments were carried out with applied currents of 80 and 8mA. For this purpose a special cell was constructed. Isotopic exchanges under electrical bias were performed on pellets with the same compositions (10mm diameter and 2mm thick) as for the equilibrium exchanges. One side of each pellet was polished flat down to 0.25μm. Half of this surface was then masked and the sample faces coated with gold by dc sputtering. This resulted in a cylindrical sample with one face half coated in gold as shown in Figure 2. A gold grid was then applied to this sputtered region using gold paste. The other side of this pellet was painted with gold paste and stuck to a gold grid, held on a circular alumina disk. Each gold grid was attached to gold wires for the current collection. To sinter the gold paste, the whole cell was annealed for 1 hour at 650°C in air with a heating and cooling rate of 4°C/min. The cell was then introduced into a silica tube and exchanges were performed under the same conditions as for the equilibrium case, but with an electrical current of 80 or 8 mA for two hours. The experimental set-up will be described in more detail in a forthcoming paper [4].

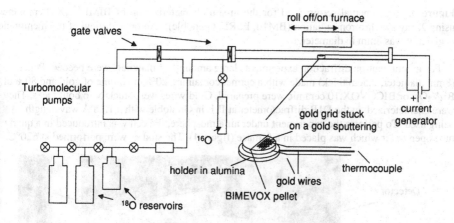

Figure 2: Schematic of the cell and the ^{18}O exchange set-up for the experiment under applied current conditions.

Pellets used for in-situ X-ray diffraction (8mm diameter and 2-2.5mm thick) were prepared from BIMEVOX powders, sintered at 710°C for 1 hour. Their faces were covered with two 50 mesh gold grids, fixed by a slurry of BIMEVOX powder in ethanol. The whole membranes were successively annealed at 740°C and 760°C for 1h30. The in-situ X-ray diffraction experiment was performed on the BM16 line of ESRF at Grenoble, Figure 3. A special cell holder was designed. The pellet was introduced between two gold contacts, connected to gold wires to allow the current density control through the membrane. Diffractograms were recorded at 620°C, for current densities up to at least $1 A/cm^2$ with a parallel beam at 0.4033Å wavelength. For each investigation, under anodic or cathodic polarization, a new pellet was systematically used and the current density stepwise increased.

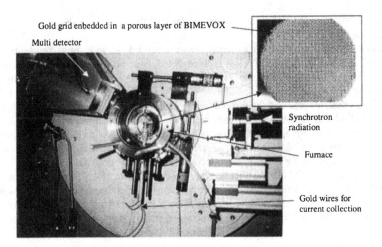

Gold grid enbedded in a porous layer of BIMEVOX

Multi detector

Synchrotron radiation

Furnace

Gold wires for current collection

Figure 3. Experimental set-up used for the in-situ characterization of BIMEVOX membranes using X-ray synchrotron radiation (BM16, ESRF, Grenoble). An optical view of the membrane is given, it was 8mm in diameter.

For in-situ neutron diffraction experiments large amounts of materials were needed. Pellets (8mm diameter, 2.5cm thick) made with a cermet containing 20% in volume of gold and 80% of $Bi_4V_2O_{11}$ or BICOVOX.10 ceramic were prepared. Gold wires were stuck onto both faces. They were characterized on the D1B diffractometer at ILL in Grenoble with a 1.28 Å wavelength being used. To perform the experiment under air atmosphere, the cell was introduced in a quartz tube open to air which was placed in a furnace (figure 4). The studies were performed at 620°C.

Detector

Furnace

Neutron beam

Figure 4. Experimental set-up used for the in-situ characterization of BIMEVOX membranes using neutrons (D1B, ILL, Grenoble). An optical view of the membrane is given, it was 8mm in diameter and 2.5cm thick.

RESULTS and DISCUSSION

Oxygen transfer in BIMEVOX materials at equilibrium

Two parameters characterize the oxygen transfer through a membrane: the surface exchange coefficient k (cm/s) which is an indication of the kinetics of oxygen transfer at the surface, and the coefficient of diffusion D* (cm²/s). These two parameters are easily deduced from the profile of concentration of the ^{18}O tracer. The solution for isotopic tracer diffusion into a semi-infinite medium has been derived by Crank [5]

$$C'(x,t) = \frac{C(x,t) - C_{bg}}{C_g - C_{bg}} = erfc\left[\frac{x}{2\sqrt{D^*t}}\right] - \left[\exp\left(\frac{kx}{D^*} + \frac{k^2t}{D^*}\right) \times erfc\left(\frac{x}{2\sqrt{D^*t}} + k\sqrt{\frac{t}{D^*}}\right)\right]$$

where C'(x,t) is the normalized concentration of ^{18}O, C(x,t) is the tracer concentration as a function of depth (x) and diffusion time (t), C_{bg} is the natural abundance of ^{18}O, C_g is the ^{18}O concentration in the gas, D* is the bulk oxygen tracer diffusion coefficient and k is the surface exchange coefficient.

For samples annealed under dry oxygen under equilibrium, ^{18}O concentrations were close to the natural background ($C_{bg} = {}^{18}O/({}^{18}O + {}^{16}O) = 0.002$) and to extract data, pellets had to be annealed for a long period of time, 12 hours. The corresponding values are given in table I. They are compared to those obtained for the classical electrolytes, stabilized zirconia (YSZ) and gadolinium doped ceria (CGO). As expected, BIMEVOX materials exhibit the best diffusion coefficients but their surface exchange coefficients are of the same order of magnitude as observed for the classical electrolytes. Moreover, because of the long annealing duration, and the risk of profile overlapping from the two surfaces (because of the high D value), it is likely that these exchange coefficients have been overestimated.

Table I: Coefficients of surface exchange k (cm/s) and of diffusion D*(cm²/s) deduced for BIBIVOX.02, BICUVOX.10 and BIBIVOX.02 compared to YSZ and CGO ones at equilibrium under dry oxygen atmosphere.

	T(°C)	D*(cm²/s)	k(cm/s)
BIBIVOX.02	705	$4(3).10^{-7}$	$0.2(2).10^{-8}$
BICUVOX.10	708	$8(5).10^{-7}$	$1.2(4).10^{-8}$
BICOVOX.10	700	$3(2).10^{-7}$	$0.5(2).10^{-8}$
YSZ [6]	700	2.10^{-8}	$0.6.10^{-9}$
CGO [7]	700	4.10^{-8}	6.10^{-8}

The oxygen transfer at the interface involves oxygen molecules, electrons and oxygen vacancies, according to the following reaction:

$$\frac{1}{2}O_2 + V_o^{\bullet\bullet} + 2e^- \rightleftharpoons O_o^X.$$

Therefore electrons are needed for the transfer to occur. Although a small electronic contribution to the conductivity was observed for BICUVOX and BICOVOX materials [8, 9], it does not seem sufficient to allow easy oxygen transfer in the absence of an electrical bias.

Oxygen transfer in BIMEVOX materials under applied current conditions

In contrast, high isotopic fractions were observed after polarization. Figure 5 shows normalized ^{18}O isotopic fractions, obtained in both the line scan mode, and from selected areas in the depth profile mode, as a function of the distance from the gold border on BIBIVOX.02. As an example, after a constant current of 80mA for one hour, an isotopic ratio of 0.20 was observed at 1 mm from the gold electrode border. The same behavior was observed for the BICUVOX.10 and BICOVOX.10 compositions. Moreover, the measured isotopic fraction was found to be a function of the applied current density. When a current of 8 mA was applied to a separate sample, a normalized isotopic fraction ten times lower was measured.

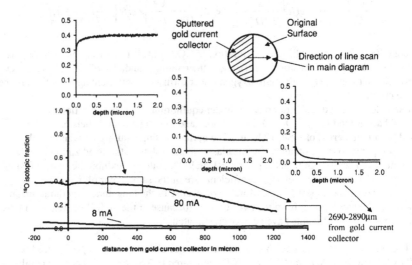

Figure 5: Normalized ^{18}O isotopic fractions obtained in both line scan mode and depth profile mode (rectangles indicate the depth profile crater positions) as a function of the distance from the gold border on BIBIVOX.02 membranes polarised at 8 mA and 80mA.

The X-ray diffractograms corresponding to a BIBIVOX.02 cathode membrane at 620°C for applied current densities up to $1A/cm^2$, are given in figure 6. The star indicates the gold Bragg peak. It remained at the same position for current densities of up to $0.4A/cm^2$, for higher current densities a Joule effect was observed. However, even for current densities lower than $0.4A/cm^2$, an evolution of the diffractograms corresponding to the BIBIVOX .02 phase was observed. The unit-cell was maintained but with a decrease of the a parameter and an increase of the c parameter. The same modifications were noticed for the cobalt and copper doped materials but with lower amplitudes.

When turning off the current, the initial pattern was recovered, indicating the apparent reversibility of the transformation.

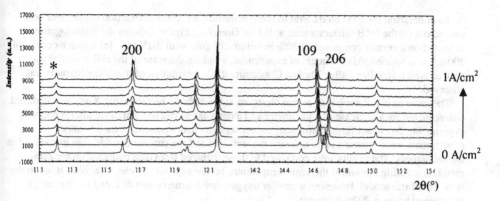

Figure 6: X-ray diffractograms corresponding to a BIBIVOX.02 cathode membrane at 620°C for applied current densities up to 1A/cm².

At the anode practically no modification was observed. However some traces of $BiVO_4$ were found after experiment on the BIBIVOX.02 membrane.

Because of grain orientations and the low X-ray diffusion factor of oxygen atoms compared to bismuth, it was not possible to extract any further data from these X-ray diffractograms.

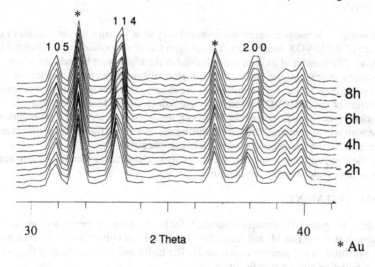

Figure 7: Difractograms of neutrons collected on a cermet composed of 20% in volume of gold with $Bi_4V_2O_{11}$, for which a current of 200mA was applied.

To understand the oxygen transfer in these materials, a neutron diffraction experiment was carried out on the D1B diffractometer at ILL in Grenoble. Figure 7 shows the difractograms collected on a cermet composed of 20% in volume of gold with $Bi_4V_2O_{11}$, for which a current of 200mA was applied. After 4 hours of experiment, a sudden decrease of the cell potential was noticed which stabilized after 2 hours. Concomitantly, an evolution of the diffractograms was observed.

The same unit-cell evolution as that observed at the cathode surface using X-ray was deduced, characterized by a decrease of a parameter and a decrease of c. They stabilized after 6 hours of experiment. To extract further structural parameters, the data measured after 6 hours of experiments were summed up to increase the statistics and the structural model was introduced in the refinement. The results were compared to those deduced from data collected on the same membrane during 3 hours at the same temperature before applying an electrical bias. It led to the same structural model. However, a smaller oxygen stoichiometry was deduced for the sample maintained under a 200mA current.

Therefore, under electrical bias, the crystal structure of BIMEVOX is maintained and a small reduction of the electrolyte, likely vanadium V^{5+} into V^{4+}, explains the oxygen transfer.

In this case the initial pattern of the BIMEVOX materials was recovered only after annealing at 700°C. No apparent modification of the cell were observed but a gray deposit was noticed on the quartz tube walls at the cathodic side. The analysis of this deposit revealed the presence of bismuth and therefore the reduction process in these membranes is probably more complicated.

CONCLUSION

$^{18}O/^{16}O$ isotopic exchange experiments revealed very slow kinetics for the transfer of oxygen at the surface of BIMEVOX membranes. This oxygen transfer is considerably enhanced under electrical bias. The transfer of oxygen is not limited to the triple point boundaries at the electrode/electrolyte/gas interface as for classical electrolytes but occurs at the electrolyte surface itself. This is explained by a slight reduction of the electrolyte under operating conditions. The crystal structure of the BIMEVOX materials is maintained and, at first glance, this transformation appeared to be reversible. However, some traces of bismuth were observed on the quartz tube wall which contained the sample during the neutron diffraction experiment and the mechanism involved in this process may be more complicated.

Finally, these experiments emphasize the necessity to characterize materials under working conditions to fully understand their behavior.

ACKNOWLEDGEMENT

The authors are grateful to Françoise Ratajack for her help for the membrane preparation. They are also grateful to Drs Michel Anne, Eric Dooryhee and Olivier Isnard for their help during the synchrotron and neutron experiments. The ESRF and ILL are thanked for providing synchrotron radiation and neutron facilities.

REFERENCES

1.   S. Lazure, C. Vernochet, R.N. Vannier, G. Nowogrocki, G. Mairesse, *Solid State Ionics*, **90**, 117 (1996).

2.   J.C. Boivin, C. Pirovano, G. Nowogrocki, G. Mairesse, P. Labrune, G. Lagrange, *Solid State Ionics*, **113-115**, 639 (1998).
3.   R.J. Chater, S. Carter, J.A. Kilner, B.C.H. Steele, *Solid State Ionics* **53-56**, 859 (1992).
4.   R.N. Vannier, S.J. Skinner, R.J. Chater, J.A. Kilner, G. Mairesse, *Solid State Ionics*, submitted.
5.   P.S. Manning, J.D. Sirman, R.A. de Souza, J.A. Kilner, *Solid State Ionics* **100**, (1997) 1.
6.   E. Ruiz-Trejo, J.D. Sirman, Yu.M. Baikov, J.A. Kilner, *Solid State Ionics* **113-115**, 565 (1998)
7.   T. Iharada, A. Hammouche, J. Fouletier, M. Kleitz, *Solid State Ionics* **48**, 257(1991).
8.   J. Fouletier, C. Muller, E. Pernot, *Electroceramics V*, Univ. of Aveiro, 37 (1996).

Mat. Res. Soc. Symp. Proc. Vol. 756 © 2003 Materials Research Society

A Study of the Oxygen Transport Kinetics in SrFeO$_{3-x}$

Jiho Yoo and Allan J. Jacobson
Center for Materials Chemistry, University of Houston,
4800 Calhoun Rd., Houston, TX, USA, 77204-5002

ABSTRACT

Experimental results from electrical conductivity relaxation (ECR) measurements on SrFeO$_{3-x}$ are described. Values of D_{chem} and k_{chem} were obtained by monitoring the variation of the time dependence of the electrical conductivity after an abrupt change in the oxygen partial pressure. Values for the oxygen ion and vacancy diffusion coefficients were calculated from the measured thermodynamic factors and compared with those of other compositions in the series La$_{1-x}$Sr$_x$FeO$_{3-x}$. The surface exchange coefficients, k_{chem} and k_{ex} were also determined.

INTRODUCTION

Solid oxides with perovskite and related structures are of interest as electrodes or interconnect materials in solid oxide fuel cells (SOFCs), as ion-permselective membrane materials, as membranes for syngas reactors, and in thin film sensors [1 ~ 3]. Even though each application has its own specific requirements, the perovskite oxide should, in general, have both high electronic and ionic conductivity and be structurally stable in reducing atmospheres. SrFeO$_{3-x}$ shows reasonable properties for practical applications when doped with other metals, and consequently has been widely investigated as a base material. Many studies have been performed to understand the structural characteristics of SrFeO$_{3-x}$ because of its interesting order-disorder transition and extremely high oxygen deficiency [4, 5]. A defect model and thermodynamic constants for compositions in the La$_{1-x}$Sr$_x$FeO$_{3-x}$ series were proposed by Mizusaki et al. [6]. A phase diagram (T vs. x) for SrFeO$_{3-x}$ was reported by Takeda et al. [7] and recently electrical conductivities and thermopowers were measured at $10^{-16} \leq pO_2 \leq 1$ atm and at $700 \leq T \leq 950$ °C by Kozhevnikov et al. [8].

Only limited data on oxygen transport kinetics are available for undoped SrFeO$_{3-x}$ [9] although bulk diffusion and surface exchange coefficients for La$_{1-x}$Sr$_x$FeO$_{3-x}$ (x = 0, 0.1, 0.25, 0.4) have been reported [10 ~ 12]. In the present work, the self-diffusion (D_O) and vacancy (D_v) diffusion coefficients were determined from the chemical diffusion coefficient (D_{chem}) of SrFeO$_{3-x}$ obtained from conductivity relaxation experiments and compared with the previous data. At the experimental conditions used for the relaxation experiments, the kinetics are controlled by both bulk diffusion and surface exchange and consequently the surface exchange coefficient (k_{chem}) was also determined [13].

EXPERIMENTAL

SrFeO$_{3-x}$ was synthesized by the self-propagating high-temperature synthesis method. The details of the method can be found elsewhere [14]. A mixture of SrO$_2$ (Aldrich) and Fe powder (Alfa Aesar, 99.998%) were ground with zirconia balls for 24 h. Three weight % of NaClO$_4$ (Sigma, 99.6%) was added to the mixture and re-mixed for 10 h. The highly exothermic NaClO$_4$ provides internal O$_2$ and heat for the initial combustion. The reactant pellet was ignited with a chemical match and then sintered for 12 h at 1250 °C. The sintered pellets were ~ 95 % dense of the theoretical density. X-ray powder diffraction (Scintag XDS 2000, Cu K_α radiation) and electron microprobe analysis (JEOL JXA-8600) were used to confirm the phase purity of SFO. The sample is more than 97% pure and chemically homogeneous.

The details of the electrical conductivity relaxation (ECR) technique are described elsewhere [15, 16]. The gas switches employed were between pO_2 = 0.5 / 1 atm, 0.15 / 0.3 atm, 0.05 / 0.1 atm, and 0.01 / 0.02 atm. The ECR data were obtained at 790, 830, 870, 910, 950, and 980 °C. The experimental data for the relative conductivity, $g(t) = \{\sigma_{expt}(t)-\sigma_{expt}(0)\}/\{\sigma_{expt}(\infty)-\sigma_{expt}(0)\}$, were fit to an analytical solution of the diffusion equation with appropriate boundary conditions with D_{chem} and k_{chem} as variables.

The variation of oxygen content of $SrFeO_{3-x}$ was investigated by thermogravimetric analysis at pO_2 = 0.15 atm and the solid-state coulometric titration technique [17]. The absolute x-values at $750 \le T \le 1040$ °C and pO_2 = 0.21 atm were taken from the previous work [7].

THEORY

In the case where the total electrical conductivity of a mixed conductor is predominantly determined by the electronic contribution, the transport is controlled by ionic diffusion. Consequently, D_{chem} can be related to vacancy- and self-diffusion coefficients using the thermodynamic factors (Γ_v, Γ_O) according to [12]:

$$D_{chem} = \Gamma_v \cdot D_v = \Gamma_O \cdot D_O, \text{ where } \Gamma_v = -\frac{1}{2} \cdot \frac{\partial \ln pO_2}{\partial \ln C_v} \text{ and } \Gamma_O = \frac{1}{2} \cdot \frac{\partial \ln pO_2}{\partial \ln C_O}$$

where C_v and C_O are the vacancy and oxide ion concentration, respectively. Since the oxide ionic conductivity is the same as the vacancy conductivity, $C_O D_O = C_v D_v$ is derived using Nernst-Einstein equation [10]:

Recently, a surface exchange model was developed to include the contribution of oxygen vacancies [18]. The vacancy concentration at the surface is assumed proportional to the vacancy concentration in the bulk. In this model, the surface exchange coefficient in the region of a small deviation from thermal equilibrium (k_{chem}), assuming a dilute solid solution, is given by:

$$k_{chem} = k_f \cdot \left(\frac{pO_2}{pO_2^o}\right)^{1/2} \cdot \frac{C_{v,e}}{C_{O,e}} \cdot (\Gamma_O + 1) = k_f \cdot \left(\frac{pO_2}{pO_2^o}\right)^{1/2} \cdot A$$

where k_f, C_h, and A are the forward rate constant of the oxygen incorporation reaction, the concentration of electron holes, and $\{1 + C_{v,e}(1/C_{O,e} + 4/C_{h,e})\}$, respectively, and the subscript e stands for thermal equilibrium. The surface exchange coefficient at thermal equilibrium (k_{ex}) is correlated with k_{chem} according to $k_{ex} = k_{chem}/(\Gamma_O + 1)$.

RESULTS AND DISCUSSION

Diffusion coefficients

The results for the chemical diffusion coefficients obtained by fitting the data for different pressure switches and at different temperatures are shown in Figures 1a and 1b. In Figure 1a, the data are shown as a function of the final pressure used in the switch. Figure 1b shows the temperature dependence of the average diffusion coefficients.

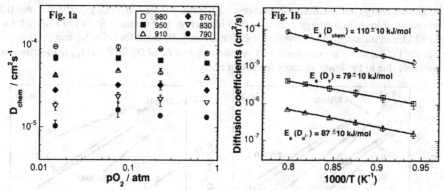

Figure 1. a) The values of D_{chem} obtained with different pressure switches; b) the temperature dependence of the diffusion coefficients.

As expected, the values of D_{chem} for SrFeO$_{3-x}$ are independent of pO$_2$ in the range $0.01 \leq$ pO$_2$ ≤ 1 atm (Fig. 1The D_{chem} values at 790 °C, for example, vary from 1.03×10^{-5} to 1.64×10^{-5} cm^2/s. The previous studies reported that D_{chem} values for La$_{1-x}$Sr$_x$FeO$_{3-\delta}$ ($x = 0.1, 0.4, 0.5$) were also independent of pO$_2$ at $0.01 \leq$ pO$_2 \leq 1$ atm [12, 17]. The activation energy (E_a) of D_{chem} is 110(10) kJ/mol for SrFeO$_{3-x}$. The E_a of D_{chem} (kJ/mol) for La$_{1-x}$Sr$_x$FeO$_{3-\delta}$ was reported to be 82(7) for $x = 0.1$, 81(12) for $x = 0.4$ [12] and, 119(13) for $x = 0.5$ [17].

Since the electrical conductivity of SrFeO$_{3-x}$ is predominantly electronic over the range of oxygen partial pressures used in these experiments, the oxygen ion and vacancy diffusion coefficients (D_v and D_O) can be obtained from D_{chem} using the relationships given above and the measured thermodynamic factors for oxygen ions and vacancies. The values of Γ_v and Γ_O were obtained by plotting pO$_2$ vs. oxygen vacancy concentration (C_v) and oxide ion concentration (C_O) using the results from thermogravimetric analysis and solid-state coulometric titration experiments reported elsewhere [17]. The results are shown in Figures 2a and 2b.

The D_v values for the series of compounds La$_{1-x}$Sr$_x$FeO$_{3-\delta}$ ($0 \leq x \leq 1$) are compared in Figure 2a. The value of D_v measured for SrFeO$_{3-x}$ is ~1.5 to 5 times smaller than observed for the lanathanum substituted compositions. Because of the high oxygen vacancy content of SrFeO$_{3-x}$, (0.35 ~ 0.47), interactions among defects to lead to defect clusters or short range order and consequently, the defect association reduces the vacancy mobility of SrFeO$_{3-x}$.

The activation for vacancy diffusion in SrFeO$_{3-x}$ is 79(10) kJ/mol (Fig. 1b). The corresponding values for La$_{1-x}$Sr$_x$FeO$_{3-x}$, with $x = 0.5, 0.4$, and 0.1, were reported to be 110(10), 80(25), and 74(12) kJ/mol, respectively [12, 17]. In addition, activation energies (E_a) of 79(25) kJ/mol for $x = 0.1$ and 114(23) kJ/mol for $x = 0.25$ were reported by Ishigaki [10]. The oxide ion diffusion coefficients (D_O) measured for SrFeO$_{3-x}$ are also compared with the previous results in Fig. 2b[10, 17, 19]. D_O is expected to be proportional to C_v/C_O if only doubly ionized vacancies exist and they are randomly distributed. On increasing the level of Sr-substitution on the La sites, the D_O values increase as expected. In the La$_{1-x}$Sr$_x$FeO$_{3-x}$ series, SrFeO$_{3-x}$ has the largest D_O value, as its vacancy concentration is much higher than the others. The E_a of D_O for SrFeO$_{3-x}$ is 87(10) kJ/mol. The E_a values of D_O for La$_{1-x}$Sr$_x$FeO$_{3-\delta}$ ($x = 0.1$ ~0.5) and La$_{0.6}$Sr$_{0.4}$Co$_{0.6}$Fe$_{0.4}$O$_{3-\delta}$ were reported to be 140 ~ 175 kJ/mol [10, 17, 19]. The E_a of D_O is given by both the ionic defect migration energy and the defect formation energy [20]. For $La_{1-x}Sr_xFeO_{3-\delta}$ (x = 0, 0.1, and 0.25), the vacancy formation energy decreases with increasing the Sr content from 137 kJ/mol for x = 0 to 62 kJ/mol for x = 0.25 [10] while the E_a for vacancy diffusion for $SrFeO_{3-x}$ and for the $La_{1-x}Sr_xFeO_{3-\delta}$ series are comparable. The relatively low E_a of D_O for $SrFeO_{3-x}$ is therefore most probably due to the lower vacancy formation energy.

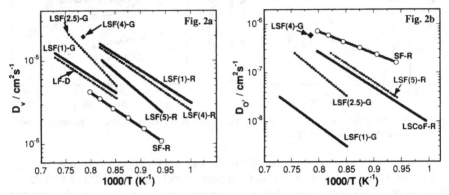

Figure 2. a) Comparison of the D_v values for the $La_{1-x}Sr_xFeO_{3-x}$ series. (a) SF-R: the present ECR data for $SrFeO_{3-x}$, (b) LSF(1)-G, LSF(2.5)-G, and LSF(4)-G: the gas phase analysis data with x = 0.1, 0.25, and 0.4, respectively [10], (c) LSF(1)-R, LSF(4)-R, and LSF(5)-R: the ECR data with x = 0.1, 0.4, and 0.5, respectively [12, 17], (d) LF-D: the depth profile data with $LaFeO_{3-x}$ [11]. b) Comparison of the D_O values of $La_{1-x}Sr_xFeO_{3-x}$ series. (a) SF-R: the present ECR data for $SrFeO_{3-x}$. (b) LSF(1)-G, LSF(2.5)-G, and LSF(4)-G: the gas phase analysis data with x = 0.1, 0.25, and 0.4, respectively [10]. (c) LSF(5)-R: the ECR data with x = 0.5 [17]. (d) LSCoF-R: the ECR data with $La_{0.6}Sr_{0.4}Co_{0.6}Fe_{0.4}O_{3-x}$ [19].

Surface exchange coefficients

The pO_2 dependence of the surface exchange coefficient, k_{chem}, of $SrFeO_{3-x}$ as a function of the final oxygen partial pressure is shown in Fig. 3. k_{chem} is found to be proportional to $(pO_2)^n$ with $0.22 \leq n \leq 0.44$. As discussed above, the value of k_{chem} is expected to be proportional to $(pO_2/pO_2^*)^{0.5+m}$, where m is determined by the pO_2 dependence of the thermodynamic factor $(\Gamma_0 + 1)$ according to the model of Kim et al. The thermodynamic factor $(\Gamma_0 + 1)$ is proportional to $(pO_2/pO_2^*)^m$ with $-0.20 \leq m \leq -0.17$ and hence k_{chem} should be proportional to $(pO_2/pO_2^*)^n$ with n = ~0.33 to 0.30. Values of n in the range 0.44 to 0.22 are obtained, in reasonable agreement with with the model. The temperature dependence of k_{chem} is shown in Figure 4. The k_{chem} activation energy varies with pO_2, indicating the pO_2 dependent kinetics. The values are 145(12) kJ/mol at pO_2 = 0.01 atm, 146(9) kJ/mol at pO_2 = 0.05 atm, 127(6) kJ/mol at pO_2 = 0.15 atm, and 98(2) kJ/mol at pO_2 = 0.5 atm. The pO_2-dependent thermal activation was also observed for $La_{0.5}Sr_{0.5}FeO_{3-x}$ in the same pO_2 range [17]. From the expression given above for k_{chem}, the change in the activation energy must arise from variations in the rate constant with pO_2 because the temperature dependence of the term $\{1+C_{v,e}(1/C_{O,e} + 4/C_{h,e})\}$ is only weakly dependent on

pO_2. This implies that the rate-determining step for the surface exchange reaction on $SrFeO_{3-x}$ varies with the oxygen partial pressure in the pO_2 range used for the measurements.

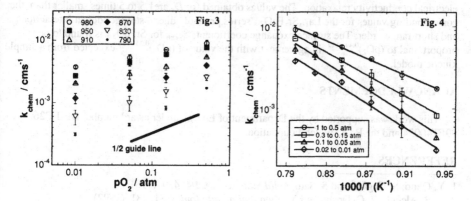

Figure 3. The pO_2 dependence of k_{chem} at 790, 830, 870, 910, 950, and 980 °C.

Figure 4. The temperature dependence of k_{chem} at $0.01 \leq pO_2 \leq 1$ atm.

The exchange coefficient at thermal equilibrium, k_{ex}, for $SrFeO_{3-x}$ was calculated using the relation $k_{ex} = k_{chem}/(\Gamma_O + 1)$. The values at $pO_2 = 0.05$ atm are compared with those of compositions in the $La_{1-x}Sr_xFeO_{3-x}$ series at $0.05 \leq pO_2 \leq 0.07$ atm in Figure 5 [10, 11, 17]. The comparison shows clearly that the magnitude of k_{ex} is closely related to the level of Sr doping. Increased acceptor doping enhances the formation of oxygen vacancies at the surface thereby leading to an increase in the magnitude of k_{ex}.

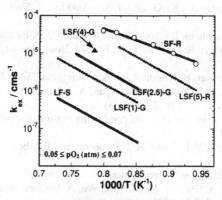

Figure 5. Comparison of the k_{ex} values. (a) SF-R: the present ECR data for $SrFeO_{3-x}$ at $pO_2 = 0.05$ atm. (b) LSF(1)-G, LSF(2.5)-G, and LSF(4)-G: the gas phase analysis data with x = 0.1, 0.25, and 0.4, respectively at $pO_2 = 0.064$ atm [10]. (c) LSF(5)-R: the ECR data with x = 0.5 at $pO_2 = 0.05$ atm [17]. (d) LF-S: the SIMS data with $LaFeO_{3-\delta}$ at $pO_2 = 0.07$ atm [11].

CONCLUSION

The oxygen diffusion and surface exchange coefficients have been measured for $SrFeO_{3-x}$ by electrical conductivity relaxation. The values obtained for D_v are 1.5 to 5 times smaller than the corresponding values for the $La_{1-x}Sr_xFeO_{3-x}$ series probably due to stronger defect interactions and short range order. The surface exchange coefficient, k_{chem}, for $SrFeO_{3-x}$ is found to be proportional to $(pO_2)^{0.44\sim0.22}$, in agreement with the value of $(pO_2)^{0.33\sim0.30}$ expected from a simple kinetic model.

ACKNOWLEDGEMENTS

This work was supported by the Department of Energy under award number DE-FC26-99FT40054 and the R. A. Welch Foundation.

REFERENCES

1. Y. Ohno, S. Nagata, and S. Sato, *Solid State Ionics*, **3/4**, 439 (1981)
2. C. B. Alcock, R. C. Doshi, and Y. Shen, *Solid State Ionics*, **51**, 281 (1992)
3. J. J. Tunney and M. L. Post, *Journal of Electroceramics*, **5**, 63 (2000)
4. P. D. Battle, T. C. Gibb, and S. Nixon, *J. Solid State Chem.*, **79**, 75 (1989)
5. J. A. M. van Roosmalen and E. H. P. Cordfunke, *J. Solid State Chem.*, **93**, 212 (1991)
6. J. Mizusaki, M. Yoshihiro, S. Yamauchi and K. Fueki, *J. Solid State Chem.*, **67**, 1 (1987).
7. Y. Takeda, K. Kanno, T. Takada, O. Yamamoto, M. Takano, N. Nakayama, and Y. Bando, *J. Solid State Chem.*, **63**, 237 (1986)
8. V. L. Kozhevnikov, I. A. Leonidov, M. V. Patrakeev, E. B. Mitberg, and K. R. Poeppelmeier, *J. Solid State Chem.*, **158**, 320 (2000)
9. K. Nisancioglu, and T. M. Gür, *Solid State Ionics*, **72**, 199 (1994)
10. T. Ishigaki, S. Yamauchi, K. Kishino, J. Mizusaki and K. Fueki, *J. Solid State Chem.*, **73**, 179 (1988).
11. T. Ishigaki, S. Yamauchi, J. Mizusaki, and K. Fueki, *J. Solid State Chem.*, **55**, 50 (1984)
12. J. E. ten Elshof, M. H. R. Lankhorst and H. J. M. Bouwmeester, *J. Electrochem. Soc.*, **144**(3), 1060 (1997)
13. H. J. M. Bouwmeester, H. Kruidhof and A. J. Burggraaf, *Solid State Ionics*, **72**, 185 (1994).
14. Q. Ming, J. Hung, Y. L. Yang, M. Nersesyan, A. J. Jacobson, J. T. Richardson, and D. Luss, *Combust. Sci. and Tech.*, **138**, 279 (1998)
15. S. Wang, *Oxygen transport in mixed conducting perovskite oxides*, PhD. Thesis, University of Houston, 2000.
16. S. Wang, A. Verma, Y. L. Yang, A. J. Jacobson, and B. Abeles, *Solid State Ionics*, **140**, 125 (2001)
17. J. Yoo and A. J. Jacobson, in preparation.
18. S. Kim, S. Wang, X. Chen, Y. L. Yang, N. Wu, A. Ignatiev, A. J. Jacobson, and B. Abeles, *J. Electrochem. Soc.*, **147**(6), 2398 (2000).
19. J. E. ten Elshof, M. H. R. Lankhorst, and H. J. M. Bouwmeester, *Solid State Ionics*, **99**, 15 (1997)
20. P. Kofstad, *Nonstoichiometry, Diffusion, and Electrical Conductivity in Binary Metal Oxides*, (Wiley, New York, 1972)

Preparation and Oxygen Permeability of La-Sr-Co-Fe Oxide Thin Films by a Chemical Solution Deposition Process

Hirofumi Kakuta[1,2], Takashi Iijima[1] and Hitoshi Takamura[3]
[1]Smart Structure Research Center, National Institute of Advanced Industrial Science and Technology, Tsukuba Central 2, 1-1-1, Umezono, Tsukuba 305-8568, Japan
[2]Core Research for Evolutional Science and Technology (CREST), Japan Science and Technology Corporation (JST), 2-1-13, Higashi-Ueno, Taito-ku, Tokyo 110-0015, Japan
[3]Department of Materials Science, Graduate School of Engineering, Tohoku University Aoba-yama02, Sendai 980-8579, Japan

ABSTRACT

Oxygen-ionic and electronic conductive thin films with the composition of $La_{0.6}Sr_{0.4}Co_{0.5}Fe_{0.5}O_{3-\alpha}$ (LSCF) were prepared on a porous alumina substrate by a chemical solution deposition (CSD) process and their oxygen permeating flux densities were measured. Thickness of the LSCF layer on the substrate was about 0.4 μm. Oxygen flux density of the LSCF sample was found to be 0.6 $\mu mol \cdot cm^{-2} \cdot s^{-1}$, however, time-dependent degradation of oxygen flux was observed. The CeO_2 barrier layer between the LSCF layer and the substrate was effective in order to improve time-dependent degradation of oxygen flux.

INTRODUCTION

Oxygen-ionic and electronic mixed conductors have been receiving much attention for the usage in a variety application such as gas sensors and fuel cell electrodes. Moreover, it can be applied to an oxygen permeating membrane [1-4]; oxygen gas can selectively pass through the membrane made of the mixed conductor from high oxygen pressure side to low oxygen pressure side at elevated temperatures. This material is expected to manufacture pure oxygen at a lower cost, however oxygen-permeating flux of bulk materials is insufficient so far. It is required to increase the oxygen flux density of mixed conductors.

One of the approaches to increase oxygen flux is to decrease the thickness of mixed conductive ceramics, because oxygen-permeating flux is in inverse proportion to the thickness of the membrane within certain limitation [3]. The bulk sample was generally cut and polished to prepare the membrane, but this method has a limitation of decreasing the thickness, since the mechanical strength decreases with decreasing the thickness. The oxygen permeating membrane should be supported by a porous substrate to permeate the gases if the membrane thickness is less than about 150 μm. It is required to prepare thinner films than 150 μm. The mixed conductive thin films have been prepared on the porous support [5-7]. However, oxygen-permeating flux was not successfully measured because it is difficult to obtain dense films.

Chemical solution deposition (CSD) process is one of the promising techniques to prepare ceramic films on dense substrates in a variety filed including superconductors [8] and ferroelectric materials [9]. This study utilized the CSD process coupled with a spin coating technique in order to obtain dense mixed conductive films.

As a mixed conductor, we made choice of the La-Sr-Co-Fe system, because bulk forms of LSCF were known to exhibit high oxygen permeability [1, 2]. The composition used in this study was $La_{0.6}Sr_{0.4}Co_{0.5}Fe_{0.5}O_{3-\alpha}$ (LSCF) with the perovskite-type structure. While we carried on this study using a porous alumina substrate, time-dependent degradation of oxygen flux was observed. A reaction between the LSCF film and the substrate was likely to occur. Therefore, a barrier layer was used between the thin film and the substrate. We chose the CeO_2 as a barrier layer, because it is hard to dissolve the constitute elements of thin film.

The purpose of this study is to prepare the LSCF thin films on the porous substrate and to measure its oxygen flux density. Moreover, the effect on the barrier layer of CeO_2 inserted between the LSCF layer and the substrate was investigated.

EXPERIMENTAL DETAILS

The porous alumina substrate named ANODISC^{TM} from Whatman was used as a substrate in this study. It has precisely controlled honeycomb-like channels with an average pore diameter of about 200 nm, but the channel narrows to about 20 nm at the bottom surface. Coating of the precursor solution was conducted at the surface having 20 nm pores.

To prepare the LSCF precursor solution, $La(NO_3)_3 \cdot 5H_2O$, $Sr(OH)_2$, $Co(NO_3)_2 \cdot 8H_2O$, $Fe(NO_3)_3 \cdot 5H_2O$ were dissolved into the 2-methoxyethanol in a concentration of 0.2 M by using a stirrer. The precursor solution of CeO_2 was also prepared with $Ce(NO_3)_3 \cdot 6H_2O$ in the concentration of 0.2 M. The sequence of spin coating and pyrolysis at 673 K was conducted three times, and then the films were fired at 973 K for 5 min. These processes were repeated several times to increase the thickness of the films.

The phases present were identified by X-ray diffractometry (XRD). Microstructure of the films was observed by a field-emission-type scanning electron microscope (SEM). Oxygen-permeating flux of the samples was analyzed by a gas chromatograph (GC) with active carbons and molecular sieves using air – He cell. Helium gas with a flow rate of 20 sccm as a carrier gas was swept to the sample. If nitrogen was detected with GC owing to leakage, the volume of oxygen leakage was calculated from the nitrogen leakage. The volume of permeating oxygen through the membrane was estimated from subtracting the volume of leakage oxygen from the volume of detected oxygen.

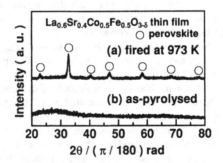

Figure 1. X-ray diffraction patterns of LSCF thin films: (a) fired at 973 K; (b) as-pyrolysed.

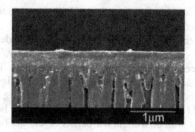

Figure 2. SEM micrograph of a cross section view of LSCF thin film on the porous alumina substrate fired at 973 K.

RESULTS AND DISCUSSION

Figure 1 shows the X-ray diffraction patterns of LSCF thin films: (a) fired at 973 K; (b) as-pyrolysed. Though the as-pyrolysed sample was not crystallized, the sample after firing 973 K was obtained as a perovskite-type single phase. All samples were fired at 973 K in this study.

Figure 2 shows SEM micrograph of a cross section view of LSCF thin film on the porous alumina substrate fired at 973 K and coated 21 times. Although a part of LSCF infiltrated into the pores of the substrate, a large portion was deposited onto the substrate. The thickness of the LSCF layer on the substrate was around 0.4 μm.

Figure 3 shows time-dependent oxygen flux of LSCF samples: (a) LSCF / CeO_2 / substrate, (b) LSCF / substrate, (c) bulk form. At first, oxygen flux of (b) LSCF thin film without CeO_2 layer is compared with that of (c) bulk form. The details of the CeO_2 layer are discussed later. Corrected oxygen flux density of the LSCF thin film with the thickness of 0.4 μm was 0.6 $\mu mol \cdot cm^{-2} \cdot s^{-1}$ at 1073 K. For comparison, the bulk sample with the same composition was prepared and measured under the same measuring condition as the LSCF films. Oxygen flux density of the bulk with a thickness of 1.42 mm was 0.02 $\mu mol \cdot cm^{-2} \cdot s^{-1}$, as seen in figure 3 (c).

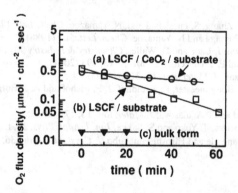

Figure 3. Time-dependent oxygen flux of LSCF samples: (a) LSCF / CeO_2 / substrate, (b) LSCF / substrate, (c) bulk form.

Oxygen flux of the thin film is higher than that of bulk form, so the effect on the decreasing the thickness was demonstrated.

However, while oxygen flux of the bulk sample is independent of elapsed time, that of the LSCF thin film deposited onto the substrate decreases gradually. This degradation seemed to be caused by reaction between the LSCF film and the substrate. Therefore, we investigated the effect on the CeO_2 layer as an intermediate layer. The barrier layer was also prepared by spin coating five times and fired at 973 K, followed by LSCF coating 21 times. Oxygen flux density of the sample with CeO_2 is shown in figure 3 (a). The difference between (a) open circles and (b) open squares in figure 3 is with or without the barrier layer, and both the LSCF layer was coated 21 times. As time-dependent degradation of the sample with CeO_2 layer was improved, CeO_2 is found to be available. It was considered that the CeO_2 layer inhibited the reaction between the LSCF film and the substrate.

CONCLUSIONS

Preparation and oxygen flux density of La-Sr-Co-Fe oxide thin films by a CSD process were studied. The LSCF dense film with the thickness of 0.4 μm was successfully fabricated onto the porous alumina substrate. Its oxygen flux density was found to be 0.6 $\mu mol \cdot cm^{-2} \cdot s^{-1}$, which was much higher than that of bulk sample with the same composition. As time-dependent degradation of oxygen flux was observed, we tried fabricating the intermediate layer of CeO_2. Since time-dependent oxygen-permeating flux was improved, this study confirmed that the barrier layer of CeO_2 was effective for preventing the reaction between the LSCF film and the substrate.

ACKNOWLEDGMENTS

This work was supported by CREST of Japan Science and Technology Corporation.

REFERENCES

1. Y. Teraoka, H. M. Zhang, S. Furukawa and, N. Yamazoe, *Chem. Lett.* 1743 (1985).
2. Y. Teraoka, T. Nobunaga and N. Yamazoe, *Chem. Lett.* 503 (1988).
3. J. Kilner, S. Benson, J. Lane and D. Waller, *Chemistry & Industry* 907 (1997).
4. S. P. S. Badwal and F. T. Ciacchi, *Adv. Mater.* **13**, 993 (2001).
5. M. Liu and D. Wang, *J. Mater. Res.* **10**, 3210 (1995).
6. C. Chen, H. J. M. Bouwmeester, H. Kruidhof, J. E. Elshof and A. J. Burggraaf, *J. Mater. Chem.* **6**, 815 (1996).
7. C. Xia, T. L. Ward and P Atanasova, *J. Mater. Res.* **13**, 173 (1998).
8. H. Zhuang, H. Kozuka, T. Yoko and S. Sakka, *Jpn. J. Appl. Phys.* **29**, L1107 (1990).
9. Takashi Iijima, Gang He and Hiroshi Funakubo, *J. Crystal Growth* **236**, 248 (2002).

Mixed Ionic – Electronic Conduction and Oxygen Permeation in Ba-In Based Oxides Doped with Transition Metals

Yusuke Aizumi, Hitoshi Takamura, Atsunori Kamegawa and Masuo Okada
Department of Materials Science, Graduate School of Engineering, Tohoku University, Aoba-yama 02, Sendai 980-8579, Japan

ABSTRACT

The electrical conductivity and oxygen permeability of transition-metal-doped $(Ba_{0.3}Sr_{0.2}La_{0.5})_2(In_{1-x}TM_x)_2O_{5+\delta}$ (TM = Fe, Co, Mn and Sn; $0 \leq x \leq 0.5$) have been investigated. The X-ray diffraction analysis revealed that all the samples had a cubic perovskite-type structure due to La^{3+} doping on the alkaline earth site. For Fe-doped specimens, the lattice parameter of 0.414 nm for $(Ba_{0.3}Sr_{0.2}La_{0.5})_2In_2O_{5+\delta}$ linearly decreased with increasing the Fe content, suggesting the incorporation of Fe into the matrix phase. Fe-doping for the In site enhanced the p-type conduction under a wide $P(O_2)$ range, and the p-type conductivity increased with increasing the Fe content without decreasing the ionic conductivity. The oxygen permeability was measured under the $P(O_2)$ difference between helium and air in the temperature range of 800 ~ 1000 °C. For the Fe-doped specimens, the oxygen flux density of $j(O_2)$ increased with increasing the Fe content, and a maximum value of 0.5 $\mu mol \cdot cm^{-2} \cdot s^{-1}$ was attained at 1000 °C for the membrane thickness of 1.0 mm.

INTRODUCTION

Mixed oxygen-ion and electronic conductors have been extensively studied because of their promising electrochemical applications such as oxygen separation membranes and electrode materials for solid oxide fuel cells [1]. Among a number of the mixed conductors, perovskite-type oxides in La-Sr-Co-Fe-O and La-Sr-Ga-Fe-O systems are well known to exhibit a high oxygen flux density, reaching to 8.2 $\mu mol \cdot cm^{-2} \cdot s^{-1}$ at 1000 °C for $La_{0.7}Sr_{0.3}Ga_{0.6}Fe_{0.4}O_3$ [2-5]. The theoretical oxygen flux density of the mixed conductors can be expressed by the following Wagner's equation:

$$j(O_2) = \frac{RT}{16F^2L} \int_{\ln P(O_2)'}^{\ln P(O_2)''} \sigma_{amb} \, d \ln P(O_2) \qquad (1)$$

$$\sigma_{amb} = \frac{\sigma_{el}\sigma_{ion}}{\sigma_{el} + \sigma_{ion}} \qquad (2)$$

where, $j(O_2)$ means the oxygen flux density, F: Faraday constant, L: membrane thickness, R: gas constant, T: temperature, $P(O_2)'$ and $P(O_2)''$: oxygen partial pressures, σ_{amb}: ambipolar conductivity, σ_{el} and σ_{ion}: electronic and oxygen-ion conductivity, respectively. To explore novel mixed conductors with a high oxygen flux density, it is important to choose a host oxide that exhibits a high ionic conductivity, since the oxygen flux density is limited by the minor charge carrier that tends to be the ionic one in many mixed conductors.

Apart from the La-Sr-Fe-based oxides, $Ba_2In_2O_5$ with a brownmillerite structure is well-known to be a good oxygen-ion conductor [6]. Recently, it is reported that $Ba_2In_2O_5$ doped with La^{3+} on the Ba site exhibited a high ionic conductivity of 0.12 S/cm at 1000 °C [7]. This La-doped $Ba_2In_2O_5$ crystallizes in the perovskite-type structure because a part of oxygen vacancy sites in the brownmillerite structure can be filled up by oxygen due to the charge compensation of $[La_{Ba}] = 2[O_i]$. As a result, the order-disorder transition that can be observed in $Ba_2In_2O_5$ at around 1203 K vanishes. Based on this La-doped $Ba_2In_2O_5$, it seems to be possible to prepare the mixed oxygen-ion and electronic conductor by doping transition metals on the In site as well as the other perovskite-type oxides. Thus, the purpose of this study is to prepare the Ba-In-based oxides doped with La and the transition metals on Ba and In sites, respectively, and to clarify the effect of the transition metal doping on the electrical conductivity and the oxygen permeability. As dopant elements, Fe, Co, Mn and Sn were selected in this study.

EXPERIMENTAL PROCEDURES

The samples of $(Ba_{0.3}Sr_{0.2}La_{0.5})_2(In_{1-x}TM_x)_2O_{5+\delta}$ (TM = Fe, Co, Mn and Sn) with $0 \leq x \leq 0.5$ were prepared by the citrate-based liquid mix technique [8,-10], where Sr was added to optimize the ionic radii on the Ba site [11]. Calcined powders were die-pressed into disc-shaped pellets with dimensions of about ϕ15mm × 1mm at 30 MPa followed by isostatic pressing at 300 MPa. The pellets were then sintered at 1260 °C ~ 1530 °C for 2 ~ 3 h in air. Phase identification was conducted by the powder X-ray diffraction. The microstructure of thermally etched samples was observed by an optical microscope and a scanning electron microscope equipped with an energy-dispersive composition analyzer. The electrical conductivity measurement was performed by means of the DC four-probe method in the temperature range of 750 ~ 1000 °C under a wide range of $P(O_2)$ controlled by CO-CO_2 gas mixtures. For the oxygen flux density measurement, disk-shaped samples with a dimension of approximately ϕ15mm × 0.5 ~ 1 mm in thickness were used. The sample was sandwiched between two quartz tubes, and the permeate side was sealed by a borosilicate glass. The oxygen flux density was measured under the $P(O_2)$ difference between He and air in the temperature range of 800~1000 °C by using a gas chromatograph and a mass spectrometer.

RESULTS AND DISCUSSION
Synthesis and crystal structure

Figure 1 shows X-ray diffraction patterns for Fe-doped samples of $(Ba_{0.3}Sr_{0.2}La_{0.5})_2(In_{1-x}Fe_x)_2O_{5+\delta}$ with $0.0 \leq x \leq 0.5$. As expected, all the samples were identified as a cubic perovskite-type structure, and reflections were found to shift towards lower angles as the Fe content increased. This peak shift seems to be reasonable based on the ionic radius difference between In and Fe cations, which are 0.80 and 0.65 nm, respectively. The lattice constant of the perovskite-type phase observed in the Fe, Co, Mn, and Sn-doped specimens were plotted as a function of the dopant content in Fig. 2. The lattice constant of Fe and Mn-doped samples linearly decreased with increasing the Fe and Mn content until x = 0.5. For example, the value of 0.406 nm for Fe-doped sample with x=0.4 was smaller by 2 % than that of 0.414 nm for $(Ba_{0.3}Sr_{0.2}La_{0.5})_2In_2O_{5+\delta}$. On the other hand, for Sn and Co-doped samples, the lattice constant variation was flattened at around x = 0.1 and 0.4, respectively, due to the formation of secondary phases. From the XRD patterns, the secondary phases were found to be mainly In_2O_3 and

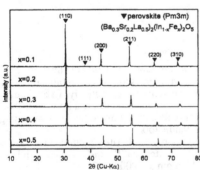

Figure 1. X-ray powder diffraction pattern of Fe-doped samples with x = 0.1 ~ 0.5.

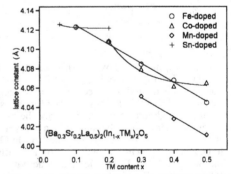

Figure 2. The lattice constants of TM-doped samples as a function of the dopant content (TM = Fe, Co, Mn and Sn.)

Figure 3. The microstructure of $(Ba_{0.3}Sr_{0.2}La_{0.5})_2(In_{0.7}Fe_{0.3})_2O_{5+\delta}$ sintered at 1530 ºC for 3h.

$La_{1-x}Sr_xCoO_3$ for Sn and Co-doped samples, respectively. Figure 3 shows the microstructure of $(Ba_{0.3}Sr_{0.2}La_{0.5})_2(In_{0.7}Fe_{0.3})_2O_{5+\delta}$ sintered at 1530 °C for 3 h. As can be seen, the sample was well densified under the sintering condition. The compositional analysis was also performed for this sample, and the observed composition, $(Ba_{0.28}Sr_{0.25}La_{0.47})_2(In_{0.68}Fe_{0.32})_2O_{5+\delta}$ showed a good agreement with the nominal one.

Oxygen permeation property

The oxygen flux density of $(Ba_{0.3}Sr_{0.2}La_{0.5})_2(In_{1-x}TM_x)_2O_{5+\delta}$, where TM = Fe, Co, Mn, and Sn was measured at a temperature range of 800 ~ 1000 °C under He. Figure 4 shows the Arrhenius plot of the oxygen flux density, $j(O_2)$, and oxygen permeability, $J(O_2)$, for the TM-doped samples. The thickness of all the samples was fixed to 1.0 mm. For the Fe-doped sample with x = 0.4, the highest oxygen flux density of 0.5 $\mu mol \cdot cm^{-2} \cdot s^{-1}$ was attained at 1000 °C. This suggests that the Ba-In-based oxides can be a mixed oxygen-ion and electronic conductors by Fe-doping. The Co-doped sample also showed a high $j(O_2)$ of 0.2 $\mu mol \cdot cm^{-2} \cdot s^{-1}$ at 1000 °C. However, the oxygen flux density of Mn and Sn-doped samples decreased by an order of magnitude, compared to that of the Fe-doped one. An activation energy of $j(O_2)$ for the Fe-doped sample with x = 0.4 was determined as 38.9 kJ/mol. The composition dependence of $j(O_2)$ was examined for the Fe and Co-doped samples as shown in Fig. 5. The oxygen flux density of $j(O_2)$ increased with increasing the Fe and Co content, and showed a maximum value at x = 0.4 and 0.3, respectively. The activation energies of Fe-doped samples obtained from Arrhenius plots (not shown) were 56.2, 43.2, and 23.1 kJ/mol for x = 0.1, 0.2, and 0.3, respectively. These activation energies will be discussed in conjunction with those for electrical conductivities later.

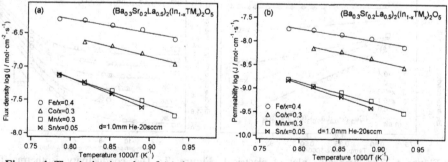

Figure 4. The Arrhenius plot of (a) the oxygen flux density and (b) the oxygen permeability for the TM-doped samples.

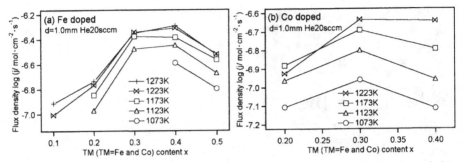

Figure 5. The composition dependence of the oxygen flux density for (a) Fe and (b) Co-doped samples.

Electrical conductivity of Fe-doped samples

The electrical conductivity measurements were performed for the Fe-doped samples, which showed a high $j(O_2)$ compared to the other TM-doped systems. Figure 6 shows the Arrhenius plots of Fe-doped samples measured in air. In the temperature range of 700 ~ 1000 °C, the data can be fitted by a single straight line, even though the sample without Fe (x = 0.0) seems to slightly exhibit the order - disorder transition. Obviously, the electrical conductivity increased and the activation energy decreased with increasing the Fe content. This may suggests that the major charge carrier switches from ionic to electronic. To clarify this, conductivity isotherm measurements were performed. Figure 7 shows the conductivity isotherms of the Fe-doped sample with x = 0.3 measured at a temperature range of 700 ~ 1000 °C. It is clearly observed that the electrical conductivity is dominated by the p-type conduction with a slope of +1/4 under a high $P(O_2)$ regime, while the $P(O_2)$ independent regime indicating the ionic conduction is observed in the low $P(O_2)$ regime. Figure 8 shows the conductivity isotherms at 750 °C for the Fe-doped samples with x = 0.0 ~ 0.4. As expected, at a given temperature, the p-type conductivity increased with increasing the Fe content. It should be also noted that, except for the sample with x = 0.2, the ionic conductivity shows the almost same value of 1.3×10^{-2} S/cm regardless of the Fe content. This indicates that the concentration of mobile oxygen ions may be determined by the following defect reaction:

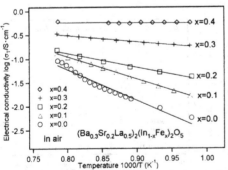

Figure 6. The Arrhenius plot of electrical conductivity for Fe-doped samples.

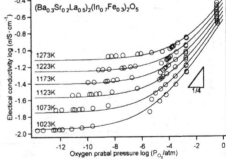

Figure 7. The conductivity isotherms for Fe-doped sample with x = 0.3.

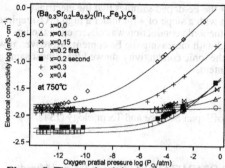

Figure 8. The conductivity isotherms at 750 °C for the Fe-doped samples with x = 0.0 ~ 0.4.

Table I. The activation energy of the p-type and oxide ion conductivities together with that of oxygen flux density for the Fe-doped samples with x = 0.1 ~ 0.3.

Fe content	ΔE_{ion} (kJ/mol)	ΔE_p (kJ/mol)	ΔE_j (kJ/mol)
x=0.1	97.3	89.7	56.2
x=0.2	111.2	50.6	43.2
x=0.3	86.3	25.1	23.1

$$[Ba_{La}] = 2[V_O^{\bullet\bullet}] \text{ (or } [La_{Ba}^{\bullet}] = 2[O_i^{''}]) ,\qquad (3)$$

while the concentration of holes can be controlled by the Fe content on In sites. The activation energy of p-type and ionic conductivities can be separated by fitting the following equation to the conductivity isotherms measured in the temperature range of 700 ~ 1000 °C.

$$\sigma_{total}(T, P(O_2)) = \sigma_{P0}\exp\left(-\frac{\Delta E_p}{RT}\right)P(O_2)^{\frac{1}{4}} + \frac{\sigma_{ion0}}{T}\exp\left(-\frac{\Delta E_{ion}}{RT}\right) \qquad (4)$$

Table 1 summarizes the activation energy of the p-type and oxide ion conductivities (ΔE_p, ΔE_{ion}) together with that of oxygen flux density (ΔE_j) for the Fe-doped samples with x = 0.1 ~ 0.3. For x = 0.2 and 0.3, ΔE_p shows a good agreement with ΔE_j, suggesting that the oxygen permeation rate is limited by the p-type conduction. However, based on the conductivity isotherms shown in Figs. 7 and 8, σ_i was the minor conductivity under the $P(O_2)$ regime between air and $10^{-2 \sim 3}$ atm, which corresponds to the $P(O_2)$ difference for the permeation measurement. This discrepancy may indicate that the rate determining process of oxygen permeation for the Fe-doped samples is not bulk diffusion but a surface exchange reaction associated with the p-type conductivity, even though further investigations will be needed to clarify the rate limiting process.

CONCLUSIONS

The preparation, electrical conductivity and oxygen permeability of transition-metal-doped $(Ba_{0.3}Sr_{0.2}La_{0.5})_2(In_{1-x}TM_x)_2O_{5+\delta}$ (TM=Fe, Co, Mn and Sn; $0 \le x \le 0.5$) have been investigated. All the samples were identified as a cubic perovskite-type structure. The lattice constant of Fe and Mn-doped samples linearly decreased with increasing the Fe and Mn content until x = 0.5. For the Fe-doped sample with x = 0.4, the highest oxygen flux density of 0.5 μmol•cm^{-2}•s^{-1} (d = 1.0 mm) was attained at 1000 °C, suggesting that the Ba-In-based oxides can be a mixed oxygen-ion and electronic conductors by Fe-doping as well as the other perovskite-type oxides. The total electrical conductivity under air increased and the activation energy decreased with increasing the Fe content. It is also observed for the Fe-doped samples that the electrical conductivity is dominated by p-type conduction with a slope of $+1/4$ under a high $P(O_2)$ regime, while the $P(O_2)$ independent regime indicating the ionic conduction was observed in the low $P(O_2)$ regime. The p-type conductivity increased with increasing the Fe content. Moreover, it was found that, except for the sample with $x = 0.2$, the ionic conductivity showed the almost same value of 1.3×10^{-2} S/cm regardless of the Fe content.

ACKNOWLEDGEMENTS

This work has been supported by CREST of Japan Science and Technology (JST).

REFERENCES

1. H. L. Tuller, "MATEIRALS DESING AND OPTIMIZATION,"*Oxygen Ion and Mixed Conductors and Their Technological Applications*, eds. H. L. Tuller et al. (Kluwer, 2000) 245-270
2. B. Ma, U. Balachandran, J.-H. Park, C.U. Segre, *Solid State Ionics* **83**, 65 (1996).
3. T. Ishihara, T. Yamada, H. Arikawa, H. Nishiguchi, Y. Takita, *Solid State Ionics* **135**, 631 (2000).
4. O. Porat, M. A. Spears, C. Heremans, I. Kosacki, H. L. Tuller, *Solid State Ionics* **86-88**, 285 (1996).
5. Y. Teraoka, T. Nobunaga, N. Yamazoe, *Chem. Lett.* 1 (1990).
6. J. B. Goodenough, J. E. Ruiz-Diaz, Y. S. Zhen, *Solid State Ionics* **44**, 21 (1990).
7. K. Kakinuma, H. Yamamura, H. Haneda, T. Atake, *Solid State Ionics* **140**, 301 (2001).
8. M. P. Pechini, *U.S. Patent* #3,330,697 (1967)
9. H. Takamura, K. Enomoto, A. Kamegawa, M. Okada, *Solid State Ionics*, **581-588**, 154-155, (2002).
10. H. Takamura, H. L. Tuller, *Solid State Ionics*, **67-73**, 134 (2000).
11. K. Kakinuma, H. Yamamura, H. Haneda, T. Atake, *Solid State Ionics*, **571-576**, 154-155, (2002).

Mat. Res. Soc. Symp. Proc. Vol. 756 © 2003 Materials Research Society

Preparation and Oxygen Permeability of Gd-Doped Ceria and Spinel-Type Ferrite Composites

Hitoshi Takamura, Masashi Kawai, Katsutoshi Okumura, Atsunori Kamegawa, and Masuo Okada
Department of Materials Science, Graduate School of Engineering, Tohoku University,
Sendai 980-8579, JAPAN.

ABSTRACT

The preparation and oxygen permeability of composites of $Ce_{0.8}Gd_{0.2}O_{2-\delta}$ (GDC) and spinel-type ferrites, MFe_2O_4 (M = Co and Mn) have been investigated. The composites of GDC - x vol% MFe_2O_4, where x ranged from 5 to 65, were prepared by a citrate-based liquid-mix technique. The composites were found to be almost fully densified by sintering at 1300 °C for 2 h. From TEM observations, the grain size of GDC and spinel-type phases was found to be less than 0.5 μm. In the case of M = Co, GDC - 25 vol% $CoFe_2O_4$ with a membrane thickness of 1.0 mm exhibited an oxygen flux density of 0.21 $\mu mol \cdot cm^{-2} \cdot s^{-1}$ under the $P(O_2)$ difference between He (20 sccm) and air at 1000 °C. Under reducing atmosphere of Ar-5%H_2, the oxygen flux density of this composite increased up to 1.3 $\mu mol \cdot cm^{-2} \cdot s^{-1}$. Moreover, under Ar-10%$CH_4$ gas flow, GDC - 15 vol% $MnFe_2O_4$ with a membrane thickness of 0.24 mm exhibited the oxygen flux density of 2 and 7 $\mu mol \cdot cm^{-2} \cdot s^{-1}$ at 800 and 1000 °C, respectively.

INTRODUCTION

Oxygen separation membranes based on mixed oxygen-ion and electronic conductors have been extensively studied because of their promising applications such as oxygen and syngas production [1]. Perovskite-type oxides in La-Sr-Co-Fe and La-Sr-Ga-Fe systems are well known to exhibit a high oxygen flux density at elevated temperatures of 750 ~ 1000 °C [2, 3]. To date, the highest value of 8.2 $\mu mol \cdot cm^{-2} \cdot s^{-1}$ has been reported for $La_{0.7}Sr_{0.3}Ga_{0.6}Fe_{0.4}O_{3-\delta}$ with a membrane thickness of 0.3 mm at 1000 °C [4]. In addition to these single-phase mixed conductors, dual-phase-type ones comprising of an ionic conductor such as yttria-stabilized zirconia (YSZ) and an electronic conductor, for example, a precious metal of Pd, have been developed as well [5]. For the composite-type mixed conductor, it is possible to choose good ionic and electronic conductors as components, and to control the mixed conductivity by adjusting the volume fraction under the restriction of the percolation theory. At the same time, since the surface oxygen exchange takes place at the three-phase-boundary (TPB) regions, fine microstructures are essential to obtain high oxygen flux densities. In addition to this, in the case of using oxide-based electronic conductors, the combination of ionic and electronic conductive phases, which does not form insulating bi-product phases, has to be carefully selected. As such ceramics-based composites, the combinations of Gd-doped CeO_2 (GDC) as the ionic conductor and Sr-doped $LaMnO_3$ (LSM) or Ca-doped $GdCoO_3$ (GCC) as the electronic conductor have been reported [6, 7]. GDC is a well-known oxygen-ion conductor and recently receiving much attention as a model material to investigate nano-scale effects [8-10]. As another class of composite-type mixed conductors based on GDC, in this study, spinel-type ferrites expressed as MFe_2O_4 (M; divalent cations) have been utilized for the electronic conductive phase. This ferrite phase is also expected to work as a sintering agent. Thus, the purpose of this study is to prepare the GDC- MFe_2O_4 (M = Co and Mn) composites with fine microstructures, and clarify the electrical conductivity and the resultant oxygen flux density. The methane conversion was also conducted by using this composite.

EXPERIMENTAL DETAILS

The composites of $Ce_{0.8}Gd_{0.2}O_{1.9}$ – x vol% MFe_2O_4 (M = Co and Mn, Ni; 5 ≤ x ≤ 65) have been prepared by the Pechini process [11, 12]. $Ce_{0.8}Gd_{0.2}O_{1.9}$ will be referred as GDC in this article. Raw materials used were $Ce(NO_3)_3 \cdot 6H_2O$, $Gd(NO_3)_3 \cdot 5H_2O$, $Co(NO_3)_2 \cdot 6H_2O$, $Mn(NO_3)_2 \cdot 6H_2O$, $Fe(NO_3)_3 \cdot 9H_2O$ as metal sources, and citric acid and ethylene glycol for chelating agents. After polymerizing, the resin was fired at 700 °C for 2 h to obtain the oxide phase. The oxide powders were die-pressed into pellets with dimensions of about $\phi 10$ mm × 1 mm at 30 MPa followed by isostatic pressing at 300 MPa. The pellets were then sintered at 1300 °C for 2 h in air. Phase identification was conducted by the X-ray powder diffraction. The chemical composition and microstructure of the composites were observed by an electron probe microanalysis (EPMA) and a transmission electron microscope (TEM). The electrical conductivity measurement was performed by the complex impedance method and the DC four-probe method in the temperature range of 750 ~ 1000 °C under a wide $P(O_2)$ range controlled by air and CO - CO_2 gas mixtures. For the oxygen flux density measurement, diamond-polished samples with dimensions of approximately $\phi 8$mm × 0.2 ~ 1 mm in thickness were used. The sample was sandwiched between two quartz tubes. The permeation and feed sides were sealed with gold rings, and the side wall of samples was also sealed by a borosilicate glass. The oxygen flux density was measured under various $P(O_2)$ difference between air and either of He (20, 40, 60 sccm), Ar-5%H_2 (60 sccm), and Ar-10%CH_4 (60 sccm) at a temperature range of 800 ~ 1000 °C. Gas concentration was determined by using a gas chromatograph and a mass spectrometer.

RESULTS AND DISCUSSION
Synthesis and microstructural observation

Phase identification and microstructural analyses for the GDC-based composites were performed by XRD, EPMA and TEM. Figure 1(a) shows the XRD patterns of the GDC – x vol% $CoFe_2O_4$ composites fired at 1300 °C for 2 h. From Fig. 1(a), as expected, the composites with x = 16.7 ~ 64.6 were found to comprise of GDC and $CoFe_2O_4$ phases. In addition, a small amount of $GdFeO_3$ was observed. The lattice constants of the fluorite-type GDC and spinel-type $CoFe_2O_4$ phases were then calculated from the XRD patterns and plotted in Fig.1 (b) as a function of the volume fraction of the $CoFe_2O_4$ phase. The lattice constants of GDC and $CoFe_2O_4$ phases in the composites decreased compared to those of pure GDC and $CoFe_2O_4$. Since EPMA revealed that the solubility of Fe and Co in the GDC phase was limited to less than 1 mol% under our preparation conditions, the formation of $GdFeO_3$, which would reduce the Gd and Fe content in the GDC and $CoFe_2O_4$ phases, respectively, may be responsible for the decrease in lattice constants.

The microstructure of the composites was observed by TEM. Figure 2 shows the TEM

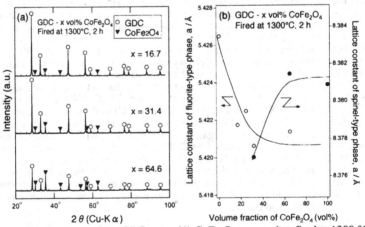

Figure 1. (a) XRD patterns of the GDC – x vol% $CoFe_2O_4$ composites fired at 1300 °C for 2 h, and (b) the lattice constants of the fluorite and spinel-type phases as a function of x.

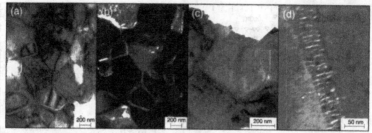

Figure 2. TEM micrographs of the GDC – 16.7 vol% CoFe$_2$O$_4$ fired at 1300 °C for 2 h. (a) bright-field image, (b) dark-field image taken from the diffraction spot of GDC, (c) rim structure and (d) its magnified image.

micrographs of the GDC – 16.7 vol% CoFe$_2$O$_4$ fired at 1300 °C for 2 h. Even though pure GDC usually requires a high sintering temperature of up to over 1600 °C for densification, from Fig. 2(a), this composite is found to be well densified by the firing at 1300 °C for 2 h. This decrease in the sintering temperature is presumably due to the presence of CoFe$_2$O$_4$ that can work as a sintering agent. As a result of the low sintering temperature, a fine grain size of less than 0.5 μm was achieved. In addition, somewhat different contrasts were observed around grain boundary regions. By taking a couple of dark-field images, for example, as shown in Fig. 2(b) that was taken from a diffraction spot of GDC, the rim of grains was found to consist of nano-sized GDC and CoFe$_2$O$_4$ grains but not GdFeO$_3$ phase. The magnified images of the rim were shown in Figs. 2(c) and (d). Further TEM observations are underway to clarify the detailed structure of the rim and the existence of GdFeO$_3$.

Electrical conductivity

Figure 3 shows the Cole – Cole plots of the GDC – x vol% CoFe$_2$O$_4$ composites, where x = 16.7, 31.4 and 64.6. The measurements were performed at 300 °C under air. For the composite with x = 16.7, three semicircles corresponding to the relaxation process of bulk, grain boundary, and electrode components were clearly observed. On the other hand, the composites with a higher volume fraction of CoFe$_2$O$_4$ (x = 31.4 and 64.6) showed two semicircles. Since the data points corresponding to low frequencies almost lie on the real-part axis, electronic conductivities seem to be enhanced for the composites with x = 31.4 and 64.6. To clarify this, conductivity isotherms were measured for the composites with x = 31.4.

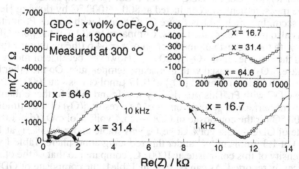

Figure 3. The Cole – Cole plots of the GDC – x vol% CoFe$_2$O$_4$ composites, where x = 16.7, 31.4 and 64.6, measured at 300 °C under air.

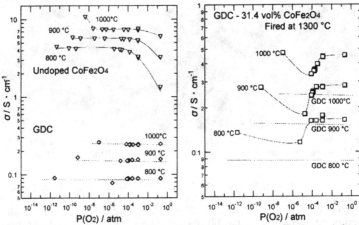

Figure 4. The conductivity isotherms of (a) GDC and undoped $CoFe_2O_4$, and (b) those of the composite of GDC – 31.4 vol% $CoFe_2O_4$ measured at 800 ~ 1000 °C.

Figure 4 shows the conductivity isotherms of (a) GDC and nominally undoped $CoFe_2O_4$, and (b) those of the composite of GDC – 31.4 vol% $CoFe_2O_4$ measured at 800 ~ 1000 °C. In Fig. 4(a), GDC and undoped $CoFe_2O_4$ exhibited typical $P(O_2)$ dependence of ionic and n-type conductors, respectively. The ionic conductivity of GDC was 9.0×10^{-2} and 2.4×10^{-1} S/cm at 800 and 1000 °C, respectively. In addition, the electronic conductivity of $CoFe_2O_4$ phase was higher than the ionic conductivity of GDC by an order of magnitude. On the other hand, the conductivity isotherms for the composite of GDC – 31.4 vol% $CoFe_2O_4$ shown in Fig. 4(b) exhibited the p-type conduction under a $P(O_2)$ regime of $> 10^{-5}$ atm. This change in the electronic carrier type may be due to the formation of $GdFeO_3$, and as a result, the compositional deviations of GDC and $CoFe_2O_4$ phases. In any case, it was confirmed that the electronic conductivity of this composite was higher than the ionic one of GDC at 800 ~ 1000 °C indicated by dashed lines.

Oxygen flux density and methane conversion

Since the composites of GDC – x vol% $CoFe_2O_4$ with relatively high electronic conductivities were prepared, the oxygen flux density was then evaluated. Figure 5 shows the oxygen flux density, $j(O_2)$, of the GDC – x vol% $CoFe_2O_4$ composites as a function of the $P(O_2)$ difference. The measurement was conducted at 800 ~ 1000 °C by flowing He gas at the rate of 20, 40 and 60 sccm. In which, $j(O_2)$ of pure GDC coated with porous Pt, imitating a mixed conductor, was also plotted as dashed lines. Compared to $j(O_2)$ of the pure GDC coated with Pt, the composite with the highest volume fraction of x = 64.6 (the membrane thickness; d = 0.67 mm) exhibited the same level of $j(O_2)$ at 800 °C. However, because of the lack of the ion conductive phase, $j(O_2)$ decreased with increasing temperature. On the other hand, the composite with x = 25.4 (d = 0.61 mm) showed $j(O_2)$ of 0.17 $\mu mol \cdot cm^{-2} \cdot s^{-1}$ at 1000 °C, which was higher that that of the GDC with Pt. This flux density was found to be further enhanced by putting Ni-catalysts on the surface of permeation side. As a result, $j(O_2)$ of 0.21 $\mu mol \cdot cm^{-2} \cdot s^{-1}$ was finally attained at 1000 °C for the composite of GDC – 25.4 vol% $CoFe_2O_4$ (d = 1.0 mm; He 20 sccm). This composite of GDC – 25.4 vol% $CoFe_2O_4$ was stable under Ar-5%H_2 at 1000 °C, and exhibited $j(O_2)$ of 1.3 $\mu mol \cdot cm^{-2} \cdot s^{-1}$ under this reducing atmosphere. Table 1 summarizes the oxygen flux density of this composite at 1000 °C, compared to that of the other GDC-based composites recently reported. As can be seen in Table1, the composite of GDC – 25.4 vol% $CoFe_2O_4$ was found to exhibit a high oxygen flux density in the class of dual-phase-type mixed

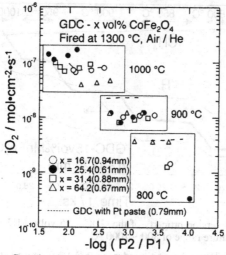

Figure 5. The oxygen flux density, $j(O_2)$, of the GDC – x vol% $CoFe_2O_4$ composites as a function of the $P(O_2)$ difference. Each membrane thickness was given in parentheses.

Table 1. The oxygen flux density of the GDC – 25.4 vol% $CoFe_2O_4$ composite with Ni-catalyst at 1000 °C.

Composition	d / mm	$jO_2/\mu mol \cdot cm^{-2} \cdot s^{-1}$	Ref.
GDC-25$CoFe_2O_4$	1.0	0.21	This work
w/ Ni-catalyst		1.3	Ar-5%H_2
GDC-LSM	1.0	0.07	@950° C[6]
GDC-GCC	1.5	0.08	[7]

conductors. Then, the partial oxidation of CH_4 was conducted by using this composite. However, since it immediately turned out that the chemical stability of $CoFe_2O_4$ was not enough high under Ar-10%CH_4, $MnFe_2O_4$ was utilized instead of $CoFe_2O_4$ for the CH_4 conversion test.

Figure 6 shows the CH_4 conversion property of the GDC – 15 vol% $MnFe_2O_4$ composite with d = 0.24 mm. In Fig. 7, the gas concentration of CH_4, CO, and H_2, and the oxygen flux density of $j(O_2)$ calculated from the reduction of CH_4 concentration (thick line) were plotted as a function of elapsed time. As can be seen, the concentration of CO and H_2 increased with increasing temperature, while that of CH_4 decreased. From the reduction of the CH_4 concentration, $j(O_2)$ was found to reach 2 and 7 $\mu mol \cdot cm^{-2} \cdot s^{-1}$ at 800 and 1000 °C, respectively. These oxygen flux densities were comparable to those of perovskite-type $La_{0.7}Sr_{0.3}Ga_{0.6}Fe_{0.4}O_{3-\delta}$ with a membrane thickness of 0.3 mm (*e.g.* 8.2 $\mu mol \cdot cm^{-2} \cdot s^{-1}$ at 1000 °C) [4].

CONCLUSIONS

The preparation and oxygen permeability of composites of $Ce_{0.8}Gd_{0.2}O_{2-\delta}$ (GDC) and spinel-type ferrites, MFe_2O_4 (M = Co and Mn), have been investigated. The composites were found to be almost fully densified by sintering at 1300 °C for 2 h. In addition to the GDC and spinel-type phase, the formation of $GdFeO_3$ was confirmed by the XRD analysis. From TEM observations, the grain size of GDC and spinel-type phases was found to be less than 0.5 μm. In the case of M = Co, GDC - 25 vol% $CoFe_2O_4$ with a membrane thickness of 1.0 mm exhibited an

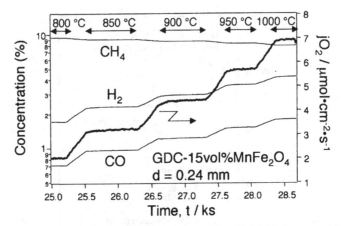

Figure 6. The CH_4 conversion property of the GDC – 15 vol% $MnFe_2O_4$ composite with d = 0.24 mm at the temperature range of 800 ~ 1000 °C.

oxygen flux density of 0.21 $\mu mol \cdot cm^{-2} \cdot s^{-1}$ under the $P(O_2)$ difference between He (20 sccm) and air at 1000 °C. Under reducing atmosphere of Ar-5%H_2, the oxygen flux density of this composite increased up to 1.3 $\mu mol \cdot cm^{-2} \cdot s^{-1}$. Moreover, under Ar-10%$CH_4$ gas flow, GDC - 15 vol% $MnFe_2O_4$ with a membrane thickness of 0.24 mm exhibited the oxygen flux density of 2 and 7 $\mu mol \cdot cm^{-2} \cdot s^{-1}$ at 800 and 1000 °C, respectively.

ACKNOWLEDGEMENTS

This work has been supported by CREST of Japan Science and Technology (JST).

REFERENCES

1. P. N. Dyer, R. E. Richards, S. L. Russek and D. M. Taylor, *Solid State Ionics* **134**, 21 (2000).
2. Y. Teraoka, H. M. Zhang, S. Furukawa, and N. Yamazoe, *Chem. Lett.*, 1743 (1985).
3. T. Ishihara, T. Yamada, H. Arikawa, H. Nishiguchi, Y. Takita, *Solid State Ionics* **135**, 631 (2000).
4. T. Ishihara, Y. Tsuruta, T. Todaka, H. Nishiguchi, Y. Takita, *Solid State Ionics*, in press.
5. C. S. Chen, B. A. Boukamp, H. J. M. Bouwmeester, G. Z. Cao, H. Kruidhof, A. J. A. Winnubst, A. J. Burggraaf, *Solid State Ionics* **76**, 23 (1995).
6. V. V. Kharton, A. V. Kovalevsky, A. V. Viskup, F. M. Figueiredo, A. A. Yaremchenko, E. N. Naumovich, F. M. B. Marques, *J. Eur. Ceram. Soc.* **21**, 1763 (2001).
7. U. Nigge, H. -D. Wiemhöfer, E. W. J. Römer, H. J. M Bouwmeester, T. R. Schulte, *Solid State Ionics* **146**, 163 (2002).
8. H. L. Tuller, A. S. Nowick, *J. Electrochem. Soc.* **126**, 209 (1979).
9. M. Mogensen, N. M. Sammes, G. A. Tompsett, *Solid State Ionics* **129**, 63 (2000).
10. P. Knauth, H. L. Tuller, *Solid State Ionics* **136-137**, 1215 (2000).
11. M. P. Pechini, U.S. Patent #3,330,697 (1967).
12. H. Takamura, H. L. Tuller, *Solid State Ionics* **134**, 67 (2000).

High-Temperature Potentiometric NO_2 and CO Sensors Based on Stabilized Zirconia with Oxide Sensing Electrodes

E. Di Bartolomeo, M. L. Grilli, N. Kaabbuathong, and E. Traversa
Department of Chemical Science and Technology, University of Rome Tor Vergata,
Via della Ricerca Scientifica, 00133 Rome, Italy

ABSTRACT

This paper reports the efforts made in our laboratory to develop electrochemical sensors that might detect NO_2 and CO at high temperatures for On Board Diagnostic (OBD) application. The non-Nernstian behaviour of zirconia-based electrochemical NO_2 sensors with various oxides as sensing electrodes was studied in the temperature range 450-700°C. Both pellets and tape-casted layers (150 μm of thickness) of yttria-stabilized zirconia (YSZ) were used for fabrication of the sensors. Pt electrodes were painted on both sides of the pellets or as two parallel fingers on one face of the layers. One of the Pt electrodes was covered with a thick-film oxide electrode. Various oxides were tested as sensing electrodes, either p- or n-type semiconductors, including WO_3 and $LaFeO_3$. The role of ionic conductivity of the oxide electrodes was investigated using Sr-doped perovskite-type oxides, such as $La_xSr_{1-x}FeO_3$, a mixed ionic-electronic conductor. The sensors were tested as potentiometric and amperometric devices. The performance of these devices was promising: fast and stable responses to different NO_2 concentrations (20-1000 ppm in synthetic air) were observed at high temperatures. The role of the metallic electrodes is also studied. The sensing mechanism of the sensors is discussed.

INTRODUCTION

Strict norms on pollution control are being enforced worldwide, especially for what concerns emissions of vehicles [1]. To satisfy the new standards on automobile emissions, the On-Board Diagnostic (OBD) system has been introduced inside the vehicles to control the main pollutants, such as NO_x, HCs and CO. The OBD is a quite complex, closed-loop system to continuously monitor the pollutant concentrations in the exhausts. At present, it consists of two solid state oxygen sensors (lambda sensors): one placed upstream the three way catalytic converter (to control the fuel/air ratio), the second one located downstream in the exhaust to control, through an electronic unit, the efficiency of the catalytic converter. The availability of new, reliable and fast NO_x, HCs and CO solid state sensors will allow a direct and precise analysis of the pollutants easily integrable in the OBD system. Moroever, NO_x sensors are required to control de-NO_x systems for fuel-lean applications in gasoline engines. Solid-state electrochemical sensors with metal oxide auxiliary phase seem to be the most suitable for applications at high temperatures and in harsh environments.

Several reports are available on solid electrolytes based sensors combined with metal (Pt, Au, etc.) and oxide electrodes for NO_x [2-16] and CO/HCs [17-19] detection. Different types of solid electrolytes have been tested in the electrochemical sensors such as: NASICON [2, 3, 6, 14], YSZ [4, 7-14, 17, 18] and ceria [19] combined with many types of semiconducting oxides: WO_3 [5, 9, 11, 15], perovskites oxides (ferrites [13, 14], cobaltites [18]) and spinel [8, 10]. A recent paper [20] has reviewed the actual trends for high temperature NO_x sensors. The major issues concerning the selectivity and the long-term stability are yet to be overcome for high temperature sensors [20]. This fact drives scientists to search for new materials, to improve the device fabrication technique and to go through the sensing mechanism. According to many authors [7, 12, 16-19], the sensing mechanism of electrochemical sensors based on coupling a solid electrolyte with semiconducting oxides can be explained using the "mixed potential" theory. The mixed potential mechanism was claimed both for NO_x [7, 12, 16] and CO/HCs [17-19] gas sensors based on similar electrochemical cells. However, a different explanation of the NO_x sensing mechanism has been proposed and named "different electrode equilibria" [22]. Different electrode equilibria is a more general concept to explain the NO_x sensitivity that is due not only to electrochemical reactions, but also to different electrocatalytic activity and/or sorption-desorption behaviour of the two electrodes [14, 15]. Mixed potential theory has also been proposed to explain the working features of single-chamber solid oxide fuel cells with electrodes exposed to the same gas atmosphere [23-25]. Given this background, in this work we report a study on yttria-stabilised zirconia (YSZ) based sensors with WO_3 (n-type), $LaFeO_3$ (p-type) and $La_xSr_{1-x}FeO_3$ oxides as sensing electrode. WO_3 and $LaFeO_3$ have been selected because of their good performances in bulk [26, 27] and thick-film [28] form as semiconductor NO_2 sensors. $La_{0.8}Sr_{0.2}FeO_3$ oxide was used in order to check whether the mixed ionic and electronic conduction could improve the sensing response. The sensors were prepared using pellets and thick layers of YSZ to get devices in "bulk" and "planar-form", respectively. The electrochemical response (EMF measurements and amperometric measurements) of the sensors was studied in the temperature range between 450-700°C in the presence of NO_2 and CO in air (20-1000 ppm). The influence of the grain size of the oxides used for the sensing electrodes on the gas response was also investigated. All sensors were wholly exposed to the same atmosphere; the reference electrode was thus not separated and this fact highly simplifies the sensor design. Moreover, the variations of oxygen concentration can be the same at both electrodes, and their influence might be compensated between the electrodes. To go through the sensing mechanism of the sensors some electrochemical measurements were also performed exposing the reference electrode to air.

EXPERIMENTAL

YSZ (8 wt.% of Y_2O_3) pellets of 10 mm in diameter were used for "bulk" sensor fabrication [14]. Pt paste was deposited on both sides of the pellets as metallic electrodes and thin gold wires were connected for current collection. YSZ tape-casted layers of 150 μm in thickness and rectangular shape (10 mm x 12 mm) were used as solid electrolyte material for "planar" sensors. Pt paste was used for the preparation of the metallic electrodes and deposited on one side of the layers in parallel fingers form (8 mm x 2 mm) at a distance of 5 mm. The firing temperatures of Pt paste were 800°C for 10 min. For the fabrication of the sensing electrode, commercial WO_3 (99.995 % purity), $LaFeO_3$ or $La_xSr_{1-x}FeO_3$ perovskite oxides, prepared in the laboratory, were mixed with a screen-printing oil and the slurry thus obtained was painted on the area of one metallic electrode and fired at 750°C for 3 hours. The metal oxide powders were deposited on one of the porous Pt electrodes making contact with the YSZ electrolyte either at edges of Pt electrode, either through the pores of the Pt.

Nano-sized $LaFeO_3$ powders were prepared by the thermal decomposition of a LaFe-hexacyanide complex at 700°C for 1 h [29]. To check the influence of the electrode grain size on the sensing response of the sensors, the $LaFeO_3$ oxide was also heated at 900°C for 4 h.

La$_{0.8}$Sr$_{0.2}$FeO$_3$ was prepared by a sol-gel route. The precursors used were: La(NO$_3$)$_2$, Sr(NO$_3$)$_2$, Fe(NO$_3$)$_3$, citric acid, ethylene glycol. The nitrates and the citric acid were mixed in the following molar ratio: La:Sr:Fe:citric acid = 0.8:0.2:1:2, while the citric acid:ethylene glycol weight ratio was 40:60. Stoichiometric amounts of salts were first dissolved into ethylene glycol. When the precursors were completely dissolved, controlled amount of citric acid was added. A complete dissolution of precursors resulted in a clear red-brown solution. The gel formation occurred at 120°C, its transparency giving an indication of an homogeneous system. The samples were dried at 120-130°C for few hours and then heated at 700°C for 3 hours. Again to check the influence of the electrode grain size on the sensing response of the devices, the La$_{0.8}$Sr$_{0.2}$FeO$_3$ oxide was also heated at 900°C for 4 h.

The X-ray diffraction (XRD) analysis was conducted for phase identification of perovskite oxides. Microstructures of the electrodes were observed by scanning electron microscopy (SEM). XRD patterns showed only the peaks of orthorhombic perovskite-type LaFeO$_3$ and La$_{0.8}$Sr$_{0.2}$FeO$_3$ phases.

Sensing experiments were carried out in a conventional gas-flow apparatus equipped with a controlled heating facility. The sensor was alternatively exposed to air and NO$_2$ or CO (200-1000 ppm in air) at the total flow rate of 100 ml/min in the temperature range between 450 and 700°C. Electromotive force (EMF) measurements were performed between the two electrodes of the sensors using a digital electrometer. During the EMF measurements, the electrode with the oxide coating was always kept at the positive terminal of the electrometer and both the electrodes were exposed to the same gas environment.

Amperometric measurements were performed applying a constant dc voltage of +1 or –1 volt to the sensing electrode and measuring the current flowing between the electrodes by switching between air and oxidising or reducing atmospheres.

RESULTS AND DISCUSSION

SEM analysis

SEM observations were performed on different oxide thick films deposited on Pt electrode. The WO$_3$ thick film showed a porous structure made of grains of sub-micrometric dimension of about 200-250 nm [30]. The LaFeO$_3$ observations were in agreement with what has been previously reported for these perovskite-type oxides [29]. The thick film prepared with powder calcined at 700°C was highly porous and consisted of large grains 2 - 10 μm in size. Each of these grains was made of soft agglomerates of homogeneous nanometric particles of 50 - 100 nm in size. The LaFeO$_3$ thick film obtained with powder calcined at 900°C showed an increase of nanosized grains up to about 200 nm [14]. The grain size of the thick film prepared with La$_{0.8}$Sr$_{0.2}$FeO$_3$ powders decomposed at 700°C was less then 50 nm as shown in figure 1. The La$_{0.8}$Sr$_{0.2}$FeO$_3$ film obtained with powder calcined at 900°C for 4 hours showed an increase in grain size up to about 150 nm (figure 2).

Figure 1. SEM micrograph of the La$_{0.8}$Sr$_{0.2}$FeO$_3$ thick film prepared with powder heated at 700°C for 3 hours.

Figure 2. SEM micrograph of the La$_{0.8}$Sr$_{0.2}$FeO$_3$ thick film prepared with powder heated at 900°C for 4 hours.

Sensing electrochemical properties of "bulk" sensors

Figure 3 shows the typical NO$_2$ response of Pt/YSZ/Pt/WO$_3$ sensor at 600°C. EMF changed quickly upon switching from air to different NO$_2$ concentrations in air. At 600°C steady-state values were observed at all measured gas concentrations. The response (at 90% of saturation EMF value) and recovery times were 40 seconds and 2 minutes respectively. The EMF response of WO$_3$ based sensors was measured in the temperature range 550-700°C. A linear correlation between the EMF saturation values and the logarithm of gas concentration was observed at different operating temperatures. The EMF response decreased at all different gas concentrations increasing the operating temperature. The largest sensitivity (18.8 mV/decade) was found at 600°C. At temperatures below 600°C, the sensing responses were unstable and did not show a linear correlation with log NO$_2$ concentrations.

Figure 4 shows that in the presence of various CO concentrations the EMF response of Pt/YSZ/Pt/WO$_3$ sensors decreased and reached negative values, unlike NO$_2$ response (always keeping WO$_3$ side at the positive terminal of the electrometer). At 600°C the EMF values at any given concentration were smaller than those measured in the presence of NO$_2$. However the sensitivity at 600°C was −15 mV/decade, comparable to NO$_2$ sensitivity at the same temperature. An attempt to decrease the CO cross sensitivity was successful when Pt electrodes were replaced with Au [30].

Figure 5 shows the sensing response at 400°C and 450°C of the Pt/YSZ/Pt/LaFeO$_3$ sensors, with LaFeO$_3$ powder calcined at 700°C. EMF changed quickly upon switching from air to 75 ppm NO$_2$ and steady-state values were observed. At higher temperature the stability of the response was improved and the response time became faster. The response of the Pt/YSZ/Pt/LaFeO$_3$ sensor was 64 mV at 400°C, and 26 mV at 450°C. The response of the same sensor, at 400°C, obtained using the LaFeO$_3$ powders calcined at 900°C was investigated and compared to the response at the same operating temperature of the sensor obtained using LaFeO$_3$ powders calcined at 700°C. A sharp decrease in EMF response (about 6 mV) was observed for the sensor prepared using the powder calcined at 900°C, which can be explained by an increase in the grain size as observed by SEM analysis [14].

Figure 6 shows the results of the amperometric measurements at 450°C of the Pt/YSZ/Pt/LaFeO$_3$ sensor when a constant dc voltage of 1 volt and −1 volt was applied to the

LaFeO₃ electrode cycling synthetic air and 300 ppm of NO₂ in air. When applying + 1 V (- 1 V) a fast increase (decrease) in the current was observed upon exposure to NO₂. The response time was less than 30 seconds both for direct and reverse polarization.

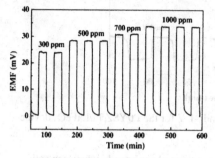

Figure 3. EMF of Pt/YSZ/Pt/WO₃ sensor at 600°C in cycling air and different concentrations of NO₂ in air.

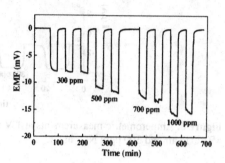

Figure 4. EMF of Pt/YSZ/Pt/WO₃ sensor at 600°C in cycling air and different concentrations of CO in air.

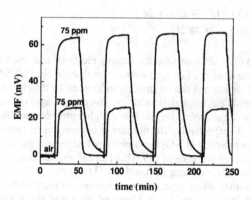

Figure 5. EMF of Pt/YSZ/Pt/ LaFeO₃ sensor at 400 and 450°C in cycling air and 75 ppm of NO₂ in air

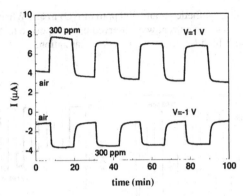

Figure 6. Amperometric measurements at 1 V and –1 V for the Pt/YSZ/Pt/LaFeO₃ device at 450°C in air and 300 ppm NO₂ in air.

The electrochemical reactions occurring at the three-phase boundary between solid electrolyte, electrode and gas, can be expressed as follows [7, 15, 18, 21]:
for NO_2

$$NO_2 + 2e \Leftrightarrow O^{2-} + NO \qquad (1)$$
$$O^{2-} \Leftrightarrow \tfrac{1}{2}O_2 + 2e \qquad (2)$$

for CO

$$CO + O^{2-} \Leftrightarrow CO_2 + 2e \qquad (3)$$
$$\tfrac{1}{2}O_2 + 2e \Leftrightarrow 2O^{2-} \qquad (4)$$

Reactions (1) and (3) mainly occur at the sensing electrode and reactions (2) and (4) at the metal electrode side. The number of oxygen ions accumulated at the Pt/YSZ and YSZ/Pt/(WO₃, LaFeO₃) interfaces is different and can be strongly influenced by the electrode material, its surface morphology and its electrocatalytic activity. The porous layer of semiconducting oxide enhances the adsorption of gas molecules and hence highly promotes reaction (1) and (3) upon exposure of NO₂ and CO, respectively. The number of oxygen ions is larger at the Pt/YSZ/(WO₃, LaFeO₃) interface in the presence of NO₂ atmosphere, while the number of oxygen ions is larger at the YSZ/Pt interface in the presence of CO. This fact can explain that the EMF response was in opposite directions to NO₂ (oxidising gas) and to CO (reducing gas).

Upon applied potential, the generated oxygen species are pumped through YSZ and give rise to a current through the solid electrolyte. Under positive bias, the increase in the current, observed in the polarization measurements upon NO₂ exposure, can be related to the decrease in the resistance at the electrolyte/oxide electrode interface due to an increase in the number of oxygen ions accumulated at the LaFeO₃/Pt/YSZ interface and available for the conduction. Under negative bias, the oxygen ions are moved from the higher to lower O^{2-} accumulation region (metal electrode side) and hence the negative current flow decreases as shown in the amperometric measurements. Upon CO exposure, the situation was just the opposite. The amperometric measurements confirmed that the Pt/YSZ/Pt/LaFeO₃ sensors show an opposite behaviour in presence of oxidising and reducing gas atmosphere. Moreover, the amperometric measurements of WO_3 based sensors behave just opposite: under positive bias the current decreases in the presence of NO_2 gas and increases under negative bias.

From potentiometric measurements, it is clearly shown that the sensing reactions (1) and (3) mainly take place at the semiconducting oxide electrodes. The amperometric measurements are consistent with the fact that WO_3 is an n-type semiconductor and $LaFeO_3$ is a p-type semiconductor.

Sensing electrochemical properties of "planar" sensors

The EMF response, performed on WO_3 based sensors at fixed temperature (550-700°C), was in opposite direction upon exposure to NO_2 and CO. Positive EMF values were measured at different NO_2 concentrations and negative EMF values at different CO concentrations, as already observed for the "bulk" sensors. The response was stable and reproducible; at 600°C the EMF magnitude was 25 mV and 17 mV upon exposure to 1000 ppm of NO_2 and CO, respectively. The response (20 seconds) and recovery (40 seconds) times were quite fast for both gas mixtures. A linear correlation was observed between the EMF values and the NO_2 and CO concentrations in logarithmic scale. The best sensitivity to both NO_2 and CO gases was observed at 600°C.

The EMF response of $LaFeO_3$ and $La_{0.8}Sr_{0.2}FeO_3$ based sensors were found in opposite directions with respect to the WO_3 based sensors response: negative EMF values upon NO_2 exposure, positive values upon CO exposure. In figures 7 and 8 the EMF response of $LaFeO_3$ based sensors to NO_2 and CO are reported.

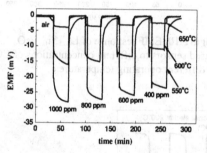 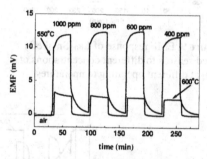

Figure 7. EMF response of $LaFeO_3$ based sensors to different concentrations of NO_2 at different operating temperatures.

Figure 8. EMF response of $LaFeO_3$ based sensors to different concentrations of CO at different operating temperatures.

Figures 9 and 10 show the EMF response of $La_{0.8}Sr_{0.2}FeO_3$ based sensors to NO_2 and CO at 550 and 600°C. At a given gas concentration, the EMF values of $La_{0.8}Sr_{0.2}FeO_3$ based sensors decreased increasing the operating temperature. At 550°C the EMF magnitude at 1000 ppm of NO_2 in air was about 30 mV, while the EMF value at the same concentration of CO was only 5 mV. The $La_{0.8}Sr_{0.2}FeO_3$ based sensors showed a reduced cross-sensitivity to CO in comparison to WO_3 and $LaFeO_3$ based sensors, especially at 550°C.

This is probably due to the mixed ionic and electronic conduction mechanism: probably oxygen ions absorbed on the surface can migrate inside the grains and so a reduced number of adsorbed oxygen ions is available for the reaction with CO. On the other hand, NO_2 adsorbs competitively with O_2 and its sensitivity is not affected by the ionic conduction of the perovskite phase. For both perovskite oxides based sensors the best sensitivity was observed at 550°C, as obtained from the slope of the linear fit of the sensitive curves.

In figure 11 the EMF of $La_{0.8}Sr_{0.2}FeO_3$ based sensors with powders decomposed at 700°C at different concentrations of NO_2 and 450°C is reported. The response is fast, stable and also reproducible. In the same figure the EMF of $La_{0.8}Sr_{0.2}FeO_3$ based sensors, prepared with powders heated at 900°C for 4 hours, at 100 ppm of NO_2 at 450°C is also shown. Increasing the grain size of the powders, the response became small, slow and unstable. So, the grain size of the sensing electrode strongly affects the sensors performances. This finding has been also previously reported for $LaFeO_3$-based sensors [14].

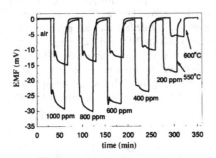

Figure 9. EMF response of $La_{0.8}Sr_{0.2}FeO_3$ based sensors to different concentrations of NO_2 at different operating temperatures.

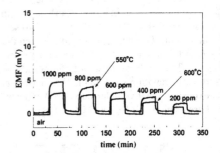

Figure 10. EMF response of $La_{0.8}Sr_{0.2}FeO_3$ based sensors to different concentrations of CO at different operating temperatures.

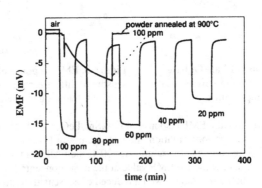

Figure 11. EMF at 450°C of $La_{0.8}Sr_{0.2}FeO_3$ based sensors with powders decomposed at 700°C and 900°C.

The results obtained from the EMF measurements on p-type based sensors cannot be explained by the occurrence of the electrochemical reactions at the three-phase boundary (solid electrolyte/electrode/gas) [7, 8, 15-18], as obtained in the case of bulk sensors. In particular, the electrochemical reaction (1), that should mainly happen at the oxide sensing electrode, as in the case of bulk sensors, cannot explain the negative EMF values of p-type semiconducting oxide based sensors in presence of NO_2 gas. This fact is probably due to the electrodes design: two close finger-electrodes exposed to the same gas atmosphere. Because of this geometry, the electrochemical reactions (1) and (2) for NO_2 detection take place with the same rate at both the electrodes and thus the adsorption mechanism characteristic of semiconductors becomes predominant.

Sensing mechanism

In order to go through the sensing mechanism of these electrochemical sensors, a different geometry of the planar sensor electrodes was tested. The reference and sensing electrodes were painted on the opposite sides of the YSZ layers. The Pt electrode was exposed to reference air and the $La_{0.8}Sr_{0.2}FeO_3$ oxide electrode was exposed to different concentrations of NO_2 in air. Figure 12 shows the EMF response of the planar sensors at 450°C and different concentrations of NO_2 in air. The EMF values were positive for all different NO_2 concentrations. These results are just the opposite of what was observed in the case of planar sensors with parallel finger electrodes exposed to the same atmosphere. Figure 13 shows the response at 450°C of the same sensor put upside down, that is exposing the oxide electrode to reference air and the Pt electrode to 100 ppm of NO_2 in air. The EMF response was positive, even if the EMF values were lower and unstable in the same operating conditions. Moreover, the EMF response in air showed a drift probably due to the oxygen ions conduction of the $La_{0.8}Sr_{0.2}FeO_3$ oxide.

These results suggest that the oxide electrode plays an important role in promoting the electrochemical reaction of NO_2. In fact, the response of the sensors with the oxide electrode exposed to the gas is much larger and stable. Moreover, the response of the sensors with reference air can be fully explained by using the mixed potential theory. For NO_2 detection, equations (1) and (2) take place at the electrode exposed to the pollutant gas, giving rise to positive EMF values. Thus, the mixed potential theory, widely accepted from many authors [7, 12, 16-19], can fully explain the behaviour of sensors with reference air. The same theory cannot explain the response of planar sensors with parallel finger electrodes or the amperometric measurements of bulk electrochemical sensors. In these cases, different electrode equilibria theory [22] and electrocatalytic activity and/or chemical sorption-desorption behaviour of the electrodes must be claimed to explain the opposite responses obtained for p-and n-type semiconducting oxide based sensors.

Two different sensing mechanisms, the "mixed potential" theory and the "different electrode equilibria" theory, are concomitant on sensors based on YSZ with oxide sensing electrodes. When the Pt electrode is separate from the sensing electrode (as in the case of planar sensors with reference air) or it is not exactly exposed to the same gas stream (as in the case of bulk sensors), the mixed potential theory is predominant. When the two electrodes are parallel and exposed to the same gas stream (as in the case of planar sensors) the different electrode equilibria can fully explain the sensors response.

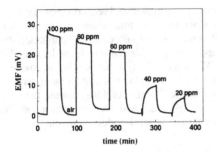

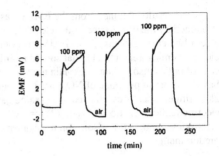

Figure 12. EMF response of La$_{0.8}$Sr$_{0.2}$FeO$_3$ based sensors with reference air at metallic electrode to different concentrations of NO$_2$ at 450°C.

Figure 13. EMF response of La$_{0.8}$Sr$_{0.2}$FeO$_3$ based sensors with reference air at the oxide electrode to 100 ppm of NO$_2$ at 450°C.

CONCLUSIONS

YSZ based sensors with oxide electrodes both in "planar" and "bulk" form are promising candidates for pollutants detection at high temperatures, up to 700°C. The sensors show fast and stable response without the use of reference air. The sensors response is strongly affected by the grain size of the powder used for the preparation of the sensing electrode.

The sensing mechanism of oxide-based sensors cannot always be explained by mixed potential theory. The EMF response can be also due to the different electrocatalytic activity and/or chemical sorption-desorption behaviour of the electrodes (different electrode equilibria). Our findings on planar type sensors suggest that, due to the electrode design (close finger-electrodes exposed to the same gas atmosphere), the electrochemical reactions take place with the same rate at both electrodes and thus the chemical adsorption mechanism of gases on semiconducting oxide is predominant. The mixed potential theory can be fully claimed only in the case of oxide-based sensors when reference air is used.

ACKNOWLEDGEMENTS

This work was partly supported by the Ministry of University and Scientific and Technological Research (MURST) of Italy and by the National Research Council of Italy (CNR), under the auspices of the Targeted Project "Special Materials for Advanced Technologies - MSTA II".Part of this work was performed during a stay of Narin Kaabbuathong at the Ehime University in Matsuyama, under the supervision of Profs. H. Aono and Y. Sadaoka.

REFERENCES

1. K. Oishi, in Proceedings of the Electrochemical Society Symposium, Vol. 93-7, Proceedings of the Symposium on Chemical Sensors II, Eds. M. Butler, A. Ricco, and N. Yamazoe, (The Electrochemical Society, Pennington, NJ,1993) p. 443.
2. Y. Shimizu, K. Maeda, *Chem. Lett.*, 117 (1996).
3. Y. Shimizu and K. Maeda, *Sensors and Actuators* **B 52**, 84 (1998).
4. Di Bartolomeo, E., Traversa, E., Baroncini, M., Kotzeva, V. and Kumar, R.V., *J. Eur. Ceram. Soc.*, **20**, 2691 (2000).
5. N. Miura, T. Shirahishi, K. Shimanoe, N. Yamazoe, *Electrochem. Commun.*, **2**, 77 (2000).
6. Y. Shimizu and N. Yamashita, *Sensors and Actuators* **B**, **64**, 102 (2000).
7. N. Miura, H. Kurosawa, M. Hasei, G. Lu, and N. Yamazoe, *Solid State Ionics*, **86-88**, 1069 (1996).
8. G. Lu, N. Miura and N. Yamazoe, *J. Mater. Chem.*, **7**, 1445 (1997).
9. G. Lu, N. Miura, and N. Yamazoe, *Ionics* **4**, 16 (1998).
10. N. Miura, G. Lu, M. Ono and N. Yamazoe, *Solid State Ionics* **117**, 283 (1999).
11. G. Lu, N. Miura, and N. Yamazoe, *Sensors and Actuators* **B**, **65**, 125 (2000).
12. N. Miura, and N. Yamazoe, in Sensors Update, Vol. 6, Eds. H Baltes, W. Göpel and J. Hesse, (WILEY-VCH, Weinheim, Germany, 2000) p. 191.
13. J.W. Yoon, M.L. Grilli, E. Di Bartolomeo, R. Polini, and E. Traversa, *Sensors and Actuators* **B 76**, 483 (2001).
14. M.L. Grilli, E. Di Bartolomeo and E. Traversa, *J. Electrochem. Soc.*, **148**, p. H98 (2001).
15. M. L. Grilli, N. Kaabbuathong, A. Dutta, E. Di Bartolomeo and E. Traversa, *J. Ceram. Soc. Jp*, **110 [3]**, 159 (2002).
16. N. Miura, S. Zhuiykov, T. Ono, M. Hasei, and N. Yamazoe, *Sensors and Actuators B*, **83**, 222 (2002).
17. T. Hibino, A. Hashimoto, S. Kakimoto, and M. Sano, *J. Electrochem. Soc.*, **148**, p. H1 (2001).
18. E.L. Brosha, R. Mukundan, D.R. Brown, F.H. Garzon, J.H. Visser, M. Zanini, Z. Zhou, E. M. Lothetis, *Sensors and Actuators* **B**, **69**, 171 (2000).
19. E.L. Brosha, R. Mukundan, D.R. Brown, F.H. Garzon, J.H. Visser, *Solid State Ionics* **148**, 61 (2002).
20. F. Ménil, V. Coillard, C. Lucat, *Sensors Actuators* **B 67**, 1 (2000).
21. Y. Shimizu, H. Nishi, H. Suzuki, and K. Maeda, *Sensors Actuators* **B**, **65**, p. 141 (2000).
22. E. D. Wachsman and P. Jayaweera, in Solid State Ionic Devices II – Ceramic Sensors, Eds. E.D. Wachsman, W. Weppner, E. Traversa, M. Liu, P. Vanysek, and N. Yamazoe, (The Electrochem. Soc. Proc. Series, Pennington, NJ, 2001) p. 298.
23. T. Hibino, A. Hashimoto, T. Inoue, J. Tokuno, S. Yoshida and M. Sano, *Science*, **288**, 2031 (2000).
24. T. Hibino, A. Hashimoto, T. Inoue, J. Tokuno, S. Yoshida and M. Sano, *J. Electrochem Soc.*, **148 (6)** A544 (2001).
25. T. Hibino, A. Hashimoto, M. Yano, M. Suzuki, S. Yoshida and M. Sano, *J. Electrochem Soc.*, **149 (2)** A133 (2002).
26. M. Akiyama, Z. Zhang, J. Tamaki, N. Miura, N. Yamazoe and T. Harada, *Sensors and Actuators* **B**, **13-14**, 619 (1993).

27. T. Inoue, K. Ohtsuka, Y. Yoshida, Y. Matsuura and Y. Kajiyama, *Sensors and Actuators B*, **24-25**, 388 (1995).
28. E. Traversa, Y. Sadaoka, M. C. Carotta, and G. Martinelli, *Sensors and Actuators B*, , **65**, 181-185 (2000).
29. E. Traversa, M. Sakamoto, and Y. Sadaoka, *J. Am. Ceram. Soc.*, **79**, 1401 (1996)
30. A. Dutta, N. Kaabbuathong, M.L. Grilli, E. Di Bartolomeo and E. Traversa, *J. Electrochem Soc.*, in press (Feb. 2003).

Mat. Res. Soc. Symp. Proc. Vol. 756 © 2003 Materials Research Society

Electrochemical NO$_x$ Sensors for Automotive Diesel Exhaust

Louis P. Martin, Ai-Q. Pham, Robert S. Glass
Lawrence Livermore National Laboratory
P.O. Box 808, L-353
Livermore, CA, 94551-0808, U.S.A.

ABSTRACT

New emissions regulations will increase the need for compact, inexpensive sensors for monitoring and control of automotive exhaust gas pollutants. Species of interest include hydrocarbons, carbon monoxide, and oxides of nitrogen (NO$_x$). The current work is directed towards the development of fast, high sensitivity electrochemical NO$_x$ sensors for automotive diesel applications. We have investigated potentiometric NO sensors with good sensitivity and fast response when operated in 10% O$_2$. The sensors consist of yttria-stabilized zirconia substrates attached with NiCr$_2$O$_4$ sensing electrodes and Pt reference electrodes. A composite NiCr$_2$O$_4$:Rh sensing electrode is shown to give significantly faster response than NiCr$_2$O$_4$ alone. The exact role of the Rh in enhancing the response speed is not clear at present. However, the Rh appears to accumulate at the contacts between the NiCr$_2$O$_4$ particles and may enhance the inter-particle electronic conduction. Ongoing testing of these sensors is being performed to elucidate the sensing mechanisms and to quantify cross sensitivity to, for example, NO$_2$.

INTRODUCTION

Increasingly stringent emissions regulations will introduce a need for compact, inexpensive sensors for monitoring and control of regulated exhaust gas pollutants including hydrocarbons, carbon monoxide, and oxides of nitrogen (NO$_x$). Significant progress as been made towards the development of electrochemical sensors using ionically conducting ceramic electrolytes, usually yttria-stabilized zirconia (YSZ), and catalytically active metal oxide sensing electrodes [1-3]. The suitability of numerous single- and mixed-metal oxides for use as NO$_x$ sensor electrodes, usually in air, has been explored in the literature [4, 5]. However, reliable, cost effective sensors suitable for diesel exhaust gas applications have not yet emerged. Improvements are still needed in sensitivity, response time, reliability, and cross-sensitivity.

The current work is directed towards the development of electrochemical NO$_x$ sensors for automotive diesel exhaust monitoring. The high oxygen content (5-15%) of the diesel exhaust suggests the use of a potentiometric sensor using the response of catalytic metal or metal-oxide electrodes on a solid ionic-conducting electrolyte. This response is most often described as a 'mixed potential' response, however alternate sensing mechanisms have been proposed in the literature [6, 7]. At present, the exact mechanism of NO response in this type of sensor does not appear to be fully understood, and there is some doubt as to the general applicability of the 'mixed potential' mechanism. The current investigation was performed by applying catalytic electrodes of NiCr$_2$O$_4$ or NiCr$_2$O$_4$:Rh to a planar O$_2$ sensor substrate. Mixed potential response to NO in 5-15% O$_2$ was monitored as a function of gas and substrate temperatures.

EXPERIMENTAL DETAILS

The NO sensors were fabricated by modification of multi-layer O_2 sensors. The O_2 sensors have Pt sensing and reference electrodes on a yttria-stabilized zirconia (YSZ) substrate, and are laminated together with an integrated resistive heater. To enhance the NO sensitivity, the sensors were modified by coating ~2 μm of sub-micron $NiCr_2O_4$ powder over the Pt sensing electrode using a colloidal spray deposition technique, and sintered at 900 °C for 1 hr. Composite $NiCr_2O_4$:Rh layers were formed by a proprietary technique which involves infiltration of small amounts of Rh into the sintered $NiCr_2O_4$ electrode. In all cases, the $NiCr_2O_4$ or composite coatings extended past the edges of the underlying Pt electrode to make direct contact with the YSZ electrolyte. The sensors are heated by applying a current to the resistive heater, and the temperature determined from the heater resistance. The heater was calibrated using independent measurements of heater resistance versus furnace temperature in the range from room temperature to 650 °C. The O_2 sensors were intended to operate with the reference electrode isolated from the test gas (i.e. air reference), however in this study both sensing and reference electrodes were exposed to the test gas stream for the NO testing. The sensor and experimental setup are shown schematically in Figure 1.

NO sensing experiments were performed in a quartz tube (inside diameter = 1.75 cm) inside a tube furnace. Pt wires were used to connect the external leads to the sensor electrodes. Unless otherwise specified, all tests were performed using 500 ppm NO in 10% O_2, balance N_2, at a flow rate of 1000 ml/min. The face velocity of the gas flowing through the reaction tube at this flow rate is ~7 cm/s, which is significantly less than in automotive exhaust. A standard gas handling system was used to deliver the gasses to the reaction tube. The configuration of a typical sensor and the testing configuration are shown in Figure 1. An electrochemical interface was used to measure open circuit potentials at 0.5 s intervals. Tests were performed with gas temperatures ranging from 250-350 °C, and sensor temperatures from 400-600 °C.

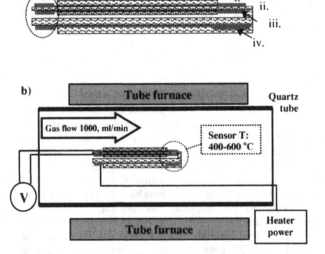

Figure 1: a) Schematic of the NO sensor showing i. sensing electrode, ii. YSZ electrolyte, iii. reference electrode, iv. resistive heater, and v. pads for making connections to the electrodes and heater. b) The sensor testing setup.

RESULTS

The NO sensors were tested in flowing gas, 10% O_2, and the response to the addition of 500 ppm NO was monitored. Figure 2 shows typical results for sensors with $NiCr_2O_4$ and composite ($NiCr_2O_4$:Rh) electrodes, with a gas temperature of 250 $^\circ$C and sensor temperature of 450 $^\circ$C. The data show the sensors to have a baseline, i.e. 10% O_2 without NO, of approximately -3 mV. With the addition of 500 ppm NO to the gas stream, the sensor responds with a significant negative potential at the sensing electrode (relative to the reference electrode). It is ofeten assumed that the NO electrochemically combines with O^{2-} ions from the YSZ at the YSZ/electrode interface. In the presence of oxygen, this reacttion combines with the electrochemical reaction for oxygen (with the YSZ) to form a local cell, with a net reduction in the overall potential. This response is most often described as a mixed potential response, and occurs as a result of the simultaneous, competing oxidation/reduction reactions at the electrode. The form of these reactions is given as [8]

$$O^{2-} + NO \Leftrightarrow NO_2 + 2e^- \qquad (1)$$
$$O_2 + 4e^- \Leftrightarrow 2O^{2-}$$

While generally accepted, this description of the sensing mechanism is probably oversimplified. There is experimental evidence that the response is not entirely determined by the properties of the gas/electrode/YSZ interface, but is also affected by the electrode thickness and microstructure, and by the nature of the electrical connection to the electrode [6, 9]. It should also be noted that under the present experimental conditions, the measured response is not strictly equivalent to the mixed potential at the sensing electrode, but rather to the difference between the mixed potentials at the sensing and reference electrodes. Ideally, these electrodes are selected so that the reference electrode is very much less sensitive to the test species, NO, than the sensing electrode. Alternatively, the reference electrode can be removed from the test gas stream and exposed to a constant atmosphere (air).

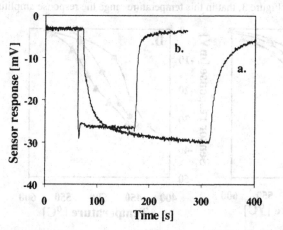

In Figure 2, both sensors exhibit responses that approach, or exceed -25 mV from the baseline, however it is immediately apparent that the composite electrode responds significantlty faster than the $NiCr_2O_4$. This is particularly evident in the time required to achieve a stable response, which takes many minutes for the $NiCr_2O_4$ sensor. Equally

Figure 2: Response to 500 ppm NO in 10% O_2 for sensors with a. $NiCr_2O_4$ and b. composite sensing electrodes. Gas: 250 $^\circ$C, sensor: 450 $^\circ$C.

important, however, is the recovery time when the NO is removed from the gas stream. Minimizing this time is crucial to the develoment of a deployable NO sensor suitable for exhaust monitoring and control. Because the sensor with the composite electrode recovers the baseline so much faster than the $NiCr_2O_4$ sensor, further testing was performed to evaluate the relative sensitivity and speed of these sensors.

Figure 3 shows the response to 500 ppm NO in 10% O_2 for each of the two sensors as a function of sensor temperature for different gas temperatures. The results show that at intermediate sensor temperatures, 425-500 °C, the sensor with the composite electrode has a sensitivity which is comparable to the $NiCr_2O_4$ sensor. At sensor temperatures below ~450 °C, the $NiCr_2O_4$ sensor had larger responses, however it was difficult to acquire reliable data due to the very long equilibration times required. Conversely, above 500-550 °C the composite sensor had essentially no response. This seems to indicate multiple, competing mechanisms in the sensing process. It is well known that Rh is highly catalytic to certain active gas species, however the quantity of Rh added to form the composite electrode is very small. In fact, during SEM analysis of the electrode microstructure it is very difficult to distinguish between the electrodes with and without the Rh. In other words, in the composite electrode nearly all the surface area of the (porous) electrode appears to be made up of the $NiCr_2O_4$ particles. Thus, it is unclear whether the added Rh contributes directly to the catalytic response, or whether it promotes the speed of the $NiCr_2O_4$ response. Very careful inspection using SEM and EDX indicates the presence of thin Rh necks forming between some of the $NiCr_2O_4$ particles. It is postulated that these metallic interparticle contacts may promote the charge transfer through the electrode and thereby quicken the response. This proposed mechanism is, at the present time, speculative and deserves further investigation since it has potential implications to a broad range of sensor development applications.

Response times (90% of time for baseline recovery upon removal of 500 ppm NO from the gas stream) are shown for the two sensors in Figure 4. The data show that the Rh addition significantly enhances the recovery speed, by an order of magnitude or more, when the sensor is operated at 450 °C. Note, from Figure 3, that in this temperature range the response amplitudes

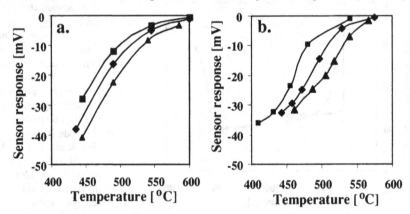

Figure 3: Sensor response to 500 ppm NO, 10% O_2, as a function of sensor temperature for sensors with a) $NiCr_2O_4$ and b) composite electrodes. Gas temperature: —■— 250 °C, —◆— 300 °C, —▲—350 °C.

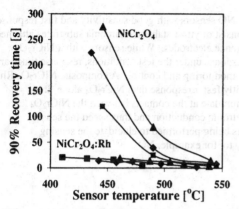

Figure 4: 90% Recovery time versus sensor temperature upon removal of 500 ppm NO from gas stream. Gas temperature: ■— 250 °C, ◆— 300 °C, ▲— 350 °C.

are comparable for the two sensors. It is interesting to note that the incorporation of the Rh into the composite electrode particularly enhances the response speed at low temperatures. This could be interpreted as consistent with the charge transfer model postulated above since the electronic conductivity of the metal and ceramic decrease and increase, respectively, with increasing temperature.

Additional measurements were performed to evaluate the sensitivity to the composite sensor to varying levels of O_2, NO and NO_2. The response to different NO concentrations in 10% O_2 (gas temperature 250 °C, sensor temperature 450 °C) is shown in Figure 5. The response is approximately logarithmic with the NO concentration from 100-500 ppm, but deviates somewhat at concentrations below ~100 ppm. This may be due to the competing effects of the catalytic activity of the Pt counter electrode, which was also exposed to the test gas in this configuration, or may be due to a systematic error in the NO concentration measurement (equivalent to ~25 ppm). Under these operating conditions, it was determined that over the range of 5-15% O_2, the sensor baseline shifted approximately +2 mV , or <10% of the response to 500 ppm NO. Over that same range of O_2 concentration, the response to 250 ppm NO, as measured from the respective baselines, was constant. Finally, it was noted that the sensitivity to NO_2 was comparable to, or slightly larger than the NO sensitivity, and opposite in sign as expected from the oxidation/reduction behavior of NO_2 relative to NO.

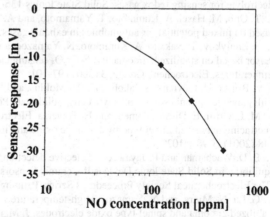

Figure 5: NO response as a function of NO concentration (in 10% O2) for the composite-electrode sensor. Gas temperature: 250 °C, sensor temperature 450 °C.

CONCLUSIONS

We have investigated potentiometric NO sensors with good sensitivity and fast response when operated in 5-15% O_2. The sensors consist of yttria-stabilized zirconia substrates attached with $NiCr_2O_4$ sensing electrodes and Pt reference electrodes. While sensors with a $NiCr_2O_4$ sensing electrode exhibit a significant NO response under the test conditions, response times are too long to be of use for automotive exhaust monitoring and control. A composite $NiCr_2O_4$:Rh sensing electrode is shown to give significantly faster response than $NiCr_2O_4$ alone. It is postulated that the Rh, which appears to accumulate at the contacts between the $NiCr_2O_4$ particles, may enhance the inter-particle electronic conduction and thus speed the sensor response. Ongoing testing of these sensors is being performed to elucidate the sensing mechanisms and to quantify cross sensitivity to, for example, NO_2.

ACKNOWLEDGMENTS

The authors wish to thank Rick Soltis, of the Ford Research Center, for supplying the O_2 sensors used in this investigation. Additional thanks go to Jim Ferrara, of the Lawrence Livermore National Laboratory, for assistance with scanning electron microscopy. This work was performed under the auspices of the U. S. Department of Energy by the University of California, Lawrence Livermore National Laboratory under Contract No. W-7405-Eng-48.

REFERENCES

1. F. Menil, V. Coillard, and C. Lucat, Critical review of nitrogen monoxide sensors for exhaust gases of lean burn engines, Sensors Actuators B 67 (2000) 1-23.
2. N. Miura, G. Lu, and N. Yamazoe, Progress in mixed-potential type devices based on solid electrolyte for sensing redox gases, Solid State Ionics 136-137 (2000) 533-542.
3. T. Ono, M. Hasei, A. Kunimoto, T. Yamamoto, and A. Noda, Performance of the NO_x sensor based on mixed potential for automobiles in exhaust gases, JSAE Review 22 (2001) 49-55.
4. S. Zhuiykov, T. Nakano, A. Kunimoto, N. Yamazoe, and N. Miura, Potentiometric NO_x sensor based on stabilized zirconia and $NiCr_2O_4$ sensing electrode operating at high temperatures, Electrochem. Comm. 3 (2001) 97-101.
5. V. Bruser, U. Lawrenz, S. Jakobs , H. -H. Mobius, and U. Schronauer, NO_x determination with galvanic zirconia solid electrolyte cells, Solid State Phen. 39-40 (1994) 269-272.
6. M. L. Grilli, E. Di Bartolomeo, and E. Traversa, Electrochemical NO_x sensors based on interfacing nanosized $LaFeO_3$ perovskite-type oxide and ionic conductors, J. Electrochem. Soc. 148 (2001) H-98-H102.
7. E. D. Wachsman and P. Jayaweera, Selective detection of NO_x by differential electrode equilibria, in Solid State Ionic Devices II-Ceramic Sensors, edited by E. D. Wachsman et al. (The Electrochemical Society Proceedings Series, Pennington, NJ, 2001), pp. 298-304.
8. G. Lu, N. Miura, and N. Yamazoe, High-temperature sensors for NO and NO_2 based on stabilized zirconia and spinel-type oxide electrodes, J. Mater. Chem. 7 (1997) 1445-1449.
9. L. P. Martin, A.-Q. Pham, R. S. Glass, Effect of Cr_2O_3 electrode morphology on the nitric oxide response of a stabilized zirconia sensor, submitted to Sensors and Actuators, B (2002).

Mat. Res. Soc. Symp. Proc. Vol. 756 © 2003 Materials Research Society   EE9.4

Kinetics of Oxygen Exchange in $Sr(Ti_{0.65}Fe_{0.35})O_3$

Th. Schneider, S. F. Wagner, W. Menesklou, E. Ivers-Tiffée
Universität Karlsruhe (TH), Institut für Werkstoffe der Elektrotechnik,
76128 Karlsruhe, Germany,
schneider@iwe.uni-karlsruhe.de

ABSTRACT

Current limiting electrochemical pumping cells (amperometric sensors) based on zirconia are commonly used for engine control applications. Fast resistive-type sensors adapted from semiconducting metal oxides are a promising alternative for future exhaust gas monitoring systems. Therefore among the interesting characteristics of the materials system $Sr(Ti_{0.65}Fe_{0.35})O_3$, including high sensitivity and temperature independence at high oxygen partial pressures ($pO_2 > 10^{-4}$ bar), a short response time ($t_{90} = 30$ ms) is obviously the most salient.

The latter is determined by the kinetics of the oxygen surface transfer and subsequent diffusion of oxygen vacancies $V_O^{\cdot\cdot}$. For thin samples and low temperatures the surface transfer is dominant, since bulk diffusion usually occurs very fast. The presented model is based on the frequency-domain analysis of amplitude and phase shift of the response signal obtained from a pO_2 modulation in a fast kinetic measurement setup. This method allows both the measurement of response times in the sub-millisecond range as well as the distinction of the behaviour either controlled by volume diffusion or by surface transfer reaction in $Sr(Ti_{0.65}Fe_{0.35})O_3$ ceramics.

INTRODUCTION

The kinetics of oxygen exchange are of primary importance for the application of semiconducting metal oxides, e.g. strontium titanate, as fast resistive oxygen sensors. The sensor's electrical conductivity reflects the equilibrium between the oxygen partial pressure pO_2 of the surrounding atmosphere and the bulk stoichiometry at high temperatures ($T > 700$ °C). Due to oxygen surface transfer and subsequent diffusion of oxygen vacancies within the sample, a pO_2 change gives rise to a change of the sample's electrical conductivity σ:

$$\sigma \propto e^{-\left(E_A/kT\right)} pO_2^{\,m} \tag{1}$$

The first r.h.s. factor describes the temperature dependence, E_A denotes an activation energy. Oxygen partial pressure dependence is described by the second factor, including m as a constant which depends on the dominant type of bulk defect ($|m| \leq + \frac{1}{4}$) and corresponds to the sensor's sensitivity. Negative values of m signify n-type, positive values p-type conduction mechanism.

The defect chemical basics which lead to the characteristic $\sigma(pO_2)$ curve of strontium titanate have been treated in several papers, e.g. [1,2], and shall therefore only be briefly summarized here. The reduction of an oxide can be described by the defect reaction

$$O_O \leftrightarrow V_O^{\cdot\cdot} + 2e^- + \tfrac{1}{2}O_2 \tag{2}$$

in which oxygen leaves its regular lattice sites (O_0) generating a doubly ionized oxygen vacancy $V_0^{..}$ thermally activated by the reduction enthalpy ΔH_{Red}, and two electrons e^- in the conduction band. The generation/recombination of electronic defects depends on the material's band gap E_g:

$$n \cdot p \propto e^{-\left(E_g/kT\right)}, \tag{3}$$

n and p being the electron and hole concentrations, respectively. The combination of eq. (3) with the mass action law of (2) yields the characteristic curve (1), leading to an activation energy

$$E_A = E_g - \frac{1}{2}\Delta H_{Red} \tag{4}$$

The perovskite $SrTiO_3$ is chemically stable up to high temperatures ($T < 1200\ °C$) and high doping levels which permit strong modifications of its sensing properties. For instance, its characteristic curve $\sigma(pO_2)$ can be rendered unambiguous over a large oxygen partial pressure range by donor doping. Whereas donor dopants inhibit the kinetics of equilibration for the bulk defects (slowly moving strontium vacancies in the cation sub-lattice), in acceptor doped material it is controlled by fast diffusion of oxygen vacancies in the anion sub-lattice only, which allows for an application as fast oxygen sensor with response times in the millisecond range [3]. The strong temperature dependence of σ – a major disadvantage for sensing applications – can furthermore be suppressed by means of a high iron (acceptor) content for a limited pO_2 range ($pO_2 = 10^{-4}$ bar - 1 bar) (figure 1). This is due to the fact that the effective band gap E_g of the materials system $Sr(Ti_{1-x}Fe_x)O_3$ depends on its iron content x, which leads to a change of sign of the activation energy E_A for $E_g \approx 2.2$ eV at $x = 0.35$ [4], according to eq. (4). Hence, the composite $Sr(Ti_{0.65}Fe_{0.35})O_3$ is a most suitable material for a temperature independent oxygen sensor [5].

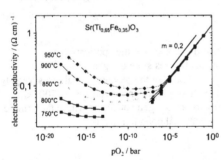

Figure 1. Temperature independence of characteristic curve $\sigma(pO_2)$ for $Sr(Ti_{0.65}Fe_{0.35})O_3$

MODELING / EXPERIMENTAL DETAILS

Modeling

In contrast to the commonly used jump method where the conductivity response to a sudden pO_2 change is measured in the time domain, in the following a technique is presented which is based on the frequency-domain analysis of magnitude and phase shift of the response signal obtained from a pO_2 modulation in a fast kinetic measurement setup (figure 2) introduced in [6,7].

The electrically contacted sample is exposed to an ambient gas atmosphere with an oxygen partial pressure $pO_2(t)$ which is periodically changed with an angular frequency ω by varying the total gas pressure $p(t)$ instead of the oxygen-gas ratio x_{O_2} so that $pO_2(t) = x_{O_2} \cdot p(t)$, $x_{O_2} = 0...100\ \%$. For low frequencies this is achieved with alternately activated valves, at higher frequencies the modulation is performed with a rotating punched disk chopping the gas stream. The frequency range runs from very low frequencies up to 1 kHz. The sample is placed in a measuring chamber within a furnace which can be heated up to 1000 °C. pO_2 is measured by a reference pressure sensor. In order to ensure linear system behaviour, the periodic pressure changes are small (in the order of $1 - 10\ \%$ of the mean pressure). The complex steady-state relationship between sensor conductivity σ and pO_2 is then given by a complex transfer function $G(i\omega)$ in the frequency domain (i: imaginary unit).

In contrast to the aforementioned jump method, this approach not only permits the measurement of response times in the sub-millisecond range but also to distinguish between kinetic behaviour either controlled by *bulk diffusion* or by *surface transfer reaction*. The kinetics of oxygen exchange in resistive oxygen sensors are determined by the rate of all consecutive reactions taking place during oxygen in- or excorporation. This process consists of various consecutive transport and electrochemical reaction steps, such as: (i) transport through a gas boundary layer surrounding the sample's surfaces, (ii) dissociation of oxygen molecules, (iii) adsorption of oxygen atoms at the surfaces, (iv) ionization, (v) transport through surface layers (space charge effects) and (vi) subsequent bulk diffusion of oxygen vacancies. It is a complex scheme of reactions which might be even more intricate (e.g. [8]).

The influence of (i) is only observed at higher modulation frequencies and/or for very thin samples when diffusion processes in the gas phase may cause a significant delay in the response times. It shall be neglected in the following. The kinetics of the elementary steps (ii) – (v) are not known in detail, therefore they are summarized as one *surface transfer reaction*. (vi) finally leads to a flow of charged carriers in the sample and thus to a conductivity change. At high oxygen partial pressures the sample's conductance depends on the concentration p of electron holes.

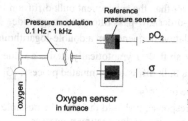

Figure 2. Schematic diagram of the fast kinetic measurement setup

The response behaviour of the bulk sample, i.e. the conductivity shift due to a harmonic pO_2 change, is decisively determined by the *bulk diffusion* of oxygen vacancies. Assuming that the sensor's kinetics are determined by this process, the concentration of oxygen vacancies $\left[V_O^{\bullet\bullet} \right]$ can be expressed by Fick's second law. Due to the fact that under oxidizing conditions $\left[V_O^{\bullet\bullet} \right]$ and p are coupled by a simple electroneutrality condition, this results in:

$$\frac{\partial p}{\partial t} = div\left(\tilde{D}_{V_O^{\bullet\bullet}}\, grad\ p \right),$$ (5)

$\tilde{D}_{V_{\ddot{O}}}$ being the chemical diffusion coefficient of oxygen vacancies. For thin samples, a one-dimensional treatment is justified. A harmonic pressure excitation of the gas atmosphere then leads to a harmonic modulation of the hole concentration $p(x,t)$ at both principal sample surfaces exposed to the gas mixture ($x = 0$ and $x = -d$). This variation propagates through the sample like a damped diffusion wave. By means of a Laplace transform to the frequency regime and subsequent integration over the sample's thickness, the conductance $\sigma(i\omega)$ and hence the transfer function $G(i\omega)$ are obtained.

Taking into account also the influence of the oxygen *surface transfer reaction* which is the rate-determining step at lower temperatures, the one-dimensional treatment is justified because surface control only occurs for thin samples. Eq. (5) is solved under the boundary conditions

$$k_1\left[O_2\right]\big|_{x=0} - k_2\, p\big|_{x=0} = -\tilde{D}_{V_{\ddot{O}}}\frac{\partial p}{\partial x}\bigg|_{x=0} \quad \text{and} \quad \frac{\partial p}{\partial x}\bigg|_{x=-\frac{d}{2}} = 0\,, \tag{6}$$

representing a continuity equation at the gas-solid interface, k_1 and k_2 being the rate constants (in m/s) for both directions of the surface transfer reaction eq. (2), and a boundary condition for symmetry reasons. A solution for the sample's conductivity σ is derived by means of a transform to the frequency regime and subsequent integration over the sample thickness d. The transfer function (frequency response) then reads:

$$G(i\omega) \propto \sinh\left(\frac{d}{2}\Omega\right) \times \left[\Omega \cdot \cosh\left(\frac{d}{2}\Omega\right) + \frac{i\omega}{k}\cdot\sinh\left(\frac{d}{2}\Omega\right)\right]^{-1}, \quad \text{with } \Omega = \sqrt{\frac{i\omega}{\tilde{D}_{V_{\ddot{O}}}}}\,, \tag{7}$$

$k \equiv k_1 = k_2$ being the surface rate constant of oxygen exchange in chemical equilibrium. If surface transfer dominates the kinetic behaviour, this corresponds to $\tilde{D}_{V_{\ddot{O}}} \to \infty$ in (7) – a "bottleneck" situation implying that surface transfer occurs much slower than the subsequent bulk diffusion. This yields a response behaviour similar to that of a first-order low-pass filter: The magnitude $|G|$ of the response signal, plotted as a function of modulation frequency on a double-logarithmic scale (Bode plot), decreases with a slope of -1, the phase shift Φ asymptotically attains a value of $-\pi/2$ for high frequencies. In contrast, in the case of a bulk diffusion dominated process, $|G|$ decreases with a slope of $-1/2$ and Φ approaches a value of $-\pi/4$.

By means of an analysis of the system's frequency response, it should therefore be possible to distinguish between bulk diffusion or surface transfer reaction as dominating processes.

Samples

Nominally undoped $SrTiO_3$ single crystals from Verneuil-grown boules were obtained from Litzenberger, Idar-Oberstein, Germany, 0.1 % and 0.3 % Fe doped $SrTiO_3$ single crystals from Commercial Crystal Laboratories, Naples FL., USA. By conductivity measurements and defect-chemical modeling the acceptor contents were determined as $N_A = 1.1\cdot10^{18}\,cm^{-3}$ for the undoped and $N_A = 5.6\cdot10^{18}\,cm^{-3}$ for the Fe doped crystals. They were cut to slabs with thicknesses varying from 40 µm to 6000 µm. These samples were polished with diamond spray, the electrode contacts applied consisted of Pt paste fired at 1200 °C in air for half an hour.

For the preparation of Sr(Ti$_{1-x}$Fe$_x$)O$_3$ thick films ($x = 0$ and 0.35), ceramic powders were prepared using a mixed oxide technique, starting with SrCO$_3$, TiO$_2$ and Fe$_2$O$_3$. A screen printing paste consisting of Sr(Ti$_{1-x}$Fe$_x$)O$_3$ powders and organic supplements only was printed onto an alumina substrate (96 %). The thick films were dried and fired at 1050 °C resulting in a layer thickness of 10 µm, grain sizes of 0.5 µm and an open porosity of 30 %. They were equipped with Pt bottom side contacts burned in at 1300 °C.

RESULTS

Kinetic measurements were performed using SrTiO$_3$ single crystals as model systems and Sr(Ti$_{0.65}$Fe$_{0.35}$)O$_3$ ceramics, since there are no single crystals available with an Fe content as high as 35 %. The ceramic samples had a high porosity in order to avoid a complicated grain boundary network for which the simple model presented above would obviously not be valid.

From the kinetic response data obtained from SrTiO$_3$ single crystals, it is possible to calculate the chemical diffusion coefficients and the surface rate constants by fitting the response curves calculated with given parameters to the measured data from the kinetic measurement setup (figure 3). For sufficiently thick samples whose kinetic behaviour is determined by bulk diffusion only, values for $\tilde{D}_{V_O^{\cdot\cdot}}$ can be determined. From further measurements on thin crystals, the kinetic behaviour of which is clearly surface reaction controlled, it is then possible to extract the surface rate constants k. The calculated values for k and $\tilde{D}_{V_O^{\cdot\cdot}}$ at different temperatures are given in table I.

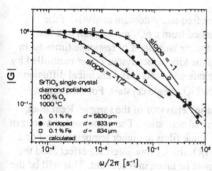

Figure 3. Measured and calculated frequency responses of SrTiO$_3$ single crystals, at $pO_2 = 1$ bar and $T = 1000$ °C. The magnitude of the transfer function, normalized to $|G| = 1$ for $\omega \to 0$, is depicted for a bulk diffusion controlled sample ($d = 5830$ µm, slope $-1/2$) and for a surface transfer dominated sample ($d = 834$ µm, slope -1). In addition, the response signal of an undoped crystal shows that bulk diffusion is reduced compared to the Fe doped case.

Table I. Surface rate constant k and chemical diffusion coefficient $\tilde{D}_{V_O^{\cdot\cdot}}$ obtained from nominally undoped and Fe doped SrTiO$_3$ single crystals

T/°C	k /(m/s)	$\tilde{D}_{V_O^{\cdot\cdot}}$ /(m²/s)	$\tilde{D}_{V_O^{\cdot\cdot}}$ /(m²/s)
	$(N_A = 1.1 \cdot 10^{18} \mathrm{cm}^{-1})$	$(N_A = 1.1 \cdot 10^{18} \mathrm{cm}^{-1})$	$(N_A = 5.6 \cdot 10^{18} \mathrm{cm}^{-1})$
750	$1.7 \cdot 10^{-5}$	$7.0 \cdot 10^{-9}$	$8.0 \cdot 10^{-9}$
850	$1.4 \cdot 10^{-4}$	$9.8 \cdot 10^{-9}$	$2.0 \cdot 10^{-8}$
1000	$3.1 \cdot 10^{-3}$	$1.5 \cdot 10^{-8}$	$4.0 \cdot 10^{-8}$

Moreover, from the run of the magnitude of the response signal in the Bode plot (e.g. figure 3) a limiting 3 dB frequency f_g can be determined from which the sensor response time t_{90} can be calculated according to the simple approximation $t_{90} \approx 1/(3 f_g)$ [5].

For thin single crystals or small grain diameters in porous thick films the kinetic behaviour is surface transfer controlled. The response times t_{90} of SrTiO$_3$ and Sr(Ti$_{0.65}$Fe$_{0.35}$)O$_3$ thick films were determined from the run of their response functions at different temperatures in this manner and compared to response times calculated for a single SrTiO$_3$ crystallite with a corresponding thickness of 0.5 µm and known k value (figure 4). It can be observed that the thermal activation behaviour of the $SrTiO_3$ crystal and thick film is obviously very similar: The slopes of the $t_{90}(T)$ curves appear to be the same, the t_{90} times of the thick film being roughly one order of magnitude larger than the single crystal value. The response times of $Sr(Ti_{0.65}Fe_{0.35})O_3$ thick films are similar to those of $SrTiO_3$ thick films at higher temperatures. In contrast, at lower temperatures their thermal activation behaviour seems to be different.

In terms of a detailed interpretation, however, it should be taken into account that there may be several sources of error: The grain sizes were only roughly estimated from SEM images ($\pm 30\%$), no inhomogeneous grain size distribution was considered. Due to the manifold influences of grain boundaries on the diffusion kinetics, porous ceramics are comparatively undefined systems within the framework of our simple model.

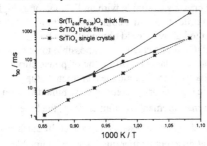

Figure 4. Sensor response times t_{90} for $SrTiO_3$ and $Sr(Ti_{0.65}Fe_{0.35})O_3$ thick films (measured) in comparison with a thin $SrTiO_3$ crystallite of the same size (theoretical). The grain sizes of the porous thick films (layer thickness 10 µm, open porosity 30 %) were 0.5 µm.

CONCLUSIONS

A model has been presented which is based on the frequency-domain analysis of the response signal of a resistive-type oxygen sensor obtained from a pO_2 modulation in a fast kinetic measurement setup. This method allows both the measurement of response times t_{90} in the sub-millisecond range as well as the distinction of the kinetic behaviour either controlled by bulk diffusion or by surface transfer reaction. Surface rate constants k and chemical diffusion coefficients $\tilde{D}_{V_O^{..}}$ were determined from well-defined $SrTiO_3$ single crystals. For thin single crystals the surface rate constant dominates the response behaviour of the sample. Real thick films of $SrTiO_3$ show slower response but similar thermal activation. The response behaviour of $Sr(Ti_{0.65}Fe_{0.35})O_3$ thick films is comparable to $SrTiO_3$ thick films at higher temperatures. For a more detailed analysis of the kinetics of $Sr(Ti_{0.65}Fe_{0.35})O_3$ thick films, interface effects and the influence of the iron content on the surface kinetics must be taken into account. This will be the subject of forthcoming investigations.

REFERENCES

1. E. Ivers-Tiffée, K. H. Härdtl, W. Menesklou, J. Riegel, *Electrochimica Acta* **47**, 807 (2001)
2. R. Moos, K.H. Härdtl, *J. Am. Ceram. Soc.* **80**, 2549 (1997)
3. W. Menesklou, H.-J. Schreiner, K. H. Härdtl, E. Ivers-Tiffée, *Sens. Actuators B* **59**, 184 (1999)
4. S. Steinsvik, R. Bugge, J. Gjønnes, J. Taftø, T. Norby, *J. Phys. Chem. Solids* **58**, 969 (1997)
5. W. Menesklou, H.-J. Schreiner, R. Moos, K. H. Härdtl, E. Ivers-Tiffée, *MRS Proc.* Vol. **604**, 305 (2000)
6. Ch. Tragut, K. H. Härdtl, *Sens. Actuators B* **4**, 425 (1991)
7. C. Tragut, *Sens. Actuators B* **7**, 742 (1992)
8. J. Maier, *Solid State Ionics* **135**, 575 (2000)

Mat. Res. Soc. Symp. Proc. Vol. 756 © 2003 Materials Research Society

Cation Transport and Surface Reconstruction in Lanthanum Doped Strontium Titanate at High Temperatures

Karsten Gömann[1], Günter Borchardt[1], Anissa Gunhold[2], Wolfgang Maus-Friedrichs[2], Bernard Lesage[3], Odile Kaïtasov[4], Horst Baumann[5]

[1]Institut für Metallurgie, Technische Universität Clausthal, Robert-Koch-Str. 42, D-38678 Clausthal-Zellerfeld, Germany
[2]Institut für Physik und Physikalische Technologien, Technische Universität Clausthal, Leibnizstr. 4, D-38678 Clausthal-Zellerfeld, Germany
[3]Laboratoire d'Etude des Matériaux Hors Equilibre, Université Paris-Sud, F-91405 Orsay Cedex, France
[4]Centre de Spectrométrie Nucléaire et de Spectrométrie de Masse, Université Paris-Sud, F-91405 Orsay Cedex, France
[5]Institut für Kernphysik, J. W. Goethe-Universität, August-Euler-Str. 6, D-60486 Frankfurt, Germany

ABSTRACT

Tracer diffusion experiments were carried out in synthetic air at 1573 K in $SrTiO_3$(100) and (110) single crystals, which were either undoped or doped with up to 1 at.% La, respectively. Tracer sources of ^{139}La and ^{142}Nd were applied by ion implantation. The resulting depth profiles were measured by SIMS. The reconstruction of the surface was monitored ex-situ using microscopic and spectroscopic methods including SEM, EPMA, and AFM. The measured tracer diffusivities show no dependency on orientation. The tracer diffusion takes place via cation vacancies. Under oxidizing conditions the dopant is compensated by Sr vacancies. Hence the diffusion is increasing strongly with La concentration. The observed time dependency of the diffusivities may be related to a space charge layer postulated by the current defect chemistry model for donor doped $SrTiO_3$. At high dopant concentrations annealing leads to segregation of bulk La to the surface. La is not significantly incorporated into the secondary crystallites at the surface which consist almost entirely of Sr and O.

INTRODUCTION

Donor doped strontium titanate $SrTiO_3$ is a promising material for resistive oxygen sensors operated at high temperature. Changing the ambient oxygen partial pressure $p(O_2)$ under high temperature leads to an undesirable surface reconstruction and the formation of secondary phases. Though an overall consistent model is still lacking, many of the phenomena can be explained by the bulk defect chemistry model of donor doped $SrTiO_3$ [1]. The defect chemistry is dominated by the following reaction (see [2] for defect notation):

$$V_O^{\bullet\bullet} + 2e' + \frac{1}{2}O_{2(g)} \Leftrightarrow O_O^x \qquad (1)$$

Upon oxidation of O deficient crystals, $V_O^{\bullet\bullet}$ and free electrons are consumed by the incorporation of O into the lattice. Finally, Sr vacancies are generated and a subsequent change in the donor compensation mechanism occurs. The excess Sr migrates to the surface where secondary SrO_x phases are formed on top of the surface (e.g. [3,4]), and Ruddlesden-Popper phases ($SrO \cdot nSrTiO_3$, [5]) are formed at the surface between the islands [6]:

$$Sr_{Sr}^x + 2e' + \frac{1}{2}O_{2(g)} \Leftrightarrow V_{Sr}'' + SrO_{sec.phase} \qquad (2)$$

The amount of V_{Sr}'' produced is fixed by the donor content

$$[D^\bullet] = 2[V_{Sr}''], \qquad (3)$$

explaining the correlation of secondary phase quantity and dopant concentration. A recent extension of the model introduces a space charge zone in the near surface region based on the strong differences in point defect mobility [7]. At low $p(O_2)$, O_2 is released into the atmosphere yielding a large number of electrons, which partly reduce Ti^{4+}, resulting in the formation of Ti^{3+} containing phases like Ti_2O_3 or $LaTiO_3$ [8-10].

For a model verification, the diffusivities of the involved species must be known. Recent studies obtain a depth-dependent O tracer diffusion coefficient which is attributed to a gradient in $[V_O^{\bullet\bullet}]$ in the surface near space charge region [7,11]. From the O diffusion data, a V_{Sr}'' diffusion coefficient was deduced, which is in acceptable accordance to a measured value [12]. Yet, except for one publication on Sr and Ti diffusion in undoped $SrTiO_3$ at 2148 K [13] and one computer simulation study [14], no data on cation diffusivities in $SrTiO_3$ is published. The work presented here is part of a larger study where diffusion experiments are combined with investigations of the topography, chemistry and electronic structure of the reconstructed surface in order to establish a kinetic model for the secondary phase formation on donor doped $SrTiO_3$. Here, we will present new results on the secondary phase growth and La and Nd diffusion experiments under oxidizing conditions.

EXPERIMENTAL

$SrTiO_3$(100) and (110) single crystals with La contents of 0, 0.02, 0.2, and 1 at.% were obtained from Crystec (Germany). The crystals were grown under reducing conditions. The samples were annealed at ambient pressure in a flow of synthetic air (80 % N_2, 20 % O_2) for up to 6 weeks at 1573 K to equilibrate the samples to a sufficient depth with the atmosphere. During the first hours the experiments were suspended several times to monitor the surface reconstruction ex situ with various methods including Scanning Electron Microscopy (SEM), Atomic Force Microscopy (AFM), and Electron Probe Microanalysis (EPMA). Additional analyses were performed at the end of the equilibration step.

Prior to tracer deposition the 5 % -doped samples were washed for 24 h at 60 °C in H_2O_{deion}, hereby removing most of the secondary phases. The tracer sources were deposited by ion implantation using a mass separated and scanned ion beam. $^{139}La^+$ ions were implanted with a dose of 1×10^{16} cm^{-2} at 40 or 120 keV, $^{142}Nd^+$ ions were implanted with a dose of 2×10^{15} cm^{-2} at 120 keV. Diffusion annealing took place under the same conditions as in the equilibration step. The depth profiles were analyzed with SIMS (Cameca IMS3f), using a scanned (250 × 250 μm^2) 10 kV O$^-$ beam at a current of 100 nA. The crater depth was measured with a surface profilometer (Tencor AlphaStep 500).

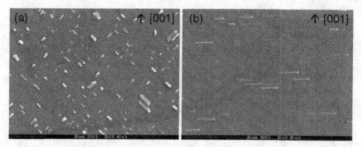

Figure 1. SEM backscattered electron images. 1 at.% La-doped SrTiO₃ annealed for 40.5 d at 1573 K in synthetic air. (a) (100) surface, (b) (110) surface.

RESULTS AND DISCUSSION

Surface Reconstruction

Annealing of donor doped SrTiO₃ in oxidizing atmosphere leads to the formation of SrO$_x$ phases on top of the surface [3,4]. At 1573 K, already after 1 hour a significant island growth is visible, which seems to be largely completed after 1 day of annealing. Fig. 1 shows backscattered electron images of the annealed surfaces. On the (100) surface the crystallites grow epitaxially in an angle of 45° to the SrTiO₃ lattice, reducing the misfit of the SrO (cubic, d = 0.51 nm) and SrTiO₃ (cubic, d = 0.391 nm) lattices to 7 %. On the (110) surface the growth is oriented perpendicular to the [001] axis, resulting in the same misfit value. EPMA investigations of 1 at.% La-doped samples confirm that the islands contain Sr, but no Ti. In contrast to earlier observations [15], also La concentrations of 0.5 at.% were measured, which corresponds to half of the initial La content of the SrTiO₃ crystal. The composition of the islands is the same on both surfaces. On the less doped samples similar secondary phases are formed, but number and size of the islands decrease substantially with the donor content. On the surface surrounding the islands of the (100) oriented crystal, terrace-like structures with step heights of typically several Å are observed with AFM [6]. In contrast, the (110) surface is heavily reconstructed, forming ridges and trenches oriented along the [001] direction perpendicular to the secondary phases.

Fig. 2 shows SIMS depth profiles after the equilibration annealing. At high dopant concentrations the thermal treatment leads to the migration of La to the surface. This segregation may result from the gradient in $V_{Sr}^{''}$ in the postulated space charge layer [7]. As the depth distribution of the dopant was considered constant in the model [7], the effect of this observation on the validity of the model will have to be evaluated.

Tracer Diffusion Experiments

Ion implanted depth profiles show a Gaussian distribution [16]. Diffusion of the implanted tracer leads to a broadening of the Gaussian profile. The diffusivities are determined by fitting the appropriate solution of Fick's second law to the profiles, which in this case is written as

$$c(x,t) = \frac{N}{\sqrt{2\pi(\Delta R_p^2 + 2Dt)}} \exp\left(-\frac{(x - R_p)^2}{2\Delta R_p^2 + 4Dt}\right), \tag{4}$$

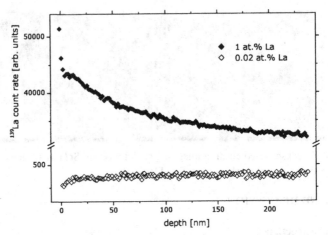

Figure 2. SIMS La depth profiles. 0.02 and 1 at.% La-doped SrTiO$_3$(100) annealed for 40.5 d at 1573 K in synthetic air.

with c being the concentration, x the depth, t the time, N the implanted dose, ΔR_p the standard deviation of the initial distribution, R_p the mean value of the projected range, and D the diffusion coefficient. Hence, the diffusion coefficient is derived directly from the increase in ΔR_p after the annealing experiment.

Fig. 3(a) shows representative La tracer depth profiles implanted at 120 keV. As a result of ion beam mixing during the SIMS analysis, the profile measured after implantation deviates slightly from the Gaussian shape. This is visible only in the low concentration region and has no significant influence on the determination of ΔR_p. After thermal treatment, the concentration maximum is shifted to lower depth values. Several explanations are possible: Predominantly Sr might evaporate from the surface, as was observed by [3]. Furthermore, material may also be consumed during the secondary phase growth. Lastly, such a shift is also observed when the tracer ions are reflected at the surface.

During implantation, radiation damage is introduced into the lattice with a depth distribution comparable to the implanted ions, but shifted in direction of the surface. On annealing this may result in extended defects like clusters, causing differing tracer diffusivities in this layer. To check for the influence of beam damage, experiments with Nd implantation were also carried out. As Nd is not contained in the samples, a lower fluence is sufficient to obtain a satisfactory dynamic range for the experiment. Like La, Nd occurs predominantly in trivalent state and has a comparable ionic radius and mass. The Nd depth profiles in Fig. 3(b) show that the Nd ions are reflected at the surface. The case of reflection is solved mathematically by adding a second exponential term to the Gaussian function:

$$c(x,t) = \frac{N}{\sqrt{2\pi(\Delta R_p^2 + 2Dt)}} \left\{ \exp\left(-\frac{(x-R_p)^2}{2\Delta R_p^2 + 4Dt}\right) + \exp\left(-\frac{(x+R_p)^2}{2\Delta R_p^2 + 4Dt}\right) \right\} \qquad (5)$$

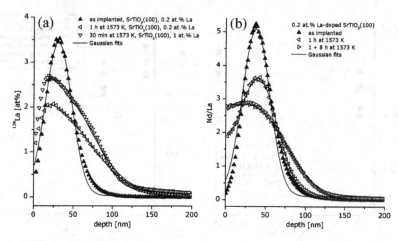

Figure 3. SIMS depth profiles of (a) La and (b) Nd. The La concentration was determined by normalizing to the natural background of the samples, which was subtracted afterwards.

Hence, under the assumption that beam damage decelerates the La tracer diffusion in the depth layer of $x \leq R_p$, only the part of $x > R_p$ of the La depth profiles was used to fit the data to Eq. 4. The Nd profiles were fitted to Eq. 5.

The results are compiled in Fig. 4. The diffusivities increase with the donor content, indicating that La and Nd diffusion takes place via Sr vacancies. As is expected for a cubic system, no dependence on the crystal orientation is found. Furthermore, a time dependence is observed. It is unlikely that this effect is related to thermal recovery, which should take place in much shorter time than the chosen experiment durations. Instead, we believe that this phenomenon is related to the space charge surface layer postulated by Meyer et al. [7]. Analogous to O diffusion, local space charge and defect concentration gradients lead to an enhancement of the La and Nd diffusivities in direction of the surface. Compared to La, the Nd diffusion coefficients are generally about an order of magnitude lower, indicating that in spite of the close relationship of the two elements, Nd may not be used as a direct analogue for La. In general, the measured diffusion coefficients are several orders of magnitude lower than the O tracer diffusion coefficients found in the literature, which range between 10^{-10} to 10^{-13} cm^2/s (T = 1573 K, x = 0 nm), depending on the dopant concentration [11]. Neglecting defect clusters, a cation vacancy mediated transport of the (dopant) tracer yields $D_{dopant} \cong 0.5 D_{V_{Sr}^-}$ (see Eq. 1 and the relation $D_i \cdot c_i = D_v \cdot c_v$, i = cation, v = vacancy), which is in good agreement with an experimental value for $D_{V_{Sr}^-} \cong 6 \times 10^{-15}$ cm^2/s at T = 1573 K in 0.2 at.% Nb doped SrTiO$_3$ [7]. This supports the assumption that La diffuses via Sr vacancies.

ACKNOWLEDGEMENTS

The authors would like to thank the Deutsche Forschungsgemeinschaft (DFG) for financial support under the DFG contracts BO 532/47 and MA 1893/2.

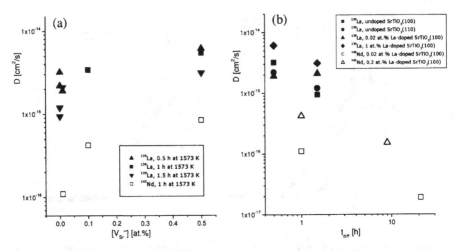

Figure 4. Values for the tracer diffusion coefficient D as a function of (a) Sr vacancy concentration (see Eq. 3), (b) annealing time.

REFERENCES

1. R. Moos and K. H. Härdtl, *J. Am. Ceram. Soc.* **80**, 2549 (1997).
2. F. A. Kröger and H. J. Vink, *Solid State Phys.* **3**, 307 (1956).
3. K. Szot, W. Speier, U Breuer, R. Meyer, J. Szade, and R. Waser, *Surf. Sci.* **460**, 112 (2000).
4. Han Wei, W. Maus-Friedrichs, G. Lilienkamp, V. Kempter, J. Helmbold, K. Gömann, and G. Borchardt, *J. Electroceram.* **8**, 221 (2002).
5. S. N. Ruddlesden and P. Popper, *Acta Cryst.* **11**, 54 (1958).
6. A. Gunhold, K. Gömann, L. Beuermann, M. Frerichs, G. Borchardt, V. Kempter, and W. Maus-Friedrichs, *Surf. Sci.* **507-510**, 447 (2002).
7. R. Meyer, R. Waser, J. Helmbold, and G. Borchardt, *Phys. Rev. Lett.* (revised).
8. K. Szot and W. Speier, *Phys. Rev. B* **60**, 5909 (1999).
9. A. Gunhold, L. Beuermann, M. Frerichs, V. Kempter, K. Gömann, G. Borchardt, and W. Maus-Friedrichs, *Surf. Sci.* (2002) (in press).
10. A. Gunhold, K. Gömann, L. Beuermann, V. Kempter, G. Borchardt, and W. Maus-Friedrichs, *Anal. Bioanal. Chem.* (submitted)
11. J. Helmbold, PhD thesis, Technische Universität Clausthal, Germany (2001).
12. F. Poignant, PhD thesis, Université de Limoges, France (1995).
13. W. H. Rhodes and W. D. Kingery, *J. Am. Ceram. Soc.* **49**, 521 (1966).
14. M. J. Akhtar, Z.-U.-N. Akhtar, R. A. Jackson, and C. R. A. Catlow, *J. Am. Ceram. Soc.* **78**, 421 (1995).
15. R. Meyer, K. Szot, and R. Waser, *Ferroelectrics* **224**, 751 (1998)
16. H. Ryssel and I. Ruge, *Ion Implantation* (Wiley, Chichester, 1986), chapter 2.1.

Mat. Res. Soc. Symp. Proc. Vol. 756 © 2003 Materials Research Society EE9.10

Initial growth stages of CeO₂ nanosystems by Plasma-Enhanced Chemical Vapor Deposition

Davide Barreca[1], Alberto Gasparotto[1], Eugenio Tondello[1], Stefano Polizzi[2], Alvise Benedetti[2], Cinzia Sada[3], Giovanni Bruno[4], Maria Losurdo[4]

[1] ISTM-CNR and CIMA Department, Padova University, Via Marzolo, 1 - 35131 Padova, Italy.
[2] Physical Chemistry Department, Ca' Foscari Venice University, Via Torino, 155/B - 30170 Venezia-Mestre, Italy.
[3] INFM and Physics Department, Padova University, Via Marzolo, 8 - 35131 Padova, Italy.
[4] IMIP-CNR, via Orabona, 4 - 70126 Bari, Italy.

ABSTRACT

Nanocrystalline CeO₂ thin films were synthesized by Plasma-Enhanced Chemical Vapor Deposition using Ce(dpm)₄ as precursor. Film growth was accomplished at 150-300°C either in Ar or in Ar-O₂ plasmas on SiO₂ and Si(100) with the aim of studying the effects of substrate temperature and O₂ content on coating characteristics. Film microstructure as a function of the synthesis conditions was investigated by Glancing Incidence X-Ray Diffraction (GIXRD) and Transmission Electron Microscopy (TEM), while surface morphology was analyzed by Atomic Force Microscopy (AFM). Surface and in-depth chemical composition was studied by X-ray Photoelectron Spectroscopy (XPS) and Secondary Ion Mass Spectrometry (SIMS).

INTRODUCTION

Cerium dioxide (CeO₂) thin films are interesting candidates for different technological applications [1], the most important being the preparation of Three-Way Catalysts (TWCs) [2]. In particular, the interest has been recently focused on oxygen-deficient ceria-based nanosystems [3]. As a matter of fact, the synergy between the peculiar nanosystem properties and the co-presence of Ce(III)/Ce(IV) might eventually allow the preparation of TWCs *without* noble metals, thus reducing their environmental impact [2].

In order to fully exploit the nanosystem potential, a major concern regards the possibility of tailoring their composition, microstructure and morphology. In this context, Plasma-Enhanced Chemical Vapor Deposition (PE-CVD) offers important advantages over other preparation techniques, allowing the synthesis of nanosystems under non-equilibrium conditions, where nucleation is predominant over the subsequent aggregate growth.

The present work is focused on the synthesis of nanocrystalline CeO₂-based thin films by PE-CVD, with particular attention to the first nucleation stages. Ce(dpm)₄ (Hdpm = 2,2-6,6-tetramethyl-3,5-heptanedione) was chosen as precursor compound and its fragmentation behavior was preliminarily investigated by Electron Impact Mass Spectrometry (EI-MS). The films were deposited on SiO₂ and Si(100) in Ar and Ar-O₂ plasmas. The most relevant results concerning the sample properties and their dependence on synthesis conditions are presented and discussed.

EXPERIMENTAL DETAILS

Ceria films were deposited by a custom-built PE-CVD reactor described elsewhere [4]. RF power (13.56 MHz) was delivered to one electrode, while the substrates were placed on a second grounded electrode. The precursor (Ce(dpm)₄, Strem Chemicals, Inc., 99.9 %) was vaporized at 170°C and introduced in the reactor in Ar flow between the two electrodes.

Film growth was performed on SiO_2 and Si(100) either in Ar or in Ar-O₂ plasmas. In the former atmosphere, RF power (20 W) and Ar flow (10 sccm) were kept constant, while the effect of substrate temperature was investigated in the range 150-300°C. In the latter atmosphere (RF power=40 W), the O₂ partial pressure (p(O₂)) was varied from 0 to 0.35 mbar, keeping the substrate temperature at 150°C in order to minimize thermal effects. Further depositions were performed in *mid-way* conditions (growth temperature=200°C, p(O₂)=0.3 mbar). In all cases, the deposition time and the total pressure were 1 h and 1.5 mbar respectively. Details concerning precursor and film synthesis and characterization can be found elsewhere [4].

DISCUSSION

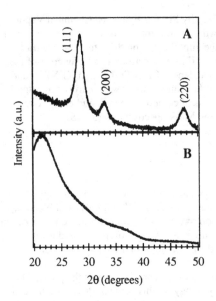

Figure 1. GIXRD patterns (incidence angle=2°) of two ceria films deposited on Si(100) (**A**) and on SiO_2 (**B**) in Ar-O₂ plasma (substrate temperature=200°C; RF power=40 W; p(O₂)=0.30 mbar).

Preliminary information on the precursor fragmentation pattern was gained by EI-MS analysis [4]. As indicated by the recorded spectra, the precursor fragmentation was characterized by a two-step ligand loss, producing [Ce(dpm)₃]⁺ (m/z=689, 1%) and [Ce(dpm)₂]⁺ (m/z=506, 100%). [Hdpm]⁺ peak (m/z=184, 16%) and several ionic/radicalic species at m/z<180 arising from its fragmentations were also detected. The formation of such species in plasmas may induce etching and reducing effects on the growing layers, [5,6], resulting in peculiar film characteristics (see below).

Fig. 1 displays the GIXRD patterns of two specimens co-deposited in Ar-O₂ plasmas on Si(100) (**A**) and on SiO_2 (**B**) at 200°C (RF power=40 W, p(O₂)=0.30 mbar). While the film supported on Si(100) displayed three reflections at 2θ≈28.5°, 33.0° and 47.5°, ascribed to (111), (200) and (220) cubic CeO₂ (cerianite) planes (average crystallite dimensions≈6 nm), no peak was observed for the sample deposited on SiO_2. This effect indicates a marked substrate influence on the microstructural characteristics of the coatings and was detected, to a different extent, for all the obtained samples.

For films deposited in Ar plasmas, cerianite peaks were clearly detectable only for T>250°C. Analyses performed by Secondary Ion Mass Spectrometry (SIMS) [4] with low-energy primary beam (3 keV) on the same samples evidenced that film thickness increased exponentially with substrate temperature, ranging from ≈3 nm at 150°C to ≈18 nm at 300°C. In our opinion, such effects could be attributed to the concomitant occurrence of ablation and growth processes in a glow discharge [7]. In fact, at higher substrate temperatures the ionic and radicalic species produced by precursor fragmentation (see above) might be more easily desorbed from the film surface, progressively favoring growth processes with respect to etching ones. As a result, both the film thickness measured by SIMS, and the XRD reflections underwent a parallel enhancement. The absence of appreciable cerianite reflections at the lower substrate temperatures might be also related to the large Ce(III) amounts (see XPS results), that are likely to prevent the cubic structure from being held.

As regards coatings deposited in Ar-O_2 plasmas, a lower XRD peak intensity was generally observed and ascribed either to the low substrate temperature (150°C), or to the presence of highly reactive species (O_2^+, O^-, O, ...) [8], both favoring etching processes and resulting in an higher microstructural disorder and a lower film thickness. In fact, for this sample set SIMS analyses revealed a consistent reduction in film thickness already at moderate oxygen partial pressure (p(O_2)). In particular, we observed that for p(O_2)=0 a ≈80 nm-thick film could be deposited, but a thickness of only ≈5 nm could be obtained for p(O_2) ≥0.05 mbar.

For Si(100)-supported films, average nanocrystal sizes from 4 to 7 nm were obtained, indicating that nucleation was predominant over the subsequent aggregate growth. Such an effect could be related to the coexistence of ablation and growth processes in PE-CVD experiments (see above) and was further investigated by TEM analyses. As an example, a plane-view micrograph for a film deposited on Si(100) at 200°C, 40 W, 0.3 mbar O_2 partial pressure (see fig. 1A) is presented in fig. 2. For some of the clusters, (111) cerianite crystalline planes (interplanar distance = 0.31 nm) were clearly discerned. For the homologous sample on silica, TEM micrographs displayed the presence of smaller nanocrystals, most of them having a lower degree of structural order and mean dimensions of ≈2 nm. Such features, in agreement with XRD results, confirmed the substrate influence on nucleation and growth events from the vapor phase. In particular, it seems reasonable to suppose that the –OH groups, present in large amounts on silica, might have strongly favored the decomposition of Ce(dpm)$_4$ and related fragments, acting as nucleation sites [9]. As a matter of fact, the high hydroxyl content and the amorphous nature of silica determine a lower mean nanocrystal size and a lower microstructural order.

Figure 2. Plane-view TEM micrograph for a ceria film deposited on Si(100) (see fig. 1A). A ceria nanocrystal evidencing (111) planes is highlighted in the box.

The surface morphology of the films was investigated by AFM. As a general trend, a low surface roughness and a regular granular-like surface texture, without cracks or inhomogeneities, was observed for all the samples, irrespective of the synthesis conditions. A representative AFM micrograph for a film deposited on silica in Ar plasma at 300°C is reported in fig. 3. It is worthwhile pointing out that such a morphology might be difficult to attain by liquid-phase synthesis techniques, that often yield a non-uniform particle distribution and/or an appreciable surface corrugation [10].

The film chemical composition was investigated by XPS analyses. In order to study Ce oxidation states and to estimate the Ce(IV)/Ce(III) ratio, particular attention was devoted to the analysis of Ce3d band shape and Binding Energy (BE) position, that display appreciable modifications on going from pure Ce(III) to pure Ce(IV) oxides.

The typical Ce3d spectrum of Ce(IV) is characterized by three main $3d_{5/2}$ features centered at 882.8 (v), 889.3 (v'') and 898.8 eV (v'''), and three main $3d_{3/2}$ peaks centered at 901.3 (u), 907.8 (u'') and 917.0 eV (u'''). Such a complex band structure is generally ascribed to the hybridization of the Ce4f states with the O2p valence band in

Figure 3. Representative AFM micrograph (2×2 μm^2) for a ceria film deposited on SiO_2 in Ar plasma (substrate temperature=300°C, Ar flow rate=10 sccm, RF power=20 W).

the ground state [1,11]. Conversely, the Ce3d region for pure Ce(III) is dominated by four bands, with the $3d_{5/2}$ contributions at 880.9 (v_0) and 885.3 eV (v_1), and the related $3d_{3/2}$ peaks at 899.7 (u_0) and 903.8 eV (u_1) respectively; so, there is partial overlap with Ce(IV) oxide signals [1]. Due to this effect, no peak deconvolution was performed and the Ce(IV) percentage was evaluated by the ratio of the u''' satellite to the total Ce3d peak area, since the u''' intensity is directly related to the Ce(IV) amount [12].

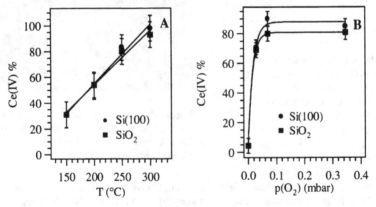

Figure 4. Dependence of Ce(IV) surface percentage: (**A**) on substrate temperature, for films deposited in Ar plasmas; (**B**) on O_2 partial pressure, for films deposited in Ar-O_2 plasmas.

Films deposited in Ar plasmas at 300°C displayed the typical Ce3d photopeak of Ce(IV) oxide. Lowering the substrate temperature resulted in a progressive decrease of Ce(IV) bands with a concomitant enhancement of Ce(III) ones, irrespective of the used substrate. Interestingly, the Ce(IV) content depended linearly on the substrate temperature (fig. 4A). This phenomenon might be ascribed to the competition between ablation and growth processes (see above) and, in particular, to the reducing action exerted on Ce by precursor-related fragments at the lowest deposition temperatures. An increase in the latter induced a progressive desorption of the above species and, in turn, a higher Ce(IV) amount.

Concerning films deposited in Ar-O_2 plasmas, an exponential dependence of Ce(IV) amount on O_2 partial pressure was observed (fig. 4B). The steep increase of Ce(IV) percentage might be ascribed to the high oxidizing power of reactive species (O_2^+, O^-, O, ...) [8], whose content is likely to increase with O_2 partial pressure. As discussed above, the same species are likely to be responsible for a reduction of film thickness on increasing $p(O_2)$.

As usually observed for metal oxides, for all samples the O1s surface peak could be fitted by three distinct components. The most intense was always detected at ≈529.5 eV and ascribed to lattice oxygen in a fluorite-type structure [11]. The other two components, located at higher BEs (≈531.5 eV and ≈533.2 eV) were attributed to the presence of hydrated/carbonated species and to adsorbed water respectively [11]. The second and third components disappeared after a mild Ar^+ sputtering. Furthermore, depth profiles confirmed that carbon presence was merely limited to the outermost sample layers, suggesting a clean conversion of the used precursor into ceria. Ce and O (fig. 5) were almost uniformly distributed throughout film thickness. The presence of a broadened film-substrate interface was confirmed by

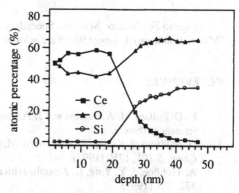

Figure 5. XPS depth profile (Ar^+, 3 kV) for a ceria film deposited on SiO_2 in Ar plasma (substrate temperature=300°C, Ar flow rate=10 sccm, RF power=20 W).

SIMS analyses and ascribed both to the continuous bombardment of ionic and neutral species during film growth [5,6,7] and to a knock-on effect induced by the primary ion beam [4]. Moreover, a careful inspection of SIMS depth profiles showed that Ce signal had a quite uniform yield at higher film thickness (>30 nm), while for thinner coatings (<20 nm) it presented an almost semigaussian shape, as expected for the convolution of a step-like film profile with the instrumental response.

CONCLUSIONS

CeO_2-based thin films were deposited by PE-CVD from $Ce(dpm)_4$ on silica and Si(100) substrates. Thanks to plasma activation, film growth was possible at temperatures (150°C) even lower than the one required for precursor vaporization (170°C). The concomitant occurrence of growth and ablation processes allowed the predominance of particle nucleation over the subsequent aggregate growth, enabling the synthesis of nanosized ceria thin films. Thin coatings with an excellent conformal coverage, a smooth and regular surface texture and average nanocrystal size lower than 5 nm were obtained. Moreover, the relative Ce(III)/Ce(IV) amount could be suitably tailored by controlled variations of the synthesis conditions, *i.e.* substrate temperature and O_2 partial pressure in Ar and Ar-O_2 plasmas respectively. Despite no compositional variations were induced by the used substrates, a significant influence of the growth surface on the precursor decomposition pathways and film microstructure was observed.

It is worthwhile highlighting that, to the best of our knowledge, no literature precedents on PE-CVD of CeO_2 thin films are available up to date.

ACKNOWLEDGEMENTS

Progetto Finalizzato "Materiali Speciali per Tecnologie Avanzate II" and Programma "Materiali Innovativi" Legge 95/95 are acknowledged for financial support.

REFERENCES

1. K. D. Pollard, H. A. Jenkins and J. Puddephatt, *Chem. Mater.* **12**, 701 (2000) and references therein.
2. G. Balducci, J. Kaspar, P. Fornasiero, M. Graziani, M. S. Islam and J. D. Gale, *J. Phys. Chem. B* **101**, 1750 (1997).
3. A. Tschöpe, J. Y. Ying, K. Amonlirdviman and M. L. Trudeau, *Mat. Res. Soc. Symp. Proc.* **351**, 251 (1994).
4. D. Barreca, A. Gasparotto, E. Tondello, C. Sada, S. Polizzi and A. Benedetti, *Chem. Vap. Deposition*, in press.
5. J. W. Lee, S. J. Pearton, C. R. Abernathy, W. S. Hobson and F. Ren, *Solid State Electronics* **39**, 1095 (1996).
6. Q. X. Guo, M. Matsuse, M. Nishio and H. Ogawa, *Jpn. J. Appl. Phys. Pt. 1* **39**, 5048 (2000).
7. J. E. Mahan, *Physical Vapor Deposition of Thin Films*, J. Wiley & Sons (Chichester, 2000).
8. D. W. Hess and D. B. Graves, in *Chemical Vapor Deposition: Principles and Applications*, edited by M. L. Hitchman & K. F. Jensen (Academic Press, London, 1993).
9. I. K. Igumenov, A. E. Turgambaeva and P. P. Semyannikov, *J. Phys. IV* **11**, Pr3-505 (2001).
10. F. B. Li and G. E. Thompson, *J. Electrochem. Soc.* **146**, 1809 (1999).
11. D. Barreca, G. A. Battiston, R. Gerbasi and E. Tondello, *Surf. Sci. Spectra* **7**, 297 (2000).
12. J. P. Holgado, G. Munuera, J. P. Espinòs and A. R. Gonzàlez-Elipe, *Appl. Surf. Sci.* **158**, 164 (2000).

Electrical Conductivity in Praseodymium-Cerium Oxide

Todd S. Stefanik and Harry L. Tuller
Crystal Physics and Electroceramics Laboratory
Department of Materials Science and Engineering, Massachusetts Institute of Technology
Cambridge, MA 02139, USA

ABSTRACT

The electrical conductivity of $Pr_xCe_{1-x}O_{2-\delta}$ (PCO) for $0 \leq x \leq 0.20$ was examined over a wide range of temperatures and oxygen partial pressures. A defect model based on multiple Pr valence states was found to be qualitatively consistent with the observed data. A unique pO_2-dependent ionic conductivity is observed at high pO_2 values in compositions containing low levels of Pr ($0 \leq x \leq 0.01$). In compositions containing higher amounts of Pr ($0.05 \leq x \leq 0.20$), formation of a Pr induced impurity band results in a significant electronic conductivity at high pO_2 values.

INTRODUCTION

Doped cerium oxide has received widespread attention owing to its high ionic conductivity at reduced temperatures and the drive to reduce solid oxide fuel cell operating temperatures below the useful limit of yttria stabilized zirconia.[1] Ceria-based electrolyte materials are typically doped with fixed valence acceptor ions such as Gd, Sm, and Y. Charge compensation is achieved through the creation of oxygen vacancies, both enhancing ionic conductivity and suppressing electron concentration. However, if a multivalent ion such as Pr is added to the ceria lattice, both electronic and ionic conductivity can be enhanced, yielding a mixed electronic-ionic conductor. Such materials are of interest in applications such as solid oxide fuel cell electrodes[2] and oxygen separation membranes.[3] The relative ease with which Pr reduces also allows for large oxygen nonstoichiometries in the PCO system.[4] This quality makes the material attractive as the sensing element in resonance-based mass sensitive sensors[5] and sorption compressors.[6] Preliminary studies in our labs have also shown PCO solid solutions to be interesting surface-effect sensor materials, potentially exhibiting enhanced selectivity over conventional sensing materials. Despite the interesting properties exhibited by PCO, relatively few studies of defects and transport in this system exist. The electrical conductivity of PCO was studied as a function of temperature in air,[2,7,8] but these studies provide little insight into the defect structure of the system. In a preliminary report, PCO with $0 \leq x \leq 0.05$ was examined as a function of temperature and pO_2 and a defect model was suggested.[5] In this study, measurements were extended to include a wider range of solid solutions and conditions of temperature and pO_2.

EXPERIMENTAL DETAILS

Powders of PCO ranging from 0 to 20% cation substitution of Ce by Pr were prepared via coprecipitation of nitrate salt solutions into oxalic acid. The resulting oxalate powders were washed, dried, and calcined to 700°C for 1 hour. The resulting powders were single phase and nanocrystalline as determined by X-ray peak broadening (crystallite size ~20nm). These powders were isostatically pressed into pellets at 275MPa then sintered to 1400°C. Higher Pr content samples cracked upon firing. It was found that firing under low vacuum (mechanical pump at approximately 5×10^{-2} Torr) helped prevent such cracking. The grain size after firing was approximately 5μm as determined by electron microscopy. Standard density measurements indicated densities of approximately 95% for low Pr content samples; 10 and 20% Pr samples did not densify as completely. Rectangular bars were cut from the sintered pellets and four point leads were painted onto the bars using Pt paint (Engelhard). Pt wires were wrapped over these electrodes and more Pt paint was applied to ensure good contact. All specimens were loaded into the sample furnace simultaneously to eliminate any measurement errors from one sample to the next. Each sample had two thermocouples attached to it, allowing for accurate temperature determination. Oxygen pressure was controlled using Ar/O_2 mixtures for high pO_2 and CO/CO_2 mixtures for low pO_2. Intermediate pO_2 values were obtained by electrochemically pumping oxygen into a flowing stream of 1000 ppm CO in CO_2, thereby increasing the otherwise low pO_2 of the buffer system. Four point DC resistance measurements were calculated from the slope on an IV sweep collected after conductivity had reached a static value at each pO_2 and temperature.

RESULTS AND DISCUSSION

Figure 1 shows representative conductivity isotherms (at 650°C) demonstrating the two general forms of behavior observed during this study. PCO behaves in a manner similar to other oxygen ion conductors, exhibiting a pO_2 independent conductivity plateau at intermediate pO_2 and increasing conductivity under reducing conditions. At

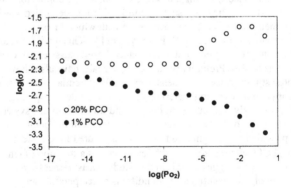

Figure 1. Conductivity of 1% and 20% $Pr_xCe_{1-x}O_{2-\delta}$ at 650°C

higher pO_2, however, PCO behaves quite differently. Depending on the Pr content, as illustrated in figure 1, the conductivity either drops steadily with increasing pO_2 (low Pr concentrations) or goes through a maximum then decreases (high Pr levels). See figure 1. In order to understand these observations, an examination of the defect chemistry of these materials is necessary.

The crystal structure of PCO is cubic fluorite. This structure is known to disorder by a Frenkel mechanism, such that

$$O_O^X \rightarrow V_O^{\bullet\bullet} + O_i^{\prime\prime}, \qquad [V_O^{\bullet\bullet}][O_i^{\prime\prime}] = K_F(T) \tag{1}$$

Generation of electronic defects is possible by the intrinsic ionization reaction:

$$nil \rightarrow e^{\prime} + h^{\bullet}, \qquad np = K_e(T) \tag{2}$$

Reduction of the host lattice may occur, yielding oxygen vacancies and charge compensating electrons:

$$O_O^X \rightarrow \tfrac{1}{2}O_2(g) + V_O^{\bullet\bullet} + 2e^{\prime}, \qquad [V_O^{\bullet\bullet}]n^2 pO_2^{\frac{1}{2}} = K_R(T) \tag{3}$$

Unlike fixed valence acceptors, Pr can exist in CeO_2 as either Pr^{4+} or Pr^{3+}. The two valences of Pr can be correlated by the following ionization reaction:

$$Pr_{Ce}^{\prime} \rightarrow Pr_{Ce}^{X} + e^{\prime}, \qquad [Pr_{Ce}^{X}]n/[\,Pr_{Ce}^{\prime}] = K_{Pr}(T) \tag{4}$$

The total amount of Pr in the system must exist as either Pr^{3+} or Pr^{4+} such that

$$Pr_{tot} = [Pr_{Ce}^{\prime}] + [\,Pr_{Ce}^{X}] \tag{5}$$

And finally, charge balance must be maintained:

$$2[O_i^{\prime\prime}] + n + [Pr_{Ce}^{\prime}] + [A^{\prime}] = p + 2[V_O^{\bullet\bullet}] \tag{6}$$

where $[A^{\prime}]$ is the concentration of background impurity acceptors in the material. While this presents a system of six equations and six unknowns, it is a complex system to work with. Since additions of Pr to CeO_2 enhance reduction in the material, it is reasonable to assume that interstitial oxygen and holes will be of low enough concentrations to be negligible. This simplifies the electroneutrality condition to:

$$n + [Pr_{Ce}^{\prime}] + [A^{\prime}] = 2[V_O^{\bullet\bullet}] \tag{7}$$

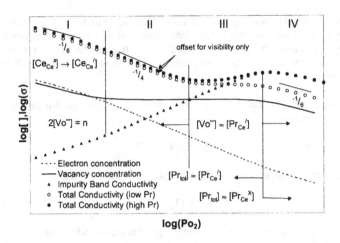

Figure 2. Predicted behavior of defect densities and electrical conductivity on oxygen partial pressure in the PCO system

Using the method of Porat and Tuller,[9] pO_2 may be solved for analytically in terms of n:

$$pO_2 = \left[\frac{2K_R}{n^2} \left(\frac{n + K_{Pr}}{(n + [A']) (n + K_{Pr}) + nPr_{tot}} \right) \right]^2 \qquad (8)$$

With these formulae, all defect concentrations can be solved as a function of pO_2 and temperature. However, appropriate values of the reaction constants and carrier mobilities must be available in order to calculate conductivity values to fit to the experimental data. Figure 2 shows the qualitative behavior of this defect model. At very low pO_2's (region I), cerium reduces to Ce^{3+} creating vacancies which are compensated for by electrons (n = $2[V_O^{\cdot\cdot}]$), yielding an electronic conductivity proportional to $pO_2^{-1/6}$. As pO_2 is increased, defect concentrations become dominated by acceptors ($[A_{Cc}'] + [Pr_{Cc}']$ = $2[V_O^{\cdot\cdot}]$, regions II and III). Acceptors include both Pr^{3+} and background impurities. For low concentrations of Pr and ignoring contributions from background acceptors, oxidation of Pr^{3+} to Pr^{4+} at high pO_2 results in a decrease in the concentration of oxygen vacancies and a corresponding decrease in ionic conductivity (region IV). This is a feature unique to systems with mutivalent dopants. In the case of fixed valence compounds, the ionic conductivity is independent of pO_2. Finally, in the case where Pr concentrations are sufficiently high, the discreet Pr levels broaden into a Pr derived impurity conducting band, resulting in elevated electronic conductivity at high pO_2. This contribution is expected to pas through a maximum when the concentration of Pr^{3+} and Pr^{4+} are equal (transition between regions II and IV). This type of impurity band contribution based on small polaron hopping is known to follow the form[10]:

$$\sigma = \frac{\sigma_0}{T} Nc(1-c)\exp\left(\frac{-E_A}{kT}\right) \qquad (9)$$

where N is the total density of atoms amongst which the electrons can hop, c is the fraction of N occupied by excess electrons (Pr^{3+}) and E_A is the hopping (migration) energy.

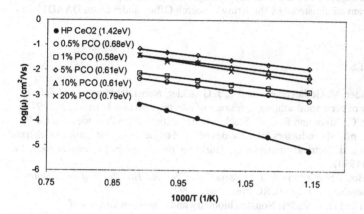

Figure 3. Ionic conductivity in PCO as a function of temperature

While the model qualitatively fits the observed conductivity data, quantitatively fitting the model remains a difficult task. There are many fitting parameters to be refined, including the cerium and praseodymium reduction coefficients, electron and ion mobilities, and impurity band mobility. Ionic mobility can be determined accurately since it is well defined in the plateau regime of each isotherm (especially at lower temperatures). Figure 3 shows the ionic conductivity as a function of temperature. It follows the expected Arrhenius behavior well, and the activation energies are in accord with data in the literature for doped ceria[11]. The highest ionic conductivity observed is in the 5 mol% Pr sample. At 800°C its ionic conductivity was 0.05 S/cm. Work is now in progress to fit the remaining parameters to the experimental data.

CONCLUSIONS

The conductivity of $Pr_xCe_{1-x}O_{2-\delta}$ for compositions ranging from 0 to 20mol% was examined. Two general forms of conductivity isotherms were identified and a defect model consistent with the observed trends in conductivity was devised. A pO2 dependent ionic conductivity was identified at high pO2's, and an impurity band conductivity was observed at sufficiently high Pr levels (5mol% and greater). Ionic mobility in the material follows the expected Arrhenius form and activation energies are consistent with those observed in other doped ceria systems. Further work is underway in order to quantitatively fit the model to the observed experimental data.

ACKNOWLEGEMENTS

This work was supported by the DoD Multidisciplinary University Research Initiative (MURI) program administered by the Army Research Office under Grant DAAD19-01-1-0566

REFERENCES

1. D. Schneider, M. Godickemeier and L. J. Gauckler, Nonstoichiometry and defect chemistry of ceria solid solutions, *Journal of Electroceramics*, **1**, 165-172 (1997).
2. M. Nauer, C. Ftikos and B. C. H. Steele, An evaluation of Ce-Pr oxides and Ce-Pr-Nb oxides mixed conductors for cathodes of solid oxide fuel cells: structure, thermal expansion and electrical conductivity, *Journal of the European Ceramic Society*, **14**, 493-499 (1994).
3. H. J. M. Bouwmeester and A. J. Burggraaf, in *The CRC Handbook of Solid State Electrochemistry* 481-534 (CRC Press, Inc., 1997).
4. P. Knauth and H. L. Tuller, Nonstoichiometry and relaxation kinetics of nanocrystalline mixed praseodymium-cerium oxide $Pr_{0.7}Ce_{0.3}O_{2-x}$, *Journal of the European Ceramic Society*, **19**, 831-836 (1998).
5. T. S. Stefanik and H. L. Tuller, Ceria-based gas sensors, *Journal of the European Ceramic Society*, **21**, 1967-1970 (2001).
6. J. A. Jones and G. D. Blue, Oxygen chemisorption compressor study for cryogenic Joule-Thompson refrigeration, *Journal of Spacecraft*, **25**, 202-208 (1988).
7. Y. Takasu, T. Sugino and Y. Matsuda, Electrical conductivity of praseodymia doped ceria, *Journal of Applied Electrochemistry*, **14**, 79-81 (1984).
8. P. Shuk and M. Greenblatt, Hydrothermal synthesis and properties of mixed conductors based on $Ce_{1-x}Pr_xO_{2-\delta}$ solid solutions, *Solid State Ionics*, **116**, 217-223 (1999).
9. O. Porat and H. L. Tuller, Simplified analytical treatment of defect equilibria: applications to oxides with multivalent dopants, *Journal of Electroceramics*, **1**, 41-49 (1997).
10. H. L. Tuller and A. S. Nowick, Small polaron electron transport in reduced CeO_2 single crystals, *Journal of Physics and Chemistry of Solids*, **38**, 859-867 (1976).
11. B. C. H. Steele, Appraisal of $Ce_{1-y}Gd_yO_{2-y/2}$ electrolytes for IT_SOFC operation at 500°C, *Solid State Ionics*, **129**, 95-110 (2000).

Mat. Res. Soc. Symp. Proc. Vol. 756 © 2003 Materials Research Society

Impedance and Mott-Schottky Analysis of a $Pr_{0.15}Ce_{0.85}O_{2-x}$ Solid Solution

R. Bouchet, P. Knauth, T. Stefanik*, H. L. Tuller*
MADIREL, Université de Provence-CNRS (UMR 6121), Centre Saint-Jérôme,
F-13397 Marseille Cedex 20, France
*Department of Materials Science and Engineering, Massachusetts Institute of Technology,
Cambridge, MA-02139, USA

ABSTRACT

The semiconductor properties of a praseodymium-cerium oxide solid solution with composition $Pr_{0.15}Ce_{0.85}O_{2-x}$ (PCO) were investigated by d.c. current-voltage and bias-dependent impedance measurements in aqueous solution. The solution data were compared with impedance values of dry cells in air. A Mott-Schottky analysis of the PCO-solution interface capacitance showed p-type semiconductivity, a flat-band potential $E_{fb} = (2.0 \pm 0.1)$ V/NHE and an ionized acceptor density $N_A = 3 \; 10^{17}$ cm^{-3}. Using these data, an electron hole mobility $\mu_h \approx 10^{-5}$ cm^2 V^{-1} s^{-1} was calculated pointing to a small polaron conduction mechanism with a hopping energy ($E_h = 0.4$ eV).

INTRODUCTION

Mixed ionic-electronic conducting (MIEC) electroceramics can be designed for various applications, including oxygen separation membranes, fuel cell electrodes, and automotive catalysts. Their function in 3-way automotive catalysts is the storage and release of oxygen during transients from reducing (fuel-rich) to oxidizing (fuel-lean) environments.[1]

Praseodymium-cerium oxides $(Pr,Ce)O_{2-y}$ (PCO) are materials with a high oxygen storage capacity, given the ready reduction-induced valence change of the praseodymium cations (Pr^{4+}/Pr^{3+}) even at elevated oxygen partial pressures.[2] The oxygen deficiency of several PCO solid solutions has been investigated using coulometric titration and was found to be extensive in agreement with the above expectation.[3,4,5] Measurements of the temperature and oxygen partial pressure dependence of the conductivity of PCO demonstrate a $p(O_2)$-dependent conductivity at high $p(O_2)$ which is predominantly ionic for Pr levels up to ~10 mol%. However, a minority electronic component becomes increasingly important as Pr levels increase due to the apparent formation of a Pr *impurity* band.[6] The relationship between Pr concentration in $(Pr,Ce)O_{2-y}$ and the effective electronic carrier density therefore becomes of interest.

This work is a part of a program for characterizing the defect and transport properties of PCO solid solutions.[3,4,5,6,7,8] Here, the semiconductor properties of polycrystalline ceramics of composition $Pr_{0.15}Ce_{0.85}O_{2-y}$ are investigated utilizing a Mott-Schottky analysis of the PCO/aqueous electrolyte junction. This analysis is capable of yielding the type and density of majority ionized dopants (acceptors or donors) and the flat-band potential of the semiconductor.[9] The bias-dependent impedance measurements in aqueous solution were complemented with a study of the temperature dependence of the complex impedance of the PCO specimen both in aqueous solution (290-360 K) and in air (290-420 K) to confirm the activated nature of electronic transport in PCO at these reduced temperatures.

EXPERIMENTAL

Powders were prepared by co-precipitation of Pr and Ce nitrate using oxalic acid solutions.[6,8] The precipitates were calcined at 700°C for 1 h and the resultant powders pressed into pellets and subsequently sintered at 1400°C for 10 h. The microcrystalline pellets were then back-contacted with silver paint and sealed into the electrode holder using epoxy resin (3M). The PCO *electrode* was then submerged into an aqueous solution containing 0.5 mol/L Na_2SO_4 at pH = 6, de-aerated by argon gas flow. Current-voltage measurements were made using a Solartron 1287 potentiostat and the bias-dependent impedance measurements were performed with a Solartron 1260 frequency response analyzer in combination with the 1287. For the impedance analysis, the amplitude was maintained between 0.02 and 0.05 V and the frequency ranged from 10^{-1} to 10^7 Hz. A saturated sulfate reference electrode (SSE: Hg/Hg_2SO_4, E = 0.642 V vs. normal hydrogen electrode (NHE)[10]) and a large area gold counter electrode were used for the experiments in solution. The gas phase measurements were performed in air using platinum paint contacts.

RESULTS

The d.c. current-voltage characteristics (Figure 1) are consistent with p-type behavior: i.e. the anodic current increases exponentially with increasing anodic bias, whereas the cathodic current remains small even for large cathodic bias.

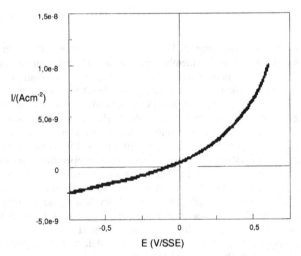

Figure 1. Current-voltage characteristics of PCO in aqueous solution at room temperature.

In the impedance spectra (Figure 2, under open circuit conditions) one distinguishes a small d.c. bias independent bulk response at high frequency, a response at intermediate frequency due apparently to blocking grain boundaries, and, at low frequency, a large strongly bias-dependent capacitive contribution of the PCO-electrolyte interface. The behavior was fitted using an equivalent circuit model consisting of three elements in series; a parallel resistor/constant phase element representing the bulk conductivity, another parallel resistor/constant phase element representing grain boundaries, and a lone constant phase element representing the blocking electrode/solution interface.

The bulk conductivity was calculated from the intercept of the high frequency arc with the real axis: at 302 K, a value $\sigma_{bulk} \approx 3 \ 10^{-7}$ S/cm was obtained. The temperature dependence between 290-360 K gave an activation energy $E_A = 0.4$ eV. Consistent with this conclusion was the derivation of the relative dielectric constant ($\varepsilon \approx 30$) from the high frequency capacitance for PCO which is close to the value reported for pure ceria[11]. Electrode corrosion was observed after extended polarization at high anodic potentials, leading to dissolution of praseodymium, as shown by ICP analysis of the solution. Therefore, an impedance study of PCO in air, using platinum contacts, was undertaken in order to check the reliability of the bulk data obtained in solution. Figure 3 shows an Arrhenius plot of the bulk resistance. The absolute values and activation energy ($E_A = 0.4$ eV) are in agreement with the solution data.

The PCO-solution interface capacitance was taken as the value of the corresponding constant phase element in the equivalent circuit (the low frequency component), the exponent of which was about 0.8. While a blocking capacitance should, in theory, result in an exponent of 1.0, electrode roughness decreases the observed value.[12] A Warburg impedance, indicative of a diffusive process, would have an exponent of 0.5, clearly not the case here. Figure 4 shows the bias dependence of the interface capacitance C in the Mott-Schottky representation.

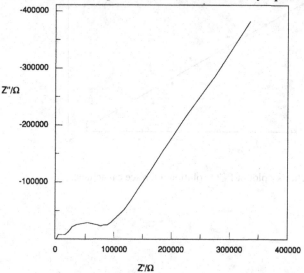

Figure 2. Typical impedance spectrum (Nyquist representation) under open circuit conditions of PCO specimen in aqueous solution (T=302K).

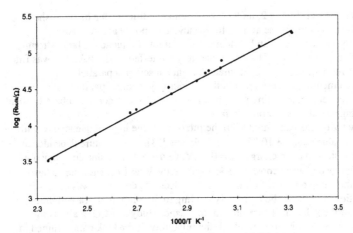

Figure 3. Arrhenius plot of the bulk resistance of PCO in air.

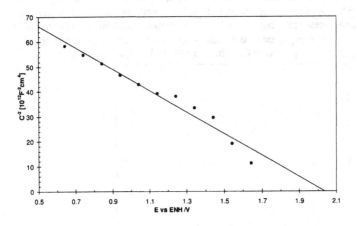

Figure 4. Mott-Schottky plot of PCO-solution interface capacitance.

DISCUSSION

According to the Mott-Schottky equation, the inverse square of C should depend linearly on the bias potential E as:[9]

$$1/C^2 = 2(E - E_{fb})/(\epsilon\epsilon_0 e_0 N_A) \qquad (1)$$

In this equation, E_{fb} is the flat-band potential, N_A is the effective acceptor concentration (for a p-type semiconductor), $\epsilon\epsilon_0$ is the dielectric permittivity and e_0 the elementary charge. From the sign and magnitude of the slope, one can conclude that PCO is a p-type semiconductor with an ionized acceptor density $N_A = 3 \times 10^{17}$ cm^{-3}. This value is markedly below the Pr concentration in PCO and therefore suggests that only a very small fraction of Pr is ionized to Pr^{3+}, while most remains Pr^{4+}. This is also consistent with expectations based on high temperature defect analysis in the PCO system.[6,8] According to recent results, the Pr level lies near mid gap in the forbidden gap of CeO$_2$.[8] A low population of electrons in the Pr level (i.e. Pr^{3+}) would be expected if the Fermi energy E_F were located sufficiently below the Pr energy level consistent with p-type behavior observed in this study. This would also be consistent with the PCO material oxidizing during cooling in air converting nearly all residual Pr^{3+} to Pr^{4+}, in agreement with high temperature defect analysis studies. Given its relatively high oxygen mobility[5], this appears to be feasible. The flat-band potential can be estimated from the intercept with the potential axis: E_{fb} = (2.0 ± 0.1) V/NHE. E_{fb} is about 6.5 V below the vacuum level, taking into account that potential of the normal hydrogen electrode is (4.5 ± 0.1) V below the vacuum level.[9] Although corrosion problems of the PCO electrode were observed at high anodic potentials, bulk impedance data obtained both in solution and in the gas phase are in good agreement. The electron hole mobility μ_h in PCO can be calculated using the relation between bulk conductivity σ_{bulk} and charge carrier density N_A:

$$\mu_h = \sigma_{bulk}/(e_0 N_A) \qquad (2)$$

Using the experimental data at 302 K, one obtains a hole mobility $\mu_h \approx 10^{-5}$ cm^{2}V^{-1}s^{-1}. This low magnitude of mobility and its activation energy for conduction are comparable to data for pure ceria[13], suggesting that a polaron-type mechanism applies as well to the PCO composition investigated here. It should be noted that in pure ceria, the electronic carrier was an electron in the Ce 4f band while here, the carrier is a hole presumably in the O 2p derived valence band.

ACKNOWLEDGMENT

The experimental help by Michal Schulz (now at University of Clausthal, Germany) is gratefully acknowledged. This work was supported in the framework of the CNRS-NSF agreement. T. Stefanik and H.L. Tuller acknowledge support from the NSF (Goalie program DMR97-01699) and the DoD Multidisciplinary University Research Initiative (MURI) program administered by the Army Research Office under Grant DAAD19-01-1-0566.

REFERENCES

[1] H. C. Yao, Yu Yao, *J. Catal.*, **86**, 254 (1984).

[2] D. Logan, M. Shelef, *J. Mater. Res.*, **9**, 468 (1994).

[3] O. Porat, H. L. Tuller, E. B. Lavik, Y.-M. Chiang, in *Nanophase and Nanocomposite Materials II*, S. Komarneni, J. Parker, H. Wollenberger, eds., Materials Research Society, Pittsburgh, 99 (1997).

[4] P. Knauth, H. L. Tuller, in *Electrochemistry of Glass and Ceramics*, S. K. Sundaram, D. F. Bickford, E. J. Hornyak, eds., American Ceramic Society, Westerville, 15 (1999).

[5] P. Knauth, H. L. Tuller, *J. Europ. Ceram. Soc.*, **19**, 831 (1999).

[6] T. Stefanik, H. L. Tuller, in *Solid State Ionics 2002*, P. Knauth, J.-M. Tarascon, E. Traversa, H. L. Tuller, eds., Materials Research Society, in press.

[7] O. Porat, H. L. Tuller, M. Shelef, E. M. Logothetis, in *Solid State Chemistry of Inorganic Materials*, P. K. Davies, A. J. Jacobson, C. C. Torardi, T. A. Vanderah, eds., Materials Research Society, Pittsburgh, 531 (1997).

[8] T. Stefanik and H.L. Tuller, *J. Eur. Ceram. Soc.*, **21**, 1967 (2001).

[9] A. J. Bard, L. R. Faulkner, *Electrochemical Methods – Fundamentals and Applications*, J. Wiley, New York (1980).

[10] A. J. Bard, R. Parsons, J. Jordan, *Standard Potentials in Aqueous Solutions*, Marcel Dekker, New York (1985).

[11] J. Lappalainen, D. Kek, and H. L. Tuller in *Electrically Based Microstructural Characterization III*, R. A. Gerhardt, A. Washabaugh, and M.A. Alim, eds., Materials Research Society, Warrendale, PA, R5.1.1-R5.1.11 (2002).

[12] J. R. Macdonald, ed, *Impedance Spectroscopy Emphasizing Solid Materials and Systems*, John Wiley & Sons, New York (1987)

[13] H. L. Tuller and A. S. Nowick, *J. Phys. Chem. Solids*, **38**, 859 (1977)

Mat. Res. Soc. Symp. Proc. Vol. 756 © 2003 Materials Research Society

Electrical Conductivity Prediction in Langasite for Optimized Microbalance Performance at Elevated Temperatures

Huankiat Seh[a,1], Harry Tuller[a], Holger Fritze[b]

[a] Crystal Physics and Electroceramics Laboratory, Department of Materials Science & Engineering, Massachusetts Institute of Technology, Cambridge MA 02139, USA

[b] Department of Physics, Metallurgy and Materials Science, Technische Universität Clausthal, D-38678 Clausthal-Zellerfeld, Germany

Abstract

The performance of the langasite-based crystal microbalance is limited due to reductions in its resistivity at high temperatures and reduced oxygen partial pressures. In this work, we utilize a recently developed defect model to predict the dependence of the ionic and electronic contributions to the total conductivity of langasite on temperature, oxygen partial pressure and acceptor and donor dopants. These results are used to select the type and concentrations of dopants expected to provide extended operating conditions for langasite-based gas sensors and crystal microbalances.

Introduction

Acoustic devices, due to their extreme sensitivity, are attractive choices for use as microbalances and for chemical and temperature sensors. The common quartz bulk acoustic wave resonators are used extensively in electrochemical studies [1-3] and also in various chemical sensors [4-6]. However, alternative materials are required for high temperature application given the destructive phase transformation that quartz undergoes at 573°C. Langasite, $La_3Ga_5SiO_{14}$ (LGS), emerges as an appealing substitute given its crystallographic stability up to its melting point of 1470°C, which, in principle, allows for operation up to that temperature. Remote high temperature operation of langasite-based SAW devices in a wireless passive mode for measuring high temperatures has been demonstrated [7]. Successful operation of a langasite oxygen sensor at temperatures as high as 600°C and as a nanobalance up to 900°C have also been demonstrated [8].

Due to the thermally activated nature of electrical conductivity in insulators such as LGS, electrical losses at elevated temperatures become greatly enhanced. These losses lead to decreased quality factors (Q) and, in turn, reduced sensor sensitivity [9]. In order to be in a position to optimize the electrical properties of LGS, studies have been initiated [10, 11] to characterize the electrical behavior of langasite at elevated temperatures and to prepare defect models based on these and related measurements. The defect model, in combination with mobility data allows one, in principal, to predict the electrical properties of langasite as functions of temperature, oxygen partial pressure and dopant levels. In this paper, we examine defect models recently developed for LGS based on preliminary experimental data as a means of predicting compositions likely to provide optimum microbalance performance at elevated temperatures.

[1] Corresponding author, Fax:+1-617-258 5748, Phone:+1-617-253 2364, Email: huankiat@mit.edu
Present address: Massachusetts Institute of Technology, Room 13-4010, 77 Massachusetts Ave, Cambridge MA 02139, USA.

Theory

A defect model based on oxygen vacancy compensation of acceptor dopants and of electrons induced during reduction has been derived and is discussed in some detail in other publications [10, 11]. The reactions and their reaction constants are tabulated in Table 1. In summary, at high and intermediate oxygen partial pressures, acceptors such as Sr^{+2} substituting on La^{+3} sites are largely compensated for by the formation of oxygen vacancies ($[2V_O^{..}] = [Sr_{La}']$) leading to mixed ionic-electronic conductivity (MIEC), as commonly observed in some fluorite (e.g. CeO_2 [12]) and perovskite ($SrTiO_3$ [13]) oxides. The minority electrons and holes, due to their higher mobilities, are expected to exhibit a conductivity dependence of the type $\log \sigma = (1/x)\log pO_2$ where $x = -4$ and $+4$ for electrons and holes respectively (see Table 2.). At sufficiently low pO_2, reduction of the oxygen lattice predominates ($[2V_O^{..}] = n$) and both vacancies and electrons follow $\log \sigma = (1/x)\log pO_2$ dependence with $x = -6$ (see Table 2.). Given the higher mobility of electrons, only the electronic contribution in this regime is normally observed.

For donor doped LGS (e.g. Nb^{5+} substituting onto Ga^{3+} sites), under somewhat oxidizing conditions, electrons are expected to compensate, e.g. $n=2[Nb_{Ga}^{..}]$ for full ionization. However, Nb being a deep donor (see below), only singly ionized Nb is considered here. This leads to a pO_2 insensitive electrical conductivity. Under reducing conditions, reduction of the oxygen lattice predominates, as above, and again a $pO_2^{-1/6}$ dependence is expected (see Table 3). Note, under sufficiently high pO_2's, another defect regime (Nb compensated by metal vacancies), n α $pO_2^{-1/4}$ is predicted but is not found to be relevant here and is therefore not considered further.

Table 1: Reactions and Reaction Constants

Reduction reaction	$O_O^x \leftrightarrow \tfrac{1}{2}O_2 + V_O^{..} + 2e'$	$K_R = pO_2^{\tfrac{1}{2}}[V_O^{..}]n^2$
Electron-hole generation	$null \leftrightarrow e' + h^{.}$	$K_i = np$
Dopant Ionization	$[Nb_{Ga}^x] \leftrightarrow e' + [Nb_{Ga}^{.}]$	$K_{ion} = n[Nb_{Ga}^{.}][Nb_{Ga}^x]^{-1}$

Table 2: Electronic Carriers Relationships for Sr-doped (acceptor) langasite.

	$n = 2[V_O^{..}]$	$[Sr_{La}'] = 2[V_O^{..}]$
$n =$	$2^{\tfrac{1}{3}}K_R(T)^{\tfrac{1}{3}}pO_2^{-\tfrac{1}{6}}$	$2^{\tfrac{1}{2}}K_R(T)^{\tfrac{1}{2}}pO_2^{-\tfrac{1}{4}}[Sr_{La}']^{-\tfrac{1}{2}}$
$p =$	$2^{-\tfrac{1}{3}}K_i(T)K_R(T)^{-\tfrac{1}{3}}pO_2^{+\tfrac{1}{6}}$	$2^{-\tfrac{1}{2}}K_i(T)K_R(T)^{-\tfrac{1}{2}}pO_2^{\tfrac{1}{2}}[Sr_{La}']^{\tfrac{1}{2}}$
	Low $\longleftarrow$	pO_2 $\longrightarrow$ High

Table 3: Electronic Carriers Relationships for Nb-doped (donor) langasite.

	$n = 2[V_O^{..}]$	$[Nb_{Ga}^{.}] = n$
$n =$	$2^{\tfrac{1}{3}}K_R(T)^{\tfrac{1}{3}}pO_2^{-\tfrac{1}{6}}$	$[Nb_{Ga}^{.}]$
$p =$	$2^{-\tfrac{1}{3}}K_i(T)K_R(T)^{-\tfrac{1}{3}}pO_2^{+\tfrac{1}{6}}$	$K_i(T)/[Nb_{Ga}^{.}]$
	Low $\longleftarrow$	pO_2 $\longrightarrow$ High

In all cases, the total conductivity is simply the sum of the electronic conductivity (both n and p contributions) and the ionic conductivity, i.e. $\sigma = \sigma_n + \sigma_p + \sigma_{ionic}$.

Experimental and Results

Mixed oxides in stoichiometric proportion were mixed, ball-milled and sintered as 1" pellets at 1450°C for 10hr. Densities of approximately 90% were obtained with a grain size on the order of 10 μm. X-ray diffraction showed the material to be langasite with no observable second phases. Pellets with effective cross sectional area of 6.55 mm^2 and length of 5.6mm were then electroded with platinum using platinum ink (Engelhard-CLAL). AC complex impedance measurements were conducted using a frequency response analyzer (Solartron 1260). The bulk conductivity values were extracted by fitting the spectra to the appropriate equivalent RC circuits. Samples were heated in a tube furnace to temperatures of 700-1000°C with the oxygen partial pressure controlled with Ar/O$_2$ (for high pO$_2$ range) and CO/CO$_2$ (for low pO$_2$ range) gas mixtures.

Figure 1 combines isotherms from nominally undoped, Sr (1 mol%) doped and Nb doped (5 mol%) langasite. In nominally undoped langasite, an oxygen partial pressure independent component is observed at higher pO$_2$, transitioning to n-type pO$_2$ dependent conductivity at lower pO$_2$ [9]. The intentional addition of the acceptor Sr increased the oxygen independent conductivity by an order of magnitude, suppressed the n-type conductivity and introduced a p-type component [11]. The addition of the Nb donor served to increase the n-type electronic conductivity compared to undoped langasite. Furthermore, n became proportional to pO$_2^{-1/6}$ at low pO$_2$ rather than pO$_2^{-1/4}$ as for undoped langasite [11].

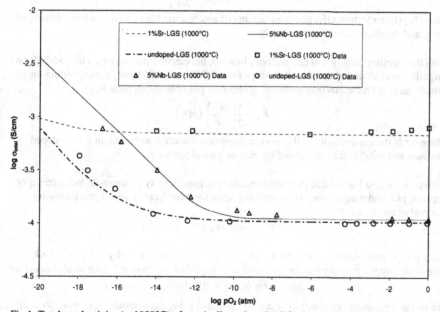

Fig 1: Total conductivity (at 1000°C) of nominally undoped, 1%Sr-doped and 5%Nb-doped langasite.

Discussion

The experimental results are in agreement with predictions of Table 2 and 3 and so key thermodynamic and kinetic data may be extracted. An activation energy of 1.17 eV, associated with ionic conduction by oxygen vacancies was obtained for conductivity in the pO_2-independent plateau region for 1%Sr-doped langasite [10]. From this data, one can estimate the mobility of oxygen vacancies (μ_i), assuming the oxygen vacancy density equals half the dopant concentration.

The pO_2-dependent conductivity of 5%Nb-doped langasite has an activation energy of 2.3eV [11]. This activation energy can be related to the reduction enthalpy defined by K_R in Table 2 (i.e. 2.3eV $\approx E_R/3$, where E_R is the reduction enthalpy). Note equating the reduction enthalpy to $E_R/3$ assumes a non-activated electron mobility. This result, together with that for nominally undoped langasite (where 3.0eV $\approx E_R/2$), gives a value of 6-6.9 eV for the reduction enthalpy E_R. The conductivity of 5%Nb-doped langasite in the oxygen independent regime, when plotted as a function of temperature, is found to have an activation energy of 0.9eV. This measured activation energy is assumed to be tied to the Nb ionization energy (Table 1) (i.e. 0.9eV = $E_{ion}/2$). To obtain the electron density in the 5%Nb-doped langasite at the oxygen independent conduction regime, we used the following equation [14]:

$$n = \sqrt{N_C N_{Nb}} \exp\left(-\frac{E_{ion}}{2kT}\right)$$

where N_C (free electron effective mass assumed) and N_{Nb} are the conduction band density of states and the donor density respectively.

With the electron density and conductivity in hand, the electron mobility (μ_e) in 5%Nb-doped langasite was calculated and found to range from 0.0051 to 0.0065 cm^2/V.s depending on temperature. With the electron mobility, it becomes possible to calculate K_R:

$$K_R = \frac{1}{2}\left(\frac{\sigma^*}{e\mu_e}\right)^3 \left(pO_2^*\right)^{\frac{1}{2}}$$

where σ^* is the conductivity in the oxygen dependent conduction regime in 5%Nb-doped langasite and pO_2^* is the corresponding oxygen partial pressure.

With μ_i, μ_e, n and K_R, we can proceed to predict the conductivity of langasite as functions of dopants, pO_2 and temperature. By combining what we have learned above, the following expression can be used:

$$\sigma = q(n\mu_e + p\mu_h + 2[V_O^{..}]\mu_i)$$

Since p-type conduction, even in Sr doped LGS, is found experimentally to be low at all accessible oxygen partial pressures, we utilize a simplified expression for the total conductivity:

$$\sigma \approx q(n\mu_e + 2[V_O^{..}]\mu_i)$$

The oxygen vacancies concentration can be estimated using the neutrality equation, $[A']=2[V_O^{..}]$, and n can be estimated by substituting K_R into suitable equations in Table 2. The detailed analysis will be provided in a future paper [15].

For the case of donor doped LGS (i.e. 5%Nb doped), since much of the data falls within the transition between $Nb_{Ga}^{\cdot}$ and $V_O^{\cdot \cdot}$ compensating for n, we solve for the electron density where both terms remain significant in the electroneutrality equation and obtain:

$$n^3 - [Nb_{Ga}^{\cdot}]n^2 - 2K_R pO_2^{-1/2} = 0$$

After solving for n, the conductivity can be determined by multiplying n by μ_e.

In modeling the predicted total conductivity, we make the simplifying assumption that carrier mobilities (electron and oxygen vacancy) are assumed independent of dopant level and type. Calculate values for the total conductivity are plotted in Fig 1 together with experimental data obtained at 1000°C for comparison. The agreement between experimental and calculated results is satisfactory. The model also allows us to predict the conductivity as a function of dopant level, temperature and oxygen partial pressure. As an example, we plot in Fig 2, the electrical conductivity as a function of net dopant level (difference between acceptor and donor levels) at 1000°C for two different oxygen partial pressures. As can be seen from the plot, to reduce the conductivity to a minimum at $pO_2 = 10^{-4}$ atm, we desire perfectly compensated langasite, i.e. [A]-[D] = 0. On the other hand, for $pO_2 = 10^{-18}$ atm, an excess of acceptor of approximately 0.1% results in a minimum in the overall conductivity.

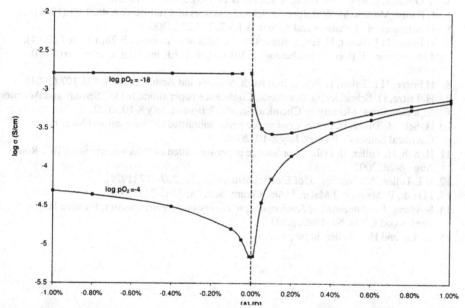

Fig 2: Predicted total conductivity as function of net dopant level, at 1000°C for two different oxygen partial pressures.

Conclusion

A model, enabling the prediction of the conductivity of langasite as a function of dopant level, temperature and oxygen partial pressure, was presented. Calculated conductivities based on the model gave satisfactory agreement to experimental data. Predicted values of total conductivity as function of net dopant concentration ($[A]-[D]$) for high and low oxygen partial pressures suggested different doping strategies to achieve minimum conductivities at high and low oxygen partial pressures. To reduce the conductivity to a minimum at $pO_2 = 10^{-4}$ atm, a perfectly compensated langasite is desired. On the other hand, for $pO_2 = 10^{-18}$ atm, an excess of acceptor of approximately 0.1% results in a minimum in the overall conductivity. These strategies will be followed to extend the operating ranges of langasite-based crystal microbalances and sensors.

Acknowledgement

This work was supported by the National Science Foundation under Grant Nos. DMR-9701699, DMR-0228787 and INT-9910012.

References

1. Z Tang, S Liu, E Wang, S Dong, Langmuir **16**(11), 4946-4952 (2000).
2. M Kunitake, Y Narikiyo, O Manabe, N Nakashima, J. Mater. Sci. **30**(9), 2338-2340 (1995).
3. Y Ohsawa, K Aoki, Sensors & Actuators B **14**(1-3), 556-557 (1993).
4. E Benes, M Groschl, W Burger, M Schmid, Sensors and Actuators A **48**, 1-21 (1995).
5. U Schramm et al, Sensors and Actuators B **67**, 219-226 (2000).
6. H Fritze, H L Tuller, H Seh, G Borchardt, Sensors and Actuators B **76**, 103-107 (2001).
7. J Hornsteiner, E Born, G Fischerauer, E Riha, Proc. IEEE Int. Freq. Control , 615-620 (1998).
8. H Fritze, H L Tuller, H Seh, G Borchardt, Sensors and Actuators B **76**, 103-107 (2001).
9. H. Fritze, O. Schneider, G. Borchardt, Conference paper submitted to "Sensors and Actuators B: International Meeting on Chemical Sensors", Boston, July 8-10, 2002.
10. H. Seh, H. Tuller, H. Fritze, Conference paper submitted to "International Meeting on Chemical Sensors", Boston, July 8-10, 2002.
11. H. Seh, H. Tuller, H. Fritze, Conference paper submitted to "Electroceramics VIII", Rome, Aug 26-28, 2002.
12. H L Tuller, A S Nowick, J. of Electrochem. Soc. **126**, 209-217 (1979).
13. I Denk, W Muench, J Maier, J. Am. Ceram. Soc. **78**, 3265-72 (1995).
14. S. Wang, *Fundamentals of Semiconductor Theory and Device Physics*, Prentice Hall, Englewood Cliffs, NJ, 1989, p.207
15. H. Seh and H.L. Tuller, in preparation.

Mat. Res. Soc. Symp. Proc. Vol. 756 © 2003 Materials Research Society   EE11.4

Thin Film Stoichiometry Determination by High Temperature Microbalance Technique

H. Fritze[1], H. Seh[2], O. Schneider[1], H. L. Tuller[2], G. Borchardt[1]
[1]Technische Universität Clausthal, Department of Physics, Metallurgy and Materials Science, Robert-Koch-Straße 42, D-38678 Clausthal-Zellerfeld, Germany, holger.fritze@tu-clausthal.de.
[2]Massachusetts Institute of Technology, Department of Materials Science & Engineering, 77 Massachusetts Avenue, Cambridge, MA 02139, U.S.A.

ABSTRACT

The in-situ determination of small mass changes of thin films became feasible with the availability of high temperature stable microbalances. With this technique, changes of the mechanical properties of thin films deposited on piezoelectric resonators are investigated at temperatures above 500 °C by monitoring the resonance behavior of the resonators. The results are valuable for fundamental understanding of the ionic and electronic transport processes in ceramic materials and for applications such as high temperature gas sensors.

This work correlates the electrical and the mechanical properties of TiO_{2-x} at different oxygen partial pressures. TiO_{2-x} films are deposited onto high temperature resonators by laser ablation and characterized by the high temperature microbalance technique as well as electrical impedance spectroscopy at 600 °C.

The oxygen partial pressure dependent resonance behavior cannot be attributed solely to mass changes of the TiO_{2-x} film. Changes of the film's mechanical stiffness have to be taken into consideration to explain the resonance behavior. The simultaneous electrical impedance measurements indicate a n-type conduction behavior of the TiO_{2-x} films.

INTRODUCTION

Piezoelectric materials such as gallium orthophosphate ($GaPO_4$), langasite ($La_3Ga_5SiO_{14}$, LGS) and related compounds are promising candidates for a wide range of new high temperature applications. The operation temperature of piezoelectric devices may, in principle, be extended up to the phase transformation at 970 and 1470 °C for $GaPO_4$ [1] and LGS [2], respectively. Our previous work demonstrated the suitability of LGS as bulk acoustic resonator up to 900 °C [3].

The application in high temperature microbalances (HTMBs) is of particular interest. Very small mass changes during (1) film deposition onto resonators or (2) gas composition dependent stoichiometry changes of thin films already deposited onto resonators can be correlated with the resonance behavior of bulk acoustic wave resonators. In addition, the mechanical properties such as density and stiffness can be extracted.

This work demonstrates the oxygen partial pressure dependent behavior of a TiO_{2-x} coated HTMB. Partial reduction of the TiO_{2-x} film is expected with decreasing oxygen partial pressure, p_{O2}. Therefore, the resonance behavior of the HTMB is governed by (1) the mass loss and (2) the decreased shear modulus of the partially reduced film.

EXPERIMENTAL

The resonators of the high temperature microbalance are prepared from polished LGS plates. They are cut from the same crystal boule and contacted on both sides with 200 nm thick key hole shaped platinum electrodes as denoted by electrode 1 and 2 in figure 1. Electrode 3 allows conductivity measurements on TiO_{2-x} films which are subsequently deposited onto some of the resonators. The deposition of the 120 nm thick TiO_{2-x} films is performed by pulsed laser ablation (LPX 325i, Lambda Physik, 20 ns pulse length).

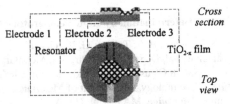

Figure 1. TiO_{2-x} coated langasite resonator with Pt electrodes.

For comparison, resonators with (TiO_{2-x} coated) and without TiO_{2-x} film (uncoated) are characterized simultaneously. Pairs of resonators are closely mounted (distance 5 mm) in an alumina support and inserted into a gas tight furnace. Oxygen is added to a 0.5 % H_2/Ar gas mixture using a zirconia oxygen ion pump (ZIROX GmbH). This allows to adjust p_{O2} in the range from 10^{-25} to 10^{-19} bar and at 10^{-4} bar. The total gas flow rate is 20 cm^3/min.

The experiments are performed at about 600 °C. Different gas compositions cause temperature changes of $\Delta T < 3$ K as determined by a thermocouple. The compensation of temperature effects is performed using equation (1) denoting X as the temperature dependent value.

$$X(T_0) = X(T) + \frac{dX}{dT}(T - T_0)$$ (1)

The temperature coefficients dX/dT result from temperature sweeps in the range from 595 to 605 °C and reverse. The values are oxygen partial pressure independent.

The resonance behavior of the LGS resonators is studied using a high speed network analyzer (HP E5100A) by monitoring the real and imaginary parts of the impedance spectra in the vicinity of the resonance frequency. The bulk properties of the resonators and the conductivity of the TiO_{2-x} films are determined by an impedance analyzer (Solartron 1260) in the range from 0.1 Hz to 2 MHz.

RESULTS AND DISCUSSION

Data evaluation

Phenomenologically, the resonance behavior can be described by the series resonance frequency and the damping of the resonator. Both properties are linked to the center frequency f_L and width w_L of the admittance peak as determinable by fitting to the Lorentz function.

The quantitative evaluation of resonators operating in transversal mode can be performed by applying a one-dimensional transmission-line model taking into consideration the material properties density ρ, shear modulus c, piezoelectric constant e and dielectric constant ε [4, 5]. To account for losses the material parameters have to be treated as complex quantities [6]. According to our investigations, the loss of the LGS resonators at high temperatures can be described satisfactorily by the mechanical and dielectric loss which are expressed by an effective viscosity η ($c" = 2\pi f\eta$) and a finite bulk resistance R_b.

Further, this one-dimensional physical model can be approximated by a modified Butterworth-van Dyke (BvD) equivalent circuit (figure 2) which corresponds to small acoustic phase shifts inside the resonator with respect to the situation without load [4, 5]. The conventional BvD circuit includes only the elements C_b, R_m, C_m and L_m.

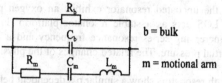

Figure 2. Extended Butterworth-van Dyke equivalent circuit.

Equations (2) explicitly connect the parameters from the equivalent circuit approach to those from the physical model, not considering the properties of the electrodes.

$$C_b = \varepsilon \frac{A}{d}, \quad C_m = \frac{8Ae^2}{N^2\pi^2 dc}, \quad L_m = \frac{\rho d^3}{8Ae^2}, \quad R_m = \frac{\eta}{cC_m} \tag{2}$$

A, d and N are the electrode area, the thickness of the resonator and the overtone, respectively. The dielectric constant $\varepsilon = \varepsilon_{22}$ applies for a y-cut resonator.

Conductivity of the TiO$_{2-x}$ film

The correlation of the electrical and mechanical properties of TiO$_{2-x}$ requires the knowledge of the TiO$_{2-x}$ film conductivity. Therefore, impedance measurements as function of the oxygen partial pressure are performed. Since electrode 1 and 3 (see figure 1) are used, the low frequency intercept of the RC semicircle in the complex impedance plane represents the resistivity of the parallel arrangement of TiO$_{2-x}$ film and LGS resonator. The conductivity of TiO$_{2-x}$ is estimated taking into account the conductivity and the dimensions of underlaying LGS. The results are shown in figure 3.

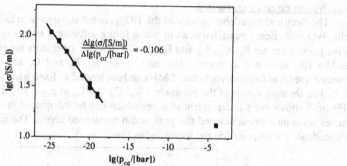

Figure 3. Oxygen partial pressure dependent conductivity of the TiO$_{2-x}$ film.

The negative slope in the lg-lg plot (figure 3) indicates a n-type conduction behavior below 10^{-19} bar. The mechanism corresponds to a partial reduction of the film and a mass loss.

Phenomenological resonance behavior

For comparison, the oxygen partial pressure dependent center frequency f_L and width w_L of the admittance peak are determined for TiO$_{2-x}$ coated and uncoated LGS resonators by fitting to the Lorentz function. Figure 4 shows the changes of these parameters in the oxygen partial pressure range from 10^{-25} to 10^{-4} bar. In the p_{O2} range from 10^{-19} to 10^{-4} bar the p_{O2} dependence is assumed to be linear.

Above 10^{-20} bar, the uncoated resonator exhibits an oxygen partial pressure independent behavior. Therefore, LGS acts as a stable resonator platform. In contrast, the TiO_{2-x} coated resonator shows an increasing series resonance frequency and a decreasing peak width with increasing oxygen partial pressure. The related changes of the physical parameters are discussed below.

Below 10^{-20} bar both resonators show a similar p_{O2} dependent behavior. The instability can be caused by changes of the LGS conductivity [7].

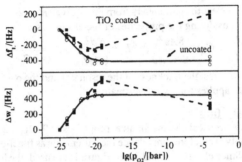

Figure 4. Oxygen partial pressure dependent changes of the center frequency f_L and width w_L of the admittance peak for the TiO_{2-x} coated and uncoated LGS resonator determined by fitting of the Lorentz function.

Equivalent circuit parameters

The electrical impedance spectra of the TiO_{2-x} coated and uncoated LGS have been fitted with the BvD equivalent circuit, using an in-house fitting software (Levenberg-Marquardt algorithm). Free parameters are R_m, C_m, L_m and R_L. The latter value represents the resistance of the sample holder (Pt wires and contacts). The bulk properties R_b and C_b are taken from impedance measurements at frequencies below 2 MHz and are treated as fixed parameters.

The absolute values of the parameters R_m, C_m and L_m at $p_{O2} = 10^{-25}$ bar are given in table 1. Figure 5 shows the p_{O2} dependent changes which can be interpreted in the same manner as the series resonance frequency and the peak width mentioned above. The implications of their p_{O2} dependence are discussed in the following section.

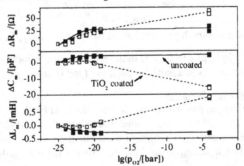

Figure 5. Oxygen partial pressure dependent changes of the parameters of the motional arm of the extended BvD equivalent circuit for the TiO_{2-x} coated and uncoated LGS resonator.

Table 1. Absolute values of the parameters of the equivalent circuit at $p_{O2} = 10^{-25}$ bar.

resonator	R_m [Ω]	C_m [pF]	L_m [mH]
TiO_{2-x} coated	592.8	308.0	19.6
uncoated	942.8	283.7	20.0

Physical properties

The material parameters effective shear modulus c, piezoelectric coefficient e and effective viscosity η are calculated using equations (2) and the equivalent circuit parameters. Figures 6 and 7 show the p_{O2} dependence of c and e for the TiO_{2-x} coated and uncoated resonators. The viscosity is not shown since both resonators behave in a similar way. For $p_{O2} > 10^{-20}$ bar, the viscosity is p_{O2} independent. The same statement is valid for the effective shear modulus and the piezoelectric coefficient of the uncoated resonator.

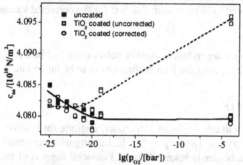

Figure 6. Oxygen partial pressure dependent changes of the effective stiffness for the TiO_{2-x} coated and uncoated LGS resonator. See text for the corrected values.

In contrast, the effective shear modulus of the TiO_{2-x} coated resonator is increasing with p_{O2}. This behavior can be described by applying p_{O2} dependent corrections of the effective density (or mass) or of the effective shear modulus in a manner that the corrected curves correspond to the undisturbed situation (i. e. to the uncoated resonator, solid line in figure 6). The correction for ρ (equation 3, left) corresponds to a density increase (or mass gain) with decreasing p_{O2}. However, the TiO_{2-x} film is partially reduced at low p_{O2}. Therefore, this correction cannot be valid.

$$\frac{d\rho}{d lg(p_{O2}/[bar])} = -1.1 \; \frac{kg}{m^3} \qquad \frac{dc}{d lg(p_{O2}/[bar])} = 8x10^6 \; \frac{N}{m^2} \qquad (3)$$

In contrast, the correction for c (equation 3, right) results in a valid statement. The effective shear modulus is decreasing with p_{O2} which corresponds to a soft (metal like) TiO_{2-x} film. It must be noted, that the approach cannot take into consideration mixed contributions of both effects.

The p_{O2} dependence of the piezoelectric coefficient shows a large difference (figure 7). An appropriate correction can be obtained by variation of the effective electrode area A, only.

$$\frac{dA}{d lg(p_{O2}/[bar])} = -1.5x10^{-7} \; m^2 \qquad (4)$$

A decrease in p_{O2} of 20 orders in magnitude yields according to equation 4 in an additional electrode area of 3 mm^2. Under low p_{O2}, the highly conductive TiO_{2-x} can be regarded as an extended electrode.

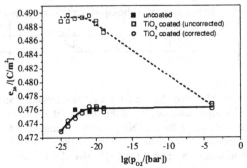

Figure 7. Oxygen partial pressure dependent changes of the effective piezoelectric constant for the TiO_{2-x} coated and uncoated LGS resonator. See text for the corrected values.

CONCLUSIONS

The oxygen partial pressure dependent resonance behavior cannot be attributed to mass changes of the TiO_{2-x} film alone. Changes of the film stiffness have to be taken into consideration to explain the resonance behavior.

ACKNOWLEDGMENTS

The authors thank Prof. Fukuda (Tohoku University, Institute for Materials Research, Japan) for providing the langasite crystal. The help of Mr. E. Ebeling with the design and machining of the samples and the sample holder is acknowledged. Financial support of the German Government (BMBF) and the German Science Foundation (DFG) made this work possible.

REFERENCES

1.   K. Jacobs, P. Hofmann, D. Klimm, J. Reichow, M. Schneider, "Structural Phase Transformations in Crystalline Gallium Orthophosphate", *J. Solid State Chem.* **149** (2000) 180-188.

2.   K. Shimamura, H. Takeda, T. Kohno, T. Fukuda, "Growth and Characterization of Lanthanum Gallium Silicate $La_3Ga_5SiO_{14}$ Single Crystals for Piezoelectric Applications", *J. Crystal Growth* **163** (1996) 388-392.

3.   H. Fritze, H. L. Tuller, H. Seh, G. Borchardt, "High Temperature Nanobalance Sensor Based on Langasite", *Sensors & Actuators B* **76** (2001) 103-107.

4.   R. Lucklum, P. Hauptmann, "Determination of Shear Modulus with Quartz Crystal Resonators", *Faraday Discuss.* **107** (1997) 123-140.

5.   C. Behling, Ph. D. Thesis, Otto-von-Guerike Universität Magdeburg, Germany, 1999.

6.   K. W. Kwok, H. L. W. Chan, C. L. Choy, "Evaluation of the Material Parameters of Piezoelectric Materials by Various Methods", *IEEE Trans. Ultrason. Ferroel. Freq. Contr.* 44 No. 4 (1997) 733-742.

7.   H. Seh, H. L. Tuller, H. Fritze, "Defect Properties of Langasite and Effects on BAW Gas Sensor Performance at High Temperatures", *Sensors and Actuators B* (in press).

Mat. Res. Soc. Symp. Proc. Vol. 756 © 2003 Materials Research Society   EE9.6

SENSOR RESPONSE OF STILBITE SINGLE CRYSTALS
UNDER "IN SITU" CONDITIONS

O. Schäf [a], H. Ghobarkar [b], P. Knauth [a]
[a] Laboratory MADIREL, University of Provence-CNRS (UMR 6121)
Centre St-Jérôme, 13397 Marseille Cedex 20, France.
[b] Free University of Berlin, Institute for Mineralogy, Berlin, Germany

ABSTRACT

The isothermal conductivity of natural stilbite single crystals depends on water content, polar organic molecule concentration and charge compensating cation species. The observed interaction processes are almost completely adsorptive at temperatures below 110°C; catalytic oxidation of the organic molecules is taking place at much higher temperatures, where zeolitic water is almost completely desorbed. A schematic model describing the observed conductivity modifications is developed.

INTRODUCTION

Zeolites are microporous materials with structure inherent channels and cage systems of sizes between ca. 0.2 to 1.4 nm [1,2]. Cations are present within the cavities for charge compensation, together with water and further polar molecules, depending on the synthesis conditions [3,4]. The water molecules are bound in a characteristic way: channel and cage size, framework charge, type and distribution of ions within the cavity system are some important factors affecting the conditions of zeolitic water bonding.

Zeolites are known to have specific catalytic, ion exchange and sorption properties [3,4]. While ion exchange takes place in aqueous solutions at temperatures mostly below 100°C, the zeolite lattice is "activated" for sorption purposes by heat treatment at temperatures above 250°C, where the maximum catalytic activity is also observed. At these temperatures, water molecules of the cavity systems are desorbed and irreversible changes of the zeolite framework lattices have often occurred [2,5,6].

Zeolites contain alumosilicate-based networks, showing ion conduction properties interesting for solid electrolyte application. However, in contrast to most conventional solid electrolytes, an open framework exists where not only the exchangeable charge compensating cations, but also adsorbed molecules with polar momentum, such as H_2O, are present. Earlier studies mostly investigated ionic motion in dehydrated zeolites: energies of defect formation and defect migration were deduced from Arrhenius plots of the electrical conductivity of chabazite (CHA) [7]. Cooperative motion by exchangeable cations in materials of different structure types was investigated [8]. The ionic conductivity of FAU-type zeolites with different Al-contents in the framework-lattice was measured and local ionic relaxation processes were observed, attributed to restricted local motion between neighboring sites [9]. Contributions of cation-framework interaction and cation-cation repulsion to the overall activation energies of FAU were investigated and activation energies for alkaline earth cations correlated with ionic radii [10]. Based on crystallographic considerations, a heterogeneous system of potential barriers caused by different sites for the cations and different occupancies for different cations was discussed. As the number of occupied sites is in the same order of magnitude as the unoccupied sites, a vacancy mechanism for cationic motion in FAU-type zeolites was postulated.

The aim of the present investigation is to check how the conduction properties of single crystal zeolites are changed by polar molecules, while a dynamic equilibrium between water in the vapor phase and water in the framework cavity system is established and to extend these investigations on zeolites in contact with liquid water. This is the foundation for developing sensors based on zeolite materials, which could be applied in the biomedical sensor, given their non-toxicity and biocompatibility.

Figure 1 shows a sketch of the crystal structure of the investigated zeolite stilbite (zeolite code STI: $NaCa_4[Al_9Si_{27}O_{72}]$ x $30H_2O$) [2]. This system presents a two-dimensional channel system along the [100] (0.49 x 0.61 nm) and [101] (0.27 x 0.56 nm) directions.

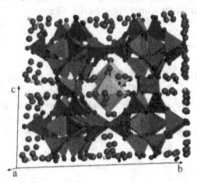

Fig. 1: View into the 10-membered ring channel system of the STI crystal structure (a-direction, single crystal data from [11]). Note that ionic radii are reduced to 20% for better visualization.

EXPERIMENTAL

Structural characterization

Natural single crystals of stilbite (STI) were visually selected. Phase purity was investigated by powder-X-ray diffraction (XRD, Philips), morphology and chemical composition by scanning electron microscopy (SEM, Leo) and energy disperse X-ray analysis (EDX) and water content by differential thermal analysis (DTA, Setaram).

Electrochemical measurements

Single crystals of STI (maximum 2 x 5 x 10 mm^3) were cut with a diamond saw. Sputtered Au electrodes of ca. 120 nm thickness were attached with Au plates and placed in a spring loaded ceramic sample holder. Two-point impedance measurements (EG&G) were made in a frequency range between 300 kHz and 50 mHz with a maximum amplitude of 20mV under a defined H_2O partial pressure using synthetic air (80.5%N_2, 20.5%O_2) as carrier gas. The water partial pressure was the respective vapor pressure, fixed by leading the gas stream through a special thermostatisized saturation bottle at different temperatures [12]. The H_2O partial pressure was verified via non disperse infrared spectroscopy (Fisher-Rosemount).

Preliminary investigations showed that at temperatures below 100°C only minor water loss of the zeolite channel system is observed. At about 80°C and 0.02 bar H_2O a sensitivity window of zeolite STI with respect to polar organic vapors exists [12]. This sensor effect was investigated using methanol, 2-propanol, and 3-pentanol as polar organic molecules with different size and dipole moment. For this purpose, the carrier gas stream was split into two equal parts of 60 SCCM (standard cubic centimeters) flow, led through two different thermostatized bottles for independent saturation with water and the respective alcohol. The partial pressures of water and liquid combustible were fixed by the thermostat temperature.

For measurements in the aqueous phase, the zeolite samples were attached with pure Ag-paint (Plano Co.). Epoxy resin DP 490 (Scotch 3M Co.) was used to separate the aqueous phase from the electrodes. Water proof and long term stability was investigated with dummy cells, using quartz crystals instead of zeolites [13].

Analysis of the gas phase

In order to determine the temperature ranges of adsorptive and catalytic interaction of organic vapors on zeolites, 50 mg of powdered STI with grain sizes smaller than 25μm were dispersed on inert quartz glass wool [14]. The gas stream (total flow: 100 SCCM) consisted of a stoichiometric mixture of N_2-O_2-carrier gas with 11 vol% O_2 and methanol vapor (7.2 vol%) obtained by gas saturation with special designed gas saturation bottles [12]. This mixture was led over the zeolite sample, while the temperature was changed in 10 K steps and the gas atmosphere was analyzed by FTIR (Biorad) after STI sample contact.

RESULTS

XRD indicated that all specimens were pure phases with the expected zeolite lattice (JCPDS 44-1479 for STI). EDX measurements exhibited no impurities, such as foreign channel cations, according to the accuracy of the method. The onset of water loss under the dynamic conditions of DTA analysis shows approximately the range of reversibility for in situ measurements [6]. Time dependent IS measurements under stationary conditions with water vapor showed that drift-free measurements are possible up to 110 °C [6, 12].

Figure 2 shows an Arrhenius plot of STI conductivity, measured in the gas phase with water vapor and in contact with liquid water. The activation energies for ionic motion are only slightly higher in the gas phase, while the total saturation of the channel system in liquid water leads to a slightly higher conductivity. Figure 3 shows the effect of water vapor and fig. 4 that of different organic molecules in the gas phase on the conductivity of single crystalline STI. With increasing $p(H_2O)$, the resistivity of the zeolite sample increases; polar organic molecules lead to a resistance decrease.

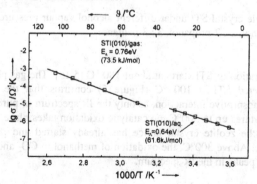

Fig. 2: Arrhenius plot of conductivity of single crystal STI. Left: contact with gas phase, right: contact with liquid water (complete saturation of the channel system; data from [13]).

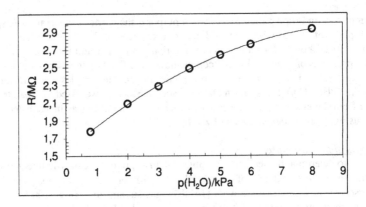

Fig. 3: Resistance of single crystal STI at different water vapor pressures.

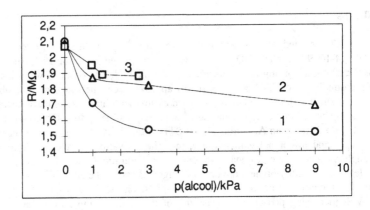

Fig. 4: Resistance of single crystal STI under different alcohol vapour pressures at constant $p(H_2O) = 2$ kPa. 1 : methanol, 2 : 2-propanol, 3 : 3-pentanol

Irreversible structural alteration of STI starts at about 180 °C [5,15]. The gas phase analysis after contact with powdered STI at 100 °C (Figure 5) confirms that the conductivity modification is based on adsorptive interaction, as only the IR spectrum of pure methanol is observed. Even at temperatures up to 300°C, no catalytic oxidation takes place, although the irreversible alteration of the zeolite crystal lattice has already started and proceeds with further rise of temperature. Above 300°C, the oxidation of methanol to CO_2 and H_2O begins as can be seen by the CO_2 peaks in the IR spectrum.

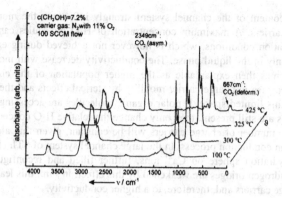

Fig. 5: IR-spectra of methanol after contact with powdered STI (grain fraction <25 μm) under oxidizing conditions. 100 °C: no catalytic oxidation; above 325 °C: formation of CO_2.

DISCUSSION

In the present gas phase investigations only slight deviations from complete filling of the zeolites channel systems with water molecules can be assumed. In the natural state, charge compensating Ca^{2+} ions occupy all possible sites and are coordinated with 8 water molecules forming a hydration sphere, Na^+ sites are only occupied for 22% for charge compensating reasons and do not show a complete hydration sphere. The effect of partial ion exchange by Cu and Ag on the activation energy of conductivity (0.76 eV after Cu exchange, but 0.54 eV after Ag exchange) was previously investigated by gas phase measurements in the temperature range 40-105°C. Na^+ ions do not take part in the ion exchange process and it can be concluded that they are rather immobile [12]. Kelemen et al. [16] stated that relatively immobile hydrated complexes are present in hydrated zeolites, and showed a relation between cation size, channel size and activation energy of conductivity. From literature and our experiments, it can be concluded that the presence of water molecules causes a sterical hindering for the cationic motion in hydrated zeolites.

Krogh-Andersen et al. [17] reviewed measurements on single and polycrystalline hydrated zeolites and concluded from their own transference number measurements that H^+ ions are the moving species during the conduction process. Resing et al. [18] concluded from NMR-relaxation investigations in FAU that the jumping rate of the water molecule at room temperature is 1/100 of the jumping rate of bulk water and that protons are at least 1000 times more mobile than the protons in ice at the melting point. They claimed zeolitic water to be an intracrystalline fluid with mobility between the crystalline and liquid state. Due to the high mobility of H_2O molecules, H^+-ion transport by a vehicle (Grotthuss) mechanism on H_2O of the hydration spheres is most likely in the present natural zeolite system. In this model, charge compensating cations are only considerably mobile during the conditions of cation exchange processes.

The conduction process is influenced by gaseous molecules accessing the zeolite void system. Effects of surface conduction are overcompensated by effects of the inner surface (channel and cage systems). In contrast to the effect of polar organic compounds, the isothermal resistivity of STI rises with increasing water vapor pressure. These experimental observations can be interpreted by taking a more precise look at the zeolite channel system (Figure 1). If only protons linked to water molecules contribute to the ionic conduction process, the water content of the channel system strongly influences the mobility and the number of charge carriers. A maximum concentration of H_2O molecules can be assumed under channel saturation conditions, which is however not achieved during experiments in the gas phase, but only in the liquid phase. The conductivity decrease with increasing water vapor pressure becomes then explainable as the higher population of the channel system causes a stronger decrease in charge carrier mobility by sterical effects and the formation of hydrogen bonds in this limited space. If polar organic molecules are accessing the channels while H_2O molecules are still present, they only change the relative H_2O concentration in the channel system. The number of charge carriers will be constant, given that alcohols do not contribute to a proton conduction process. In the large channel system of STI, H_2O molecules form a complete hydration sphere on Ca^{2+} ions, further filled and reconfigured by polar MeOH, while the hydrogen bridges are weakened by this effect. In sum, this leads to a higher mobility of the charge carriers and, therefore, to a higher conductivity.

ACKNOWLEDGMENT

The authors gratefully acknowledge the catalytic measurements done by F. Adolf.

REFERENCES

[1]   W.M. Meier, D.H. Olson and C. Baerlocher, eds. Atlas of Zeolite Structure Types, 4th rev. edn., Elsevier, London, Boston, Singapore, Sidney, Toronto, Wellington (1996)

[2]   G. Gottardi, E. Galli, Natural Zeolites. Springer, Heidelberg, N.Y., Tokyo (1985)

[3]   H.G. Karge, J. Weitkamp (eds.), Molecular Sieves, Science and Technology Vol. 1, Synthesis. Springer, Berlin, Heidelberg, N.Y. (1998)

[4]   J. Weitkamp, L. Puppe (eds.), Catalysis and Zeolites, Fundamentals and Application, Springer Verlag, Berlin, Heidelberg, N.Y., London, Paris, Tokyo (1999)

[5]   M.H. Simonot-Grange, Clays and Clay Minerals 27(6) (1979) 423-428

[6]   H. Ghobarkar, O. Schäf and U. Guth, Progr. Solid State Chem. **27** (1999) 29-73

[7]   I.R. Beattie, Trans. Faraday Soc. 50 (1954) 581-587

[8]   D.C. Freeman, D.N. Stamires, J. Chem. Physics 35(3) 1961) 799-806

[9]   F.J. Jansen, R. Schoonheydt, J. Chem. Soc. Farad. Trans. I (1973) 1338-1355

[10]   W.J. Mortier, R. Schoonheydt, Prog. Solid State Chem. 16 (1985) 1-125

[11]   E. Galli, Acta Cryst. B 27 (1971) 833-841

[12]   O. Schäf, H. Ghobarkar and U. Guth, Ionics 3 (1997) 282-288

[13]   O. Schäf, H. Ghobarkar, A.C. Steinbach, U. Guth, Fresenius J. Anal. Chem. 367 (2000) 388-392

[14]   O. Schäf, H.Ghobarkar, F. Adolf, P. Knauth, Solid State Ionics 143(3/4) (2001) 433-444

[15]   O. Schäf, H. Ghobarkar, Proceedings Conference on Basic Science and Advanced Technology, BSAT-II, Nov. 5-7, 2000 Assiut, Egypt, Vol.1, 35-64

[16]   G. Kelemen, W. Lortz, G. Schön, J. Mat. Sci. 24 (1989) 333-338

[17]   E. Krogh-Andersen, I.G. Krogh Andersen, E. Skou, Proton Conduction in Zeolites in: Proton Conductors, P. Columban (ed), Cambridge University Press, Cambridge (1992)

[18]   H.A. Resing, J.K. Thomson in: Molecular Sieve Zeolites I, M. Flanigen et al. (eds). Adv. Chem. Series 101, American Chemical Society (1971)

Mat. Res. Soc. Symp. Proc. Vol. 756 © 2003 Materials Research Society   EE3.16

PREPARATION AND CHARACTERIZATION
OF ANATASE (TiO₂) NANOCERAMICS

A.WEIBEL, R.BOUCHET, P.KNAUTH
Laboratory MADIREL, University of Provence-CNRS (UMR 6121)
Faculty of Sciences of Marseille St-Jérôme, 13397 Marseille Cedex 20, France.

ABSTRACT

We have systematically investigated the influence of various parameters (pressure, temperature, time, sample size) on the hot pressing process of TiO₂ nanoceramics. This study is backed by thermodilatometric experiments and microscopic observations of the microstructure. Impedance spectra of TiO₂ nanoceramics with different porosity are also presented.

INTRODUCTION

Nanocrystalline titanium dioxide is important for many applications, including dye-sensitized solar cells [1], photocatalytic decomposition of organic pollutants in waste water [2] and solid state gas sensors [3]. For all these applications, a well-defined, large surface area and porosity are important. Following many studies on the preparation of nanoporous and nanocrystalline thin layers of TiO₂ [4], a fundamental study of dense nanocrystalline TiO₂ is worthwhile to check the influence of a nanometric grain size on the electrical properties.

The usual technique for preparation of conventional ceramics, low temperature forming followed by high temperature sintering [5] is not applicable to nanocrystalline anatase ceramics, because the high temperature treatment leads invariably to large grain growth and phase transition [6]. The objective of nanocrystalline oxide ceramics is not attainable by this way. The method of choice for preparation of nanocrystalline ceramics is hot-pressing (HP). Here, compaction of nanocrystalline precursor powders and sintering are performed during the same treatment by application of a high pressure at moderate temperature. HP was previously explored for nanocrystalline ceria [7] and titania [8], but the influence of the experimental parameters was not fully investigated.

The objective of this paper is to present a study of density versus the physical parameters temperature, pressure and time for nanocrystalline TiO₂ (anatase) ceramics. Furthermore, the influence of the sample dimensions is explored and preliminary impedance spectra are presented.

EXPERIMENT

Pellets of different relative densities are obtained by HP phase-pure anatase powder (Bayer, Germany) calcined at 300°C for 1 hour. This powder is prepared by the sulfate process and is nominally undoped (only residual impurities from the preparation process are present, essentially Na⁺). The precursor particles have an average size of 17.5 nm, determined from X-ray diffraction (XRD) and BET adsorption measurements [9].

The hot-press is a prototype built in collaboration with Cyberstar S.A.; it permits to work up to 5 tons weight and temperatures up to 1100°C. This apparatus is equipped with a digital comparator Sylvac S229 with ±1µm precision, which allows to follow the dilatation of sample and sample holder. The dyes are from pure alumina with internal diameters of 4, 6 and 12.7 mm (SOTIMI S.A.). The dilatometric curves are obtained with a heating rate of 1 K/min. The HP procedure is to apply first the pressure and to increase temperature with a rate of 5 K/min. After reaching the desired temperature, the sample is held at this temperature for a certain time. The pressure is then released and the sample cools down at the intrinsic cooling rate of the hot press, in order to relax, at least partially, strains and stresses in the nanoceramics.

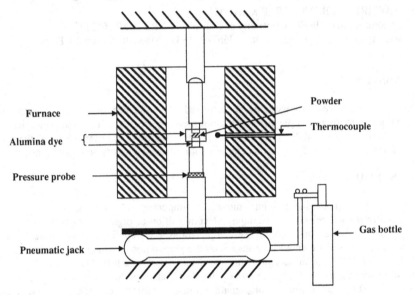

Figure 1. Schematic representation of the hot-press.

XRD was used to determine the phase purity of the samples and the mean grain size from Scherrer's equation, using the full width at half-maximum intensity of the (101) and (200) diffraction peaks of the anatase phase. These tests were carried out on a Siemens D5000 diffractometer with conventional Bragg-Brentano (θ-2θ) geometry and CuKα radiation (λ = 0.15406 nm).

The microstructural evolution following compaction was analyzed by high resolution Scanning Electron Microscopy. Electrical conductivities of two sample pellets with different porosity were measured in air at 600°C using a Solartron 1260 Frequency Response Analyzer. The impedance spectra were recorded with 50 mV a.c. amplitude in the range of 10^6 to 0.1 Hz using gold electrodes.

RESULTS AND DISCUSSION

Hot-pressing

Figure 2 shows the relative density, i.e. the ratio experimental/theoretical density (3,84 g/cm^3 for anatase), as a function of time for 3 different pressures at constant temperature (600°C)

We notice an important influence of the time for t < 120 min. As a consequence, we systematically applied a plateau time of 2 hours to study the role of temperature and pressure on sample densification.

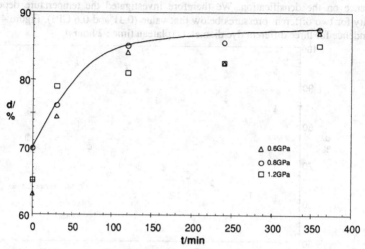

Figure 2. Time dependence of the density for three applied pressures at 600°C (6 mm dyes).

The following curve (figure 3) shows the pressure dependence of the relative density at constant temperature and time (600°C, 2 hours) for three dye diameters.

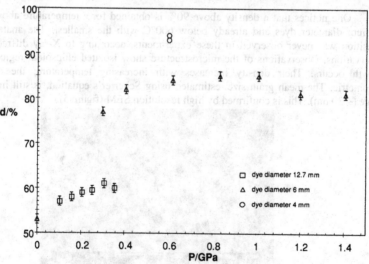

Figure 3. Pressure dependence of the relative density for three dye diameters at 600°C during 2h.

The first part of the curves has a similar shape; the relative density depends on the pressure following a power law with small exponent. A similar result was found by Jak using dynamic compaction of TiO_2 nanoparticles [10]. Above 0.6 GPa, the applied pressure has no major influence on the densification. We therefore investigated the temperature dependence of the density for two different pressures below that value (0.31 and 0.6 GPa). Figure 4 represents this dependence for three different dye diameters (plateau time : 2 hours).

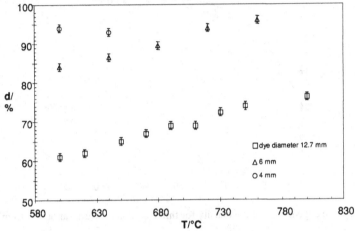

Figure 4. Temperature dependence of the relative density for three different dye diameters and two applied pressures (0.31 GPa for 12.7 mm and 0.6 GPa for 6 and 4 mm).

One notices that a density above 90% is obtained for a temperature above 680°C using medium diameter dyes and already below 600°C with the smallest. The anatase-rutile phase transition was never observed in these experiments, according to X-ray diffraction and SEM observations. Observations of the microstructure show isolated ellipsoidal regions where grain growth occurs. Their density increases with increasing temperature; the matrix remains nanometric. The mean grain size, estimated using Scherrer's equation, is still in the nanometer range (~ 30 nm). This is confirmed by high resolution SEM (figure 5).

Figure 5 : High resolution electron micrograph of a TiO_2 nanoceramic (92% relative density).

Dilatometry

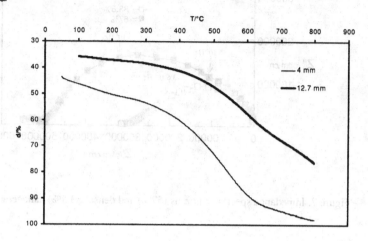

Figure 6. Comparison of dilatometric curves with two different dye diameters.

The dilatometric curve (Figure 6) shows directly sample density vs. temperature. For the smaller dyes, sintering begins at about 350°C and 90% density is already attained at 600°C, where pore shrinkage is observed and the slope changes. For the larger dyes, sintering starts at higher temperature ($\approx$ 450°C) and densification continues above 600°C. These results confirm the previously discussed temperature dependence of the density using different dye diameters.

Impedance spectra

Normalized impedance spectra of samples with considerably different density are shown in figure 7. The fitting of the spectra requires two (R//CPE) impedance elements for the porous sample and three (R//CPE) impedance elements for the dense one. The high frequency part represents the bulk response, as usual. The calculated bulk resistivity is the same for dense and porous samples, as expected. The low frequency response includes contributions by pores, for both samples, and grain boundaries, for the dense nanoceramic. The temperature and $P(O_2)$ dependence of the impedance is currently under investigation.

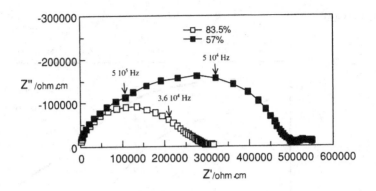

Figure 7. Impedance spectra of porous (57%) and dense (83.5%) nanoceramics.

REFERENCES

[1] Ch. J. Barbe, F. Arendse, P. Comte, M. Jirousek, F. Lenzmann, V. Shklover, M. Grätzel, J. Am. Ceram. Soc., **80**, 3157 (1997).

[2] R. M. Alberici, M. C. Canela, M. N. Eberlin, W. F. Jardim, Appl. Catalysis B, **30**, 389 (2001).

[3] E. Traversa, O. Schäf, E. DiBartolomeo, P. Knauth, in *Nanocrystalline Metals and Oxides – Selected Properties and Applications*, P. Knauth and J. Schoonman, ed., Kluwer, Boston, 2002, p. 189-207.

[4] T. M. Wang, S. K. Zheng, W. C. Hao, C. Wang, Surface and Coatings Technology, **155**, 141-145 (2002).

[5] W. D. Kingery, H. K. Bowe, D. R. Uhlmann, *Introduction to Ceramics*, ed., John Wiley & Sons, New York, 1976, p. 474.

[6] H.G. Kim, K.T. Kim, Acta Mat., **47**, 3570 (1999).

[7] Y. M. Chiang, E. B. Lavik, I. Kosacki, H. L. Tuller, J. Y. Ying, J. Electroceramics, **1**, 7-14 (1997).

[8] P. Knauth, H. L. Tuller, J. Appl. Phys., **85**, 897-902 (1999).

[9] P. Knauth, R. Bouchet, O. Schäf, A. Weibel, G. Auer, "Functionalized TiO_2 Nanoparticles for Pigments, Photoelectrochemistry, and Solid State Chemical Sensors" in *Synthesis, Functionalization and Surface Treatments of Nanoparticles*, M.-I. Baraton, ed., American Science Publ., Stevenson (2002).

[10] M. J. G. Jak, in *Nanostructured Materials : Selected Properties and Applications*, P. Knauth, J. Schoonman, ed., Kluwer, Boston, 2002, p. 66.

Phase Equilibria in the In_2O_3-WO_3 System

Annette P. Richard and Doreen D. Edwards
School of Ceramic Engineering and Material Science, Alfred University,
Alfred, NY, 14802, U.S.A.

ABSTRACT

The subsolidus phase relationships in the In_2O_3-WO_3 system at $800-1400^\circ C$ were studied by X-ray diffraction. Two binary oxide phases – $In_2(WO_4)_3$ and In_6WO_{12} – are stable in air over the temperature range of $800-1200^\circ C$. Preferential volatilization of WO_3 prevented the determination of phase equilibria above $1300^\circ C$.

INTRODUCTION

Numerous phases have been reported in the In_2O_3-WO_3 system, including several tungsten-bronze phases [1-5], $In_2(WO_4)_3$[6-8], and In_6WO_{12} [8-11]. Interest in many of these materials has arisen because of their similarity to rare-earth tungstates and indium-containing defect-fluorites and their potential use as optical and electronic materials [6-11]. $In_2(WO_4)_3$ is a trivalent ion conductor that is isostructural with $Sc_2(WO_4)_3$.[7] In_6WO_{12} has been investigated as possible electrochromic material.[8] While numerous phases have been reported in this system, the phase relationships in the system are not well understood. This report summarizes our efforts to determine the phase stability of the high-indium content compounds in this system.

EXPERIMENTAL

Samples were prepared with compositions ranging from $x_{In} = 0.16$ to $x_{In} = 0.98$, where x_{In} is defined on a cation basis, i.e. $x_{In} = [In]/ ([In]+[W])$. Most samples were prepared by solid-state reaction from commercially obtained powders: In_2O_3 (Indium Corporation of America) and WO_3 (>99.99% purity, Aldrich Chemical Co.). Pre-weighed amounts of powders were moistened with acetone and mechanically mixed in an alumina mortar and pestle. Samples with $x_{In} = 0.4$ were also produced using a solution method in which a 1M aqueous solution of $In(NO_3)_3$ (Indium Corporation of America) and a 1M aqueous solution of H_2WO_4 (Kodak) were mixed in a porcelain crucible at a cation ratio of 2In:3W. Samples prepared by the solution method were calcined at $750^\circ C$ to remove water and nitrate. Powders prepared by both methods '////were ground in an alumina mortar and pestle and then unniaxially pressed into 12.5 mm diameter pellets. Pellets were fired in alumina crucibles between $800^\circ C$ and $1200^\circ C$ at $100^\circ C$ increments for 50 hours. Pellets were placed atop a sacrificial pellet and covered with either a second sacrificial pellet or powders of the same composition to limit volatilization of the constituent oxides. After firing, the samples were quenched in dry air, massed and ground for phase identification through x-ray diffraction analysis. The samples were then re-pressed and re-fired until consecutive X-ray diffraction patterns were identical. Select samples were heated to $1400^\circ C$ in an attempt to identify the upper limits of phase stability.

X-ray diffraction analysis was conducted using a Phillips XRG 3100 diffractometer (Phillips Inc., USA) with Cu$K\alpha$ radiation (40kV, 20mA). Reacted powders were mounted on a zero-background holder for phase identification and lattice-parameter measurements. Silicon was used as an internal standard. Diffraction patterns were collected between 10°-70° 2Θ at a stepped scan rate of 0.04° per step for a count time of 4 seconds per step. Commercial software (Jade 6.0, Material Data, Inc.) was used to calculate the positions of the reflections with respect to the silicon standard and to determine lattice parameters via a least squares method. Figures 1 – 3 show as-collected x-ray diffraction data, i.e. the background has not been removed.

DISCUSSION

Figure 1 shows the x-ray diffraction patterns of samples prepared at 1000°C. Based on comparisons to powder diffraction files [12], four distinct phases are evident in the x-ray diffraction patterns, including WO_3, $In_2(WO_4)_3$, In_6WO_{12}, and In_2O_3.

A few samples within the composition range reported for indium tungsten bronzes, i.e. x_{In} ≤ 0.28, were prepared at 1000 °C and examined (Figure 1b). These samples were dark green in color, which is in direct contrast to literature reports which describe tungsten bronzes as dark blue with metallic luster [1-5]. The luster of samples prepared at $x_{In}<0.28$ increased as the In to W cation ratio increased, but the samples remained dark green. X-ray diffraction patterns of the samples made with $x_{In}\leq 0.28$ indicated biphasic mixtures WO_3 and $In_2(WO_4)_3$.

Samples prepared with $x_{In} = 0.33$ and heated at 1000°C, 1100°C, and 1200°C were white and green suggesting a biphasic mixture. With increasing temperature, the amount of the green

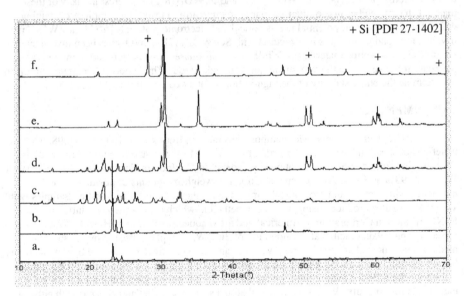

Figure 1. X-ray diffraction patterns of samples prepared in the pseudo-binary WO_3-$InO_{1.5}$ system at 1000°C: a. WO_3, b. $x_{In} = 0.28$, c. $x_{In} = 0.40$ or $In_2(WO_4)_3$, d. $x_{In} = 0.67$, e. $x_{In} = 0.86$ or In_6WO_{12}, f. In_2O_3.

phase around the edge of the pellet appeared to decrease. The x-ray diffraction patterns of these samples indicate the presence of $In_2(WO_4)_3$ (the white phase) as well as additional reflections, which were presumed to belong to WO_3.

Figure 2 chronicles the phase evolution of the samples prepared with $x_{In} = 0.40$. Samples at 800°C were yellow and green, suggesting the presence of more than one phase. The X-ray diffraction pattern for the 800°C sample shows peaks corresponding to $In_2(WO_4)_3$ (PDF # 49-0337) as well as several additional peaks which do not correspond to any known phases in the In_2O_3-WO_3 system. The yellow color is characteristic of unreacted In_2O_3, but this phase was not detected in the X-ray diffraction pattern.

Samples prepared at 1000°C and 1200°C were white with no apparent secondary phases. The X-ray diffraction patterns for samples prepared at 1000°C and 1200°C are generally consistent with PDF # 49-0337 for $In_2(WO_4)_3$, which reports reflections over the range 10 to 34° 2-Θ Cu$K\alpha$ [12]. Samples heated at 1300°C and 1400°C exhibited excessive weight loss, presumably due to the preferential vaporization of WO_3. Samples prepared at 1400°C consisted of a white interior pellet covered with a relatively thick layer of needle-like gray material. The x-ray diffraction patterns of the 1300°C sample and of the interior of the 1400°C sample indicated the presence of $In_2(WO_4)_3$ with no clear evidence of a secondary phase. X-ray diffraction analysis of the gray material on the 1400°C sample indicated In_6WO_{12}.

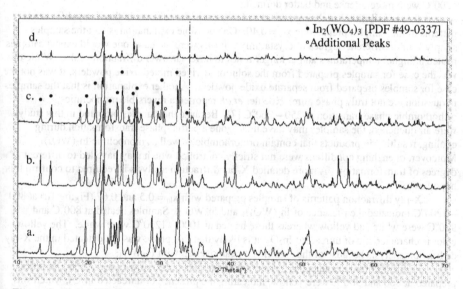

Figure 2. X-ray diffraction patterns of $In_2(WO_4)_3$ samples prepared by solid-state reaction at: a. 800°C, b. 1000°C, and c. 1200°C, and d. 1400°C.

While samples of $x_{In} = 0.40$ prepared at 1000°C and 1200°C have been identified as nominally phase-pure $In_2(WO_4)_3$, there are a number of features in the X-ray diffraction patterns that require further discussion. First, both patterns possess two well-defined, low-intensity peaks at ~25.5° and ~30.4° that are not reported in the powder diffraction file for $In_2(WO_4)_3$ or those of any other reported WO_3-In_2O_3 phases. Second, there are notable differences in the intensity and widths of the higher angle reflections of the two patterns.

Because the powder diffraction file PDF # 49-0337 only reports reflections for 10 to 34° 2-Θ (Cu$K\alpha$ radiation), it is difficult to comment on the significance of these differences. In general, most of the higher angle peaks are broad and ill defined, which is typical of samples that have not completely reacted or that are not well crystallized. However, prolonged heating did not result in better peak definition.

In an attempt to develop a better understanding of the $In_2(WO_4)_3$ phase, mixed oxide powders prepared using a solution method were reacted and equilibrated at temperatures ranging from 800 – 1200°C. The X-ray diffraction patterns of the samples prepared from the solution-derived mixed-oxide powders were similar to those prepared from separate oxide powders with two major exceptions. First, the sample prepared with the mixed-oxide powders and heated at 800°C did not have any evidence of unreacted In_2O_3, i.e. it was not yellow. Second, the X-ray diffraction patterns of samples heated at 800 and 1000°C contained broad, ill-defined peaks at higher angles, whereas the higher-angle reflections in the pattern of the sample prepared at 1200°C were more intense and better defined.

Two possibilities are being considered to explain the features in the X-ray diffraction patterns of the samples prepared at $x_{In} = 0.40$. One possible explanation is that the samples simply exhibit different degrees of crystallinity. If this were the case, one would expect samples heated at higher temperatures and at longer times to have narrower diffraction peaks. While this was the case for samples prepared from the solution-derived mixed oxide powders, it was not the case for samples prepared from separate oxide powders. Another explanation is that the samples in question are not truly phase pure. Koehler et al. reported a reversible monoclinic-orthorhombic phase transition at 250 – 260°C [7]. Because the samples prepared in this study were air quenched, the samples may have undergone a partial phase transformation during cooling, resulting in products that contain orthorhombic as well as monoclinic $In_2(WO_4)_3$. Moreover, quenching conditions were not strictly controlled which may have led to different degrees of transformation. A more detailed X-ray diffraction analysis is required to confirm this hypothesis.

X-ray diffraction patterns of samples prepared with $x_{In} = 0.5$ and 0.67 (Figure 1d) at 800 – 1200°C indicated the presence of $In_2(WO_4)_3$ and In_6WO_{12}. Samples heated at 800°C and 900°C were white and yellow whereas those heated at 1000 – 1200°C were white. The yellow color is characteristic of unreacted In_2O_3; it is likely that this phase was not detected in the X-ray diffraction analysis because of the significant peak overlap with In_6WO_{12}.

Figure 3 chronicles the phase evolution of samples prepared with $x_{In} = 0.86$. Samples prepared at 800 and 900°C were yellow, indicating the presence of unreacted In_2O_3, which was confirmed by X-ray diffraction. The samples heated at 1000-1200°C were white, and the corresponding X-ray diffraction patterns indicated the presence of In_6WO_{12}. The lattice parameters for the samples heated at 1200°C were determined to be a_R=6.2323 (1) and α= 99.029(3) [hexagonal axes: a_H=9.4803 (4) and c_H=8.9414 (4)], which is in agreement with the single-crystal lattice parameters reported by Michel and Kahn [9]. The X-ray diffraction patterns of samples heated at 1300°C and 1400°C showed evidence of In_2O_3 in addition to In_6WO_{12} and

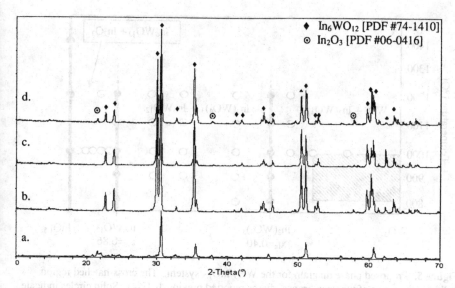

Figure 3. X-ray diffraction pattern of In_6WO_{12} powders fired at a. 800°C, b. 1000°C, c. 1200°C and d. 1300°C.

were greenish-white throughout the samples. These results as well as excessive weight loss at 1400°C again indicate the preferential volatilization of WO_3 above 1300°C. Samples equilibrated at 1200°C were annealed at 800 and 900°C for several days with no evidence of dissociation, indicating that In_6WO_{12} is stable at those temperatures.

Samples with $x_{In} = 0.92-0.98$ were prepared and characterized to determine whether or not W could be incorporated into the bixbyite structure to form an In_2O_3 solid solution. The cubic lattice parameter of the In_2O_3 phase in the sample prepared at $x_{In} = 0.92$ was measured to be 10.112 (4)Å compared to 10.152 (2)Å for pure In_2O_3. While this decrease in lattice parameter may indicate the substitution of some W^{6+} (r = 0.60 Å for octahedral coordination) for In^{3+} (r = 0.80Å for octahedral coordination) [13], the solubility of WO_3 in In_2O_3 is definitely less than 2% on a cation basis because In_6WO_{12} was noted in all of the x-ray diffraction patterns of samples prepared with $x_{In} = 0.92-0.98$.

Figure 5 is a proposed subsolidus phase diagram for the In_2O_3-WO_3 system based on our work and that reported for the indium-tungsten bronzes [1-4].

CONCLUSIONS

Subsolidus phase equilibria in the WO_3-$InO_{1.5}$ system were investigated by x-ray diffraction in the temperature range of 800 – 1400°C. Our work has shown that two binary oxide phases – $In_2(WO_4)_3$ and In_6WO_{12} – are stable in air over the temperature range of 800 – 1200°C. Preferential volatilization of WO_3 prevented the determination of phase equilibria above 1300°C.

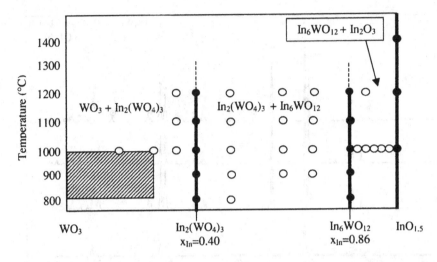

Figure 5. Proposed phase diagram for the WO_3-$InO_{1.5}$ system. The cross-hatched region indicates the range of tungsten bronze phases reported previously [2-4]. Solid circles indicate nominally phase-pure materials identified in this study. Open circles indicate biphasic samples identified in this study.

ACKNOWLEDGEMENTS

Annette Richard was supported through a graduate fellowship provided by the New York State College of Ceramics at Alfred University and the State University of New York.

REFERENCE

1. D. J. M. Bevan and P. Hagenmuller, *Non-Stoichiometric Compounds: Tungsten Bronzes, Vanadium Bronzes and Related Compounds*, Vol. 1 (Pergamon Press, Toronto, 1973) p. 542.
2. R. J. Bouchard and J. L. Gillson, Inorg. Chem.**7**, 969 (1968).
3. T. Ekstrøm, M. Parmentier, K. A. Watts, R. J. D. Tilley, J. Solid State Chem. **54**, 365 (1984).
4. Labbé *et. al.*, Acta Cryst. B. **35**, 1557 (1979).
5. S. K. Srivastava, J. Matter. Sci. Lett. **13**, 832(1994).
6. K Nassau, H. J. Levinstein and g. M. Loiacono, J. Phys. Chem. Solids. **26**, 1805 (1965).
7. J. Köhler, N. Imanaka, G. Adachi, Z. anorg. allg. Chem. **625**, 1890 (1999).
8. V.K. Trunov and L.M. Kovba, Vestn. Mosk. Univ. Ser. II. **22**, 91 (1967).
9. D. Michel and A. Kahn, Acta. Cryst. B. **38**, 1437 (1982).
10. W.S. Dabney, N.E. Antolino, B.S. Luisi, A.P. Richard, and D.D. Edwards, Thin Solid Films. **411**, 192 (2002).
11. T. Gaewdang, J.P. Chaminade, A. Garcia, C. Fouassier, M. Pouchard, P. Hagenmuller and B. Jacquier, Mater. Letters. **18**, 64 (1993).
12. Powder Diffraction File, International Centre for Diffraction Data, Newtown Square, PA, 19073-3273, USA.
13. R.D. Shannon, *Acta Cryst.*, **A32**, 751 (1976).

Cathode Materials for
Lithium Batteries

Mat. Res. Soc. Symp. Proc. Vol. 756 © 2003 Materials Research Society   EE7.1

Next Generation Positive Electrode Materials Enabled by Nanocomposites:
-Metal Fluorides-

F. Badway[a], N. Pereira[a,b], F. Cosandey[b], and G.G. Amatucci[a,z]

[a]Telcordia Technologies, Red Bank, NJ 07701 USA
[b]Rutgers University, Piscataway, NJ 08855 USA

[z]e-mail: gamatucc@rci.rutgers.edu

ABSTRACT

Through the use of nanostructures and nanocomposites, the electrochemical activity of metal fluoride materials was opened as potential candidates as next generation high energy density positive electrodes for Li batteries. This class of materials, utilizing FeF_3 as an example, is shown to exhibit good reversible behavior of approximately 200 mAh/g in the 3V region. The specific capacity is extended to 600 mAh/g when the discharge is extended to take into account the additional specific capacity associated with a 2V plateau. Through the use of XRD, SAED and high resolution TEM, the 2V reaction mechanism was associated to a reversible metal fluoride conversion mechanism. It is shown that LiF + Fe nanocomposite can be utilized as initial components in order to make the technology suitable for Li-ion applications. Although exhibiting relatively poor rate capabilities at this initial stage, reversible conversion metal fluorides enable for the first time the utilization of all the redox states of the constituent metal in a reversible manner in the positive electrode. This translates to 4X the specific capacity and double the energy density of today's state of the art $LiCoO_2$.

INTRODUCTION

The Li-ion battery is the premiere high-energy rechargeable energy storage technology of the present day. Unfortunately, its high performance still falls short of energy density goals in applications ranging from telecommunications to biomedical. Although a number of factors within the battery cell contribute to this performance parameter, the most crucial ones relate to how much energy can be stored in the positive and negative electrode materials of the device.

The positive electrode of Li-ion batteries is dominated by the layered Li intercalation compound, $LiCoO_2$[1]. $LiCoO_2$ has a practical reversible specific capacity of 150 mAh/g. Alternative electrode materials such as compounds and solid solutions consisting of $LiNiO_2$[2] and $LiMn_2O_4$[3][4] have been introduced in the past. Although the capacity of these materials do not exceed that of $LiCoO_2$ by a great extent, they are lower in cost and the latter is more environmentally acceptable. For the past decade there has been an extensive effort for search for new positive electrode materials. Current focus is on layered manganese compounds of the general formula $LiMnO_2$[5] and phosphate materials of the general formula $LiMePO_4$[6] and $Li_3Me_2(PO_4)_3$[7] where Me is a transition metal. Although operating at a lower voltage and close to the same capacity as present day $LiCoO_2$, these materials are of interest due to their low cost and safety. However, little new ground has been revealed in the quest for positive electrode materials of higher energy density.

Metal fluorides have been largely ignored as positive electrodes for lithium batteries. This is due to their insulative nature brought about by their characteristic large bandgap. Iron trifluoride (FeF$_3$) was first reported as showing limited electrochemical activity by Arai et al.[8] Arai reported a capacity of 80mAh/g from FeF$_3$ in a discharge voltage region from about 4.5V to 2V, involving the Fe^{3+} to Fe^{2+} redox transition. The poor electronic conductivity combined with a questionable ionic conductivity result in the disparity between 80mAh/g and the theoretical (1e$^-$ transfer) capacity of 237mAh/g. Since then, the fluorides have been discarded as useful positive electrode materials.

We have recently introduced the use of carbon metal fluoride nanocomposites (CMFNCs) to enable the electrochemical activity of metal fluorides[9]. We have shown over a 200% improvement (relative to Ref. 8) in the electrochemical activity of FeF$_3$ in the 4.5 to 2.5V region such that we could recover 99% of the FeF$_3$ theoretical capacity (235 vs. 237) in the 4.5 to 2.5V region with a total CMFNC specific capacity of approximately 200 mAh/g. The capacity and voltage were consistent with the reduction:

$$Li^+ + e^- + Fe^{3+}F_3 \rightarrow LiFe^{2+}F_3 \quad (1)$$

We also recently presented evidence that FeF$_3$ CMFNCs offered excellent reversible specific capacity through a second reaction occurring at 2V[10]. The combined specific capacities resulted in an exceptional total capacity of approximately 600 mAh/g. It was reported that the metal fluoride reaction was due to a reversible fluoride-based conversion reaction and could be utilized for positive electrode technology. Herein, we present an overview of this work.

EXPERIMENTAL:

CMFNCs were fabricated by the high energy milling of FeF$_3$ (Alfa), and either expanded graphite (Superior), carbon black (MMM Super P), or activated carbon (Norit A-supra). Stoichiometric mixtures (metal fluoride: C, 85:15 wt%) were placed inside a hardened steel milling cell along with hardened steel media. All cell assembly was done in He atmosphere. The milling was performed for the designated times in a high energy mill (Spex 8000). As formed CMFNC samples were removed from the milling cell in a He-filled glovebox. Unless otherwise noted, 85/15% FeF$_3$:C CMFNCs fabricated by high energy milling for 1h with activated carbon were utilized for general characterization.

The structure of the materials was characterized by x-ray diffraction (XRD), utilizing a Scintag X2 diffractometer with CuKα radiation. The material's microstructure was analyzed by transmission electron microscopy (TEM), utilizing a Topcon OO2B microscope. Dark field (DF) imaging techniques were used to image the size and distribution of the various phases. In addition, selected area electron diffraction patterns (SAED) from various areas were obtained to determine the structure of the phases present.

Electrodes for electrochemical characterization were prepared by adding Poly(vinylidene fluoride-co-hexafluoropropylene) (Kynar 2801, Elf Atochem), carbon black (Super P, 3M) and dibutyl phthalate (Aldrich) to the CMFNC in acetone. The slurry was tape cast, dried for 1 hour at 22ºC, and rinsed in 99.8% anhydrous ether (Aldrich) to extract the dibutyl phthalate plasticizer. The electrodes, 1cm^2 disks typically containing 57±1% CMFNC and 12±1% carbon black were tested electrochemically versus Li metal (Johnson Matthey). The Swagelock (in-house) or coin (NRC) cells were assembled in a He-filled dry box using Whatman GF/D glass fiber separators saturated with 1M LiPF$_6$ in ethyl carbonate: dimethyl carbonate (EC: DMC 1:1 in vol.) electrolyte (Merck). The cells were controlled by Mac-Pile (Biologic) or Maccor battery cycling system. Cells were cycled under a constant current of 7.58 mA/g at 22°C, unless noted otherwise.

RESULTS and DISCUSSION

a. Structure

X-ray patterns revealed a systematic broadening of the Bragg peaks of the FeF_3 as a function of milling time. Scherrer calculations showed a systematic decrease in primary crystallite size down to approximately 20nm within 1h (Fig. 1) from an original crystallite size in excess of 150nm. TEM analysis was utilized to determine whether the peak breadth was directly related to small crystallite size and whether a true nanocomposite formed. A representative bright field image of a FeF_3 based CMFNC with activated carbon is shown in Fig.2a. Much textural development was found to occur with domains of FeF_3 (determined by SAED) particles < 30nm but widely scattered from 1-20nm as evidenced by the dark field image of Fig. 2b. Many domains were very fine on the order of 1-3nm. Interspersed were areas of carbon and porosity (light regions in bright field, dark in dark field, Fig. 2a and b, respectively) thereby confirming the formation of a nanocomposite and supporting the dramatic reduction in crystallite sizes as determined by XRD.

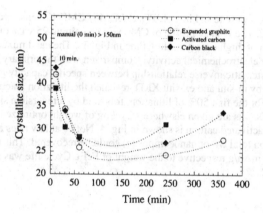

Fig. 1 Crystallite size as a function of milling time for FeF_3:C CMFNCs based on activated carbon (squares), carbon black (diamonds), expanded graphite (circles)

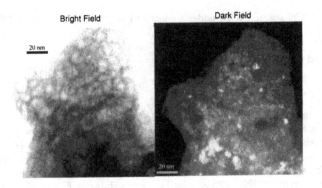

Bright Field   Dark Field

20 nm

Fig. 2 Bright field (a) and dark field (b)TEM images of 85/15 wt% FeF$_3$/ activated carbon CMFNC fabricated by high energy milling for 1h

b. Electrochemistry

i. 2.5-4.5V

The CMFNCs were tested in Li/LiPF$_6$ EC:DMC/ CMFNC cells, cycling from 2.5V to 4.5V at 7.58 mAh/g. The voltage profile of CMFNCs containing 25 wt% of carbon black is shown as a function of high energy milling time in Fig. 3a. The hand mixed FeF$_3$ materials exhibited very little electrochemical activity. Comparison with the primary crystallite size of Fig. 1 realizes a systematic inverse relationship between specific capacity and primary crystallite size. Initial studies by in-situ and ex-situ XRD revealed the lithiation reaction proceeded in a two phase manner for the first 50% of lithiation, followed by a reversible single phase topotactic insertion process. Details are given elswhere[9]. Cycling of weight optimized 85:15 wt% CMFNCs based on FeF$_3$ and activated carbon is shown in Fig. 4. Notice at low rates approximately 200mAh/g (based on total wt. of nanocomposite) could be recovered. This translates to approximately 235 mAh/g respective to the weight of FeF$_3$. Cycle life was shown to be respectable at 22°C.

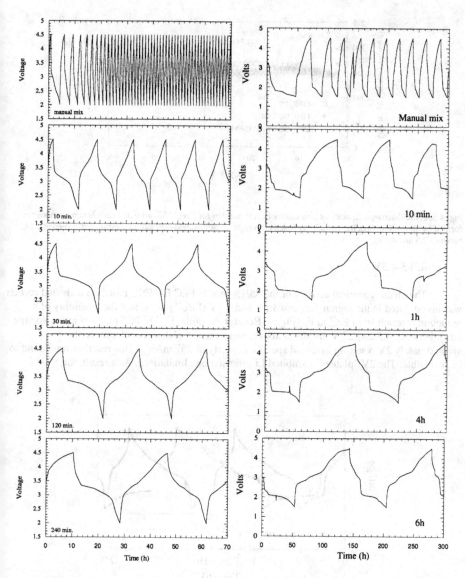

Fig. 3 Voltage vs. time for FeF3 based CMFNCs (a) fabricated by high energy milling with 25% carbon black and tested at 22°C at 7.58 mA/g from 2V to 4.5V and (b) fabricated by milling with 15% activated carbon and tested at 7.58 mA/g from 1.5V to 4.5V at 70°C.

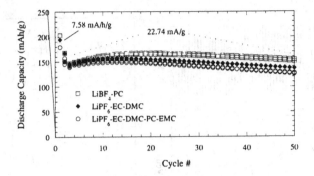

Fig. 4 Specific discharge capacity as a function of cycle number for FeF$_3$/ 15% activated carbon nanocomposites formed by 1h milling. Specific capacity is based on weight of nanocomposite material, testing performed at 22°C between 2.5 and 4.5V.

ii. 1.5-4.5V

The electrochemical activity of the 85/15 wt% FeF$_3$:C CMFNC based on activated carbon was investigated in the region between 4.5V and 1.5V (Li/Li$^+$) to explore the feasibility of a reduction reaction from Fe^{3+} to Fe (Fig.5). Besides the initial Fe^{3+} to Fe^{2+} reaction in the voltage range between 4.5 and 2.5V discussed above, there exists a plateau development at approximately 2V resulting in a total specific capacity of 367 mAh/g. This reaction was found to be reversible. The 2V "plateau" exhibited Li insertion rate limitations. As a result, the

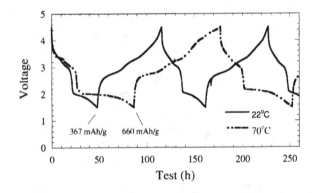

Fig. 5 Comparison of discharge/charge voltage profiles of 85:15 wt%FeF$_3$:C nanocomposites (1h activated carbon) at 7.58 mA/g.

electrochemical activity was investigated at 70°C to increase the kinetics of the reaction. At 70°C, the specific capacity of the CMFNC increased significantly to 660 mAh/g. Evidence clearly showed that the process was at least partially reversible with at least 2 distinct oxidation reactions occurring at approximately 3.0V. In addition, the voltage vs. capacity plot also showed that the 3.8-2.8V ($Fe^{3+} \rightarrow Fe^{2+}$) reduction reaction that was observed for the virgin FeF_3:C nanocomposite was reversible. This, combined with the coulombic efficiency of the charge process with respect to the discharge, suggest the complete reoxidation to Fe^{3+}.

The FeF_3 based CMFNC samples were initially discharged to 1.5V and subsequently charged to 4.5V at 70°C at 7.58mA/g. The effect of the milling time to form the CMFNC on the electrochemical behavior of the activated carbon-based CMFNC samples is shown in the representative voltage profiles of Fig. 3b. The manually-milled sample exhibited poor performance. A dramatic yet systematic increase in the electrochemical activity develops after 10 min. of milling to form the CMFNC, and reaches an optimum for the 1 and 4 h sample. Total CMFNC capacities after the third cycle exceeded 540 mAh/g (631 mAh/g of FeF_3) in a number of the samples. The development of the electrochemical activity paralleled the reduction in particle size (Fig. 1) in similar fashion to the 3V activity discussed above. Cycle life for FeF_3 CMFNCs milled for 6h is shown in Fig. 6. Reasonable reversibility was shown for the 6h milled carbon black sample at a reversible CMFNC specific capacity of approximately 540 mAh/g at 70°C.

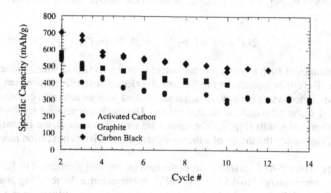

Fig. 6 Composite specific discharge and charge capacities for FeF_3:C CMFNCs formed by milling for 6h. Cells were cycled between 1.5 and 4.5V at 7.58mAh/g at 70°C.

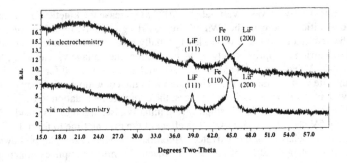

Fig. 7 XRD patterns of ex-situ analysis of reaction products post electrochemical and mechanochemical reduction of FeF_3:C CMFNCs with Li^+ or Li (see text).

Analysis of the discharge voltage profiles of Fig. 5 show approximately one third of the total capacity was recovered in the upper voltage area between 2.5 and 4.5V. This behavior is consistent with a 1 e^- process at high potential ($Fe^{3+} \rightarrow Fe^{2+}$) vs. the 2$e^-$ process at low potential ($Fe^{2+} \rightarrow Fe^0$). Analysis by ex-situ XRD of a FeF_3 CMFNC discharged to 1.5V at 70°C revealed two phases, LiF and Fe Fig. 7 This is consistent with a thermodynamic driven conversion reaction of the type:

$$2Li^+ + 2e^- + LiFe^{2+}F_3 \rightarrow 3LiF + Fe \ (2)$$

The thermodynamics of such a reaction is consistent with the voltage at which the reaction was found to occur. To further support the mechanism, the electrochemical reaction was replicated through the use of mechanochemistry. Li metal was milled with an activated carbon:FeF_3 CMFNC in a 3:1 molar ratio under a He atmosphere. The mechanochemical results are plotted vs. the electrochemical results (Fig. 7). The results are virtually identical, the formation of Fe and LiF, therefore, supporting the theory of a thermodynamically-driven conversion mechanism.

High sensitivity ex-situ SAED analysis was performed on one cell discharged to 1.5V ($Li_{2.58}FeF_3$) and one recharged to 4.5V ($Li_{0.596}FeF_3$), to characterize the resulting phases. The SAED pattern of the fully lithiated 1.5V sample (not shown) showed diffuse rings associated to the full conversion product Fe and LiF; d-spacings derived from SAED spectra taken from two fields of the 1.5V sample are shown in table 1. Field 1 had d-spacings which all agreed well with that of LiF and possibly small quantities of Fe. The second field had d-spacings which agreed well with that of Fe. Both results were consistent with that of the ex-situ powder XRD results and the conversion reaction of equation (2). The full recharge to 4.5V showed a development of crystallinity and crystallite size as represented by more spot development on the SAED pattern (not shown). Listing of the d-spacings of the SAED spectra from the sample recharged to 4.5V is shown in table 2 for two different fields of analysis. The peaks from both fields differ significantly from the fully discharged sample (table 1). Analysis shows they agree well with FeF_2.

The peak at approximately 2.0(3) Å was associated to a composite of residual Fe from lack of full conversion, LiF which would be left as a 1 mole constituent (see equation (3)), and the (210) Bragg reflection from FeF_2. The slight d-spacing mismatch combined with an attenuated (110) peak suggests that the FeF_2 - type phase may be a partially lithium substituted FeF_2 of $Li_xFe_{(1-x)}F_2$ origins.

$$3LiF + Fe \rightarrow FeF_2 + LiF \ (3)$$

1.5V Field 1	LiF - Fm3m	1.5V Field 2	Fe - Im3m
2.301	2.325 (111)		
2.011	2.013 (200)	2.014	2.027 (110)
1.420	1.424 (220)	1.429	1.433 (200)
1.207	1.214 (311)		
1.156	1.163 (222)	1.163	1.170 (211)
		1.001	1.013 (220)
		0.898	0.906 (310)

Table 1. d-spacings derived from SAED analysis of two fields of a FeF_3:C CMFNC discharged to 1.5V. JCPDS standards for LiF and Fe shown for reference.

4.5V Field 1	4.5V Field 2	FeF_2 - P42/mnm
		3.320 (110)
2.661	2.694	2.700 (101)
2.339	2.377	2.344 (111)
2.034	2.014	2.101 (210) LiF 2.013 (220) Fe 2.014 (110)
1.792		1.773 (211)
	1.706	1.660 (220)
1.530	1.541	1.488 (310) 1.482 (112)
1.433	1.433	1.417 (310)
1.219		1.213 (321)
1.165	1.160	1.173 (222)

Table 2. d-spacings derived from SAED analysis of two fields of a FeF_3:C CMFNC recharged to 4.5V. JCPDS standard for FeF_2 is shown for reference.

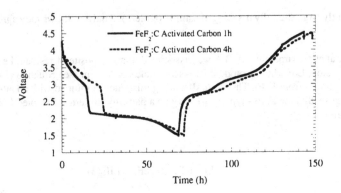

Fig. 8 Comparison of second cycle voltage curves for $FeF_3:C$ and $FeF_2:$ C CMFNCs showing distinct similarities. All cells were cycled at 7.58 mA/g at 70°C in $LiPF_6$ EC:DMC and were precharged to 4.5V

The formation of FeF_2 is consistent with a reversible metal fluoride conversion process. However, a discrepancy arises as subsequent cycles show that close to $3e^-$ are reversibly consumed as opposed to $2e^-$ that we expect for an FeF_2 phase. In addition, the reversible profile clearly shows the 3.5-2.8V region associated with the $Fe^{3+} \rightarrow Fe^{2+}$ reduction and the conversion at approximately 2V associated with the $Fe^{2+} \rightarrow Fe^{o}$ reduction. The former reduction reaction would not be expected to occur in $Fe^{2+}F_2$. We have found that to our surprise a similar behavior is shown for CMFNCs based on FeF_2 (Fig. 8).The basis of this reaction will be discussed in the a future publication.

The cathodic reduction of the FeF_3 material results in the products of LiF + Fe. A positive electrode comprised of a CMFNC of FeF_3 is only really suitable for use in Li based batteries containing negative electrode with prelithiated or lithium negative electrodes. Therefore, one could utilize the starting components of LiF + Fe as the positive electrode. In this case the reaction would begin on anodic oxidation. Fig. 9 shows a 3LiF/Fe : C nanocomposite positive electrode commencing on charge showing good electrochemical activity at 70°C.

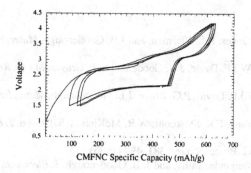

Fig. 9 Voltage vs. specific capacity curve for nanocomposite of 3LiF:Fe cycled at 7.48 mA/g between 1.5V and 4.2V at 70°C.

SUMMARY

The reaction mechanism we have proposed for these positive electrodes is similar to the reversible conversion reaction for negative electrodes first identified by Tarascon et al. [11] for oxides and sulfides and later on by our group for nitrides [12]. Assuming a Gibbs-free energy of $Fe^{2+}F_2$ to be approximately equivalent with that of $LiFe^{2+}F_3$ the calculation of the free energy of (2) is 2.44V. Experimentally, a slightly lower potential (approx. 2V) was found for the reaction to proceed. This may be nestled in the slow kinetics of the reaction as suggested by its temperature dependence.

Most FeF_3 CMFNC samples cycled in the 540 mAh/g range resulting in a corrected specific capacity of 621 mAh/(g FeF_3)). These capacities are approximately 400% greater than present day $LiCoO_2$. Even though average voltage is low (approx. 2.3V), energy densities based on the smaller 540 mAh/g specific capacity number result in 1242 Wh/kg, a factor of 2 greater than $LiCoO_2$'s 600 Wh/kg. This is the highest specific capacity non-sulfur rechargeable positive electrode reported to date. We have found that this reaction mechanism is prevalent among most of the transition metals when formed into nanocomposites. Although the numbers look promising there is a long path ahead in the development of such compounds for practical cells. Namely, major improvements must be made in kinetics and cycling stability of the conversion process and voltage profile. However, it is another beacon of hope in the continuing struggle to improve the energy densities of Li based systems in a non-evolutionary step.

ACKNOWLEDGEMENTS

The authors at Telcordia would like to sincerely thank the US Government for its support of this research project.

REFERENCES

[1] K. Mizushima, P.C. Jones, P.J. Wiseman, and J.B. Goodenough, *Mater. Res. Bull.*, **15**, 783 (1980)

[2] M.G.S.R. Thomas, W.I.F. David, J.B. Goodenough, P. Groves, *Mater. Res. Bull.*, **20**, 1137 (1985)

[3] M.M. Thackeray, W.I.F. David, P.G. Bruce, J.B. Goodenough, *Mater. Res. Bull.*, **18**, 461 (1983)

[4] J.M. Tarascon, E. Wang, F.K. Shokoohi, W.R. McKinnon, S. Colson, *J. Electrochem. Soc.* **138**, 2859 (1991)

[5] A.R. Armstrong, P.G. Bruce, *Nature*, **381**, 499 (1996)

[6] A. K. Padhi, K. S. Nanjundaswamy, and J. B. Goodenough, *J. Electrochem. Soc.*, **144**, 1188 (1997)

[7] A.K. Padhi, K.S. Nanjundaswamy, C. Masquelier, S. Okada, *J.B. Goodenough*, **144**, 1609 (1997)

[8] H. Arai, S. Okada, Y. Sakurai, J. Yamaki, *J. Pow. Sources*, **68**, 716 (1997)

[9] F. Badway, N. Pereira, F. Cosandey, G.G. Amatucci, Submitted *J. Electrochemical Soc.*

[10] F. Badway, F. Cosandey, N. Pereira, G.G. Amatucci, Submitted *J. Electrochem. Soc.*

[11] P. Poizot, S. Laruelle, S. Grugeon, L. Dupont, J.M. Tarascon, *Nature*, **407**, 496 (2000)

[12] N. Pereira, L.C. Klein, G.G. Amatucci, *J. Electrochem. Soc.*, **148**, A262 (2002)

Mat. Res. Soc. Symp. Proc. Vol. 756 © 2003 Materials Research Society   EE5.2

Surface properties of disordered γ MnO$_2$ at the solid-electrolyte interface.

Bénédicte Prélot[1,2], Christiane Poinsignon[*1], Fabien Thomas[2], Frédéric Villiéras[2]

[1] Laboratoire d'Electrochimie et de Physico-Chimie des Matériaux et des Interfaces-CNRS UMR 5631 INPG-ENS Electrochimie Electrométallurgie Grenoble, BP 75, F-38402 St Martin d'Heres Cedex

[2] Laboratoire Environnement et Minéralurgie CNRS UMR 7569 INPL-ENS Géologie, BP 40, F-54501 Vandoeuvre-les-Nancy Cedex.

ABSTRACT

Relationships between lattice parameters of manganese dioxides (γ/ε-MD) and their surface properties at the solid-aqueous solution interface were investigated. The studied series ranged from orthorhombic ramsdellite to tetragonal pyrolusite and encompassed disordered MD samples. The structural model used takes into account two structural defects which affect the orthorhombic network of ramsdellite: Pr (rate of pyrolusite intergrowth) and Tw (rate of microtwinning). Water adsorption isotherms showed that the cross sectional surface area of water molecules is linearly correlated to Pr: from 6.3 Å^2 (Pr=0.2) to 13.1 Å^2 (Pr=1). Titration of their surface charge evidenced a linear relationship between PZC and Pr starting from ramsdellite (Pr = 0, Tw = 0, PZC = 1) to pyrolusite (Pr = 1, Tw = 0, PZC = 7.3). γ-MD with intermediate values of Pr (0.2 to 0.45) have increasing PZC values. For similar Pr values (0.45), high Tw percentage (0.3 and 1) makes the PZC to increase. The experimental results are compared with data collected in the literature for dioxides of transition elements with tetragonal structure. Surface titration leads to the determination of electrochemically active surface area at alkaline pH.

INTRODUCTION

The most active manganese dioxides used in dry and alkaline batteries are the synthetic γ/ε-MD. The structural disorder, necessary to a good electrochemical activity [1], is described by two types of defects, which affect the orthorhombic network of ramsdellite: Pr, a random intergrowth of tetragonal pyrolusite structural blocks and Tw: microtwinning of this interspersed structure [2]. Both parameters quantify the electrochemical activity of these synthetic MDs [3]. To better understand their role on the reduction process, kinetics studies are carried out and lead to calculate the proton diffusivity. Then appears the difficulty to determine the electrochemically active surface area. Relationships between structural disorder parameters of γ/ε-MD and their surface properties at the solid-aqueous solution interfaces were investigated. The studied series ranged from ramsdellite to pyrolusite and encompassed disordered MD samples having an increasing percentage of structural defects.

EXPERIMENTAL DETAILS

Materials

Five MnO$_2$ samples were examined. Two synthetic pyrolusite samples with different unit grain size, respectively coarse-sized β-MnO$_2$ (Touzart and Matignon, France), and nanometric-sized "nano-β-MnO$_2$" obtained by spray-vapor deposition [4]. A synthetic ramsdellite (S-Ramsd), a Chemical Manganese Dioxide, "WSA" grade (CMD WSA), purchased from Sedema, an Electrodeposited Manganese Dioxide (EMD Delta) purchased from DELTA Company.

Analytic methods

Experimental methods are extensively described in an extended paper [5]. XRD patterns of the samples were recorded on a Siemens D5000 diffractometer equipped with a diffracted-beam monochromator. Specific surface areas were determined using the BET method, from nitrogen adsorption isotherms at 77K . Water vapor adsorption was performed at 30°C on a home-made gravimetric apparatus [6]. The adsorption isotherms were exploited by the BET method to calculate the monolayer capacity of the samples. A statistical cross-sectional area of adsorbed water molecules was calculated according to the N_2-BET surface area. The titration set-up is described in [5]. The surface charge was calculated by difference in consumed titrant between the MD suspension and the electrolyte solution.

RESULTS AND DISCUSSION

Characterization of the samples

XRD patterns of the studied samples are presented on Figure 1. Pr and Tw are calculated among the protocol defined by Pannetier [2] and presented in Table I.

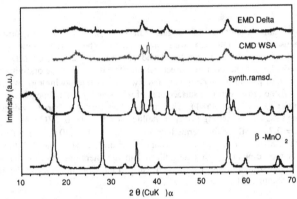

Figure 1. XRD patterns of synthetic ramsdellite, β-MnO_2, CMD WSA and EMD Delta.

The CMD and EMD samples show differences only in their rate of microtwinning. The end-members pyrolusite and natural ramsdellite are the references for Pr and Tw. The two pyrolusite samples displayed very different surface areas, according to their unit particle sizes, respectively 4.0 61.2 m^2 g^{-1} for β-MnO_2 and nano-β-MnO_2.

The calculated mean cross sectional areas of adsorbed water are linearly correlated with Pr (Fig. 2). The walue calculated on pyrolusite, 13.1 $Å^2$ (Table I) falls within commonly measured values ranging from 10.6 to 14.8 $Å^2$ [10]. The surface of pyrolusite presents very narrow channels of one (MnO_6) octahedron width and depth, which can not accommodate more than 1 water molecule. Furthermore sp2 hybridization confers the apical oxygens in the channels a hydrophobic character. Water molecules may then experience only physisorption involving weak normal and lateral interaction, thus allowing for high molecular area. In contrast, rhamsdellite containing samples displayed unusually low molecular areas, suggesting that adsorbed water molecules experience strong intermolecular interaction or strong normal interaction with the surface. On ramsdellite, the width (on the 100 faces) or depth (on the 010 faces) of the surface structural channels is equivalent to two octahedra: 2 to 3 water molecules can be accommodated in such space, which results in narrow confinement and strong intermolecular interactions. Furthermore, the oxygens in the ramsdellite channels are characterized by two types of hybridization, sp2 (planar) and sp3 (pyramidal) [11]. Water molecules are likely to be dissociated when adsorbed in the first layer, which would explain their extremely low average molecular area on ramsdellite. The two samples CMD WSA and EMD Delta of intermediate structure, present intermediate cross section of water molecules, compared with the end-members pyrolusite and ramsdellite. It appears thus that the surface structure of adsorbed water molecules is dominated by the rate of pyrolusite intergrowth clearly shown in Figure 2.

Table I: Structural and surface parameters of the MD samples, compared with literature data.

Material	System	Pr	Tw	Surface area (m^2.g^{-1})	Area water* (Å^2)	PZC	Additional references
natural ramsdellite	Orth.	0	0	n.d.	n.d.	nd	[7]
Synthetic-ramsdellite	Orth.	0.20	0.20	15.2	6.3	<2.5	[8]
CMD WSA	Orth.	0.45	0.35	40.4	9.8	3.7	[2]
EMD Delta	Hexag.	0.45	1	38.7	8.6	6.7	-
β-MnO$_2$ Pyrolusite/nano-β	Tetrag.	1	0	4.0/61.2	13.1	7.5/7.3	[9]
IC1 EMD γ	Hexag.	nd	nd	nd	nd	4.15	[12]
IC12 EMD γ	Hexag.	nd	nd	nd	nd	3.76	[12]
IC22 EMD γ	Hexag.	nd	nd	nd	nd	4.72	[12]

* Mean cross-sectional area of adsorbed water molecules at the monolayer capacity
Orth.: orthorhombic, Hexag: hexagonal, Tetrag.: tetragonal, nd: not determined
The high percentage of microtwinning in the EMD Delta sample requires indexation of the pattern in a pseudo hexagonal symmetry.

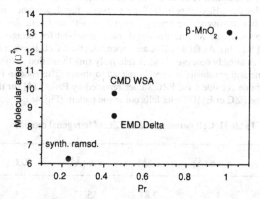

Figure 2. Dependence of the cross sectional area of adsorbed water molecules on Pr.

Relationships between PZC and structural-chemical disorder

From the curves on Figure 3, it appears that the samples display different PZC over a broad pH range, as already shown in literature [18,19]. The corresponding points of zero charge (PZC) are given in Table I. The end-members ramsdellite and pyrolusite display respectively the lowest and highest PZC, whereas the CMD and EMD display intermediate values.

If one refers to the successive models predicting the PZC of oxides, it appears that two directions were basically investigated. On one hand, it has been shown that the value of the PZC is correlated with intrinsic properties of the metal atom such as ionization potential or electronegativity [12, 13]. On the other hand, following the formal bond valence [14], atomic distance parameters were combined with the charging properties of the metal ion to predict the position of the PZC of oxides[15-18]. Healy et al. [18] established a correlation between PZC and the ability of the electrostatic field of the solid lattice to polarize and dissociate water molecules at the interface. They calculated the electrostatic field strength using a lattice parameter r_a the cation-anion spacing, and a surface parameter r_c the adsorbate-adsorbent ionic distance.

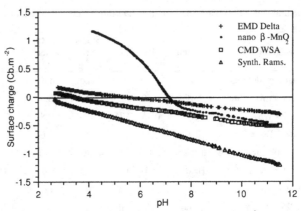

Figure 3. Charging curves of the MD samples.in KNO_3 $10^{-1}M$

In the present work the electronic field strength parameter [18] was used to compare the studied MD series to other divalent metal oxides of tetragonal symmetry. Its calculation relies on a the following equation: $pH_{PZC} = (A.V^{-2/3})+B$, where V is the mean volume of the unit cell per metal ion, A and B are constants. The constant A contains the Hückel equation for the calculation of the volume field strength [18]. Plotting pH_{PZC} versus $V^{-2/3}$ then reveals the dependence of the Point of Zero Charge on the metal ion density in the crystallographic pseudo unit cell. An approximately linear variation of PZC with the mean volume per metal ion is observed for tetragonal dioxides of rutile structure (Table II and Fig. 4a). As far as MD's are concerned, their PZC is positively correlated to the volume field strength, as already observed [18]. A relatively small variation in mean volume per metal ion between ramsdellite and pyrolusite is linearly related to strong difference in PZC. This great PZC range is unexpected if one considers the PZC values reported by Parks [14] for different forms of Fe and Al oxides. Also, the PZC of EMD Delta falls out of the pattern (Fig. 4a).

Table II: Cell parameters and PZC of tetragonal dioxides

Material	a = b (Å)	c (Å)	Vol (Å³)	PZC	ref
β-MnO₂ Pyrolusite	4.39	2.87	55.57	7.5	[9]
SiO₂ stishovite	4.179	2.66	46.42	2	[12]
TiO₂ rutile	4.59	2.96	62.42	5.8	[12]
PbO₂ plattnerite	3.92	3.37	83.27	9.5	[12]
SnO₂ cassiterite	4.738	3.12	71.59	7.3	[12]

These observations clearly show average relationships between lattice parameters and surface properties. However, inverse trends indicate that other parameters that average ones must be investigated.

Therefore, taking into account the nature and rate of structural disorder in the solid should provide a better understanding of the PZC of manganese dioxides. Figure 5 shows the effect of pyrolusite intergrowth (Pr) and microtwinning (Tw) on the PZC of the studied MD's. It appears clearly that pyrolusite intergrowth is the leading factor, since it is linearly related to the PZC.

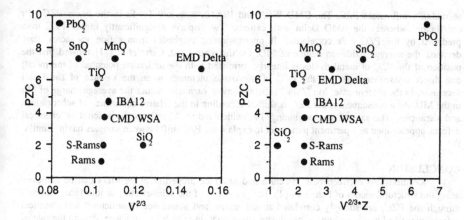

Figure 4. Dependence of PZC on crystallographic parameters for rutile structure dioxides and MD's .a) the mean volume of the unit cell per metal ion; b) the amount of electrons in the mean volume of the unit cell per metal ion.

The two end-members ramsdellite and pyrolusite display very different PZC, respectively < 2.5 and 7.3 (extrapolation of the linear variation of PZC with Pr for Pr = 0 gives PZC = 1 for ramsdellite). Materials such as EMD's or CMD's can be considered after the intergrowth model as mixtures of both end-members ramsdellite and pyrolusite, and thus display intermediate PZC. The IBA12 sample of Tamura *et al.* [12] fits very well in this correlation.

The calculated molecular areas of adsorbed water molecules (Fig. 2) clearly parallel the PZC. The lowest PZC is obtained on ramsdellite, on which water occupies the lowest molecular area (6.3 $Å^2$). This unusually low value is attributed (section 1) to the dissociative adsorption of the first water layer, which may result in lability of the proton, and thus in low protonation constant of surface groups. Following this approach, the high PZC of pyrolusite can be related to high protonation constant originating from non-dissociative physisorption on weak affinity surface sites, revealed by the high water molecular area (13.1 $Å^2$). The intermediate samples CMD WSA and EMD Delta display intermediate PZC and water molecular area.

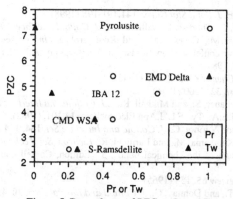

Figure 5. Dependence of PZC on Pr and Tw

The effect of microtwinning is less straightforward: no direct correlation with PZC is observed. However, it can be seen (Fig. 5) that for the samples with the same Pr, microtwinning secondarily influences PZC. The CMD WSA and IBA12, with low Tw fit in the linear PZC / Pr correlation, whereas the EMD Delta with extreme Tw displays a significantly higher PZC than predicted by the PZC / Pr correlation. If microtwinning originates from cation vacancies, they decrease the average oxidation degree of MD. On the other hand, Carre et al [11] showed that the position of the PZC of metal oxides is linearly correlated with the ionization potential of the metal: one direct consequence is that the PZC of the oxide increases when the valence of the metal decreases. In the present case, Mn^{3+}/Mn^{4+} substitutions or vacancies reduce the average charge of Mn in the MD. As a consequence, the PZC is shifted according to the relative abundance of substitutions and vacancies. The rate of microtwinning Tw, which indirectly reflects the amount of chemical defects, appears then as a pertinent parameter to explain the PZC drift between samples having similar Pr.

CONCLUSION

Relationships between bulk structure and interface properties of γ/ϵ-MD's can be achieved using the structural model of Pannetier [2]. The properties of the solid-electrolyte interface, analyzed through the PZC, are strongly correlated to the nature and abundance of structural and chemical defects, described by the amount of pyrolusite intergrowth in ramsdellite structure, Pr, and the rate of microtwinning Tw. A linear relationship is evidenced between PZC and Pr, starting from ramsdellite (Pr = 0, Tw = 0, PZC = 1) to pyrolusite or β-MD (Pr = 1, Tw = 0, PZC = 7.3). γ-MD with intermediate values of Pr (0.2 to 0.45) have increasing PZC values ranging between 2.5 and 5.6. For similar Pr values (0.45), high Tw percentage (0.3 and 1) makes the PZC to increase from 3.7 to 5.7. The experimental results, compared with data collected in the literature for dioxides of transition elements with tetragonal structure, shows a surprising difference between PZC of both MnO_2 forms: 7.3 for the quadratic pyrolusite, 1 for orthorhombic ramsdellite.

ACKNOWLEDGEMENT

EBC Company is acknowledged for funding this work and Dr. Elisabeth Djurado (LEPMI Grenoble) for her help in preparing nanocrystalline pyrolusite.

REFERENCES

1. Poinsignon, C., Tedjar, F., Amarilla, J.M., J. Mater.Chem.. 12, 1227 (1993).
2. Chabre,Y., and Pannetier, J., Prog. Solid St. Chem. 23, 1 (1995).
3. McLean, L., Poinsignon, C., Amarilla, J.M., Lecras, F., and Strobel, P., J. Mater.Chem. 8, 1183 (1995).
4. Djurado, E., Meunier E. J. Solid Stat Chem. 141,191-198,(1998).
5. Prélot, B., Poinsignon, C., Thomas, F., Villiéras, F., J. Colloid Interface Sci. in press (2002).
6. Poirier, J.E., François, M., Cases, J.M., and Rouquerol, J., In: Proceedings of the second engineering foundation conference on fundamental adsorption. A.I. Liapis (ed.). New York. AIChE Pub., 473-482, 1987.
7. Byström, A.M., Acta Chem. Scand. 3, 163 (1949).
8. Baur, W.H., Acta Cryst. 32, 2000 (1976).
9. Hagymassy, J. Jr., Brunauer, S., and Mikhail, R.S., J. Colloid and Interface Sci. 29, 485 (1969).
10. Maskell W.C., Shaw J.E.A., Tye F.L. J.App.Electrochem. 12,101, (1982).
11. Carre, A., Roger, F., and Varinot, C., J. Colloid and Interface Sci. 154, 174 (1992).
12. Tamura, H., Oda, T., Katayama, M., and Furuichi, R., Environ. Sci. Technol. 30, 1198 (1996).
13. Pauling, L., "The nature of the chemical bond". 3rd Edition, Cornell University Press, Ithaca, N.Y. 1960.
14. Parks, G.A., Chem. Review 65, 153 (1965).
15. Yoon, R.H., Salman, T., and Donnay, G., J. Colloid and Interface Sci. 70, 483 (1979).
16. Bleam, W., J. Colloid and Interface Sci. 159, 312 (1993).
17. Hiemstra, T., de Witt, J.C.M., and Van Riemsdjik, W.H., J. Colloid Interface Sci. 133, 91 (1989).
18. Healy, T.W., Herring, A.P., and Fuerstenau, D.W., J. Colloid Interface Sci. 21, 435 (1966).

Cation Ordering in Substituted $LiMn_2O_4$ Spinels

P. Strobel [1], A. Ibarra-Palos [1] and C. Poinsignon [2]

[1] Laboratoire de Cristallographie CNRS, BP 166, 38042 Grenoble Cedex 9, France
[2] LEPMI, ENSEEG, BP 75, 38402 Saint-Martin d'Hères Cedex, France

ABSTRACT

In order to overcome the capacity fading of $LiMn_2O_4$ in lithium batteries, various substitutions for Mn have been proposed. The structural implications of substitution in $LiMn_{2-x}M_xO_4$ with x = 0.5, i.e. with exactly 1/4 octahedral (16d-site) cations replaced, are investigated here. For this stoichiometry, cationic ordering was known previously for M = Mg and Zn, resulting in a super-structure with primitive cubic symmetry. Given the poor chemical contrast of X-ray diffraction between Mn and Co, Ni or Cu, $LiMn_{1.5}M_{0.5}O_4$ samples were studied by neutron diffraction and IR spectroscopy. Both techniques show the occurence of cationic ordering for M = Ni and Cu, but not for Co or Ga. In the case of M = Zn, further complication due Li/Zn ordering on the tetrahedral (8a) site is well resolved by FTIR. This investigation shows that the main driving force for octahedral cation ordering is the charge difference between Mn and M atoms.

INTRODUCTION

The spinel compound $LiMn_2O_4$ is one of the major candidates for positive electrode materials in lithium batteries [1, 2]. However, its practical use is beset by capacity fading, that numerous groups have sought to reduce by partial substitution of manganese by transition metals M [3]. The presence of M elements such as Cr, Fe, Co, Ni or Cu gives rise to an extra charge plateau at high potential and gives interest to compositions with relatively high substitution levels [4]. An especially interesting stoichiometry is $LiMn_{1.5}M_{0.5}O_4$ (or $Li_2Mn_3MO_8$), where the replacement of exactly 1/4 of the manganese yields strictly tetravalent Mn if M is divalent.

On the basis of X-ray diffraction data, most electrochemical studies on such materials assumed a random substitution of M on the manganese sites (octahedral 16d sites in the spinel structure). However, the 3/4-1/4 ratio can favour ordering of the 16d cations, as in the well-known compound $Li_2Mn_3MgO_8$ [5]. This has also been predicted from ab initio studies for other dopants [6]. In this work, a number of $LiMn_{1.5}M_{0.5}O_4$ compounds containing various divalent or trivalent M species are investigated. Using neutron diffraction and FTIR spectroscopy, we will show that compositions are indeed ordered for M = Ni and Cu, whereas the chemical contrast in X-ray diffraction is too weak to establish this by X- ray diffraction. The general trends of random vs. ordered 16d site occupation in lithium manganese oxide spinels as a function of dopant charge and size is discussed.

EXPERIMENTAL DETAILS

Samples were prepared by solid state reaction using Li_2CO_3, highly divided MnO_2 and appropriate oxides or carbonates of Mg, Co, Ni, Cu, Zn and Ga. Mixtures were ground together and heated repeatedly at 700-850 °C in air with intermittent re-grinding and furnace cooling. For neutron diffraction samples, quantities in the range 10-15 g were prepared. Synthesis details for M = Mg, Co, Ni and Cu have been reported previously [7,8].

Fired samples were analyzed by X-ray diffraction using a Siemens D-5000 diffractometer equipped with a diffracted beam monochromator (Cu $K\alpha$ radiation). Counting times higher than 20 s/step were systematically used to provide good statistics for Rietveld refinements. Neutron diffraction (ND) was carried out at Institut Laue-Langevin, Grenoble, France, using the powder diffractometers D2B (λ = 1.5940 Å) or D1B (λ = 1.2806 Å). Structural refinements were carried out using the Rietveld method with the WinplotR/Fullprof package.

Infrared spectra were recorded on pellets prepared by diluting 1 mg sample in 150 mg TlBr and pressed under 8 tons/cm2. A Nicolet 710 FTIR spectrometer was used in transmission mode with a resolution of ± 2cm^{-1}. The mirror velocity of the interferometer was reduced to 20 cycles/mn to increase the beam intensity; 200 accumulations were carried out to record well resolved spectra.

RESULTS

Composition and diffraction analysis

The compositions studied are listed in table I. $LiMn_{1.5}M_{0.5}O_4$ spinels have been prepared with divalent (Mg, Ni, Cu, Zn) and trivalent M elements (Ga, Co). In the cobalt case, the charge distribution ($Mn^{+3.67}/Co^{+3}$ rather than Mn^{+4}/Co^{+2}) was established from magnetic susceptibility measurements [8]. In addition, two "$Li_{0.5}Mn_{1.5}MO_4$" compositions were also prepared. In these, partial substitution of Li by M on the tetrahedral site allowed to vary the manganese valence at constant octahedral site contents, as between sample pairs 25 and 53 (Mg), or 112 and 111 (Zn).

Table I. Nominal composition and symmetry for $AMn_{1.5}M_{0.5}O_4$ samples studied (in order of decreasing M valence).

sample	formula	R_M Å [a]	cell parameter [b]	space group	experimental evidence [b]	reference
54	$LiMn_{1.5}Ga_{0.5}O_4$	0.62	8.211	Fd3m [c]	ND, FTIR	this work
03	$LiMn_{1.5}Co_{0.5}O_4$	0.525	8.138	Fd3m	ibid.	this work
07	$LiMn_{1.5}Ni_{0.5}O_4$	0.69	8.167	P4$_3$32	ibid.	this work
96	$LiMn_{1.5}Cu_{0.5}O_4$	0.73	8.189	P4$_3$32	ND	this work
25	$LiMn_{1.5}Mg_{0.5}O_4$	0.67	8.187	P4$_3$32	XRD	[5]
112	$LiMn_{1.5}Zn_{0.5}O_4$	0.74	8.182	P2$_1$3	ibid.	[9]
53	$(Li_{0.5}Mg_{0.5})Mn_{1.5}Mg_{0.5}O_4$		8.279	Fd3m	XRD, FTIR	this work
111	$(Li_{0.5}Zn_{0.5})Mn_{1.5}Zn_{0.5}O_4$		8.268	Fd3m	ibid.	this work

[a] Shannon's octahedral ionic radius of substituting species
[b] e.s.d. < 0.001 Å in all cases
[c] normal spinel-type space group.

Figure 1 shows typical XRD diagrams. Unreacted MO oxide impurities were detected for M = Ni and Cu (1.7 % and 6 % from 2-phase Rietveld refinements, respectively). Additional reflections are clearly seen for M = Mg (and for Zn, not shown). They correspond to a primitive cubic cell, i.e. a lowering of symmetry due to 16d-site cation ordering. Such extra reflections are not observed for M = Ni or Cu.

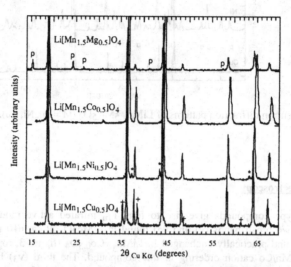

Figure 1. XRD patterns of members of the $LiMn_{1.5}M_{0.5}O_4$ series (M = Mg, Co, Ni, Cu). Primitive cubic cell reflections are marked "P" (* NiO, + CuO impurities).

When turning to neutrons, the picture is markedly different (see figure 2). The differences in coherent neutron scattering lengths between Mn (-3.73 fm) and the neighbouring elements (Co: +2.50, Ni: +10.3, Cu: +7.72 fm) [10] give access to clear evidence of superstructure reflections for Ni and Cu, and *not* for Co.

Refinement details will be given elsewhere [11]. Full Rietveld refinements show that the Ni and Cu phases are ordered. Superstructure lines are weaker for Cu; their presence is remarkable in view of the fact that (i) this sample contains a significant CuO fraction, so that the spinel stoichiometry is not exactly Cu/Mn = 1/3, (ii) it exhibits partial Li/Cu inversion. The structural formula of sample 96 is thus $(Li_{0.9}Cu_{0.1})[Mn_{1.5}Cu_{0.25}Li_{0.25}]O_4$. This result is not surprising in view of the numerous $Cu^{2+}M^{3+}_2O_4$ spinel phases known with tetrahedral Cu^{2+} [5].

Note also the difference in space group between the Mg ($P4_332$) and Zn ($P2_13$) cases (table 1). In the latter, the strong tetrahedral preference of Zn^{2+} induces an additional cation ordering on the tetrahedral (8a) sites, yielding the structural formula $(Li_{0.5}Zn_{0.5})[Mn_{1.5}Li_{0.5}]O_4$. This also affects the $Li_{0.5}Mn_{1.5}MO_4$ case with M = Zn : for sample 111, the Rietveld refinement gives the cation distribution $(Zn_{0.83}Li_{0.17})$ $[Mn_{1.5}Li_{0.33}Zn_{0.17}]$ O_4, i.e. an almost complete displacement of lithium by zinc from the tetrahedral sites.

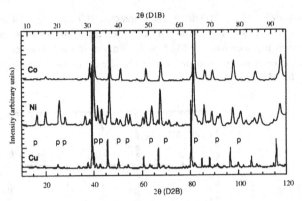

Figure 2. Neutron diffraction patterns of $LiMn_{1.5}M_{0.5}O_4$ (M = Co, Ni, Cu). P = $P4_332$ reflections.

Infrared spectroscopy

Spinel-type compounds give rise to four T_{1u} infrared active modes [12,13]. The most intense are two broad bands (v_1 and v_2) found in $LiMn_2O_4$ at ca. 610 and 500 cm^{-1} [14-16]. These are found practically unchanged in $LiMn_{1.5}Co_{0.5}O_4$ (figure 3, top left), confirming the absence of Mn/Co cation ordering in this compound. The third (v_3) band is observed as a shoulder on the low-energy side of the v_2 one at 435 cm^{-1}. A similar spectrum was obtained for $LiMn_{1.5}Ga_{0.5}O_4$ (not shown).

For nickel substitution (figure 3, top right), FTIR shows a dramatic increase in the number of infrared active modes, in agreement with previous results [13]. This feature is readily explained by group theory analysis, which predicts an increase from 4 to 21 infrared active modes when the symmetry decreases from Fd3m to $P4_332$ [17]. The Co-Ni comparison illustrates the power of vibration spectroscopy in determining symmetry changes undetectable by X-Ray diffraction.

Regarding Mg-substituted samples, $LiMn_{1.5}Mg_{0.5}O_4$ is known to possess cation ordering, and its infrared spectrum was reported previously [13]. For $(Li_{0.5}Mg_{0.5})Mn_{1.5}Mg_{0.5}O_4$ (sample 53 in table I), XRD did not detect any superstructure in spite of a very similar composition. The absence of cation ordering in this sample is indeed confirmed by FTIR (figure 3, bottom left).

The last panel of figure 3 (bottom right) shows the FTIR spectrum of the zinc analog. It shows two additional peaks at 583 and 685 cm^{-1}, which can be attributed to the peculiar cation distribution revealed by XRD in this sample: this is the only case in the series under study where the tetrahedral 8a site is occupied mostly by a heavy atom (Zn). The occurence of extra infrared bands due to distribution changes involving cations with different masses has been reported previously [18]. The high-frequency feature has been assigned to the 8a site contribution [19]. It is actually present as a shoulder already in sample 53, which also contains Zn on the 8a site (figure 3, bottom left).

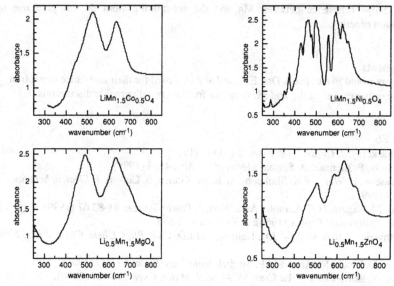

Figure 3. FTIR Spectra of selected substituted $LiMn_{1.5}M_{0.5}O_4$
(clockwise from top left: samples 03, 07, 111 and 53 from table 1).

DISCUSSION

An examination of ionic radii and structure types (table 1) seems to show a correlation between the ionic radius of the substituting element M and the occurence of cationic ordering. However, the last two samples contradict this trend. The main difference between samples 25 and 112 on one hand, samples 53 and 111 on the other, is a change in manganese average valence (+3.67 instead of +4), while the octahedral site contents remain unchanged. Now, a decrease in Mn valence produces an increase in Mn ionic radius. As a result, samples 53 and 111 (not ordered) exhibit lower diffe-rences in *both* charge and size between Mn and the substituting elements. This seems to indicate that a high valence difference ($\Delta v = 2$) is required for the occurence of cation ordering in substituted Li-Mn-O spinels.

CONCLUSIONS

The present study addresses the question of cation ordering in Li-Mn-O spinels with 1/4 substitution on the octahedral sites. To overcome the inability of XRD to detect such ordering when the cations involved have undistinguishable scattering powers, neutron diffraction and FTIR were used. These techniques unambiguously showed the occurence of octahedra site ordering for Ni and Cu substitution, while trivalent substituents (Ga, Co) or compositions with unsufficient charge difference between Mn and the substituting atom do not give rise to octahedral cation ordering.

Acknowledgments

The authors would like to thank Drs. E. Suard and O. Isnard for their assistance in neutron diffraction measurements at I.L.L., and M. Anne for fruitful crystallographic discussions.

REFERENCES

1. M. M. Thackeray, *J. Ceram. Soc. Amer.* **82**, 3347 (1999).
2. M. Broussely, P. Biensan, B. Simon, *Electrochim. Acta* **45**, 3 (1999).
3. J. M. Tarascon, E. Wang, F.K. Shokoohi, W.R. McKinnon, S. Colson, *J. Electrochem Soc.* **138**, 2859 (1991).
4. H. Kawai, M. Nagata, H. Tukamoto, A.R. West, *J. Power Sources* **81-82** 67 (1999).
5. G. Blasse, *Philips Res. Rept.* 1 (1964); JCPDS card 32-573.
6. J.S. Braithwaite, C.R.A. Catlow, J.H. Harding and J.D. Gale, *Phys. Chem. Chem. Phys.* **2** 38 (2000).
7. F. Le Cras, D. Bloch, M. Anne and P. Strobel, *Solid State Ionics* **89** 203 (1996).
8. P. Strobel, A. Ibarra Palos, F. Le Cras, M. Anne, *J. Mater. Chem.* **10** 429 (2000).
9. J.C. Joubert and A. Durif, *C.R. Acad. Sci. Paris* **258** 4482 (1964); JCPDS card 74-1260.
10. V.F. Sears, *AECL Report* 8490 (1984).
11. P. Strobel, C. Poinsignon, A. Ibarra Palos and M. Anne, *J. Solid State Chem.* (submitted).
12. W.B. White and B.A. DeAngelis, *Spectrochim. Acta* **23A** 985 (1967).
13. J. Preudhomme, *Ann. Chim. (France)* **9** 31 (1974).
14. T.J. Richardson, S.J. Wen, K.A. Striebel, P.N. Ross and E.J. Cairns, *Mater. Res. Bull.* **32** 609 (1997).
15. C. Julien, M. Massot, E. Haro-Poniatowski, G.A. Nazri, A. Rougier, *MRS Proc.* **496** 415 (1998).
16. B. Ammundsen, G.R. Burns, M.S. Islam, H. Kanoh and J. Rozi□re, *J. Phys. Chem. B* **103** 5175 (1999).
17. G.C. Allen and M. Paul, *Appl. Spectrosc.* **49** 451 (1995).
18. V.A.M. Brabers, *Phys. Stat. Sol. A* **12** 629 (1972).
19. M. Laarj, S. Kacim and B. Gillot, *J. Solid State Chem.* **125** 67 (1996).

Mat. Res. Soc. Symp. Proc. Vol. 756 © 2003 Materials Research Society

The Syntheses and Characterization of Layered LiNi$_{1-y-z}$Mn$_y$Co$_z$O$_2$ Compounds

J. Katana Ngala, Natasha A. Chernova, Luis Matienzo, Peter Y. Zavalij, M. Stanley Whittingham[*]
Chemistry Department and the Institute for Materials Research,
State University of New York at Binghamton,
Binghamton, New York 13902-6016, U.S.A.

ABSTRACT

The layered compounds of formula LiNi$_{0.4}$Mn$_{0.6-y}$Co$_y$O$_2$, for y=0.2, 0.3, and 0.4 and LiNi$_{0.7-y}$Mn$_{0.3}$Co$_y$O$_2$ for y=0.3, and 0.1 were synthesized at 800°C. X-ray powder diffraction indicates layered structure of R$\bar{3}$m symmetry similar to α-NaFeO$_2$. Rietveld refinement data shows that Mn and Ni increase the tendency of transition metal ions to migrate into the interlayer sites relative to LiCoO$_2$. Both magnetic susceptibility and XPS data support a 2+ oxidation state for Ni and 4+ and 3+ for Mn and Co, respectively. The layered compound LiNi$_{0.4}$Mn$_{0.4}$Co$_{0.2}$O$_2$ shows a high initial capacity of about 200mAh/g when cycled between 2.5V and 4.3 V at 20°C.

INTRODUCTION

Suitable cathode materials need to be found to replace LiCoO$_2$ in commercial lithium rechargeable batteries. The compound LiCoO$_2$ is expensive and toxic. The new material needs to be cheaper, less toxic and, preferably, have a higher capacity than LiCoO$_2$.

Much interest has been on the layered LiMnO$_2$ compound for its prospect of providing not only a cheap but also an environmentally benign cathode material [1-3]. However, it easily converts to the, thermodynamically stable, spinel structure. The conversion is promoted by Jahn-Teller distortion, which occurs due to the $t_{2g}^3 e_g^1$ configuration of the Mn^{3+} ion.

Two approaches to stabilizing the layered LiMnO$_2$ have been taken. In the geometric stabilization approach, "pillars" between the layers provide the stabilization. We reported [4, 5] on the compounds, KMnO$_2$ and (VO)$_y$MnO$_2$, which are examples of such structures. The two compounds have reasonable stability towards cycling, particularly at low current densities.

In the electronic stabilization approach, the electron configuration is increased by substitution of the Mn with more electronegative elements such as Co [6-9] and Ni [10-14]. The more electronegative element helps by keeping the Mn above 3+ oxidation state. For instance, the compound LiMn$_{0.5}$Ni$_{0.5}$O$_2$ has indicated a high capacity without formation of the spinel [6].

The layered compounds, LiNi$_{1-y-z}$Mn$_y$Co$_z$O$_2$, have demonstrated remarkable electrochemical behavior [15,16] such that they are suitable candidates as cathodes for lithium rechargeable batteries. They have smooth discharge curves, which are devoid of plateaus, over a wide potential range. Thus, they are able to deliver high capacities without going through irreversible phase changes.

In this paper we report on the syntheses and characterization of some LiNi$_{1-y-z}$Mn$_y$Co$_z$O$_2$ compounds. We report, for the first time, on the magnetic susceptibility and XPS analyses on LiNi$_{0.4}$Mn$_{0.4}$Co$_{0.2}$O$_2$.

---

[*] Contact author: stanwhit@binghamton.edu

EXPERIMENTAL

The mixed hydroxide of the transition metals was prepared. Stoichiometric amounts of the soluble salts: $Mn(OAc)_2 \cdot 6H_2O$, $Co(OAc)_2 \cdot 6H_2O$, and $Ni(NO_3)_2 \cdot 4H_2O$ were mixed in a beaker and dissolved using de-ionized water. Excess NaOH solution was added to this solution, while stirring, to precipitate the mixed hydroxide. The precipitate was filtered out and rinsed with deionized water, in a Buchner funnel. Complete removal of Na^+ ions from the precipitate was ensured by returning the precipitate into the beaker and thoroughly stirring it in more deionized water. Filtration and rinsing of the precipitate repeated followed by the drying of the precipitate in an oven at $50°C$ for about 12 hours.

Lithium was introduced into the mixed hydroxide by adding $LiOH \cdot nH_2O$ to the dried precipitate. The mixed hydroxide precipitate was ground and from it an appropriate amount was weighed out. To the mixed hydroxide powder was added a suitable amount of $LiOH \cdot nH_2O$. The resulting mixture was thoroughly ground and mixed before being made into a pellet.

The reaction was allowed to take place in two steps. In the first step the pellet was placed in the furnace and heated at $450°C$ for about 12 hours. At the end of this time it was removed, ground and made into a pellet. The pellet was placed in the furnace and heated at $800°C$ for about 8 hours. At the end of this time, the pellet was removed from the furnace and rapidly cooled by placing it in the hood.

The product formed was characterized using X-ray powder diffraction. Cu $K\alpha$ radiation on a Scintag XDS2000 θ-θ diffractometer equipped with a Ge(Li) solid state detector was used. The data was collected from $2\theta=15°$ to $2\theta=80°$ in 0.02 steps and at 7 seconds per step.

The samples were tested as cathodes versus Li anode. The bag cell type was used and the electrochemistry performed at $20°C$ using Mac Pile II Instrument. The sample was thoroughly mixed with carbon black and teflon in the ratio 80:10:10, respectively. The sample mixture was dehydrated by heating it on a hot press, at about $180°C$, for about 2 hours before transferring it into the glove box and setting up the cell. The electrolyte was $LiPF_6$ in a mixture of dimethyl carbonate (DMC) and ethylene carbonate (EC). The electrolyte was from EM industries. Cycling of the samples was performed using 0.1 mA/cm^2 and 2.5 V- 4.3 V potential range.

Magnetic properties of $LiNi_{0.4}Mn_{0.4}Co_{0.2}O_2$ were studied using Quantum Design MPMS XL SQUID magnetometer. Temperature dependence of the magnetic susceptibility was measured over the temperature range from 1.8 to 298 K in the magnetic field of 0.1 T. Magnetization curves were measured at 298 K and 10 K in the magnetic field up to 2 T.

X-ray photoelectron spectroscopy (XPS) of $LiNi_{0.4}Mn_{0.4}Co_{0.2}O_2$ was performed using a modified PHI model 5500 multi-probe spectrometer. Monochromatized Al $K\alpha$ X-ray excitation was produced by a 7 mm Al filament at 14 keV and 350 mA. Survey and high-resolution spectra were collected at $65°$ to the detector with pass energy resolutions of 187.5 eV and 11.75 eV, respectively. All measured binding energies were referenced to the C 1s line at 284.8 eV.

RESULTS AND DISCUSSION

Synthesis and X-Ray Analysis

The compounds were synthesized using solid state reaction. Mixed hydroxide precursors were heated at $800°C$ in air. This is the first report of the synthesis to be carried out at this

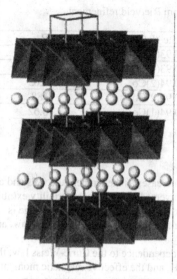

Figure 1. Layered structure of $LiNi_{1-y-z}Mn_yCo_zO_2$.

temperature. Previously, higher temperatures only have been used for this type of compounds [15, 16]. At the end of the heating, the product was rapidly cooled. The rapid cooling was performed since temperature affects the occupancies of the metal ion sites.

The powder x-ray patterns of the compounds indicate layered structure of $R\bar{3}m$ symmetry, similar to α–$NaFeO_2$. The structure, which consists of a cubic close packed arrangement of oxide ions, is shown in Fig. 1. The transition metal ions in the structure occupy alternating layers in octahedral sites. A typical x-ray pattern is given in Fig. 2a.

Rietveld refinement was performed to investigate the distribution of the metal ions in the 3a and 3b sites. The refinement was performed by taking the average atomic number of the metal ions in the main framework to be equal to that of cobalt. Nickel was assumed to be the migrating ion. Fig. 2b shows the refined pattern for $LiNi_{0.4}Mn_{0.4}Co_{0.2}O_2$ and Table I gives the summary of the refinement data for the series $LiNi_{1-y-z}Mn_yCo_zO_2$. The refinement always shows the presence of transition metal ions in the lithium sites. The amount of transition metal ions in the lithium sites is inversely proportional to the cobalt content and proportional to the manganese content.

Thus, we expect the Mn-rich compounds to show higher irreversible capacities than the Co-rich ones.

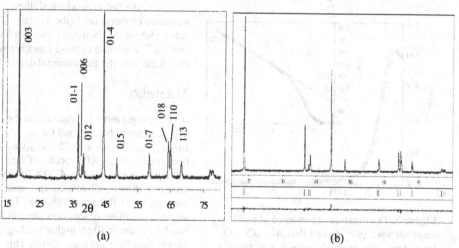

(a)

(b)

Figure 2. X-ray pattern (a) and refinement fit (b) of $LiNi_{0.4}Mn_{0.4}Co_{0.2}O_2$.

Table I. Metal ions occupation for $LiNi_{1-y-z}Mn_yCo_zO_2$, from Rietveld refinement.

y	z	3a site (0,0,0)	3b site (0,0,0.5)	R_w
0.4	0.2	0.960(4)Li+0.040(4)Ni	0.93(1)Co	10.71
0.3	0.3	0.963(5)Li+0.037(5)Ni	0.93(1)Co	8.83
0.2	0.4	0.981(9)Li+0.019(9)Ni	0.94(2)Co	12.63
0.3	0.4	0.991(9)Li+0.009(9)Ni	0.90(2)Co	9.88
0.3	0.1	0.913(6)Li+0.087(6)Ni	0.94(1)Co	10.62

Magnetic Susceptibility

The magnetic susceptibility of $LiNi_{0.4}Mn_{0.4}Co_{0.2}O_2$ increases as the temperature falls and a small down-turn is observed at about 25 K (Fig. 3). The inversed magnetic susceptibility exhibits Curie-Weiss behavior only above 150 K, the magnetization curve in this temperature range is linear. Below 150 K magnetic susceptibility deviates significantly from the Curie-Weiss law, and a non-linear magnetization is observed.

From the fit of the high-temperature part of $1/\chi_M(T)$ dependence to the Curie-Weiss law, the value of the asymptotic Curie temperature is $\Theta = -86\pm0.6$ K, and the effective magnetic moment per transition metal atom is $\mu = 2.61$ μ_B. Negative Curie temperature indicates rather strong antiferromagnetic exchange. The obtained value of μ may be used to determine charge states of Ni, Mn, and Co. Two reasonable combinations of the charge states can be assumed. In the first one, all the elements are in 3+ state, and this yields magnetic moment per atom of 2.65 μ_B, assuming 1.73 μ_B per low spin Ni^{3+}, 4.90 μ_B per high spin Mn^{4+} and non-magnetic low-spin Co^{3+}. The other possibility of Ni^{2+} (2.83 μ_B), Mn^{4+} (3.87 μ_B), and Co^{3+}, results in $\mu=2.68$ μ_B. Thus, the observed magnetic moment per atom in $LiNi_{0.4}Mn_{0.4}Co_{0.2}O_2$ is consistent with both scenarios, the first case where all the transition elements are in the 3+ state, the second where the compound contains Ni^{2+} and Mn^{4+}; a mixture of these cases is also consistent with the experimental data.

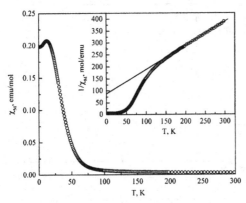

Figure 3. Temperature dependence of molar magnetic susceptibility of $LiNi_{0.4}Mn_{0.4}Co_{0.2}O_2$. Inset shows the inversed magnetic susceptibility, the straight line is the best fit to the Curie-Weiss law.

XPS studies

In order to get more information on the charge states of Ni, Mn, and Co in $LiNi_{0.4}Mn_{0.4}Co_{0.2}O_2$ an XPS study was performed. The 2p XPS spectra of these elements are given in Fig. 4. The 2p Co spectrum shows well-defined $2p_{3/2}$ and $2p_{1/2}$ components at 780.3 and 795.1 eV. The shake-up satellites of low intensity can be resolved at about 10 eV higher binding energy than the main components. This picture is consistent with the one reported for $LiCoO_2$ [17], and allows us to confirm

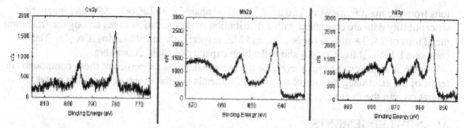

Figure 4. 2p XPS spectra for Co, Mn, and Ni in LiNi$_{0.4}$Mn$_{0.4}$Co$_{0.2}$O$_2$.

the 3+ oxidation state for Co in our compound. The Mn 2p spectrum shows only main 2p$_{3/2}$ and 2p$_{1/2}$ peaks at 642.3 and 654.1 eV, respectively. The 2p$_{3/2}$ binding energy is closer to the values reported for Mn^{4+} in β-MnO$_2$ (642.2 to 642.6 eV), than those for Mn^{3+} in α-Mn$_2$O$_3$ (641.8 to 641.9 eV) [18]. Thus, the Mn^{4+} charge state is very likely in LiNi$_{0.4}$Mn$_{0.4}$Co$_{0.2}$O$_2$, which means that Ni should be in the 2+ state. In the Ni 2p spectrum the main 2p$_{3/2}$ and 2p$_{1/2}$ components are at 855.0 and 872.6 eV, respectively. Well-resolved wide satellites are observed at about 7 eV higher energies than the main peaks. This structure is consistent with the one known for NiO, however, a complicated satellite structure typical for this compound hinders the peaks expected for Ni^{3+}, which makes identification of the Ni^{3+} state very difficult [19].

Electrochemistry

The electrochemical behavior of the compounds was studied. Here we report the electrochemical behavior of LiNi$_{0.4}$Mn$_{0.4}$Co$_{0.2}$O$_2$, which has indicated the best results among the compounds that we studied. The sample was cycled between 2.5 V and 4.3 V, at a current density of 0.1 mA/cm^2 as shown in Figure 5. The cycling was performed at 20°C. The initial capacity obtained was about 200 mAh/g, which corresponds to about 0.7 Li per formula unit. About 15% of the initial capacity was lost in the first cycle. The profile is smooth and devoid of any plateaus associated to phase changes in the potential range. The shape of the profile is consistent with the Ni in the +2 oxidation state being the electroactive species.

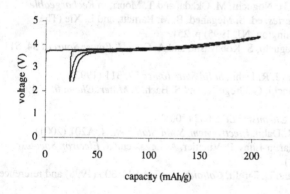

Figure 5. Electrochemical behavior of LiNi$_{0.4}$Mn$_{0.4}$Co$_{0.2}$O$_2$.

CONCLUSIONS

The layered compounds, LiNi$_{0.4}$Mn$_{0.6-y}$Co$_y$O$_2$, for y=0.2, 0.3, and 0.4 and LiNi$_{0.7-y}$Mn$_{0.3}$Co$_y$O$_2$ for y=0.3, and 0.1, were synthesized at 800°C. Rietveld refinement shows that Mn and Ni in the structure increase the tendency of the migration of transition metal ions from the main framework into the Li sites, compared with $LiCoO_2$. XPS and magnetic susceptibility data are consistent with a distribution of the oxidation states among the transition metal ions of 2+, 3+ and 4+ for Ni, Co, and Mn, respectively in $LiNi_{0.4}Mn_{0.4}Co_{0.2}O_2$. The compound, $LiNi_{0.4}Mn_{0.4}Co_{0.2}O_2$ showed a high capacity of about 200mAh/g.

Work is in progress to fully evaluate the electrochemical properties of these compounds in terms of the electronegativities of the transition metals and the extent of transition metal ion migration into the 3a sites.

ACKNOWLEDGEMENTS

We thank the Department of Energy, through the BATT program at LBNL for the support of this work (DE-AC03-76SF00098).

REFERENCES

1. C. Delmas and F. Capitaine, in Extended Abstracts of 8th Int. Meet. Lithium Batteries, Nagoya, Japan, 1996, p.470.
2. R. Chen, P. Zavalij and M. S. Whittingham, *Chem. Mater.* **8**, (1996) 1275.
3. A.R. Armstrong, A. J. Paterson, A. D. Robertson and P. G. Bruce, *Chem. Mater.* **14**, 710 (2002).
4. R. Chen and M. S. Whittingham, *J. Electrochem. Soc.* **144**, L64 (1997).
5. F. Zhang, K. Ngala and M. S. Whittingham, *Electrochem. Commun.* **2**, 445 (2000).
6. M. E. Spahr, P. Novak, Bernhard Schnyder, Otto Haas, and R. Nesper, *J. Electrochem Soc.* **145**, 1113 (1998).
7. C. Delmas, I. Saadoune, and A. Rougier, *J. Power Sources* **43/44**, 595 (1993).
8. E. Zhecheva and R. Stoyanova, *Solid State Ionics* **66**, 143 (1993).
9. T. Ohzuku, H. Komori, K. Sawai, and T. Hirai, *Chem. Express* **5**, 773(1990).
10. M. Fujiwara, S. Yamada, and M. Kanada, in Extended Abstracts of 34th Battery Symposium, Nagoya, Japan, 1993, p.135.
11. M. Yoshi, K. Yamato, J. Itoh, H. Noguchi, M. Okada, and T. Mouri, in *Rechargeable Lithium and Lithium-Ion Batteries*, ed. S. Megahed, B. M. Barnett, and L. Xie (The Electrochemical Society, Pennington, NJ, 1995) p. 251.
12. Y. Nitta, K. Okamura, K. Haraguchi, S. Kobayashi, and A. Ohta, *J. Power Sources* **54**, 511 (1995).
13. E. Rossen, C. D. W. Jones, and J. R. Dahn, *Solid State Ionics* **57**, 311 (1992).
14. D. Caurant, N. Baffier, V. Bianchi, G. Grégoire, and S. Bach, *J. Mater. Chem.* **6**, 1149 (1996).
15. T. Ohzuku and Y. Makimura, *Chemistry Lett.* 642 (2001)3.
16. Z. Lu, D. D. MacNeil and J. R. Dahn, *Electrochem. Solid-State Lett.* **4**, A201 (2001).
17. J. P. Dupin, D. Gonbeau, I. Martin-Litas, P. Vinatier, A. Levasseur, *J. Electron Spectosc. Relat. Phenom.* **120**, 55 (2001).
18. S. Ardizzone, C. L. Bianchi, and D. Tirelli, *Colloids Surf., A* **134**, 305 (1998) and references therein.
19. A. F. Carley, S. D. Jackson, J. N. O'Shea, M. W. Roberts, *Surf. Sci.* **440**, L868 (1999) and references therein.

Mat. Res. Soc. Symp. Proc. Vol. 756 © 2003 Materials Research Society   EE5.7

Crystal Chemistry of Chemically Delithiated Layered Oxide Cathodes of Lithium Ion Batteries

A. Manthiram and S. Venkatraman
Materials Science and Engineering Program, ETC 9.104
The University of Texas at Austin
Austin, Texas 78712

ABSTRACT

The structural and chemical stabilities of layered $Li_{1-x}CoO_{2-\delta}$, $Li_{1-x}Ni_{0.85}Co_{0.15}O_{2-\delta}$ and $Li_{1-x}Ni_{0.5}Mn_{0.5}O_{2-\delta}$ $(0 \leq (1-x) \leq 1)$ cathodes have been investigated by chemically extracting lithium from the corresponding $LiMO_2$ with the oxidizer NO_2BF_4 in acetonitrile medium. While $Li_{1-x}CoO_{2-\delta}$ and $Li_{1-x}Ni_{0.85}Co_{0.15}O_{2-\delta}$ begin to form a P3-type and a new O3-type (designated as O3') phases, respectively, for $(1-x) < 0.5$ and $(1-x) < 0.3$, $Li_{1-x}Ni_{0.5}Mn_{0.5}O_{2-\delta}$ maintains the initial O3-type structure without forming any second phase. Chemical analysis with a redox titration indicates that the $Li_{1-x}CoO_{2-\delta}$, $Li_{1-x}Ni_{0.85}Co_{0.15}O_{2-\delta}$, and $Li_{1-x}Ni_{0.5}Mn_{0.5}O_{2-\delta}$ systems begin to lose oxygen from the lattice, respectively, for $(1-x) < 0.5$, < 0.3 and < 0.4, which is accompanied by an onset of a decrease in the c parameter. The oxygen loss signals chemical instability and the trend in instability correlates with the charging voltage profiles of the cathodes.

INTRODUCTION

Commercial lithium-ion cells presently use the layered $LiCoO_2$ as the cathode, but only 50% of its theoretical capacity (140 mAh/g) could be practically utilized. In contrast, the layered lithium nickel oxide with a partial substitution of Co for Ni ($LiNi_{0.85}Co_{0.15}O_2$) shows a much higher reversible capacity of 180 mAh/g, which corresponds to 65% of its theoretical capacity. Also, the layered $LiNi_{0.5}Mn_{0.5}O_2$ has recently been found to exhibit higher capacities of 150 – 200 mAh/g depending on the synthesis conditions [1]. However, the reason for the differences in capacities among the three systems has not been fully understood in the literature although they have the same O3-type structure. Most of the studies in this regard have focused invariably on the structural characterization of the electrochemically charged cathodes. Despite the recognition that the highly oxidized redox couples such as $Co^{3+/4+}$ and $Ni^{3+/4+}$ are characterized by a near-equivalence of the metal:3d and O^{2-}:2p energies particularly in the case of perovskite oxides, little attention has been paid in the literature to the possible oxidation of O^{2-} ions during the charge/discharge process and the consequent chemical instability leading to oxygen loss from the lattice. One of the reasons for the lack of such information is the contamination of the electrochemically charged samples by carbon, binder, and electrolyte, and the consequent difficulty in analyzing the oxidation states and oxygen contents by wet-chemical analysis.

To overcome this difficulty, our group has focused recently on synthesizing bulk samples of $Li_{1-x}MO_2$ (M = Co, Ni, $Ni_{0.5}Mn_{0.5}$) free from carbon, binder, and electrolyte by chemically extracting lithium from $LiMO_2$ with an oxidizer in non-aqueous media [2,3]. We present here the structural and chemical characterizations by X-ray diffraction and wet-chemical redox titrations of the $Li_{1-x}CoO_2$, $Li_{1-x}Ni_{0.85}Co_{0.15}O_2$, and $Li_{1-x}Ni_{0.5}Mn_{0.5}O_2$ systems for $0 \leq (1-x) \leq 1$ and a correlation of the differences in the reversible capacities of the three systems to the observed chemical and structural instabilities.

EXPERIMENTAL

LiCoO$_2$ was synthesized by firing Li$_2$CO$_3$ and Co$_3$O$_4$ in air at 900 °C for 24 h. LiNi$_{0.85}$Co$_{0.15}$O$_2$ was prepared by firing a co-precipitated hydroxides of Ni and Co with lithium hydroxide in a flowing oxygen atm at 725 °C for 24 h. LiNi$_{0.5}$Mn$_{0.5}$O$_2$ was synthesized by firing the co-precipitated hydroxides of Ni and Mn with lithium hydroxide in air by a two-step heating process [1] at 480 °C for 3 h and 900 °C for 3 h. Chemical extraction of lithium was carried out by stirring the LiMO$_2$ powders in an acetonitrile solution of various quantities of NO$_2$BF$_4$ for 2 days under argon atm using a Schlenk line:

$$LiMO_2 + x\ NO_2BF_4 \rightarrow Li_{1-x}MO_2 + x\ NO_2 + x\ LiBF_4 \qquad (1)$$

The products formed after the reaction were washed several times with acetonitrile under argon to remove LiBF$_4$, dried under vacuum at ambient temperature, and stored in an argon-filled glove box to avoid reaction with the ambient. The lithium contents were determined by atomic absorption spectroscopy. The oxidation state of transition metal ions and the oxygen content were determined by iodometric titration [4]. Structural refinements and lattice parameter determinations were carried out by analyzing the X-ray diffraction data with the Rietveld method using the DBWS-9411 PC program [5].

RESULTS AND DISCUSSION

Figure 1 shows the evolution of the X-ray diffraction patterns of Li$_{1-x}$CoO$_{2-\delta}$ and the variations of the hexagonal lattice parameters with lithium content. The corresponding data for both Li$_{1-x}$Ni$_{0.85}$Co$_{0.15}$O$_{2-\delta}$ and Li$_{1-x}$Ni$_{0.5}$Mn$_{0.5}$O$_{2-\delta}$ are shown, respectively, in Figures 2 and 3. The Li$_{1-x}$CoO$_2$ system maintains the parent O3-type structure with an oxygen stacking sequence of ...ABCABC... along the c axis for $0.5 \leq (1-x) \leq 1.0$ and a new phase begins to form around $(1-x) = 0.45$ as indicated by the formation of a shoulder to the right of the (003) reflection. The new phase grows with decreasing lithium content and the end member CoO$_{2-\delta}$ consists of reflections corresponding to only the new phase, which could be indexed on the basis of a P3-type structure [6] having an oxygen stacking sequence of ...ABBCCA... along the c axis. The P3-type phase is formed from the O3-type phase by a gliding of the CoO$_2$ sheets. In the region $0 < (1-x) \leq 0.45$, the O3-type and the P3-type phases coexist. The observation of a P3-type structure for chemically synthesized CoO$_{2-\delta}$ is in contrast to the O1-type structure (...ABABAB... stacking) reported for the electrochemically synthesized samples [7, 8]. However, the P3-type structure is metastable and it transforms slowly to the O1-type structure with increasing reaction time of > 2 days with the oxidizer NO$_2$BF$_4$ [6] due to a further gliding of the CoO$_2$ sheets. The formation of the metastable P3-type phase during chemical delithiation is due to the rapid extraction of lithium in < 15 min, which translates into a high charging rate of > 4C [6], compared to the slow electrochemical charging (< C/10 rate).

The Li$_{1-x}$Ni$_{0.85}$Co$_{0.15}$O$_{2-\delta}$ system, on the other hand, maintains the parent O3-type structure for a wider range of $0.3 \leq (1-x) \leq 1.0$ and the new phase begins to form at a much lower lithium content of around $(1-x) = 0.25$ (Figure 2a) as indicated by the small shoulder to the right of the (003) reflection. The new phase grows with decreasing lithium content as in the case of Li$_{1-x}$CoO$_2$ and the end member Ni$_{0.85}$Co$_{0.15}$O$_{2-\delta}$ consists of only the new phase, which could also be indexed on the basis of an O3-type structure, but with a smaller c parameter. We designate the new phase as O3' phase. Both the O3 and O3' phases coexist in the region $0 < (1-x) \leq 0.25$. In contrast, the $Li_{1-x}Ni_{0.5}Mn_{0.5}O_{2-\delta}$ system (Figure 3) maintains the parent O3-type structure for all values of lithium content $0 \leq (1-x) \leq 1$ without showing the formation of a new phase or a two-phase region. The reason why the $Li_{1-x}Ni_{0.5}Mn_{0.5}O_{2-\delta}$ system does not show a two-phase region while both the $Li_{1-x}CoO_{2-\delta}$ and $Li_{1-x}Ni_{0.85}Co_{0.15}O_{2-\delta}$ systems exhibit is not clear at present.

In the cases of both $Li_{1-x}CoO_{2-\delta}$ and $Li_{1-x}Ni_{0.85}Co_{0.15}O_{2-\delta}$ systems, the c parameter increases initially with decreasing lithium content in the region where the initial O3-type structure is maintained (Figure 1b and 2b). One could visualize this increase to be due to the increasing electrostatic repulsion across the van der Waals gap between the MO_2 sheets with decreasing lithium content. However, the new phases – P3-type or O3'-type – have smaller c parameters than the O3-type phases. In the case of $Li_{1-x}Ni_{0.5}Mn_{0.5}O_{2-\delta}$ system, the c parameter increases initially with decreasing lithium content as in the cases of $Li_{1-x}CoO_{2-\delta}$ and $Li_{1-x}Ni_{0.85}Co_{0.15}O_{2-\delta}$ in the region $0.4 \leq (1-x) \leq 1$, but decreases with further decrease in lithium content for $(1-x) < 0.4$ although the parent O3-type structure is maintained for the entire lithium content of $0 \leq (1-x) \leq 1$. The observed decrease in the c lattice parameters at lower lithium contents in all the three systems may signal an O-O interaction across the van der Waals gap between the MO_2 sheets of the layered structure. Also, all the three systems begin to show an increase in a parameters at the lithium contents where the c parameters begin to decrease.

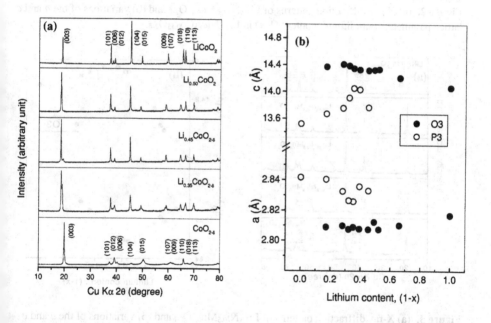

Figure 1. (a) X-ray diffraction patterns of $Li_{1-x}CoO_{2-\delta}$ and (b) variations of the a and c lattice parameters with lithium content $(1-x)$ in $Li_{1-x}CoO_{2-\delta}$.

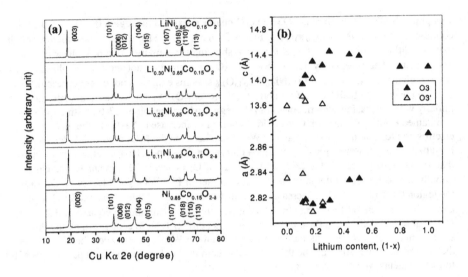

Figure 2. (a) X-ray diffraction patterns of $Li_{1-x}Ni_{0.85}Co_{0.15}O_{2-\delta}$ and (b) variations of the a and c lattice parameters with lithium content (1-x) in $Li_{1-x}Ni_{0.85}Co_{0.15}O_{2-\delta}$.

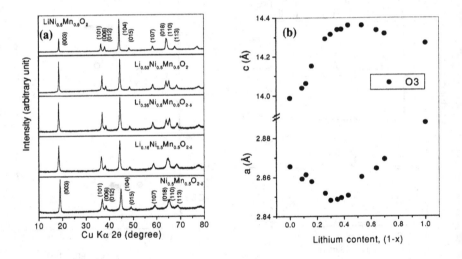

Figure 3. (a) X-ray diffraction patterns of $Li_{1-x}Ni_{0.5}Mn_{0.5}O_{2-\delta}$ and (b) variations of the a and c lattice parameters with lithium content (1-x) in $Li_{1-x}Ni_{0.5}Mn_{0.5}O_{2-\delta}$.

Figure 4 compares the variations of the oxidation state of the transition metal ions and the oxygen content with lithium content (1-x) for the three systems $Li_{1-x}CoO_{2-\delta}$, $Li_{1-x}Ni_{0.85}Co_{0.15}O_{2-\delta}$ and $Li_{1-x}Ni_{0.5}Mn_{0.5}O_{2-\delta}$. In the case of $Li_{1-x}CoO_{2-\delta}$, the oxidation state of Co increases initially with decreasing lithium content as one would expect and the oxygen content remains close to 2 for $0.5 \leq (1-x) \leq 1$. However, for $(1-x) < 0.5$, the oxidation state of Co remains nearly constant and the charge compensation is achieved by a loss of oxygen from the lattice. The end-member $CoO_{2-\delta}$ that was obtained after a reaction time of 2 days has an oxygen content of 1.72 with $\delta = 0.28$. In contrast, the $Li_{1-x}Ni_{0.85}Co_{0.15}O_{2-\delta}$ system loses oxygen at a much lower lithium content of $(1-x) < 0.3$, and the end-member $Ni_{0.85}Co_{0.15}O_{2-\delta}$ has a much smaller $\delta = 0.1$. Interestingly, the lithium content $(1-x)$ at which the oxygen loss begins to occur coincides with the lithium content at which the new phase P3 or O3' begins to form in both the systems. The oxygen loss behavior of $Li_{1-x}Ni_{0.5}Mn_{0.5}O_{2-\delta}$ is in between those of the $Li_{1-x}CoO_{2-\delta}$ and $Li_{1-x}Ni_{0.85}Co_{0.15}O_{2-\delta}$ systems. The $Li_{1-x}Ni_{0.5}Mn_{0.5}O_{2-\delta}$ system loses oxygen from the lattice for $(1-x) < 0.4$ and the end-member $Ni_{0.5}Mn_{0.5}O_{2-\delta}$ has $\delta = 0.25$. In this system, the oxygen loss begins to occur at the lithium content where the c lattice parameter begins to decrease.

The observed loss of oxygen from the $Li_{1-x}MO_2$ lattices is due to an overlap of the metal:3d band with the top of the O^{2-}:2p band and a consequent oxidation of the oxide ions at deep lithium extraction. The value of lithium content at which the oxygen loss begins to occur will depend on the extent of overlap between the two bands and the position of the Fermi energy E_F [2, 6]. In fact, the discharge/charge voltage profiles would be a reflection of the relative positions of E_F [9] in the three systems and Figure 5 compares the charge profiles collected at a slow charging rate of C/100. The data illustrate that $LiCoO_2$ with a higher voltage for a given lithium content would have a greater lowering of the metal:3d energy and a greater overlap between the metal:3d and O^{2-}:2p bands compared to that in $LiNi_{0.85}Co_{0.15}O_2$ and consequently a greater tendency to lose oxygen from the lattice. $LiNi_{0.5}Mn_{0.5}O_2$ with an intermediate voltage profile exhibits a behavior in between those of $LiCoO_2$ and $LiNi_{0.85}Co_{0.15}O_2$.

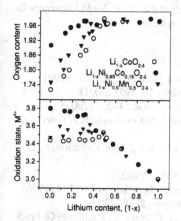

Figure 4. Variations of the oxidation state of M^{n+} and oxygen content with lithium content $(1-x)$ in $Li_{1-x}MO_{2-\delta}$.

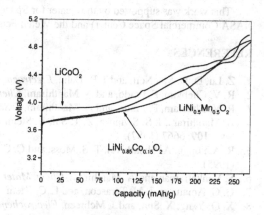

Figure 5. Charging voltage profiles of the cathodes at C/100 rate.

CONCLUSIONS

Bulk samples of $Li_{1-x}CoO_{2-\delta}$, $Li_{1-x}Ni_{0.85}Co_{0.15}O_{2-\delta}$, and $Li_{1-x}Ni_{0.5}Mn_{0.5}O_{2-\delta}$ free from carbon and binder have been synthesized for $0 \leq (1-x) \leq 1$ by chemically extracting lithium from the respective parent oxides $LiMO_2$ and systematically analyzed for structural and chemical instabilities. While the $Li_{1-x}CoO_{2-\delta}$ and $Li_{1-x}Ni_{0.85}Co_{0.15}O_{2-\delta}$ systems begin to form, respectively, a P3-type phase for $(1-x) < 0.5$ and a new O3-type phase – designated as O3' phase – for $(1-x) < 0.3$ that have lower c parameters than the original O3 phase, the $Li_{1-x}Ni_{0.5}Mn_{0.5}O_{2-\delta}$ system maintains the original O3-type structure for the entire value of lithium content $0 \leq (1-x) \leq 1$. The $Li_{1-x}CoO_{2-\delta}$, $Li_{1-x}Ni_{0.5}Mn_{0.5}O_{2-\delta}$, and $Li_{1-x}Ni_{0.85}Co_{0.15}O_{2-\delta}$ systems begin to lose oxygen from the lattice, respectively, for $(1-x) < 0.5$, 0.4 and 0.3. The beginning of oxygen loss from the lattice is accompanied by a decrease in the c lattice parameter in all the three systems signaling the introduction of holes into the $O^{2-}:2p$ band and a beginning of a stronger O-O interaction across the van der Waals gap at the lithium content where oxygen loss begins to occur.

We believe the much higher practical capacity (180 mAh/g) of $LiNi_{0.85}Co_{0.15}O_2$ compared to that of $LiCoO_2$ (140 mAh/g) is due to a better chemical stability of the former without losing any oxygen down to a lower lithium content of 0.3. Although the chemical instability leads to an actual loss of oxygen during our chemical lithium extraction experiments, neutral oxygen may not be evolved in actual lithium-ion cells. Instead, the chemical instability may lead to a stronger reaction of the cathode with the electrolyte and a consequent degradation of the cathode and capacity fade. One way to suppress such a chemical instability is to develop a more stable electrode-electrolyte interface. In fact, recent experiments indicate that a coating of $LiCoO_2$ with inert oxides such as ZrO_2, TiO_2, and Al_2O_3 provides remarkable capacity retention to much higher voltages possibly due to the suppression of the chemical instability [10].

ACKNOWLEDGEMENT

This work was supported by the Center for Space Power at the Texas A&M University (a NASA Commercial Space Center) and the Welch Foundation Grant F-1254.

REFERENCES

1. Z. Lu, D. D. MacNeil, and J. R. Dahn, *Electrochem. Solid-State Lett.*, **4**, A191 (2001).
2. R. V. Chebiam, F. Prado, and A. Manthiram, *Chem. Mater.*, **13**, 2951 (2001).
3. R. V. Chebiam, F. Prado, and A. Manthiram, *J. Solid State Chem.*, **163**, 5 (2002).
4. A. Manthiram, J. S. Swinnea, Z. T. Sui, H. Steinfink, and J. B. Goodenough, *J. Amer. Chem. Soc.*, **109**, 6667 (1987).
5. R. A.Young, A. Shakthivel, T. S. Moss, and C. O. Paiva Santos, *J.Appl. Crystallogr.*, **28**, 366 (1995).
6. S. Venkatraman and A. Manthiram, *Chem. Mater.*, **14**, 3907 (2002).
7. G. G. Amatucci, J. M. Tarascon, and L. C. Klein, *J. Electrochem. Soc.*, **143**, 1114 (1996).
8. X. Q. Yang, X. Sun, and J. Mcbreen, *Electrochem. Commun.*, **2**, 100 (2000).
9. C. Delmas, M. Menetrier, L. Croguennec, S. Levasseur, J. P. Peres, C. Pouillerie, G. Prado, L. Fournes, and F. Weill, *Int. J. Inorg. Mater.*, **1**, 11 (1999).
10. A. M. Kannan, L. Rabenberg, and A. Manthiram, *Electrochem. Solid-State Lett.*, **6**, A16 (2003).

Mat. Res. Soc. Symp. Proc. Vol. 756 © 2003 Materials Research Society   EE5.9

Lithium electrochemical deintercalation from O2-LiCoO2: structural study and first principles calculations

D. Carlier[a,b], A. Van der Ven[a], G. Ceder[a], L. Croguennec[b], M. Ménétrier[b] and C. Delmas[b]

[a] Department of Materials Science and Engineering, Massachusetts Institute of Technology, 77 Massachusetts Ave. Cambridge, MA, 02139 (USA)

[b] Institut de Chimie de la Matière Condensée de Bordeaux-CNRS and Ecole Nationale Supérieure de Chimie et Physique de Bordeaux
87 av. Dr A. Schweitzer, 33608 Pessac cedex (France)

Abstract

We present a detailed study of the O2-LiCoO$_2$ phase used as positive electrode in lithium batteries. This phase is a metastable form of LiCoO$_2$ and is prepared by ionic exchange from P2-Na$_{0.70}$CoO$_2$. The O2-LiCoO$_2$ system presents interesting fundamental problems as it exhibits several phase transformations upon lithium deintercalation that imply either CoO$_2$ sheet gliding or lithium/vacancy ordering. Two unusual structures are observed: T$^{\#}$2 and O6. The T$^{\#}$2 phase was characterized by X-ray, neutron and electron diffraction, whereas the O6 phase was only characterized by XRD.

In order to better understand the structures and the driving forces responsible for the phase transformations involved in lithium deintercalation, we combine our experimental study of this system with a theoretical approach. The voltage-composition curve at room temperature is calculated using Density Functional Theory combined with Monte Carlo simulations, and is qualitatively in good agreement with the experimental voltage curve over the complete lithium composition range. Pseudopotential and thermodynamic calculations both show that two tetrahedral sites have to be considered for Li in the T$^{\#}$2 structure. The calculated voltage curve thus exhibits a two-phase O2/T$^{\#}$2 region that indicates that this phase transformation is driven by the entropy maximization and not by a non-metal to metal transition. We also predict two ordered phases for Li$_{1/4}$CoO$_2$ (O2) and Li$_{1/3}$CoO$_2$ (O6) and show that the formation of the O6 phase is not related to Li staging or Co^{3+}/Co^{4+} charge ordering.

I.   Introduction

LiCoO$_2$ can exhibit two types of layered structures: O3 and O2. The thermodynamically stable structure is prepared by solid-state reaction and exhibits an O3 stacking where the LiO$_6$ and CoO$_6$ octahedra share only edges. The O2-type LiCoO$_2$ is metastable and was prepared for the first time by Delmas et al. by Na$^+$/Li$^+$ exchange from the P2-Na$_{0.70}$CoO$_2$ phase [1]. In the O2 structure, the LiO$_6$ octahedra share edges but also faces with the CoO$_6$ ones (Fig. 1). The Rietveld refinement of the neutron diffraction pattern of O2-LiCoO$_2$ (S.G. P6$_3$mc) indicated that the repulsion between lithium and cobalt ions through the common face of their octahedra is strong enough to displace them from the center of their octahedra [2]. Whereas O3-LiCoO$_2$ has been extensively studied as positive electrode in lithium batteries, to our knowledge, very few papers deal with the O2-LiCoO$_2$ system

[3-4]. During lithium deintercalation, this system exhibits several reversible voltage plateaus, which are associated with phase transformations involving either CoO_2 sheet gliding or lithium/vacancy ordering. We recently investigated the structural and physical properties of this system and showed that several unusual structures in layered oxides are observed: $T^\#2$, $T^\#2$' and O6 [5]. All these phases are named following the packing designation proposed previously by Delmas et al. for the layered oxides: the P, T or O letter describes the alkali ion site (Prismatic, Tetrahedral or Octahedral, respectively) and the number 1, 2, 3... indicates the number of slabs within the hexagonal cell. The symbol was used for the $T^\#2$ phase, which is stable for $0.52 < x \leq 0.72$, because this phase exhibits new oxygen stacking, in which the oxygen ions do not occupy positions of the same hexagonal lattice, similar to the recent observations in T2-$Li_{2/3}Ni_{1/3}Mn_{2/3}O_2$, and T2-$Li_{2/3}Co_{2/3}Mn_{1/3}O_2$ [6-7].

In order to better understand these unusual structures and the driving forces responsible for the phase transformations involved in lithium deintercalation, we have performed a first principles study of the phase diagram of the layered Li_xCoO_2 phases deriving from O2-$LiCoO_2$. It has been demonstrated how accurate phase stability information can be obtained through a combination of accurate first principles total energy calculations and statistical-mechanics techniques such as Monte Carlo simulations [8-10]. Such a method was already applied to investigate the phase diagram of the layered Li_xCoO_2 phases derived from O3-$LiCoO_2$ [11-12] and more recently the phase diagram of the Li_xNiO_2 system [13].

In this paper, we present and discuss the structural transitions occurring during Li deintercalation from O2-$LiCoO_2$ by analyzing the experimental and calculations results.

II.   Experimental section and first principles method

Preparation of O2-$LiCoO_2$

O2-$LiCoO_2$ was prepared by ion exchange reaction from P2-$Na_{0.70}CoO_2$. The $Na_{0.70}CoO_2$ precursor material was prepared by solid state reaction of a mixture of Na_2O (Aldrich, 5% excess) and Co_3O_4 (obtained from $Co(NO_3)_2.6H_2O$ Carlo Erba 99% min. at 450°C for 12 h under O_2) at 800°C under O_2 for 48 h. The mixture was ground and pelletized in a dry box (under argon) before the heat treatment, and was then quenched into liquid N_2 in order to fix the oxygen stoichiometry. Ion-exchanging sodium for lithium was performed in a LiCl / LiOH (1:1) solution in water (5 M) during 24 h with an intermediate washing, drying and grinding. An excess of lithium was used (Li/Na = 10) and the exchanged phase was washed and dried under vacuum at 100 °C for around 14 h.

Electrochemical studies

Electrochemical studies were carried out with Li/liquid electrolyte/O2-$LiCoO_2$ cells. $LiPF_6$ ethylene carbonate - dimethyl carbonate - propylene carbonate (Merck, ZV1011 Selectipur) was used as electrolyte. The cells were assembled in an argon-filled dry box. The positive electrode consisted of a mixture of 86% by weight of active material, 4% of PTFE (polytetrafluorethylene) 10% of carbon black and a C/40 current density was used (i.e. 40 h are needed to remove 1 mole of lithium).

First principles investigation

A more detailed theoretical background to obtain a phase diagram from first principles calculations can be found in some previous papers [11-12]. Essentially, the approach consists of

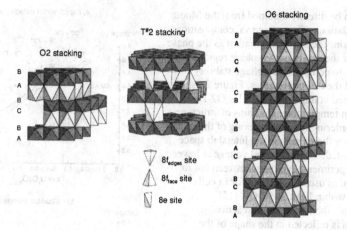

Figure 1 : The O2, T#2 and O6 stackings of Li_xCoO_2. The cobalt and the lithium ions are situated respectively in the dark and light gray polyhedra. The distorted tetrahedral sites available for the lithium ions in the T#2 structure are displayed.

using first principles calculations to parameterize the energy as a function of the occupation of possible Li sites by Li or vacancies. The parameterization takes the form of a generalized lattice Hamiltonian (the cluster expansion). The interactions in this expansion are determined by fitting the first principles calculated energy of a large number of configurations. A separate cluster expansion needs to be constructed for the O2, O6 and T#2 hosts. After the Monte Carlo simulations, the comparison of the free energy as a function of lithium concentration and temperature for each structure, allows one to build the complete phase diagram. As we investigated in the metastable phase diagram of O2-LiCoO2, we restrict ourselves solely to hosts that can be derived from it by sheet gliding, such as O6 and T#2. Other forms of Li_xCoO_2 such as O1, O3 and H1-3 require the rearrangement of the Co-O bonds and hence are not included in our phase diagram.

All energies used in the parameterization are calculated with the pseudopotential approximation to Density Functional Theory as implemented in the Vienna ab initio Simulation Package (VASP) [14]. Exchange correlations is approximated in the local density approximation (LDA), and the total-energy calculations are unpolarized.

III.   Results and discussion

The Voltage curve

Electrochemical lithium deintercalation shows a good reversibility in the overall composition range (Fig. 2a). Several structural transformations involving either reversible rearrangement of the CoO2 layers or lithium/vacancy ordering occur.

In order to compare the calculations with the experimental results, it is convenient to compare the Li insertion voltage as a function of composition. The Li chemical potential and hence the voltage can be directly obtained from the Monte Carlo simulations. The voltage vs. composition curve contains the same information as the phase diagram for the single or two-phase regions, and allows also to compare the voltage stability domains of the different phases. Figure 2b shows the calculated voltage curve of a Li//O2-LiCoO$_2$ cell at room temperature. Domains of structures with a disordering (dis.) or ordering of the Li and vacancy over the Li sites in the interslab space are represented. The calculated voltage is lower than the experimental voltage (between 0.4 to 1 V below), as usually obtained for Li cells calculated within LDA and GGA [15]. Nevertheless, the intercalation reaction mechanism is reflected in the shape of the voltage curve.

A good general agreement is observed, as the relative phase stability of the three hosts as a function of the lithium concentration is well reproduced. Basically, O2 is more stable for very high and low Li amounts, whereas T$^{\#}$2 and O6 are stable for intermediate concentrations. We discuss the phase transitions below following the Li deintercalation from O2-LiCoO$_2$:

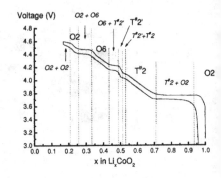

a) Experimental voltage curve

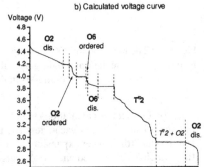

b) Calculated voltage curve

The T$^{\#}$2 structure

For x < 0.94, the system undergoes an O2 → T$^{\#}$2 phase transition that is associated with a large plateau. The calculated voltage curve exhibits also the two-phase O2/T$^{\#}$2 region indicating that this phase transformation is driven by the entropy maximization and not by non-metal to metal transition, as such transitions are rarely reproduced by LDA calculations [11].

Figure 2: experimental (a) and calculated voltage curves of a Li//O2-LiCoO$_2$ cell. the experimental results the higher and lo⁻ curves respectively represent Li removal : insertion.

In the T$^{\#}$2-Li$_x$CoO$_2$ phase the oxygen ions do not occupy the classical A, B, C positions of a triangular lattice and lithium ions are situated in distorted tetrahedra. The orthorhombic Cmca spa￼ group is used to describe the cell. The lithium ions can be situated in three different distorted tetrahedral sites: one 8e and two 8f shown in Fig. 1. The neutron diffraction characterization reve: that Li is mainly located in the 8e site, but some nuclear density is also observed in the 8f$_{edges}$ site: Calculations show that even though the 8f$_{edges}$ sites are less stable than the 8e ones for dilute Li content, they can be occupied at room temperature and are essential for the correct reproduction c the phase diagram as they add an important amount of extra configurational entropy.

For x = 0.5, the T$^{\#}$2 phase is ordered at room temperature with Li only in the more stable si￼ (8e) and corresponds to the larger Li-Li distances, so that the electrostatic Li-Li repulsion is minimized. This structure might correspond to the $T^{\#}2$' domain observed experimentally around $x = 0.5$ (Fig. 2a). For higher Li concentration, placing the lithium ions in the two types of sites ($8e$ and $8f_{edges}$) allows the Li-Li distances to be larger than those of the structures with only $8e$ site occupation, so that the higher site energy of the $8f_{edges}$ site compared to the $8e$ one is compensated for by the reduction of the Li-Li electrostatic repulsions. From the relative formation energies calculated for $T^{\#}2$, we expect an increase of the occupation of the $8f_{edges}$ sites as the Li concentration increases from $x = 0.5$ (no $8f_{edges}$ occupied) to $x = 0.75$ (experimental limit of the $T^{\#}2$ stability).

Furthermore, some of us recently performed electron diffraction of the $T^{\#}2$ crystals and showed that the large $T^{\#}2$ domain does not correspond to a single phase but to a series of structures with several complex Li/vacancy orderings. It was shown that within one $T^{\#}2$ crystal, three $T^{\#}2$ superstructures could coexist in different regions. While we did see a tendency to form complex ordered structures in the $T^{\#}2$ region in the Monte Carlo simulations as shown by the shape of the calculated voltage curve in the $T^{\#}2$ domain (Fig. 2b), more detailed work on the calculated phase diagram in the $T^{\#}2$ region is required to understand these observations.

The O6 structure

Further lithium deintercalation leads to the O6 phase ($0.33 \leq x < 0.42$) (S.G. R-3m), that was first characterized with XRD by Mendiboure et al. [3]. This stacking exhibits two types of CoO_2 layers and the possibility of Co^{3+}/Co^{4+} ordering over these layers has been inferred from XRD Rietveld refinement [5]. Spin-polarized calculations of O6-$Li_{0.5}CoO_2$ and O6-$Li_{0.33}CoO_2$ do not indicate such a charge separation so that the Co^{3+}/Co^{4+} ordering does not seem essential for the formation of the O6 stacking.

Calculations predict an ordered O6-$Li_{1/3}CoO_2$ phase to be stable at room temperature, whereas the existence of this phase was not reported. The phase transition from this ordered structure to the O6 disordered one was found to be second order, so that no two-phase domain is predicted between these phases. Actually with a more carefully look at the experimental voltage curve, we can see a small voltage slope around $x = 1/3$ (Fig. 2a). This experimental observation may therefore correspond to the formation of the ordered O6 phase.

Back to the O2 structure

For Li concentration lower than $x = 1/3$, the calculated voltage composition curve is in quite good agreement with the experimental results, but the two-phase domain associated with the O6 → O2 phase transition is predicted to be smaller: $0.29 < x < 0.33$ instead of $0.25 \leq x < 0.33$ experimentally observed. Calculations show that an O2 ordered phase is expected for $x = 1/4$ at room temperature, whereas no evidence of it has been reported. However, two different O2-Li_xCoO_2 phases were evidenced by XRD, respectively for $0.21 \leq x < 0.25$ [4-5] and $x < 0.18$ [5]. One may actually correspond to the ordered O2-$Li_{0.25}CoO_2$ and the second O2 phase to the disordered O2-Li_xCoO_2. Indeed, in the voltage curve (Fig. 2a), this first O2 phase exhibits a small stability domain, not centered exactly on $x = 1/4$ but close to this value.

IV.   Conclusions

The Li deintercalation from O2-$LiCoO_2$ is associated with several phase transformations that imply either CoO_2 sheet gliding or lithium/vacancy ordering. The synergy between the experimental characterization of the deintercalated phases (X-ray, neutron or electron diffraction), and the first principles investigation of the zero-K and finite temperature phase stability allowed us to better understand the series of unusual structures experimentally observed during Li deintercalation.

For the $T^{\#}2$ structure, we showed that, even though the $8f_{edges}$ sites are less stable than the $8f$ ones for dilute Li content, they can be occupied, and are essential for the correct reproduction of the phase diagram as they add an important amount of extra configurational entropy. We also predict several stable ordered compounds in the $T^{\#}2$ domain, especially for $x = 0.5$ that may correspond to the phase previously named $T^{\#}2'$.

We indicate that what was previously thought of by us as a single phase O6 region actually consists of an ordered region near $x = 1/3$ separated from a disordered phase region by a second order phase transition. Our calculations indicate that the formation of the O6 phase is not linked with a Li staging and is not driven by a Co^{3+}/Co^{4+} ordering on the two different cobalt layers as previously proposed.

Finally, we predict an ordered O2 phase for $x = 1/4$ and find that the O2 structure should remain stable for CoO_2 (with respect to $T^{\#}2$ and O6).

Acknowledgements

The authors wish to thank I. Saadoune, E. Suard and Y. Shao-Horn for their contribution in the experimental work and fruitful discussions on the results presented here, as well as the French ministry of foreign affairs (Lavoisier fellowship), the NSF/CNRS exchange grants (NSF-INT-0003799) and the Center for Materials Science and Engineering of MIT for financial support.

References

[1] C. Delmas, J. J. Braconnier and P. Hagenmuller, *Mat. Res. Bull.*, **17**, 117 (1982).

[2] D. Carlier, I. Saadoune, E. Suard, L. Croguennec, M. Ménétrier and C. Delmas, *Solid State Ionics*, **144**, 263 (2001).

[3] A. Mendiboure, C. Delmas and P. Hagenmuller, *Mat. Res. Bull.*, **19**, 1383 (1984).

[4] J. M. Paulsen, J. R. Mueller-Neuhaus and J. R. Dahn, *J. Electrochem. Soc.*, **147**(2), 508 (2000).

[5] D. Carlier, I. Saadoune, M. Ménétrier and C. Delmas, *J. Electrochem. Soc.*, **149**(10), A1310 (2002).

[6] J. M. Paulsen, R. A. Donaberger and J. R. Dahn, *Chem. Mater.*, **12**, 2257 (2000).

[7] Z. Lu, R. A. Donaberger, C. L. Thomas and J. R. Dahn, *J. Electrochem. Soc.*, **149**(8), A1083 (2002).

[8] D. De Fontaine, *Solid State Physics*, Academic, New York (1994).

[9] G. Ceder, A. F. Kohan, M. K. Aydinol, P. D. Tepesch and A. Van der Ven, *J. Am. Ceram. Soc.* **81**(3), 517 (1998).

[10] A. Zunger, *Statistics and Dynamics of Alloy Phase Transformations*, Plenum, New York (1994).

[11] A. Van der Ven, M. K. Aydinol, G. Ceder, G. Kresse and J. Hafner, *Phys. Rev. B*, **58**(6), 2975 (1998).

[12] G. Ceder and A. Van der ven, *Electrochem. Acta*, **45**(1-2), 131 (1999).

[13] M. E. Arroyo y de Dompablo, A. Van der Ven and G. Ceder, *Phys. Rev. B*, **66**, 064112 (2002).

[14] G. Kresse and J. Furthmuller, *Comp. Mat. Sci.*, **6**, 15 (1996).

[15] S. K. Mishra and G. Ceder, *Phys. Rev. B*, **59**(9), 6120 (1999).

Mat. Res. Soc. Symp. Proc. Vol. 756 © 2003 Materials Research Society

New Iron (III) Hydroxyl-Phosphate with Rod-packing Structure as Intercalation Materials

Yanning Song, Peter Y. Zavalij and M. Stanley Whittingham
Department of Chemistry and Institute for Materials Research, State University of New York at Binghamton, Binghamton, NY 13902-6016

ABSTRACT

A new iron hydroxyl-phosphate, $H_2Fe_{14/3}(PO_4)_4(OH)_4$ has been synthesized under hydrothermal conditions. In this compound, perpendicular chains formed by the face-sharing FeO_6 form rod-packing structure. Only about 60% of the chain sites are occupied by iron atoms; other metals, such as manganese, nickel, zinc, can be incorporated into the chain either by filling in the vacancies and/or replacing some of the iron atoms. Reversible insertion and extraction of lithium into this compound shows it to be an excellent cathode material. At current density of 0.1 mA/cm^2, 90 % of the theoretical capacity (176 mAh/g) can be obtained. The utilization was reduced to about 70 % on a ten-fold increase of current density. The electrochemical behavior is attributed to the 3-dimensional rod packing structure, where lithium can move freely even at high current densities inside the 3-dimensional framework without altering the host structure. Two of the protons in the lattice may be exchanged by lithium giving $Li_2Fe_{14/3}(PO_4)_4(OH)_4$. These lithium atoms are not removable in electrochemical cycling and similar electrochemical property was found for these two compounds, suggesting an ion-exchange process for the lithiation.

INTRODUCTION

Iron phosphate materials are pervasive in nature. They have long been used as catalyst in the synthesis of organic acids and in steel and glass industry. Recently, they are used as cathode materials in lithium batteries. They are advantageous because of the low-cost, environment-friendly properties. With a NASICON-related-3D framework, $Li_3Fe_2(PO_4)_3$ [1,2] can be electrochemically inserted 1.8 lithium atoms per formula unit at 0.05 mA/cm^2 (~116 mAh/g). And when the current density is increased to 1.0 mA/cm^2, the capacity fades to 60 mAh/g. The more successful material is $LiFePO_4$ with the olivine structure [3,4]. About 0.6 lithium atoms (~110 mAh/g) can be inserted into the structure at a current density of 2 mA/g. The excellent reversibility is due to the striking similarity of the charged and discharged materials. However, the capacity fades quickly also at higher current density due to the low diffusion rate of ions and electrons [5]. $FePO_4$ is also investigated as a candidate [6,7]. Although the amorphous phase shows practical capacity, the possibility of commercialization has been limited by the low density and low voltage.

Here we report our study on an iron hydroxyl phosphate with "rod-packing" structure.

EXPERIMENTAL DETAILS

Synthesis

The reaction mixture for synthesis contains 0.001mol $FeCl_3$ (J. T. Baker), 0.004 mol H_3PO_4 (Fisher) and 0.2 mol water. Methylamine (Aldrich) was used to adjust pH. The mixture was loaded in a Teflon-lined stainless steel autoclave and heated at 170 °C for 3 days. The resulting product, $H_2Fe_{14/3}(PO_4)_4(OH)_4$, was filtered off, thoroughly washed with an excess of deionized water and dried in oven at 55 °C for 15 hours. Some other metal chlorides, $MnCl_2$, $NiCl_2$, and $ZnCl_2$ with different ratios to iron were also added to substitute some of the iron. The reaction with LiI was done by adding $H_2Fe_{14/3}(PO_4)_4(OH)_4$ to a solution of LiI (Aldrich) in acetonitrile (Burdick & Jackson) in a N_2 glove box. After 5 days the ion-exchanged product, $Li_2Fe_{14/3}(PO_4)_4(OH)_4$, was washed several times with acetonitrile before it was dried in vacuum.

Characterization techniques

Powder diffraction patterns were obtained on a Scintag XDS2000 θ-θ powder diffractometer equipped with a Ge(Li) solid state detector (CuK_α radiation). Thermal analysis was performed on a Perkin-Elmer TGA in an oxygen atmosphere. Electrochemical studies were conducted in a helium-filled glove box using a Macpile galvanostat. In all cases, the sample was mixed with 20 wt.% carbon black and 10 wt.% PTFE powder; around 20 mg/cm^2 was hot-pressed into a stainless steel Exmet™ grid for one hour at about 120 °C. A bag cell configuration was used with a 1M solution of $LiPF_6$ in 1:1 DEC:EC (EMI, LP30) as the electrolyte, pure lithium as the anode, and Celgard 2400 (Hoechst Celanese Corp.) for the separator. The morphologies of the compounds were studied on a JEOL 8900 SEM.

DISCUSSION

Synthesis and Structure

It was found out that pH is critical in the synthesis of these compounds. The highly pure $H_2Fe_{14/3}(PO_4)_4(OH)_4$ can only be obtained in the pH range of 4.5 to 4.8. At lower pH, monoclinic $FePO_4.2H_2O$ was obtained. At higher pH, spheniscidite $Fe_2(NH_4)(OH)(PO_4)_2.2H_2O$ [8] was obtained. The manganese, nickel and zinc substituted compounds were also successfully synthesized in the pH range of 4.5 to 5.0.

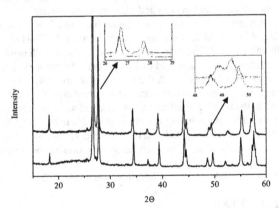

Figure 1. XRD patterns of $H_2Fe_{14/3}(PO_4)_4(OH)_4$ (bottom) and $Li_2Fe_{14/3}(PO_4)_4(OH)_4$ (top); insets magnify to show the difference.

Table 1. Cell parameters of the substitution for $H_2Fe_{14/3}(PO_4)_4(OH)_4$ after [10].

	a (Å)	c (Å)	Volume (Å³)	Occupation Factor
$H_2Fe_{14/3}(PO_4)_4(OH)_4$	5.180 (1)	13.034(1)	349.78(1)	0.588(2)
$Li_2Fe_{14/3}(PO_4)_4(OH)_4$	5.211(1)	12.914(1)	350.65(1)	-
Ni:Fe=1:4*	5.189(1)	13.016(2)	350.5(2)	0.662(2)
Zn:Fe=1:4	5.194 (1)	13.015(1)	351.1(1)	0.682(4)
Mn:Fe=1:2	5.226(1)	13.122(2)	358.4(1)	0.670(4)
Mn:Fe=1:1	5.244(1)	13.155(1)	361.71(2)	0.684(4)

* The molar ratio of M to Fe (M=Ni, Zn, Mn)

The XRD patterns of $H_2Fe_{14/3}(PO_4)_4(OH)_4$ and $Li_2Fe_{14/3}(PO_4)_4(OH)_4$ are quite similar as shown in figure 1. The magnified patterns show the difference. Both patterns were refined in space group $I4_1/amd$ and the cell parameters are a=5.180(1) Å, c=13.034(1) Å for $H_2Fe_{14/3}(PO_4)_4(OH)_4$ and a=5.211(1) Å, c=12.914 (1) Å for $Li_2Fe_{14/3}(PO_4)_4(OH)_4$. The structure is described elsewhere [9]. Basically, the iron octahedra share two opposite faces with two nearest neighbors and form infinite chains along [100] and [010] directions. The distance (2.6 Å) of the centers for the neighboring octahedra in the chain is too short for Fe^{3+}-Fe^{3+} due to strong electrostatic repulsion; at most 2/3 of the chains can be occupied. That means at least one of the neighbors of the octahedra is empty. In the case of Fe-A, 58.8 % of the chain sites are occupied. Therefore, it is possible to incorporate other cations into the chain. The cell parameters for the substitutions with other cations are shown in table I. Indeed, in all of the substitutions the occupation factor goes up to around 2/3. For $Li_2Fe_{14/3}(PO_4)_4(OH)_4$, it is supposed that lithium ions replace protons via ion-exchange as discussed later. We are presently attempting to determine the lithium ion position in the ion-exchanged compound.

Morphology study

SEM study (figure 2) shows that very uniform spherical materials can be obtained under hydrothermal conditions. The size of $H_2Fe_{14/3}(PO_4)_4(OH)_4$ particles is about 10 μm, while the particles for the nickel and zinc substituted compounds are smaller, 4 μm and 3 μm respectively. Meanwhile, the particles for zinc substituted compound are mostly incomplete spheres. The substitution for the iron has changed the morphologies.

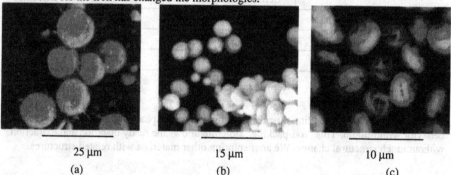

25 μm   15 μm   10 μm

(a)   (b)   (c)

Figure 2. SEM of (a) $H_2Fe_{14/3}(PO_4)_4(OH)_4$, (b) ratio Ni:Fe=1:4 and (c) ratio Zn:Fe=1:4 [10].

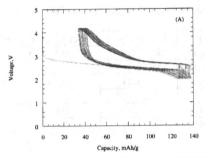
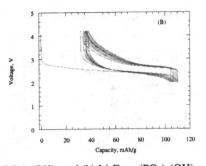

Figure 3. Electrochemical behavior of (a) $H_2Fe_{14/3}(PO_4)_4(OH)_4$ and (b) $Li_2Fe_{14/3}(PO_4)_4(OH)_4$ at 0.5 mA/cm^2 after [10].

Electrochemical study

Good electrochemical behavior is found in the hydroxyl phosphate as shown in figure 3. The capacity is about 140 mAh/g in the first cycle at current density of 0.5 mA/cm^2, with an irreversible capacity lost of about 40 mAh/g in the following cycles. This is found to be due to kinetic effects. More than 90% of the theoretical capacity is reversible when cycled at a lower current density of 0.1 mA/cm^2 [10]. Even when cycled at a current density of 1mA/cm^2, more than 70% of the capacity can be maintained. The good electrochemical behavior is believed to be the result of the stability of the "rod-packing" structure. Between the chains, lithium ions can be easily inserted or extracted. The high similarity of the electrochemical behavior of $H_2Fe_{14/3}(PO_4)_4(OH)_4$ and $Li_2Fe_{14/3}(PO_4)_4(OH)_4$ suggests that the iron in $Li_2Fe_{14/3}(PO_4)_4(OH)_4$ is also Fe^{3+}. This suggests an ion-exchange mechanism for the chemical lithiation.

The lattice expands as lithium is intercalated into the structure, as shown in table 2 for a sample after cycling 10 times. The volume change on lithium insertion is only 2.31%. This is much smaller than in $LiFePO_4$ (7.11%).

Table 2. Cell parameters of $H_2Fe_{14/3}(PO_4)_4(OH)_4$ before and after 10 cycles after [10].

	Before cycling	10 cycles, discharged state
a (Å)	5.180(1)	5.249(1)
c (Å)	13.034(1)	12.990(2)
Volume (Å^3)	349.8(1)	357.9(1)

CONCLUSION

We have shown for the first time that iron hydroxyl phosphate exhibits excellent redox behavior with lithium. This "rod-packing" structure allows the ready insertion and extraction without much structural change. We are exploring other materials with related structures.

ACKNOWLEDGEMENTS

The authors would like to thank the National Science Foundation for support of this work through grant DMR-9810198. We also thank William Blackburn for the SEM study.

REFERENCES

1. K. S. Nanjundaswamy, A. K. Padhi, J. B. Goodenough, S. Okada, H. Ohtsuka, H. Arai and J. Yamaki, *Solid State Ionics* **92**, 1(1996).
2. C. Masquelier, A. K. Padhi, K. S. Nanjundaswamy and J. B. Goodenough, *J. of Solid State Chem.* **135**, 228 (1998).
3. A. K. Padhi, K. S. Nanjundaswamy and J. B. Goodenough, *J. Electrochem. Soc.* **144**, 1188 (1997).
4. A. K. Padhi, K. S. Nanjundaswamy, C. Masquelier, S. Okada and J. B. Goodenough, *J. Electrochem. Soc.* **144**, 1609 (1997).
5. A. S. Andersson, B. Kalska, L. Haggstrom and J. O. Thomas, *Solid State Ionics* **130**, 41 (2000).
6. Y. Song, S. Yang, P. Y. Zavalij and M. S. Whittingham, *Mater. Res. Bull.* **37**, 1249 (2002).
7. Y. Song, P. Y. Zavalij, M. Suzuki and M. S. Whittingham, *Inorg. Chem.* **41**, 5778 (2002).
8. M. Cavellec, G. Ferey, J. M. Greneche, *J. Magnetism and Magnetic Mater.* **167**, 57 (1997).
9. Y. Song, N. A. Chernova, P. Y. Zavalij and M. S. Whittingham, *Chem. Mater.* (submitted).
10. Y. Song, P. Y. Zavalij and M. S. Whittingham, *Electrochem. Comm.* (submitted).

Synthesis and Electrochemical Behaviour of Ramsdellite LiCrTiO$_4$

F. García-Alvarado, M. Martín-Gil and A. Kuhn
Departament of Chemistry, Universidad San Pablo CEU, E-28688 Boadilla del Monte, Madrid
(Spain); e-mail: flgaal@ceu.es

Abstract

A ramsdellite with composition LiCrTiO$_4$ has been obtained by heating the spinel of same composition to high temperature. The new ramsdellite has been investigated in view of its possible use as an electrode material in lithium rechargeable batteries. Lithium can be partially extracted from ramsdellite LiCrTiO$_4$ and further intercalated into, by contrast to the spinel of same composition. The average operating voltage during lithium extraction is 4 Volts vs. lithium, and the process produces a specific capacity of 90 mAh/g at 0.1 mA/cm^2. On the other hand, upon reduction from open circuit voltage, lithium can be reversibly intercalated into the ramsdellite polymorph at ca. 1.5 V vs. lithium yielding a rechargeable capacity of 110 mAh/g at 0.1 mA/cm^2.

Introduction

Among the compounds that intercalate lithium at low voltage, several titanium oxides have been found [1-5]. A common feature in all these materials is partial or complete reduction of Ti^{4+} to Ti^{3+} at voltages close to 1.5 Volts. This value can be regarded relatively high when compared to carbonaceous or tin-based compounds. However, the great advantage is that Ti-O framework sustains a good cycle life. For the particular case of Li$_2$Ti$_3$O$_7$ we demonstrated that the titanate suffers a low internal stress upon intercalation, which was directly related to the observed good cycling behaviour [5]. Another example can be found in Li$_4$Ti$_5$O$_{12}$ [6]. Recently, new titanium-based ramsdellites have been reported: ramsdellite-like LiTi$_2$O$_4$ was obtained by heat treatment of the spinel of same composition [7] and lithium can be either chemically or electrochemically removed [8,9]. Furthermore, ramsdellite-like LiTi$_2$O$_4$ has been investigated in view of its potential use as negative electrode [9-11] in lithium ion batteries. It has been shown that ramsdellite LiTi$_2$O$_4$ shows excellent electrochemical behaviour in both reduction and oxidation, i.e. it can undertake both lithium intercalation (Li$_{1+x}$Ti$_2$O$_4$) and lithium extraction reactions (Li$_{1-x}$Ti$_2$O$_4$) [8,9]. It has been pointed out that only part of the total capacity would be useful for negative electrode use in rechargeable lithium ion batteries [9]. This part corresponds to the capacity involved in the process characterised by a low voltage plateau located at 1.5 V vs. Li, which corresponds to the region where LiTi$_2$O$_4$ transforms to Li$_2$Ti$_2$O$_4$ through a biphasic region. Regarding lithium extraction from LiTi$_2$O$_4$, a continuous variation of voltage, from 3 to 1.5 V vs. Li$^+$/Li is observed. This explains why the maximum useful capacity as negative electrode of both ramsdellites Li$_2$Ti$_3$O$_7$ and LiTi$_2$O$_4$ are finally very similar (190 mAh/g). In the same way, a study of the complete solid solution series of ramsdellite phases Li$_{1+x}$Ti$_{2-2x}$O$_4$ has been recently performed regarding their possible application as electrode material in lithium batteries [10].

It is noteworthy to recall that, since in the spinel LiTi$_2$O$_4$ half of the titanium ions are formally as 3+ and the other half as 4+, i.e. can be formulated as LiTi^{3+}Ti^{4+}O$_4$, partial or complete substitution of Ti(III) by other transition metals with stable trivalent state has been widely described: by vanadium [12,13], chromium [13,14] or iron [13-15], leading to the corresponding spinel solid solutions $LiM^{3+}_xTi^{3+}_{1-x}Ti^{4+}O_4$ (or $LiM_xTi_{2-x}O_4$). In this paper we report that $LiCrTiO_4$ ramsdellite, obtained at a temperature high enough to overpass the stability of the corresponding spinel phase, has an interesting electrochemical behaviour regarding its performance as possible electrode material for lithium ion batteries. We have also investigated the transformation to ramsdellite and electrochemical behaviour of other mixed spinels $LiTiMO_4$ with M=V, Fe and Mn [16].

Experimental

The low temperature (spinel) $LiCrTiO_4$ has been obtained following the route reported by Edwards et al. [14]. The ramsdellite form of $LiCrTiO_4$ was obtained when the spinel was heated during 24 hours at 1250°C in air and afterwards quenched to room temperature. This procedure was repeated twice. The X-ray phase purity was checked on a Philips X'Pert diffractometer using Cu $K\alpha$ radiation. Unit cell parameters were refined with the Fullprof program [17] by means of Rietveld profile fit [18]. In addition, crystal symmetry has been checked by means of electron diffraction on JEOL 2000 electron microscope.

Magnetic measurements on ramsdellites were performed by the Zero Field Cooling method under a magnetic field of 50 Oe in a conventional SQUID magnetometer.

Electrochemical measurements were made in Swagelok^{TM} test cells with the following configuration:

(-) Li/LiPF$_6$(1M) in EC+DMC (1:1)/LiCrTiO$_4$+carbon black + binder (+)

Lithium intercalation and extraction was evaluated by constant current discharge/charge experiments in the above described lithium cells. The positive electrode (15-20 mg) consisted of a mixture containing 85 w % $LiCrTiO_4$, 10 w % carbon black and 5 w % binder, which was pressed to 8 mm diameter pellets. For measurements under equilibrium conditions a current pulse of 50 μA was applied during 30 minutes. Cells were then allowed to stay during 6 hours in order to reach equilibrium potential before the next current pulse. Chronoamperommetric experiments were performed at ±10 mV / 30 min. at 25°C; however for this type of experiment the working electrode consisted of 70 w % oxide, 25 w % carbon black and 5 % binder. Selectipur LP30® (Merck) was used as the liquid electrolyte. Kynarflex®, kindly provided by Elf-Atochem, was used as a binder. Cells were both monitored and controlled using a MacPile II system.

Results and discussion

Characterisation

X-ray diffraction pattern of low-temperature $LiCrTiO_4$ was indexed on the basis of cubic spinel with a=8.3139 (1) Å, and space group Fd-3m, in good agreement with the literature values [13,14]. After treatment at high temperature, the corresponding pattern was indexed with the space group P bnm, the orthorhombic lattice parameters being: a = 4.9818 (8) Å, b = 9.503 (1) Å, c = 2.9264 (2) Å. Both space group and lattice dimensions are consistent with a ramsdellite phase and similar to those described for $LiTi_2O_4$ [7,8]. This structure consists of chains along c-direction form by edge-sharing [M-O$_6$] octahedra with Cr and Ti ions located at their centre (see

Figure 1). These units are further linked together in the ab-plane by sharing both corners and edges to build up a rather open 3D structure. The lithium atoms are located in the tunnel.

The magnetic measurements performed on the ramsdellite LiCrTiO$_4$ in the temperature range 200-300 K showed a Curie-like paramagnetic behaviour. The value of the effective magnetic moment, 3.78 B.M., is close to the expected spin-only value for localised paramagnetic Cr^{3+} ions (3.87 B.M.). A similar value has been reported [14] for the corresponding spinel form and it was also attributed to localised paramagnetic Cr^{3+} ions. Then we propose the formulation LiCr^{3+}Ti^{4+}O$_4$ as appropriate for the new ramsdellite reported in here.

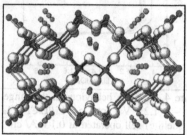

Figure 1.- Perspective view of the LiCrTiO$_4$ (R) structure along the tunnel direction. Large spheres: oxygen atoms; small spheres: Cr/Ti atoms. Small spheres located in the tunnel represent the lithium atoms.

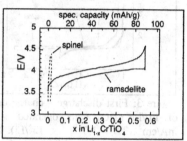

Figure 2.- First charge and discharge curves of Li/LiCrTiO$_4$ cells operated at 0.1 mA/cm^2 between 3.5 and 5.0 V vs. Li$^+$/Li. Solid line: ramsdellite. Dotted line: spinel. No appreciable oxidative reaction is observed.

Electrochemical study

In order to know whether ramsdellite LiCrTiO$_4$ can be oxidised to Li$_{1-x}$CrTiO$_4$, the electrochemical cells were tested in the potential range between 3.5 and 4.6 V vs. Li$^+$/Li. The first charge and discharge curves of a Li//LiCrTiO$_4$ (ramsdellite) cell operated at 0.1 mA/cm^2 are shown in Figure 2 (solid line). For comparison, the first oxidation – reduction cycle of the spinel polymorph, obtained under identical current density, is presented (dotted line). For ramsdellite, a reversible behaviour involving almost 0.6 Li/formula unit (specific capacity of about 90 mAh/g) is observed albeit we observed a complete capacity loss in the following two cycles. On the other hand lithium extraction is not possible from the spinel polymorph (dotted line), in agreement with a previous report [19].

The first discharge – charge curves in the range 3.5 – 1.0 V vs. Li$^+$/Li for a Li/LiCrTiO$_4$ (ramsdellite) cell operated at 0.1 mA/cm^2 current density are presented in Figure 3 (solid line). The operating voltage quickly drops down to 2.1 V vs. Li$^+$/Li. Between 2.1 and 1.5 V, a relatively small amount of lithium is inserted. From there on, a relatively flat discharge curve is obtained at 1.5 V vs. Li$^+$/Li. Up to 0.7 Li ions per formula are inserted under these experimental conditions. The 1st discharge capacity of ramsdellite (solid line) is 110 mAh/g and as it can be observed in Figure 4, capacity is well maintained during subsequent discharge-charge cycles. After 15 cycles capacity remains close to 100 mAh/g. The origin of the observed reduction process is clear since it is a common feature in titanium oxides able to intercalate lithium, that reduction of Ti^{4+} to Ti^{3+} occurs at voltages close to 1.5 Volts [1-5]. One example is the spinel form of $LiCrTiO_4$, which has been reported to reversibly undergo a one-electron reduction [19]. This is also shown in Figure 3 although we have to note that insertion proceeds a little bit further than the expected limit for $LiCr^{3+}Ti^{4+}O_4$, i.e 1 Li/formula [19]. The maximal theoretical capacity of $LiCrTiO_4$ is then calculated to be 157 mAh/g, assuming that all Ti^{4+} is reduced to

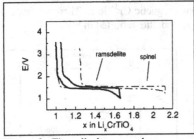

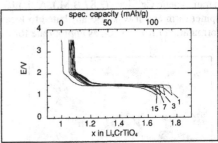

Figure 3: First discharge - charge curves of $Li//LiCrTiO_4$ cells operated at 0.1 mA/cm^2 (3.5 - 1.0 V vs. Li^+/Li). Solid line: ramsdellite. Dotted line: spinel.

Figure 4.- Cycling behaviour in the range 3.5 – 1.2 V for a $Li//LiCrTiO_4$ (ramsdellite) cell operated at 0.1 mA/ cm^2 current density

Ti^{3+}. We have obtained that maximal theoretical capacity for the spinel form under a current density of 0.1 mA $\cdot$ cm^{-2} while same did not happen for ramsdellite (see Figure 3).

In order to know whether the new ramsdellite $LiCrTiO_4$ can reach in certain conditions 157 mAh/g, and hence if the same redox mechanism (complete reduction of 1 Ti^{4+}/formula unit) is operating, discharging runs were performed under equilibrium conditions. As it can be seen in Figure 5, ramsdellite now does reach the theoretical capacity corresponding to 1 Li/Ti. We are presently paying attention to the slight excess of capacity of $LiCrTiO_4$ observed for our both samples spinel and ramsdellite. Either composition or any change in oxidation state distribution due to lithium and chromium loss (for example $Li_{1-x}(Cr^{3+})_{1-x}(Ti^{4+})_{1+x}O_4$) may be at the origin of the overpass. A chromium deficiency due to segregation of Cr_2O_3 is possible due to the fact that we used Al_2O_3 crucibles. At high temperature diffusion of Cr_2O_3 into the crucible to form the well known $Al_{2-x}Cr_xO_3$ solid solution is likely. In fact, during the synthesis of the ramsdellite (at 1250 ºC) a red colour, like that of the ruby $Al_{2-x}Cr_xO_3$, developed on the wall of the crucible. However, other possibility is the loss of lithium ($Li_{1-2x}Cr_xTiO_{4-x}$) that would reduce the formula weight of our active material which would be reflected in our experiments as if extra lithium ions were intercalated. In good agreement with these two possibilities, analysis results indicate that lithium content is lower than nominal. The oxidation state of cations are being presently investigated by Electron Energy Loss Spectroscopy (EELS). The results will be reported elsewhere.

A complete overview of differences between spinel and ramsdellite can be extracted from Figure 6 where a first charge-discharge chronoamperogram of $Li//LiCrTiO_4$ cell operated in the whole stability range between 1.0 and 4.5 V vs. Li^+/Li have been obtained at a scan rate of ±10 mV / 30 min. The main difference between the two polymorphs is found in the 3.5 – 4.5 V voltage region. Ramsdellite (solid line) presents an oxidative peak with its maximum at 4.2 V vs.

Li, whereas the homologous reductive signal, indicating reversibility of the process, appears at 3.9 V vs. Li. On the contrary, no appreciable current is detected in the spinel form of $LiCrTiO_4$ in the high voltage region up to 4.5 V vs. Li.

It is suitable to note that in layered $Li_{1.2}Cr_{0.4}Mn_{0.4}O_2$ oxidation of Cr^{3+} to Cr^{6+} has been proved to occur in the potential range from 3.5 to 4.4 V [20]. Besides, it has been demonstrated that oxidation is accompanied by migration of Cr from octahedral to tetrahedral position and that this process is reversible and provide good cycling life [21]. Regarding our system, results from magnetic measurements confirm the presence of localised Cr^{3+} in the parent compound $LiCrTiO_4$ and oxidation occurs in the same potential range as that of $Li_{1.2}Cr_{0.4}Mn_{0.4}O_2$. Then, it is likely that in our case the pair Cr^{6+}/Cr^{3+} is the responsible of the observed high voltage process.

To find out if oxidation to Cr^{6+} does occur we performed magnetic measurements on oxidised samples. The result has not helped since the observed magnetic behaviour, that we have not interpreted up to date, indicates the presence of strong cooperative effects between ions different from diamagnetic Cr^{6+}. The EELS study that is now in under progress may finally clarify the origin of the high voltage process.

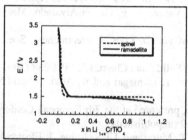

Figure 5: Open circuit voltage measurement of Li/LiCrTiO₄ cell. Solid line: ramsdellite. Dotted line: spinel.

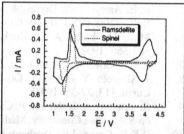

Figure 6.- Chronoamperogram of Li/LiCrTiO₄ cells obtained at a scan rate of ±10 mV / 30 min. (1- 5 V vs. Li⁺/Li). Solid line: ramsdellite. Dotted line: spinel.

Conclusions

A high temperature form of $LiCrTiO_4$ with ramsdellite structure has been obtained by simply heating the analogous spinel of same composition at 1250 °C.

On the contrary to the spinel form, the ramsdellite $LiCrTiO_4$ undergoes a reversible oxidation (lithium extraction) reaction through an almost constant voltage region situated at high voltage (4 Volts vs. Li⁺/Li). A reversible capacity of approximately 90 mAh/g at 0.1 mA/cm² is developed albeit complete capacity loss is observed after two cycles.

Furthermore, ramsdellite-like $LiCrTiO_4$ undergoes lithium intercalation at an average potential of approximately 1.5 Volts vs. Li⁺/Li. The reversible discharge capacity of ramsdellite $LiCrTiO_4$ is close to 100 mAh/g at 0.1 mA/cm² although a capacity of 157 mA/g can be reached under equilibrium conditions. On the other hand, a good cycling stability is observed in this low potential range.

In summary the oxide $LiCrTiO_4$ presents two well defined and potential-separated redox processes. The first, at 1.5 V, attributed to the redox couple Ti^{4+} / Ti^{3+} and the second, at 4 V, due to oxidation of Cr^{3+} although the final state of chromium has not been yet determined.

Acknowledgements

This work has been financially supported by both Ministerio de Ciencia y Tecnología through the project MAT2001-3713-C04-01 and Universidad San Pablo CEU.

References

1. B. Zachau-Christiansen, K. West, T. Jacobsen and S. Skaarup, Solid State Ionics **53-56**, 364-369 (1992).
2. F. García-Alvarado, M.E. Arroyo y de Dompablo, E. Morán, M.T. Gutiérrez, A. Kuhn and A. Várez, J. Power Sources **81-82**, 85 (1999).
3. L.D. Noailles, C.S. Johnson, J.T. Vaughey and M.M. Thackeray, J. Power Sources **81-82**, 259-263 (1999)
4. K.M. Colbow, J.R. Dahn and R.R. Haering, J. Power Sources **26**, 397-402 (1989).
5. M.E. Arroyo y de Dompablo, E. Morán, A. Várez and F. García-Alvarado, Mat. Res. Bull. **32**(8), 993-1001 (1997).
6. M.M. Thackeray, A. De Kock, M.H Rossouw and D. Liles, J. Electrochem. Soc. **139**, 363-366 (1992).
7. R.K.B. Gover, J.T.S. Irvine and A.A. Finch, J. Solid State Chem. **132**, 382-388 (1997).
8. J. Akimoto, Y. Gotoh, Y. Oosawa, N. Nonose, T. Kumagai and K. Aoki, J. Solid State Chem. **113**, 27-36 (1994).
9. R. Amandi, A. Kuhn and F. García-Alvarado, presented at the 7th European Conference on Solid State Chemistry, Madrid (Spain) 1999 (unpublished).
10. R.K.B. Gover, J.R. Tolchard, H. Tukamoto, T. Murai and J.T.S. Irvine, J. Electrochem. Soc. **146**(12), 4348-4353 (1999).
11. A. Kuhn, R. Amandi and F. García-Alvarado, J. Power Sources **92**, 221-227 (2001).
12. T. Hayakawa, D. Shimada and N. Tsuda, J. Phys. Soc. Jap. **58**(8), 2867-2876 (1989).
13. G. Blasse, J. Inorg. Nucl. Chem. **25**, 230 (1963).
14. P.M. Lambert, P.P. Edwards, and M.R. Harrison, J. Solid State Chem. **89**, 345 (1990).
15. S. Scharner and W. Weppner, J. Solid State Chem. **134**, 170-181 (1997).
16. F. García-Alvarado, A. Kuhn and M. Martín-Gil, Bol. Soc. Esp. Cer. Vidrio **41**, 385-392 (2002)
17. J. Rodríguez-Carvajal, *Fullprof Manual*, Institute Laue-Langevin, Grenoble France (1992).
18. H.M. Rietveld, Cryst. **22**, 151 (1967); J. Appl. Cryst. **65**, 2 (1969).
19. T. Ohzuku, K. Tatsumi, N. Matoba and K. Sawai, J. Electrochem. Soc. **147**(10), 3592-3597 (2000).
20. B. Ammundsen, J. Paulsen, I. Davidson, R. Liu, C. Shen, J. Chen and J. Lee, J. Electrochem. Soc. **149**, A431-A436 (2002).
21. M. Balasubramanian, J. McBreen, I.J. Davidson, P. S. Whitfield and I. Kargina, J. Electrochem. Soc. **149**, A176-A184 (2002).

Electrochemical intercalation of lithium in ternary metal molybdates $MMoO_4$ (M = Cu, Zn)

Th. Buhrmester, N. N. Leyzerovich, K. G. Bramnik, H. Ehrenberg, H. Fuess

University of Technology Darmstadt, Department of Materials Science, Structure Research, Petersenstrasse 23, D-64287 Darmstadt, Germany
Fax: +49 6151 16-60 23; E-mail: buhrmester@tu-darmstadt.de

ABSTRACT

Ternary oxides with general formula $MMoO_4$ (where M is a $3d$-transitional metal) were characterized as cathode materials for lithium rechargeable batteries by galvanostatic charge-discharge technique and cyclic voltammetry. The significant capacity fading after the first cycle of lithium insertion/removal takes place for different copper molybdates (standard α-$CuMoO_4$ and high-pressure modification $CuMoO_4$–III) corresponding to the irreversible copper reduction and formation of Li_2MoO_4 during the first discharge. X-ray powder diffraction data reveal the decomposition of pristine $ZnMoO_4$ by electrochemical reaction, lithium zinc oxide with the NaCl-type structure and Li_2MoO_3 seem to be formed.

INTRODUCTION

Among the systems which are intriguing in the rechargeable lithium ion technology molybdenum (VI) compounds remain of significant potential. The idea to take an advantage of the charge couple Mo^{+6}/Mo^{+4} where the metal-redox oxidation state can change by two units is very attractive for the development of batteries with high capacity and consequently, with high energy density. Since more than 20 years the various molybdenum oxides have been tested as cathode materials in lithium rechargeable batteries [1-5]. However, the significant capacity fading after several cycles accompanied with irreversible structure transformations during the first discharge (lithium intercalation) has been reported.

Mixed-molybdenum oxides are interesting as matrices for lithium intercalation while the combination of two metals in oxide-matrices produces the materials with new structure and chemical properties compared to that of binary oxides. Additionally, because in many cases the bimetallic oxides can form different crystal structures or phases depending on the synthetic conditions and compositions, a large variety of systems could be modelled and tested. Vanadium-based compounds like $MeVO_4$ (Me = In, Cr, Fe, Al, Y, Ni) were shown to be good candidates for anode materials in lithium batteries [6,7]. However, up to now there are very few investigations concerning lithium intercalation into the same type of molybdenum compounds. Quite early the copper molybdate was probed as cathode for primary lithium batteries, but the reversibility of intercalation was not discussed [8,9]. Recently, the low potential lithium insertion in manganese-molybdenum oxide and its potential application as anode material in lithium batteries has been described [10].

In this work, we report our results of electrochemical insertion of lithium into ternary molybdates α-$MeMoO_4$ (Me = Cu, Zn). α-$CuMoO_4$ is known to crystallize in the triclinic system (space group $P1$), where three molybdenum atoms on different sites are tetrahedrally coordinated by oxygen [11]. For comparison, the high pressure modification $CuMoO_4$-III (space group P1) [12, 13] with molybdenum atoms octahedrally coordinated by oxygen was included in this investigation. $ZnMoO_4$ is isotypic with α-$CuMoO_4$.

EXPERIMENTAL DETAILS

Ternary metal α-molybdates were synthesized by solid-state reaction of a stoichiometric mixture of MoO_3 (99.99 % Aldrich) and MeO (Me = Cu, Zn) (99.99% Aldrich) as described previously [11]. The high pressure modification CuMoO4-III was prepared following the procedure described by Tali et al. [12]. Phase identification was performed by X-ray powder diffraction.

Electrochemical studies were carried out with a multichannel potentiostatic-galvanostatic system VMP (Perkin Elmer Instruments, USA). Swagelok^{TM}-type two-electrode cells were assembled under argon using a lithium metal negative electrode and glass-fibre separator soaked with a electrolyte solution. The electrolyte was Selektipur^{TM}-30 (Merck, Germany, 1M $LiPF_6$ in EC:DMC-1:1 vol. %), which was used as received. The composite cathode was fabricated as follows: 80% active material, 15% acetylene carbon black and 5% polyvinylidene fluoride as polymer binder were intimately mixed, ground in an agate mortar and subsequently about 30mg were pressed into each pellet.

A three electrode glass-cell with lithium as reference and counter electrodes was used for the cyclic voltammetry experiments. In this case the cathode was prepared by pressing of the cathode mix (~3 mg) onto Ni-mesh. Then, the cell was mounted and operated in a dry argon atmosphere. All the experiments were performed at room temperature.

Structure characterization of cathode materials was made after galvanostatic charge/discharge of the cells to the desired depth of intercalation. The cathode was picked out from the cell and studied separately by X-ray powder diffraction. The data were collected with a STOE STADI/P powder diffractometer (CoKα_1 radiation, curved Ge (111) monochromator, transmission mode, step width 0.02^0 (2θ), linear PSD counter). For structure refinements the RIETAN-97 program was used [14], based on the Rietveld method with a modified pseudo-Voigt profile function.

DISCUSSION

α-CuMoO4 and CuMoO4-III

The evolution of cell voltage with composition (x mol Li pro formula unit of copper molybdate) was obtained for two modifications of CuMoO4 from the galvanostatic cycling in potential range 0.5-3 V at the current density 0.5 mA/cm^2 (Fig.1).

During the first discharge down to 0.5 V both compounds intercalate the same number of lithium ions (~5) per formula unit, that corresponds to about 600 mAh/g deliverable capacity. Nevertheless, the character of the first discharge curves is quite different. The first lithium insertion into the α-form proceeds through four flat areas at the E=f(x) curve in the ranges 2.9, 2.7, 1.6 and 1-0.5 V, respectively. For CuMoO4-III the first and second plateaus are identified at lower voltage (2.7 and 2.4 V) and, then, after insertion of 1 Li equivalent per mol CuMoO4-III, the voltage changes to 1 V with the reasonably high slope dV/dx. For both compounds, about 3 mol Li per formula unit cannot be removed from cathode upon the following charge up to 3 V. This indicates that the cathode material undergoes an irreversible structure transformation at the first lithiation.

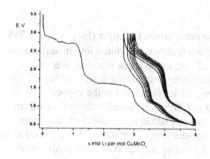 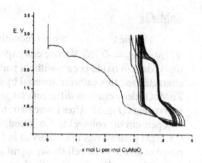

x mol Li per mol CuMoO₄   x mol Li per mol CuMoO₄

Fig. 1 Discharge-charge curves of the cells with α-CuMoO₄ and CuMoO₄-III as cathode materials at the rate 0.5 mA/cm² in 1M LiPF₆ solution (cutoff voltages 0.5V and 3V were chosen)

In order to obtain the information about structural changes of pristine copper molybdates we performed XRD studies of charged cathode materials after 8 cycles. No difference in the phase compositions of the samples could be observed, both diffraction patterns containing diffraction peaks of metallic copper and Li_2MoO_4. Metallic copper can be visually identified at the surface of the sample after extracting cathodes from the cell. Thus, during the first lithiation the irreversible reduction of the copper, accompanied with the formation of Li_2MoO_4 takes place for both copper molybdates. XRD-study for two samples of α-CuMoO₄ with different depths of the first discharge (x=0.9 and x=2.5) shows the presence of metallic Cu, Li_2MoO_4 and some amount of pristine CuMoO₄ at the potential 2.7V (the end of the second plateau at the voltage-composition curve) (Fig.2).
If the system is discharged to x=2.5 (the third plateau) metallic Cu and Li_2MoO_4 were identified only. Consequently, after reduction of Cu the following insertion of Li into Li_2MoO_4 proceeds via changing of stoichiometry without any structural phase transition.

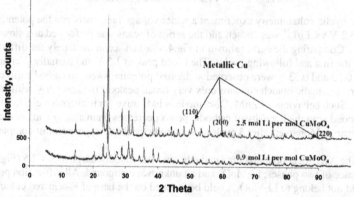

Fig. 2 Powder X-ray diffraction patterns of composite α-CuMoO₄-cathodes discharged to x = 0.9 and x = 2.5.

It is evident, the weak difference in the behaviour of the two forms of CuMoO₄ after the first discharge is due to the same initial reduction leading to formation of the new insertion compound Li_2MoO_4. Lithium intercalation occurs here at relatively low voltages so that copper molybdates cannot be assumed as good cathode materials for lithium rechargeable batteries.

263

ZnMoO₄

ZnMoO$_4$

While zinc is a stronger reduction agent in comparison to copper (E⁰ $_{Cu^{2+}/Cu}$ = 0.345 V, E⁰ $_{Zn^{2+}/Zn}$ = -0.762 V), one can expect that the intercalation of lithium into mixed zinc-molybdenum oxide occurs without participation of zinc in the redox reaction. We characterized this cathode material by cyclic voltammetry and galvanostatic cycling. ZnMoO$_4$ demonstrates different charge-discharge profiles compared to the copper molybdates (Fig.3). After intercalation of 0.02 Li per ZnMoO$_4$ the cell voltage drops from 3.2 V (open-circuit voltage) to 1.6 V and, then, the voltage plateau is identified up to x approx. 0.8. This plateau was not observed in further charge-discharge operations. Being cycled in the range of 0.6-3 V the cell shows significant capacity fading from 260 and 255 mAh/g in the first and second cycle, respectively, to 70 mAh/g in the seventh cycle.

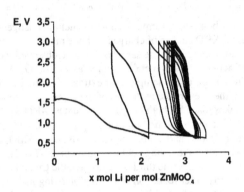

Fig. 3 Galvanostatic cycling for ZnMoO$_4$ at the current density 0.5 mA/cm^2 in the voltage range 0.6-3.0 V (electrolyte 1M LiPF$_6$ in EC-DMC=1:1).

In cyclic voltammetry experiment a wider voltage range between the potential limits 0.0 and 3.2 V vs. Li/Li$^+$ was chosen and the series of scans was performed at a slow scan rate 0.1 mV/s. Comparing the curves shown in Fig.4, one can see immediately the difference between the first and following cycles. The broad peak at 1.5 V and partially overlapped peaks at 0.52 and 0.33 V were observed at the first potential sweep in cathodic direction. However, the anodic branch contains only very broad peaks at 0.7 and 1.6 V with small intensity. Such behaviour of ZnMoO$_4$-cathode is indicative for irreversible electrode reaction. In the second scan the intensity of cathodic peaks decreases dramatically and anodic peaks are not visible. Moreover, after 5 cycles the cathodic peaks disappear almost completely.

The XRD-data obtained for a cathode after the first discharge to x = 3.5 (Fig.5) reveal the presence of two phases: Li$_2$MoO$_3$ and an unknown compound. All diffraction peaks, which did not belong to Li$_2$MoO$_3$, could be indexed on the base of F-centered cubic cell with cell parameter a = 4.197(2) Å.

Two-phase Rietveld refinement of X-ray powder data for the sample obtained at the intercalation of 5 Li per pristine ZnMoO$_4$ was also carried out (Fig.5). Due to the similiraty between the cell parameters of the unknown compound and the already described compound Li$_{0.21}$Co$_{0.79}$O [15], which has NaCl-type structure (space group $Fm3m$), the crystal structure of Li$_{0.21}$Co$_{0.79}$O was used as the initial structure model for the refinement. The atomic parameters for the second phase, Li$_2$MoO$_3$, was taken from [16]. After sequential iterations,

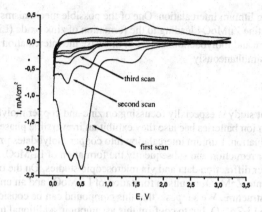

Fig. 4 Cyclic voltammogramms of $ZnMoO_4$ (electrolyte 1M $LiPF_6$ in EC-DMC) at the scan rate 0.1 mV/s.

good agreement between experimental and calculated patterns was achieved: $R_I = 0.028$, $R_P = 0.039$, $R_{wP} = 0.051$. ($R_I = 0.047$ for Li_2MoO_3).

To resume, the electrochemical intercalation of Li into $ZnMoO_4$ leads to the formation of a two-phase system: Li_2MoO_3 and Zn-containig phase, which possesses rock salt structure and was not previously described. Recently, the formation of the intermediate compound with NaCl-type structure during electrochemical lithiation of $MnMoO_4$ was reported by Kim et al. [10]. The authors observed this compound as the only product at transformation of the pristine $MnMoO_4$ upon the discharge down to 0.25 V, after that the amorphization of the sample occurs. It was shown [10], that the lattice constant of this compound was 4.30 Å and this value is in good agreement with the theoretical lattice constant of Li_2MnMoO_4 (4.27 Å) calculated based on the ionic radii reported by Shannon [17]. In our case, the presence of Li_2MoO_3 and an unknown compound with NaCl-structure was established in the discharged sample. Up to now, it is diffucult to imagine the simultaneous formation of Li_2MoO_3 and

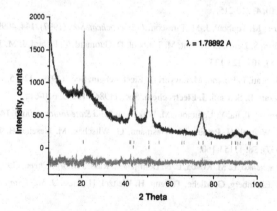

Fig. 5 Experimental, calculated and difference X-ray powder diffraction
pattern for cathode intercalated 3.5 Li per mol pristine $ZnMoO_4$.

265

Li_2ZnMoO_4 during lithium intercalation. One of the possible mechanisms can include the destruction of pristine ZnMoO4 leading to the formation of zink oxide ($Li_xZn_{1-x}O$) with NaCl-type structure and molybdenum oxide. In this case the intercalation could proceed into these two phases simultaneously

CONCLUSIONS

The current study is especially focussing on zinc and copper molybdates as cathode material in lithium ion batteries because they exhibit an irreversible phase transformation during the first lithiation. Lithium intercalation into copper molybdates proceeds through an irreversible copper reduction and subsequently the formation of Li_2MoO_4. Metallic copper is observed in powder diffraction data and via microscopic studies. On the other hand, the lithium insertion into $ZnMoO_4$ results in formation of Li_2MoO_3 and an unknown compound with NaCl - type structure. We suppose, that this compound can be considered as mixed lithium zinc oxide $Li_xZn_{1-x}O$, but to confirm this assumption, additional methods have to be applied and are currently being carried out on this system. All studied compounds were shown to be unsatisfactory as cathode materials in lithium rechargeable batteries, but may be of interest for general aspects of decomposition of ternary oxides induced by intercalation and for investigations on the degradation mechanisms in cathode materials used in lithium ion batteries.

REFERENCES

1. P. A. Christian, J .N. Carides, F. J. DiSalvo, J. V. Waszczak. *J. Electrochem. Soc.* (1980), **127**, 11, 2315-2319

2. J. O. Besenhard, J. Heydecke, E. Wudy, H.P.Fritz, W.Foag. *Solid State Ionics*, (1983), **8**, 61-65

3. T. Tsumura, M. Inagaki. *Solid State Ionics*, (1997), **104**, 183-189

4. A. Yu, N. Kumagaj, Z. Liu, J. Lee. *Solid State Ionics*, (1998), **106**, 11-18.

5. R. H. Sanchez, L.Trevino, A. F. Fuentes, A. Martinez-de la Cruz, L. M. Torres-Martinez. *J. Solid State Electrochem.* (2000), **4**, 210-215.

6. S. Denis, E. Baudrin, M. Touboul, J.-M. Tarascon. *J. Elecrochem. Soc.* (1997), **144**, 4099

7. F. Orsini, E. Baudrin, S. Denis, L. Dupont, M. Touboul, D. Guomard, Y. Piffard, J.-M. Tarascon. **Solid State Ionics**, (1998), **107**, 123-133.

8. B. Di Pietro, B. Scrosati, F. Bonino, M. Layyari. **J. Electrochem. Soc.**, (1979), **126**, 5, 729-731.

9. F. Bonino, M. Lazzari, B. Scrosati. **J. Electrochem. Soc.**, (1984), **131**, 3, 610-612.

10. S.-S. Kim, S. Ogura, H. Ikuta, Y. Uchimoto, M. Wakihara. *Solid State Ionics*, (2002), **146**, 249-256.

11. H. Ehrenberg, H. Weitzel, H. Paulus, M. Wiesmann, G. Wltschek, M. Geselle, H. Fuess. *J. Phys. Chem. Solids*, (1997), **58**, 1, 153-160.

12. R. Tali, V. V. Tabachenko, L. M. Kovba, L. N. Dem'janets. Russ. *J. Inorg. Chem.*, (1991), **36**, 927.

13. M. Wiesmann, H. Ehrenberg, G. Miehe, T. Peun, H. Weitzel, H. Fuess. *J. Solid State Chem.*, (1997), **132**, 88-97

14. F. Izumi, in "The Rietveld Method" (R.A.Young, Ed.), Chap.13. Oxford Univ. Press, Oxford, (1993)

15. W. D. Johnston, R. R. Heikes, D. Sestrich. *J. Phys. Chem. Solids*, (1958), **7**, 1-13

16. A. C. W. P. James, J. B. Goodenough. *J. Solid State Chemisty*, (1988), **76**, 87-96

17. R. D. Shannon, *Acta Crystallogr., Sect. A*, (1976,) **32**, 751

Mat. Res. Soc. Symp. Proc. Vol. 756 © 2003 Materials Research Society

X-ray Absorption Spectroscopy study of lithium insertion mechanism in $Li_{1.2}V_3O_8$

N. Bourgeon, J. Gaubicher, D. Guyomard, G. Ouvrard
Institut des Matériaux Jean Rouxel,
2 rue de la Houssinière, BP32229, 44322 Nantes, France

ABSTRACT

X-ray absorption spectroscopy (XAS) measurements were performed to thoroughly understand lithium insertion mechanism in $Li_{1.2}V_3O_8$. The evolution of the absorption pre-edge and edge corresponding to the local environment of the vanadium in the bulk has been examined by ex-situ XAS measurement at the vanadium K edge, during the first discharge-charge cycle. The results show a regular and reversible evolution of the pre-edge intensity, the edge position and the vanadium environment toward nearly perfect VO_6 octahedra.

INTRODUCTION

Lithium trivanadate, $Li_{1.2}V_3O_8$, has been investigated as a positive electrode material for rechargeable lithium batteries during the past decade [1-6]. Much research work has been focused on the structural characterization and cyclability of this compound. The crystal structure of $Li_{1.2}V_3O_8$ was first reported by Wadsley [7]. Thackeray et al. [8] confirmed that $Li_{1.2}V_3O_8$ has a monoclinic structure with the space group $P2_1/m$ and is composed of two basic structural units, VO_6 distorted octahedra and VO_5 distorted trigonal bipyramids. These two structural units share edges and corners to form layers. Lithium ions are inserted between layers in two different sites, octahedral and tetrahedral. The lithium insertion mechanism has been reported by Thackeray [8] and Kawakita [9]. When it is used as a positive in a lithium battery, the $Li_{1.2}V_3O_8$ compound can accommodate up to 3.8 lithium ions in the host structure with an operating voltage between 3.7V and 2V [8]. For $0 \leq x \leq 1.7$, lithium is inserted in a single-phase reaction process. The voltage plateau observed for $1.7 \leq x \leq 2.8$ is attributed to the coexistence of two phases; the first is $Li_3V_3O_8$, and the second is a defect rock salt structure with a nominal composition $Li_4V_3O_8$. It was found that the electrochemical properties of $Li_{1.2}V_3O_8$, such as discharge capacity and cyclability, depend on the synthesis method of the compound and the preparation conditions of the positive electrode [10-13].

In this paper, we focus on the characterization of the lithium insertion process during the first discharge-charge cycle. The local structure determination of the various phases and the modification generated by lithium insertion are investigated by XAS in order to get better understanding of the electrochemical behavior of $Li_{1.2}V_3O_8$.

EXPERIMENTAL

The lithium trivanadate is prepared by solid state reaction of Li_2CO_3 and V_2O_5 with a molar ratio of 1.2/3 at 580°C in air. The positive electrodes are prepared by mixing the active material, carbon black (Super P from Chemetal) and binder (polyvinylidene difluoride: PVDF) with the massic ratio (85:10:5), and coating the mixture onto an Al disk serving as the current collector, according to [14]. Such electrodes were vacuum-dried at 100°C for 1 h before entering a dry Ar glove box. Two-electrode Swagelok™ test cells [14] using these positive electrodes, a porous paper soaked with the electrolyte as the separator, and metallic lithium as the negative electrode, were assembled in the glove box. A 1 M solution of $LiPF_6$ in ethylene carbonate (EC) and dimethyl carbonate (DMC) (70:30) (Merck) was used as the electrolyte. All voltages given in the text are reported vs. Li^+/Li. Electrochemical Li insertion and cell cycling were performed at 22°C, monitored by a Mac-Pile™ system [15] in galvanostatic and potentiodynamic mode. The voltage range used was 3.7-2 V unless otherwise noted.

Samples for XAS studies were prepared with voltage scanning (±10mV/h) to predetermined potential values situated before and after each electrochemical process. Each potential was maintained during the time needed for insertion current to become lower than the current corresponding to 1 lithium per formula unit in 1000 h. These potentials are shown on the incremental capacity versus potential curve of figure 1, obtained in galvanostatic mode at a rate corresponding to the insertion of 1 lithium per formula unit in 10 hours. The electrochemically lithiated samples were removed from the preparation battery, washed with DMC and then mixed with boron nitride in an argon filled glove box.

XAS experiments were carried out at the D44 station at the DCI storage ring of LURE (Orsay, France). X-ray absorption measurements were performed at the vanadium K-edge (5465eV). XANES spectra were recorded in using a Si (311) double crystal monochromator in transmission mode with 0.25eV energy steps in the energy range 5445-5495eV and an integration time of 3 s. EXAFS data were collected with a Si(111) double crystal monochromator between 5350 to 6150eV with 1eV steps and an integration time of 2s per point. The treated data are obtained by averaging three successive data collection for each sample.

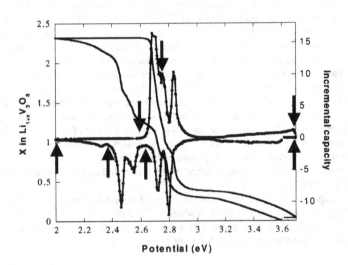

Figure 1 : Incremental capacity and number x of inserted lithium versus potential for $Li_{1.2+x}V_3O_8$.

RESULTS

XANES (X-ray Absorption Near Edge Spectroscopy)

Figure 2 shows the absorption edge evolution for various compounds in the $Li_{1.2+x}V_3O_8$ system. The pre-edge intensity is closely related to the environment symmetry. The decrease of the pre-edge intensity when the lithium content increases, confirms the symetrization of the vanadium environment, as shown by XRD [8,9]. The main edge shifts toward smaller energies as the lithium content increases, in agreement with vanadium reduction.

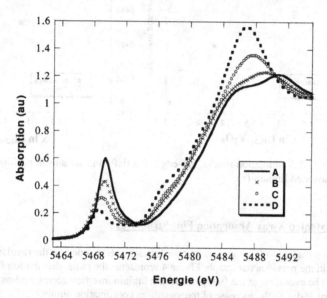

Figure 2 : XAS spectra evolution on discharge of $Li_{1.2+x}V_3O_8$ samples. (A: $Li_{1.2}V_3O_8$ (x=0), B : $Li_{2.36}V_3O_8$ (x=1.16), C : $Li_{3.35}V_3O_8$ (x=2.15), D : $Li_{4.14}V_3O_8$ (x=2.94)).

To quantify the absorption spectra, we have simulated the absorption edge between 5460 to 5495 eV with simple functions : one arctangent which describes the edge jump and three gaussian functions, one of which characterizing the pre-edge. The pre-edge intensity variations are represented in figure 3, left. During lithium insertion, the pre-edge area regularly decreases. This means that the local environment around vanadium becomes more and more symmetric without notable discontinuity. The structural modification detected by XRD [9] can then be attributed to the modification of the long distance ordering of the coordination polyhedra. These changes can be imposed by either their symetrization or the steric hindrance of adjacent lithium ions or the filling of available different sites for lithium ions. Figure 3, right, shows the evolution of the arctangent position versus the number of inserted lithium, which reveals a regular shift to smaller energy. This result corresponds to a regular and progressive reduction of vanadium by electrons transferred during lithium insertion. During the charge, all variations measured during discharge are fully reversible.

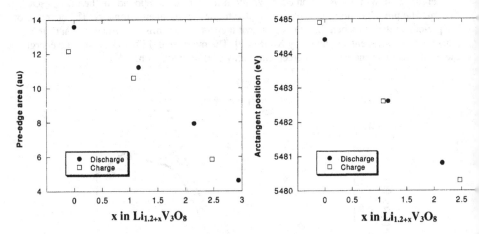

Figure 3 : XANES simulation results. Pre-edge area (left) and arctangent position (right) versus number of inserted lithium (x).

EXAFS (Extented X-ray Absorption Fine Structure)

We have studied the first vanadium coordination shell to confirm the regular evolution mentioned in the previous paragraph. Figure 4 represents the radial distribution functions of vanadium. The evolution of the first peak during lithium insertion corresponding to V-O contributions is due to the increase of the vanadium coordination number.

Two different vanadium-oxygen distances have been considered for the refinements. For each oxygen type, distance R and coordination number N have been determined. Figure 5 shows the N and R variations versus number of inserted lithium. We observe that the initial vanadium coordinence of 5 (two short (R1,N1) and three long bonds (R2,N2)) gradually tends toward a standardization of the distances, with a large decrease of the number of oxygen atoms at short distance. However, we never observe a coordinence equal to 6 which would correspond to a perfect octahedral environment. This point is in total agreement with the presence of a pre-edge which is not supposed to exist for a perfect octahedral environment. The distance variations are very regular. The average distance increases with the increase in vanadium size due to reduction

Thus, the results obtained by EXAFS confirm the regular transformations undergone by vanadium during the reversible reaction with lithium.

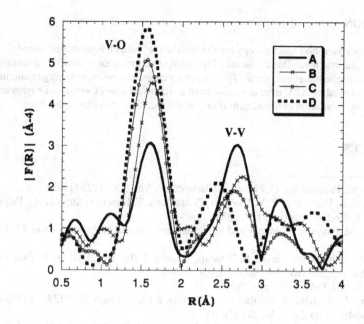

Figure 4 : Moduly of the fourier transforms of vanadium K-edge EXAFS for various lithium compositions. (A: $Li_{1.2}V_3O_8$ (x=0), B : $Li_{2.36}V_3O_8$ (x=1.16), C : $Li_{3.35}V_3O_8$ (x=2.15), D : $Li_{4.14}$ V_3O_8 (x=2.94)).

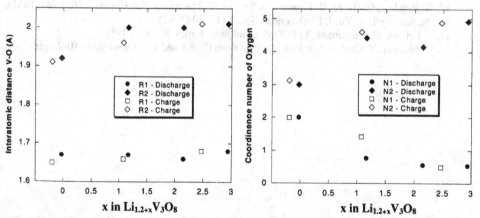

Figure 5 : Interatomic distance V-O (left) and coordinence (right) vs number of inserted lithium (x).

CONCLUSION

The X-ray absorption spectroscopy results, obtained ex-situ on equilibrated samples, indicate a regular and reversible evolution of the local structure around vanadium toward a nearly perfect octahedral environment. They suggest that the macroscopic average structure modification viewed by XRD does not come from a sharp variation of vanadium environment but probably by a variation of the lithium filling of the available cristallographic sites.

REFERENCES

1. S. Panero, M. Pasquali and G. Pistoia, J. Electrochem. Soc. **130**, 1225 (1983)
2. Y. Geronov, B. Puresheva, R. V. Moshtev, P. Zlatilova, T. Kosev, Z. Stoyvov, G. Pistoia and M. Pasquali, J. Electrochem. Soc. **137**, 3338 (1990)
3. G. Pistoia, M. Pasquali, V. Manev and R. V. Moshtev, Journal of Power Sources **15**, 13 (1985).
4. K. West, B. Zachou-Christiansen, S. Skaarup, Y. Saidi, J. Barker, I. I. Olsen, R. Pynenberg and R. Kokshang, J. Electrochem. Soc. **143**, 820 (1996).
5. B. Scrosati, Brit. Poly. J. **20**, 219 (1988)
6. F. Bonino, M. Ottaviani, B. Scrosati and G. Pistoia, J. Electrochem. Soc. **135**, 12 (1988)
7. A. D. Wadsley, Acta Cryst. **10**, 261 (1957).
8. L. A. de Picciotto, K. T. Adendorff, D. C. Liles and M. M. Thackeray, Solid State Ionics **62**, 297 (1993).
9. J. Kawakita, Y. Katayama, T. Miura, T. Kishi, Solid State Ionics **107**, 145-152 (1998)
10. S. Friberg, K. Roberts, in: G. Lindner, K. Nyberg, (eds.), *Environmental Engineering*, Reidel, Dordrecht, 1973.
11. L. Eriksson, B. Alm, Mater. Sci. Technol. **28**, 203 (1993).
12. G. Renders, G. Broze, R. Jerome, Ph. Teyssic, J. Macromol. Sci. Chem. A **16**, 1399 (1981).
13. N. Kumagai, A. Yu, J. Electrochem. Soc. **144**, 830 (1997).
14. F. Leroux, D. Guyomard, Y. Piffard, Solid State Ionics **80**, 30 (1995).
15. C. Mouget, Y. Chabre, MacPile, licensed from CNRS and UJF Grenoble to BioLogic Co.

Mat. Res. Soc. Symp. Proc. Vol. 756 © 2003 Materials Research Society   EE7.10

In Situ Electrochemical Scanning Probe Microscopy of Lithium Battery Cathode Materials: Vanadium Pentoxide (V_2O_5)

Joseph W. Bullard III and Richard L. Smith
Department of Materials Science and Engineering, Massachusetts Institute of Technology
Cambridge, Massachusetts 02139, USA

ABSTRACT

Atomic force microscopy was used to characterize the structural evolution of the $V_2O_5(001)$ surface during the electrochemical cycling of lithium. With Li insertion, nanometer-scale pits develop at the $V_2O_5(001)$ surface. The pits first appear as the composition of the crystal approaches $Li_{0.0006}V_2O_5$. Pit nucleation and growth continue through further discharge, resulting in a micro-porous (001) surface morphology. During subsequent Li extraction, cracks develop along the V_2O_5 <010> axis. Surface regions in the vicinity of these cracks "swell" during ensuing lithiation reactions, suggesting that the cracks locally facilitate Li uptake.

INTRODUCTION

Vanadium pentoxide (V_2O_5) readily takes part in reversible topotactic redox reactions and has attracted interest for a number of applications, including electrochromic windows and secondary batteries [1-5]. The oxide's ability to reversibly intercalate ionic species derives from its open layered structure, which is composed of sheets of VO_5 square pyramids that link by sharing corners and edges (Figure 1) [6,7]. Adjacent layers are held together by only van der Waals forces, enabling the structure to reversibly adjust to accommodate guest ions in the interlayer interstitial sites [1,2]. The electrochemical cycling of Li has been of particular interest, due to its relevance to secondary battery applications [1-5]. The phase relations in the Li/V_2O_5 system have been established, as have the structures of the various $Li_xV_2O_5$ phases [2,3]. Across the compositional range from x = 0 to x = 1, there are three distinct phases, with compositions of $x \leq 0.1$, $0.35 \leq x \leq 0.5$, and $0.9 \leq x \leq 1.0$ [3]. Lithium uptake is generally accompanied by structural changes, such as changes in the interlayer spacing and the symmetry of the host crystal. In the initial solid solution phase $Li_xV_2O_5$ ($x \leq 0.1$), Li^+ ions are accommodated with only subtle increases in the b and c lattice parameters, which vary from those of V_2O_5 at x = 0 (Figure 1 caption) to $b = 3.565$ Å and $c = 4.386$ Å for $x \approx 0.1$ [3].

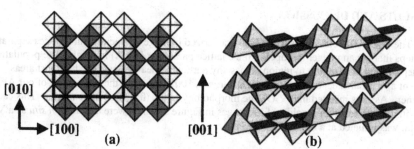

Figure 1. The V_2O_5 structure. V_2O_5 has orthorhombic symmetry (space group = Pmmn) and lattice parameters of $a = 11.496$ Å, $b = 3.551$ Å, and $c = 4.357$ Å [6].

While a number of studies have examined the bulk structural chemistry and electrochemical behavior related to the Li/V_2O_5 system, little is known about the surface microstructural changes that accompany Li cycling. Such knowledge could shed light on a number of issues, for example the local mechanics of interfacial Li exchange. A microstructure-level picture might also aid in the interpretation of macroscopic performance characteristics, such capacity fade [5]. Moreover, with the growing interest in nanostructured and thin film cells, interfacial processes are likely to assume an increasingly important role in battery performance and technology, necessitating a better understanding of the interfacial structure and properties of electrode materials. This paper highlights initial efforts to understand interface structure and dynamics in the Li/V_2O_5 system. The objective was to characterize the structural evolution of the $Li_xV_2O_5(001)$ surface during Li cycling between $x = 0$ and $x \approx 0.003$. Lithium insertion and extraction have been observed *in situ* using atomic force microscopy (AFM) and model cells based on a V_2O_5 single crystal electrode.

EXPERIMENTAL

Single crystals of V_2O_5 were grown by chemical vapor transport [6,7]. Individual mm-sized crystals were mounted in the (001) orientation on steel discs (d = 1.5 cm) using Apiezon wax. Each crystal was then cleaved to form a fresh (001) surface. An electronic contact was then painted between the steel disc and the surface with colloidal Ag (Ted Pella, Redding, CA). *In situ* atomic force microscopy (AFM) observations were made with a commercial microscope (Digital Instruments, Santa Barbara, CA) within a stainless steel glove box with a catalytically purified Ar atmosphere. AFM images were acquired in the contact mode with pyramidal Si_3N_4 tips. A fluid cell supplied by the AFM manufacturer was used for the experiments. The liquid electrolyte was contained within a silicone O-ring that was compressed between the steel sample disc and a glass plate, on which the AFM probe was mounted. The glass plate had inlet and outlet ports, through which liquid could be transferred, as well as ports to accommodate the Li metal (99.9%, Alfa Aesar) reference and counter electrodes. Once the cell was assembled, the 0.005 M $LiClO_4$ (99.99%, Aldrich) in propylene carbonate (PC, 99%, Avocado) electrolyte was introduced via a syringe. The electrochemical current between the V_2O_5 working electrode and the Li counter electrode was controlled by a Keithley 2400 Sourcemeter (Cleveland, OH), which also facilitated measurement of the potential between the V_2O_5 crystal and the reference electrode. In the Results and Discussion section, we describe in detail the structural evolution of a single 20 x 20 μm^2 region of one $V_2O_5(001)$ surface as Li was inserted and extracted. These results are characteristic of a number of similar experiments. In this particular example, Li was first inserted for 33 min at a current density of ~ 45 $\mu A/cm^2$. The current was then reversed for 15 min to extract Li. Finally, Li was cycled back into the crystal for 10 min. The (001) surface of the V_2O_5 crystal was imaged with AFM continuously throughout this cycle.

RESULTS AND DISCUSSION

The $V_2O_5(001)$ cleavage surface is characterized by large atomically flat terraces separated by steps that are an integer multiple of the c lattice parameter (4.4 Å) in height. The population of steps varies from region to region across any particular cleavage surface, but most areas present a singular (i.e. step-less) terrace as large as 100 x 100 μm^2, the upper limit of our AFM scanner. As Li is inserted into V_2O_5, the morphology of the (001) terraces changes. This can be seen in the series of topographic AFM images in Figure 2, which were acquired *in situ* as a V_2O_5 crystal was lithiated at a rate of ~ 45 $\mu A/cm^2$.

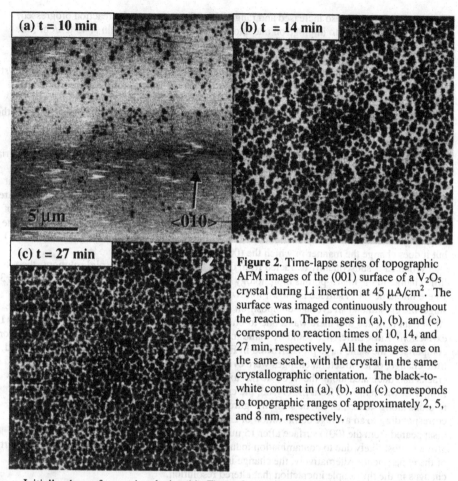

(a) t = 10 min

(b) t = 14 min

5 μm

<010>

(c) t = 27 min

Figure 2. Time-lapse series of topographic AFM images of the (001) surface of a V_2O_5 crystal during Li insertion at 45 $\mu A/cm^2$. The surface was imaged continuously throughout the reaction. The images in (a), (b), and (c) correspond to reaction times of 10, 14, and 27 min, respectively. All the images are on the same scale, with the crystal in the same crystallographic orientation. The black-to-white contrast in (a), (b), and (c) corresponds to topographic ranges of approximately 2, 5, and 8 nm, respectively.

Initially, the surface region depicted in Figure 2 was simply a singular terrace. No changes were detected in terrace morphology over the first ~ 8 min of discharge. Within ~ 10 min (Figure 2a) however, nanometer-sized pits began to appear at the surface. The pits had lateral dimensions on the order of 100 – 250 nm but were shallow, with depths of not more than 1 nm. Based on the discharge rate and the elapsed time, the pits nucleated as the average composition of the crystal approached $Li_{0.0006}V_2O_5$. Note that the upper portion of the area depicted in Figure 2a appears to have a higher density of pits than the lower portion. This derives from the fact that it takes a finite period of time (~ 1 min) to scan an area with the AFM tip and acquire a single AFM image. Within the reference frame of Figure 2a, the tip was scanned from bottom to top as the image was acquired. Thus, the bottom was imaged at an earlier point in time.

With increasing discharge time and Li insertion, pits continued to nucleate at the V_2O_5(001) surface. Figure 2b shows the same region as in Figure 2a after an additional 4 min at 45 $\mu A/cm^2$. Within that time frame, pre-existing pits grew laterally, though few exceeded 1 μm in diameter.

The depths of the pits also increased and ranged from 3 – 4 nm. As the reaction progressed, neighboring pits began to impinge on one another. After a total of 20 min at 45 $\mu A/cm^2$, the surface had taken on a rough and "mossy" appearance that did not change substantially with further Li insertion. The general surface morphology is illustrated in Figure 2c with an AFM image acquired at 27 min. At this point, the surface pits had depths of up to 6 nm, and the crystal had an average composition of approximately $Li_{0.0016}V_2O_5$. The micro-porous morphology persisted as the crystal was taken to a final composition of about $Li_{0.002}V_2O_5$. Additional experiments with other crystals demonstrated that this porosity dominates the morphology of the (001) surface to compositions of at least $Li_{0.003}V_2O_5$.

The nucleation and growth of pits during Li intercalation indicates that material is being removed from the V_2O_5(001) surface. The only viable mechanism that can explain such material removal is that the (001) surface is slowly dissolving in the electrolyte solution. Control experiments demonstrated that pitting did not occur in the absence of Li electrodes and current flow; no changes were detected in the morphology of (001) terraces over the course of exposures in excess of 30 min. After extended (several hours) treatments however, there was evidence of local V_2O_5 dissolution. Most notably, dislocation etch pits were formed in select regions of the single crystal surfaces. Thus, it is not clear that Li intercalation enhances the solubility of V_2O_5, but it does change the manner in which the (001) surface dissolves.

When Li is subsequently extracted from dilute $Li_xV_2O_5$ ($x \leq 0.003$) solid solutions, there is little change in the micro-porous morphology of the (001) terraces. This can be seen in Figure 3a, where the same (001) surface is shown following 5 min of Li extraction at 45 $\mu A/cm^2$. At this stage, the crystal had an average composition of approximately $Li_{0.0017}V_2O_5$. Inspection of the AFM data revealed that there was a one-to-one correspondence between pits prior to and after Li extraction. Little difference could be found in the location, morphology, or size of the pits. For comparison, the same large pit is arrowed in Figures 2c and 3a. Although the porosity showed little change, other structural changes were evident with Li extraction. Within 5 min, straight cracks had nucleated along the V_2O_5 <010> axis. One such crack is arrowed in Figure 3a. As further Li was removed from the crystal, the cracks propagated, often extending for tens of microns along <010>, as can be seen in Figure 3b. This image was recorded after 15 min, corresponding to an average composition of $Li_{0.0011}V_2O_5$. Seemingly, the micro-pores had disappeared from the (001) surface after 15 min. However, we believe that this was an image artifact, most likely due to contamination induced "dulling" of the tip during this particular part of the experiment. Alternatively, the change in cell polarity may have induced electrostatic changes in the tip-sample interaction that altered resolution.

After extracting Li for 15 min, the current was again reversed such that Li was re-inserted into the crystal. The same V_2O_5(001) surface is shown again in Figure 3c following 10 min at 45 $\mu A/cm^2$, corresponding to an approximate composition of $Li_{0.0017}V_2O_5$. Note that the surface porosity is once again apparent. Furthermore, there is a close correspondence between the pits observed at this stage of the cycle and those observed previously. Hence, the porosity introduced during the initial insertion reaction was little affected by subsequent cycling. Not surprisingly, the cracks introduced during Li extraction persisted through the insertion cycle. However, the surface "swelled" upward along the lengths of the cracks, such that the topographic heights of surface areas along the cracks were up to 50 nm higher than areas far (~ 5 μm) from a crack. The heights along the cracks increased with increasing reaction time.

The nucleation and growth of pits and cracks during Li insertion and extraction can be qualitatively explained with the aid of the schematic illustrations in Figure 4. During the initial intercalation of V_2O_5 (Figure 4a), Li enters at the surfaces of the crystal and diffuses into the

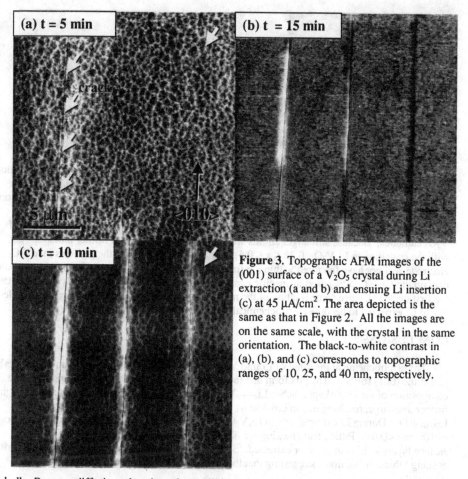

Figure 3. Topographic AFM images of the (001) surface of a V_2O_5 crystal during Li extraction (a and b) and ensuing Li insertion (c) at 45 $\mu A/cm^2$. The area depicted is the same as that in Figure 2. All the images are on the same scale, with the crystal in the same orientation. The black-to-white contrast in (a), (b), and (c) corresponds to topographic ranges of 10, 25, and 40 nm, respectively.

bulk. Because diffusion takes time, there will be a time dependent gradient in Li concentration from the surface to the bulk along any particular crystallographic direction. This concentration gradient will result in a gradient in the b and c lattice parameters, with b and c expanded at the (001) surface relative to the pure interior. Assuming the surface is saturated with Li ($x \approx 0.1$), b and c would be expanded by 0.25% and 0.67%, respectively. Regardless of the exact concentration, the (001) surface will experience a compressive load as it expands against the underlying crystal along <010>. The strain energy associated with this load would be expected to enhance dissolution, which may explain the pitting observed during Li insertion. When Li is then extracted from the crystal, it is first removed from the surface, which likely results in a concentration profile qualitatively similar to that in Figure 4b. Now, the surface experiences a tensile load, as it contracts along <010> against the underlying Li-rich portion of the crystal. Under the conditions studied, this load was apparently sufficient to instigate cracking.

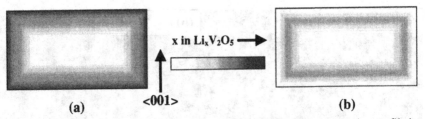

x in $Li_xV_2O_5$ →

$<001>$

(a)   (b)

Figure 4. Schematic representations (cross-sectional) of the Li concentration profile in a $Li_xV_2O_5$ crystal following Li insertion (a) and during subsequent Li extraction (b).

The topographic "swelling" observed along the $<010>$ cracks during subsequent Li insertion indicates that interior crystal regions underneath or adjacent to cracks are lithiated to a greater extent than the rest of the crystal. Presumably, this is because the cracks provide a fast transport or insertion path for incoming Li^+ ions. Although these AFM observations span a limited compositional range, the observed cracking may provide some insight into the performance of V_2O_5 cathodes. Because still larger strains are expected to develop with further lithiation (e.g. with the transition to $Li_xV_2O_5$, $0.35 \leq x \leq 0.5$), it is hard to imagine that the particles in a V_2O_5 electrode will not have fractured within the first full discharge-charge cycle. Such fracture could explain the large (~ 50%) drop in specific capacity that has been observed in V_2O_5 cathodes after the first cycle [5], since some fraction of the material that is fractured from the primary particles would be expected to become electrochemically isolated from the remainder of the electrode.

CONCLUSIONS

The electrochemical insertion of Li into V_2O_5 results in the development of nanometer-scale pits at the (001) surface. The pits form due to dissolution of the surface and first appear as the composition of the crystal approaches $Li_{0.0006}V_2O_5$. Pit nucleation and growth continue through further discharge, resulting in a micro-porous (001) morphology that persists to at least $Li_{0.003}V_2O_5$. During Li extraction from $Li_xV_2O_5$ ($x \leq 0.003$), cracks nucleate and propagate along $<010>$ and $<001>$. Pitting and cracking are likely driven by the strain that develops in the surface layers as Li is inserted or extracted. Surface regions in the vicinity of cracks swell during ensuing lithiation reactions, suggesting that the cracks locally facilitate Li uptake.

ACKNOWLEDGEMENTS
This work was supported by the National Science Foundation under DMR-MRSEC-9808941.

REFERENCES
1. R. Schöllhorn, *Agnew. Chem. Int. Ed. Engl.*, **19**, 983 (1980)
2. A. J. Jacobson in <u>Solid State Chemistry Compounds</u>, Eds. A.K. Cheetham and P. Day (Clarendon, Oxford, 1992), p. 182
3. P.G. Dickens, S.J. French, A.T. Hight, and M.F. Pye, *Mat. Res. Bull.*, **14**, 1295 (1979)
4. K.West, B.Zachau-Christiansen, T.Jacobsen, and S.Skaarup, *Solid State Ionics*, **76**, 15 (1995)
5. P.P.Prosini,Y.Xia,T.Fujieda,R.Vellone,M.Shikano,T.Sakai, *Electroch. Acta*, **46** 2623 (2001)
6. R.L. Smith, W.Lu, and G.S. Rohrer, *Surface Science*, **322**, 293 (1995)
7. R.L. Smith, G.S. Rohrer, K.S. Lee, D.-K. Seo, and M.-H. Whangbo, **367**, 87 (1996)

Anode Materials for
Lithium Batteries and
Polymer Electrolytes

Mat. Res. Soc. Symp. Proc. Vol. 756 © 2003 Materials Research Society

THE ELECTROCHEMISTRY OF GERMANIUM NITRIDE VERSUS LITHIUM

N. Pereira [a, b], M. Balasubramanian [c], L. Dupont [d], J. McBreen [c], L.C. Klein [b] and
G.G. Amatucci [a]

[a] Telcordia Technologies, Red Bank, NJ 07701, USA
[b] Rutgers University, Piscataway, NJ 08854, USA
[c] Brookhaven National Laboratory, Upton, NY 11973, USA
[d] Laboratoire de Réactivité et Chimie des Solides, Université de Picardie Jules Verne,
80039 Amiens, France

E-mail: npereira@telcordia.com, gamatucc@telcordia.com

ABSTRACT

Germanium nitride (Ge_3N_4) was examined as a potential negative electrode material for
Li-ion batteries. The electrochemistry of Ge_3N_4 versus Li showed high reversible capacity
(500mAh/g) and good capacity retention during cycling. A combination of ex-situ and in-situ x-
ray diffraction (XRD), ex-situ transmission electron microscopy (TEM) and ex-situ selective
area electron diffraction (SAED) analyses revealed evidence supporting the conversion of a layer
of Ge_3N_4 crystal into an amorphous Li_3N+Li_xGe nanocomposite during the first lithiation. The
nanocomposite was electrochemically active via a reversible Li-Ge alloying reaction while a core
of unreacted Ge_3N_4 crystal remained inactive. The lithium/metal nitride conversion reaction
process was kinetically hindered resulting in limited capacity. Mechanical milling was found to
improve the material capacity.

INTRODUCTION

Intensive effort has aimed at the development of alternative negative electrode materials
which improve on the performance of graphite currently utilized in commercial Li-ion batteries.
Although most work has concentrated on alloys and oxides, nitrides have been shown to exhibit
interesting electrochemical properties.

The ternary lithium transition metal nitrides of general formula $Li_{3-x}M_xN$ (M= Cu, Ni,
Co) [1-6], isostructural to the layered hexagonal Li_3N, were investigated for their
electrochemical properties. $Li_{2.6}Co_{0.4}N$ was found to exhibit the best performance with high
reversible capacity, 700mAh/g, and good cycling stability. However, this type of materials are
moisture sensitive and need to be pre-delithiated before use as negative electrode in Li-ion
batteries containing lithiated transition metal oxide such as $LiCoO_2$.

Binary nitrides such as Sn_3N_4 [7-8], InN [8] and Zn_3N_4 [7] were proposed to undergo an
irreversible conversion reaction resulting in the generation of a Li_3N matrix and an
electrochemically active metal M (Eq.1) which subsequently react with Li via an alloying
reaction (Eq.2).

$$M_xN_y + 3y\ Li^+ + 3y\ e^- \rightarrow x\ M + y\ Li_3N \qquad (1)$$
$$M + z\ Li^+ + z\ e^- \leftrightarrows Li_zM \qquad (2)$$

Our investigation of the reaction mechanism of Zn_3N_4 with Li revealed a first irreversible conversion reaction into $LiZn+\beta Li_3N$ was followed by a reversible conversion reaction of $LiZn+\beta Li_3N$ into LiZnN [9]. We believe the electromechanical grinding of the alloy to be the main cause responsible for the Zn_3N_4 poor cycle life.

Finally, we identified Cu_3N, where Cu does not alloy with Li, as an interesting negative electrode material candidate which exhibited improved cycling stability and excellent rate capabilities [10]. The chemistry of this system is complex with several processes, one of which consisted in the reversible lithium/metal nitride conversion reaction (Eq.3).

$$Cu_3N + 3Li^+ + 3e^- \leftrightarrows 3Cu + Li_3N \qquad (3)$$

Herein, we investigated the reaction mechanism of Ge_3N_4 with Li and report experimental evidence supporting a conversion reaction process. High-energy milling was considered to improve the reaction kinetics which limited the capacity of the material.

EXPERIMENTAL

Electrochemical characterization

Swagelock and coin cells were controlled using Mac Pile galvanostats (Biologic, Claix, France). The two-electrode cells were assembled in a He-filled dry box and consisted of a Li metal counter electrode (Johnson Matthey), a glass fiber separator (GF/D, Whatman) saturated with 1M $LiPF_6$ in (EC: DMC) (1:1 in vol.) ethylene carbonate: dimethyl carbonate electrolyte (Merck) and a $1cm^2$ Ge_3N_4 working electrode. The working electrodes were fabricated from a mixture of poly(vinylidene fluoride-co-hexafluoropropylene (Kynar 2801, Elf Atochem), carbon black (Super P, MMM) and dibutyl phthalate (Aldrich) in acetone. The slurry was cast and dried at room temperature. After extraction of the dibutyl phthalate plasticizer with 99.8% anhydrous ether (Aldrich), the working electrodes typically contained 50±1% Ge_3N_4 and 18±1% carbon black.

Structural characterization

Ge_3N_4 was characterized using ex-situ XRD, ex-situ TEM and SAED. The measurements were performed in a X2 Scintag diffractometer using $CuK\alpha$ as radiation source and in a Philips CM12 microscope, respectively. The electrodes stopped at various states of discharge and charge were rinsed in anhydrous DMC to remove residual salts and deposited on a glass slide covered with Kapton film (Spex Certiprep) sealed with a layer of silicon grease to avoid contamination with air during analysis. The electrodes for the XRD analyses consisted of plastic tape whose fabrication was described above while a mixture of 85% Ge_3N_4 powder and 15% carbon black was utilized for the TEM measurements.

The Ge_3N_4 electrode structural changes were also investigated by XRD during cycling using a hermetically sealed in-situ cell utilizing a beryllium window, which is in contact with the working electrode, as current collector.

RESULTS AND DISCUSSION

Germanium nitride was found to react with lithium reversibly (Fig.1). The reaction with Li occurs at low potential, below 0.3V, which constitutes a desirable voltage range for use as

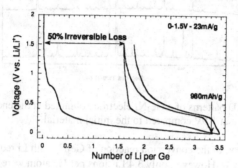

Figure 1. First two cycle voltage profile of a Ge_3N_4 electrode cycled vs. Li metal under a constant current of 23mA/g.

negative electrode. Despite a 50% irreversible capacity loss in the first cycle, it exhibited a large reversible capacity of about 500mAh/g and good cycling stability. An electrode cycled between 0 and 0.7V vs. Li metal retained 85% of the second cycle capacity after 70 cycles (Fig.2).

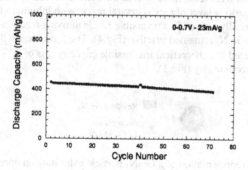

Figure 2. Cycle life of a Ge_3N_4 electrode cycled between 0 and 0.7V vs. Li metal under a constant current of 23mA/g.

XRD analyses indicated that two phases, denoted α- and β-Ge_3N_4, constituted the initial material. A mixture of these two phases is generally obtained under normal synthesis conditions. The investigation of the structural change of Ge_3N_4 using in-situ XRD, and reconfirmed ex-situ, revealed the structure of the α- and β-Ge_3N_4 phases were maintained during cycling (Fig.3). Both phases exhibited a substantial decrease, about 40%, in the x-ray reflection integrated intensity during the first discharge associated to the destruction of a layer of Ge_3N_4 crystal to form a nanocomposite of Li_3N+Li_xGe amorphous to XRD.

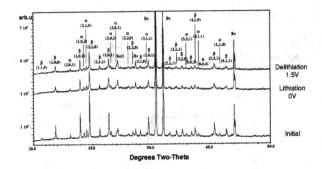

Figure 3. In-situ XRD patterns of a Ge_3N_4 electrode obtained at the end of lithiation at 0V and at the end of delithiation at 1.5V compared to the initial material.

The completion of the conversion reaction of Ge_3N_4 with Li requires in theory 8.4 Li^+ ions per Ge atom (Eq.4). However, only 3.4 Li^+ ions per Ge atom were inserted experimentally

$$1/3\ Ge_3N_4 + 8.4Li^+ + 8.4e^- \rightarrow Li_{4.4}Ge + 4/3Li_3N \quad (4)$$

$$Li_{4.4}Ge + 4/3Li_3N \leftrightarrows 4.4Li^+ + 4.4e^- + Ge + 4/3Li_3N \quad (5)$$

during lithiation which accounts for 40% of the theoretical capacity of the material (equivalent to 2470mAh/g). As a result, the destruction of 40% of the initial material upon reaction with Li agrees well with the 40% decrease in integrated intensity observed by XRD during the first lithiation. In this scenario, a layer of Ge_3N_4 crystal reacts with lithium to form a nanocomposite which is electrochemically active via a reversible Li-Ge alloying reaction (Eq.5) while a core of unreacted crystalline Ge_3N_4 remained inactive (Fig.4). The formation in the first cycle of a Li_3N inert matrix is associated to a theoretical irreversible capacity loss of 50%, which is consistent with the Ge_3N_4 electrochemistry (Fig.1).

Residual Ge_3N_4

Amorphous
$Li_xGe + Li_3N$

Figure 4. Schematic representation of a Ge_3N_4 particle exhibiting an unreacted Ge_3N_4 core surrounded by an electrochemically active converted "shell".

While no new phase was observed to emerge during cycling using x-ray diffraction techniques even with a beam of higher energy, diffuse rings not present in the initial material emerged in SAED patterns at the end of lithiation and delithiation. While the identification of the diffuse rings obtained at the end of the lithiation was not conclusive, the ones obtained at the end of delithiation were identified as Ge metal and Li_3N using a process diffraction software that converts the rings into intensity distributions similar to XRD patterns (Fig.5). The identification of Ge metal and Li_3N at the end of the delithiation is consistent with the dealloying reaction depicted in equation 5. Moreover, the generation of crystalline $Li_{22}Ge_5$ by high energy milling of

Li metal with Ge_3N_4 further supported the hypothesis of a thermodynamically favorable conversion reaction mechanism.

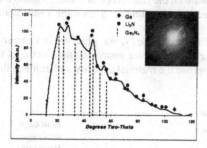

Figure 5. Ex-situ SAED pattern (inset) of a β-Ge_3N_4 rod obtained at the end of the first delithiation and its corresponding intensity distribution obtained using a Process Diffraction software.

We showed the Ge_3N_4 conversion reaction was limited as only 40% of the initial material reacted with Li in the first lithiation. This limitation may be rooted in the combination of several factors: 1) non-equilibrium conditions, 2) rate-limited reaction process and 3) low reaction potential (about 0.1V) too close to the lithiation cut-off voltage set at 0V. The voltage that gradually decreases with lithium insertion reaches the lithiation cut-off voltage before the Li^+ ions can react with the core of the particles. At higher rate, the voltage profile exhibited a larger underpotential and a more sloped decrease in voltage which resulted in a decrease in lithiation capacity from 980mAh/g to 800mAh/g. The large increase in lithiation capacity at elevated temperature further supported the reaction of Ge_3N_4 with Li is kinetically hindered.

In order to improve the kinetics of the reaction process of Ge_3N_4 with Li and therefore the material capacity, the initial Ge_3N_4 powder was milled for periods of time ranging from 10 minutes to 8 hours. The structural characterization of the resulting materials revealed a peak

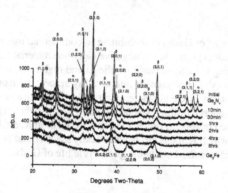

Figure 6. Ex-situ XRD pattern of Ge_3N_4 after high energy milling for various periods of time. The pattern of Ge_2Fe synthesized by energy milling is also provided.

broadening indicative of a decrease in size and crystallinity with milling time (Fig.6). Contamination of the initial Ge_3N_4 powder with the stainless steel vial and milling media was observed for long milling times with the emergence of a second phase identified as Ge_2Fe. Pure Ge_2Fe was synthesized by mechanical milling for 8 hrs using a stoichiometric mixture of Ge and Fe.

The initial capacity of the milled Ge_3N_4 was observed to increase with milling time which was consistent with a decrease in particle size improving the kinetics of the reaction (Fig.7). Although we succeeded to improve the material capacity with milling, the cycling stability

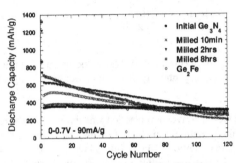

Figure 7. Cycle life of the initial Ge_3N_4 and the milled Ge_3N_4 electrodes cycled *vs.* Li metal between 0 and 0.7V at 90mA/g.

deteriorated with milling time. An increase of the low cut-off voltage and an increase in rate resulted in improved cycle life. We believe the degradation of the cycle life to be rooted in an increase of the nitride surface area in contact with the electrolyte,

The good cycling stability of crystalline Ge_3N_4, unusual for a material based on a reversible lithium alloying reaction, is believed to result from the nano-structural nature of the electrochemically active composite and its adhesion to a Ge_3N_4 substrate.

SUMMARY

Although the cost of Ge_3N_4 is prohibitive to commercial use, it was shown to offer promising electrochemical properties as negative electrode material for Li-ion batteries. It maintained good cycling stability versus metallic Li and high reversible capacity, about 500mAh/g. Ge_3N_4 was shown to convert into an x-ray amorphous nanocomposite during the first lithiation that contains Li-Ge alloys remaining electrochemically active in the subsequent cycles. Finally, high-energy milling was proposed to increase the material capacity limited by the slow kinetics of the conversion process leaving a core of unreacted Ge_3N_4. However, the core of crystalline Ge_3N_4, which remains electrochemically inactive and acts as a conductive substrate, was believed to contribute to the good cycle life of Ge_3N_4.

REFERENCES

1. M. Nishijima, T. Kagohashi, N. Imanishi, Y. Takeda, O. Yamamoto and S. Kondo, *Solid State Ionics*, **83**, 107 (1996).
2. T. Shodai, S. Okada, S. Tobishima and J. Yamaki, *Solid State Ionics*, **86-88**, 785 (1996).
3. T. Shodai, S. Okada, S. Tobishima and J. Yamaki, *J. Power Sources*, **68**, 515 (1997).
4. T. Shodai, Y. Sakurai and T. Suzuki, *Solid State Ionics*, **122**, 85 (1999).
5. Y. Takeda, M. Nishijima, M. Yamahata, K. Takeda, N. Imanishi and O. Yamamoto, *Solid State Ionics*, **130**, 61 (2000).
6. J. Yang, Y. Takeda, N. Imanishi and O. Yamamoto, *Electrochimica Acta*, **46**, 2659 (2001).
7. J.B. Bates, N.J. Dudney, B. Neudecker, A. Ueda and C.D. Evans, *Solid State Ionics*, **135**, 33 (2000).
8. B.J. Neudecker and R.A. Zuhr, *Intercalation Compounds for Battery Materials, Proceedings of the Electrochemical Society*, **99-24**, 295 (2000).
9. N. Pereira, L.C. Klein and G.G. Amatucci, *J. Electrochem. Soc.*, **149**, A262 (2002).
10. N. Pereira, L.C. Klein and G.G. Amatucci, *J. Electrochem. Soc.*, submitted.

Mat. Res. Soc. Symp. Proc. Vol. 756 © 2003 Materials Research Society

Carbon Coated Silicon Powders As Anode Materials For Lithium Ion Batteries

M. Gulbinska[1], F. S. Galasso[1], S. L. Suib[1], S. Iaconetti[2], P. G. Russell[2], and J. F. DiCarlo[2]

[1]Dept. Chemistry, University of Connecticut, Storrs, CT 06269
[2]Lithion Inc., 82 Mechanic St., Pawcatuck, CT 06379

ABSTRACT

Novel lithium ion battery anode materials consisting of carbon-coated silicon were prepared. Chemical vapor deposition (CVD) methods were used in the syntheses of these composite materials with toluene being used as the precursor for carbon coatings on silicon powder. The temperature of carbon deposition was 950°C and the deposition time was 30 min for powdered substrates. Carbon-coated silicon powders were analyzed by HRSEM, TEM, and Raman spectroscopy. Coin cells were made and cycled to evaluate the electrochemical performance of carbon-coated silicon powders. Coated materials performed well during the initial coin cell testing. Silicon wafers (orientation <111>) were also used as the substrates for carbon coatings. Silicon wafers were coated for 40 seconds at 950°C. Auger spectroscopy and XPS depth profiling were used to analyze the thin films of toluene-derived carbons on silicon wafers.

INTRODUCTION

Silicon-based anodes provide alternatives to carbonaceous materials used presently as anodes in lithium ion batteries. However, carbon-based anodes do not have very large energy capacities (the typical carbon-based anode discharge capacities are 290 mAh / g) [1]. Another reason for the limited value of carbons as anode materials is the slow diffusion of Li^+ between graphite layers. The Li/Si ratio is high in lithium silicides (like $Li_{4.4}Si$) formed in-situ in silicon-based anodes. Therefore, the theoretical capacity of silicon is exceptionally high (4000 mAh/g). In addition, lithium ions diffuse faster in $Li_{4.4}Si$ alloys that are formed *in-situ* in Si-based anodes.

A major obstacle in commercializing silicon as anode material in lithium ion batteries is the poor reversibility of the alloying reaction between lithium and silicon at room temperature [2, 3]. Volume changes in silicon during the repeated charging and discharging of the anodes cause mechanical degradation of the anode. Pulverization of the silicon powder eventually leads to the loss of electrical contact between the current collector and the silicon-based anode material [4]. Thus, the excellent initial capacities (over 1000 mAh / g) offered by silicon-based anodes decrease dramatically during subsequent cycles. The addition of carbon to silicon powder improves its cycleability [5, 6] and prevents capacity fading that is observed in pure silicon anodes. The goal of this study was to obtain materials containing silicon of sub-micron particle size and silicon-carbon composite materials.

EXPERIMENTAL DETAILS

Si powder (previously ball milled) was ground in an agate mortar. About 0.5 g of sample was loaded into the reactor. The temperature was raised to 950°C. Upon reaching 950°C, argon carrier gas was passed through a bubbler filled with liquid toluene at a flow rate of 100 mL/min. After 30 min the carrier gas flow was stopped and the reactor was allowed to cool down.

Carbon-coated Si wafer samples were also prepared in order to perform the AES studies. Si wafer (orientation <111>, boron-doped) was cleaned in 1% HF solution and rinsed with deionized water. The wafer sample was loaded into the reactor. The temperature was raised to 950°C. Upon reaching 950°C, argon carrier gas was passed through the bubbler filled with liquid toluene at a flow rate of 100 mL/min. After 40 sec carrier gas flow was stopped and the reactor was allowed to cool down.

A Renishaw Ramascope Raman spectrometer equipped with 514 nm and 785 nm lasers was used to obtain the Raman spectra of the coated powders. A Zeiss EM910 120 keV high resolution scanning electron microscope was used to examine the surface morphology. A Philips 420 transmission electron microscope equipped with STEM and EDX units was used to examine the carbon deposits. An Arbin cycler was used to perform cell cycling experiments.

RESULTS AND DISCUSSION

Raman spectroscopy results for silicon powders are shown in Figure 1. Two characteristic first-order Raman peaks attributed to carbon are present in the carbon-coated silicon powder spectrum (top spectrum). The 1600 cm^{-1} band is associated with the short-range sp^2 order in carbon deposits. The 1356 cm^{-1} band (also called a "D-band") is observed in highly disordered sp^2 bonded carbons [7].

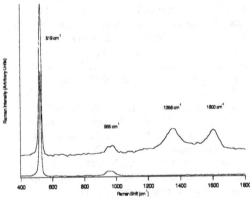

Fig.1. Raman spectra of pure silicon powder (bottom spectrum) and carbon-coated Si powder (top spectrum).

(a)   (b)

Fig. 2. HRSEM micrographs of Si powders: Si powder reference (left) and carbon-coated Si powder (right).

HRSEM micrographs of silicon powders are shown in Figures 2 and 3. Figure 2 a shows pure silicon powder reference and carbon-coated silicon powder is shown in Figure 2 b. The silicon powder particles are surrounded by sphere-shaped particles of carbon deposits. Other forms of carbon deposits are shown in Figure 3. Carbon nanofibers are deposited onto silicon powder agglomerates.

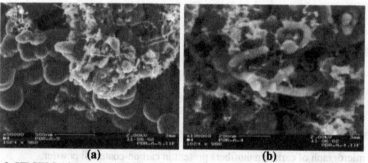

(a)   (b)

Fig. 3. HRSEM micrographs of coated Si powders: various types of deposited carbon are present.

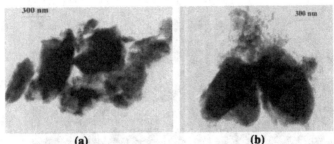

(a)   **(b)**

Fig. 4. TEM micrographs of Si powders: Si powder reference (a) and carbon-coated Si powder (b).

TEM micrographs of silicon powders are shown in Figure 4. Growth of nanofibers on silicon powder particles is illustrated in Figure 4. The carbon coating on silicon powder is non-uniform and its thickness is difficult to establish. A TEM micrograph of isolated carbon nanofibers (Figure 5) shows their hollow structure and channels running through the fibers.

Fig. 5. TEM micrograph of carbon nanofibers present in carbon-coated Si powder.

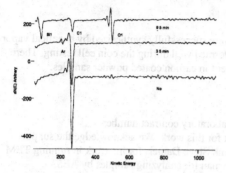

Fig. 6. AES analysis of carbon-coated Si wafer: as-synthesized sample (bottom), after 3.5 min sputter (middle), after 9.5 min sputter (top).

Figure 6 shows results of AES analysis of a silicon wafer coated with carbon (40 seconds deposition time). Auger electron spectra were obtained on an as-synthesized sample, a sample sputtered with argon ions for 3.5 minutes and a sample sputtered for 9.5 minutes. Spectrum after 3.5 minutes shows a decrease in amount of carbon coating thickness resulting from sputtering. The argon Auger peak appears in the spectra of both samples subjected to sputtering. A silicon Auger peak is visible along with a carbon peak in a sample analyzed after 9.5 minutes of sputtering. The presence of the oxygen on the silicon wafer can be attributed to silicon surface oxidation occurring during the heating of the sample.

Figure 7 illustrates coin cell cycling results performed at capacity cutoff 500 mAh/g and in the voltage range of 0.01 V to 1.3 V. No capacity fading is observed in preliminary capacity profiles. Research work is continuing to further characterize carbon-coated silicon powders.

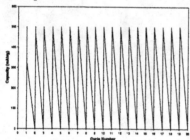

Fig. 7. Capacity profile for cell containing carbon-coated silicon powder material.

CONCLUSIONS

Carbon coated Si-based anode materials were successfully synthesized by chemical vapor deposition methods. Coated materials performed well during the coin cell testing. There are various types of deposited carbon present in carbon coated powder samples.

ACKNOWLEDGMENTS

We would like to thank Army Research Laboratory contract number DAAD17-01-C-0044 for financial support for this work. We acknowledge the support from Yardney Technical Products. We thank Steve Daniels for his work in running TEM samples. We also acknowledge Dan Goberman for analyzing samples by AES.

REFEREENCES

1. M. Wakihara, *Materials Science and Engineering*, **R33,** 109 (2001).
2. W. J. Weydanz, M. Wohlfahrt-Mehrens, R. A. Huggins, *J. Power Sources*, **81,** 237 (1999).
3. S. Bourdeau, T. Brousse, D. M. Schleich, *J. Power Sources*, **81,** 233 (1999).
4. H. Li, X. Huang, L. Chen, Z. Wu, Y. Liang, *Electrochemical and Solid State Letters*, **2,** 11 (1999).
5. M. Yoshio, H. Wang, K. Fukuda, Y. Hara, Y. Adachi, *J. Electrochem. Soc.*, **147,** 1245 (2000).
6. C. S. Wang, G. T. Wu, X. B. Zhang, Z. F. Qi, W. Z. Li, *J. Electrochem. Soc.*, **145,** 2751 (1998).
7. P. C. Eklund, J. M. Holden, and R. A. Jishi, *Carbon*, **33,** 7 (1995).

Mat. Res. Soc. Symp. Proc. Vol. 756 © 2003 Materials Research Society

Sn and SnBi Foil as Anode Materials for Secondary Lithium Battery

Shoufeng Yang, Peter Y. Zavalij and M. Stanley Whittingham
Institute for Materials Research, SUNY-Binghamton University, Binghamton, NY 13902

ABSTRACT:

In order to better understand the cycling mechanism of metal alloy anodes, and to mitigate the capacity fade observed in lithium battery use a study of simple systems was initiated. Tin foil and tin-bismuth mixtures were chosen because there is no need for conductive diluents or binders so that the intrinsic behavior could be observed. A pure tin foil was found to react rapidly with lithium, ≥ 3 mA/cm^2, and with no capacity fade for over 10 cycles. This is better than tin powder or electrodeposited tin. After the first cycle, the foil reacts with Li following a stepwise formation of different alloys as dictated by the thermodynamics. Incorporation of bismuth into the foil increased the capacity fade after the first few cycles, with the eutectic composition $Sn_{0.57}Bi_{0.43}$ having better capacity retention than the $Sn_{0.5}Bi_{0.5}$ composition. XRD and SEM-EDS shows that bismuth is rejected from the tin rich phase during lithium insertion and is not reincorporated on lithium removal, just as expected from the phase diagram.

INTRODUCTION

The use of Li-Ion batteries is very popular due to their small size and high capacity, but new electronics demand even more capacity at affordable price. The current graphite anode cannot meet the higher demand because its capacity is only 372 mAh/g [1], and higher capacity carbons can lead to unsafe lithium formation. Sn based materials have received much attention in the past 30 years due to their high capacity and energy density, 2.7 and 9 times of graphite, respectively. [2-5] However, pure Sn has not performed very well due to the large volume changes during reaction with lithium, so previous studies mainly dealt with near nano-sized Sn or Sn intermetallic powder in order to mitigate the capacity fading. [6-8] Alternatively the Sn phase may be evenly dispersed into an electronically conductive matrix system [9,10], which system can buffer the large volume changes resulting in better performance [11-13].

We have undertaken a study to better understand the reactions occurring on lithiation, so as to find a means to mitigate the capacity fading found in electrode performance. Our initial studies used pure tin foil, without any conductive additives or binder to complicate the data interpretation. We also report here the effect of control of particle size by addition of a second phase, in particular bismuth which is also electrochemically active. Tin and bismuth form a eutectic composition at $Sn_{0.57}Bi_{0.43}$. Throughout the paper, discharge refers to the lithiation of the foil and recharge refers to the delithiation process.

EXPERIMENTAL DETAILS

Sn foil (average thickness: 25-30 μm) was purchased from Aldrich, Sn-Bi binary alloys were prepared by arc-melting a mixture of Sn (Aldrich) and Bi shots (Alfa Aesar) at 600°C in a water-cooled copper hearth under an argon atmosphere. The material was rolled to 12-15 μm with a roll mill (International Rolling Mill). To prepare the cell, a thin foil with an average surface area of 0.5 cm^2 was cut and used as the positive electrode, and Li foil (Aldrich) mounted on a stainless steel disk as the negative electrode. The electrolyte was $LiPF_6$ (EC:DMC = 1:1, LP30) from EM Industries and circular Celgard 2400 (Hoechst Celenese Corp.) were used as the separator. The cell was assembled in a helium glove box with a Swagelok cell [14], and cycled with a MacPile potentiostat at constant current at ambient temperature with cut-off voltages of 0.1-1.0V for Sn and 0.005-1.3V for Sn_yBi_{1-y}. X-ray diffraction of the materials at various stages of reduction, protected from air by a Mylar film, was obtained on a Scintag XDS2000 diffractometer.

DISCUSSION

Fig. 1 shows the first 1.5 cycles for Sn and $Sn_{0.5}Bi_{0.5}$. There is an initial small voltage drop in every cell, probably associated with the nucleation of a new phase. One long plateau equivalent to 2.33Li/Sn can be seen from the Sn foil during the first lithium insertion, but only after this first insertion are the thermodynamically expected phases seen for x<3. In contrast, the SnBi system shows the thermodynamic phase Li_3Bi and then the tin phases even on the first

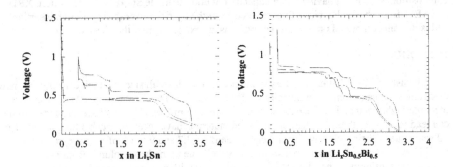

Figure 1. First one and a half cycles for Sn and $Sn_{0.5}Bi_{0.5}$ at 0.1 mA/cm^2.

lithium insertion form the thermodynamically expected phases unlike pure tin foil. Both materials have plateaus due to the formation of each alloy except $Li_{3.5}Sn$ and $Li_{4.4}Sn$, which is probably due to the lack of a significant driving force at these low potentials, and because all the Sn-Sn bonds must be broken to form $Li_{4.4}Sn$. The apparent capacity loss for the tin is 16% capacity during the first cycle, however this is recovered on subsequent cycles. The $Sn_{0.5}Bi_{0.5}$, however, has a larger capacity maintenance during the first recharge.

Fig.2 shows the capacity vs cycle number for Sn and Sn_yBi_{1-y} foils as a function of current density and composition. At 0.8 mA/cm^2, Sn foil has an initial capacity of 700 mAh/g, almost twice that of graphite and had only small fading for the first 7 cycles. However, it dropped quickly from 566 mAh/g to 312 mAh/g from the 9th to the 10th discharge. The capacity then dropped slowly and stabilized at 130 mAh/g. While the initial capacity at 3.0 mA/cm^2 was only 563 mAh/g and dropped to 434 mAh/g in the second cycle, it recovered to an average of 550 mAh/g for 10 cycles, finally it also dropped to only 120 mAh/g. The high rate capability is expected from the high lithium diffusion coefficients [15]. The Sn-Bi foil readily reacted with Li, with the eutectic composition $Sn_{0.57}Bi_{0.43}$ foil having an initial capacity of 500 mAh/g, but decreasing almost linearly to 150 mAh/g after 10 cycles. The capacity of the $Sn_{0.5}Bi_{0.5}$ foil dropped to 250 mAh/g in the 5th cycle and to 100 mAh/g in the 10th cycle. This latter foil had a range of particle sizes with some large tin particles, whereas the eutectic composition had a uniform and small particle size.

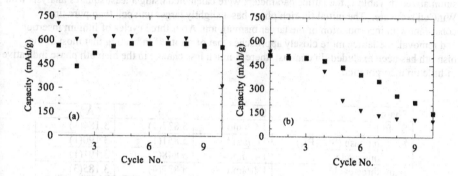

Figure 2. (a) Discharge capacity vs cycle number for Sn foil at 3.0 mA/cm^2 (■) and 0.8 mA/cm^2 (▼). (b) Discharge capacity vs cycle number for $Sn_{0.57}Bi_{0.43}$ (■) and $Sn_{0.5}Bi_{0.5}$ (▼) at 0.5 mA/cm^2.

To determine the phases formed, an X-ray study of the electrochemically reacted materials was carried out. In order to understand the behavior of the Sn-Bi binary system, the temperature of formation of the compounds and of the reaction with lithium must be considered. An inspection of Figure 3 shows that on solidification from the melt, pure bismuth and a tin rich phase are expected. The composition of this tin rich phase is strongly dependent on the temperature; thus, on initial formation from the melt it can be as high as 20 mass % bismuth. However, at room temperature, where the electrochemistry is performed the equilibrium bismuth

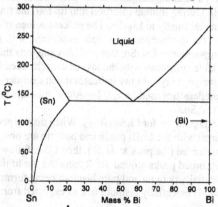

Figure 3.   Calculated phase diagram for Sn-Bi system.

content is expected to be around an order of magnitude lower. Thus, on reaction with lithium this tin rich phase is expected to expel the bismuth forming Li_3Bi and $Li_{0.4}Sn$. On the subsequent lithium removal cycle, much less if any bismuth will be incorporated back into the tin phase.

The X-ray data of some materials after lithiation and complete lithium removal is summarized in Table 1; the lattice parameters were calculated using a least squares analysis with WinCell software. The initial tin-rich phase has a slightly larger unit cell than pure tin, consistent with incorporation of the larger bismuth ion. After three cycles of lithium insertion and removal, the lattice more closely approaches that of tin itself indicating that most of the bismuth has been excluded. In contrast, there is much less change in the bismuth phase indicative of little tin incorporation.

Phase	S.G	a (Å)	c (Å)
Sn foil	I 4_1/amd	5.837(1)	3.184(1)
Sn rich phase ($Sn_{1-y}Bi_y$)	I 4_1/amd	5.851(1)	3.190(1)
After one cycle	I 4_1/amd	5.848(2)	3.191(1)
After three cycles	I 4_1/amd	5.837(4)	3.185(3)
Bi (ICDD 44-1246)	R-3m	4.547(1)	11.861(3)
Bi rich phase ($Bi_{1-y}Sn_y$)	R-3m	4.546(1)	11.881(2)
After one cycle	R-3m	4.536(3)	11.888(8)
After three cycles	R-3m	4.541(2)	11.886(5)

Table 1. Standard and calculated lattice parameters of the Sn and Bi rich phases.

X-ray patterns for the lithium containing materials were also obtained to determine which lithium phases were being formed. Fig. 4 shows the X-ray patterns for Li_xSn and "$Li_xSn_{0.5}Bi_{0.5}$" as a function of the lithium content, x. The first four X-ray patterns, from the bottom, in each case represent the first insertion of lithium, and the second four the removal of the lithium. Each X-ray pattern was obtained on a separate sample and was made about 12 hours after removal from the electrochemical cell. During lithium insertion into tin foil, the tin converts to LiSn, then to the phases $Li_{2.33}Sn$-$Li_{2.6}Sn$ and finally to $Li_{3.5}Sn$. The process is then reversed on lithium removal, with mostly pure tin being observed on complete recharge, with just a trace of $Li_{0.4}Sn$ even though the nominal composition is $Li_{0.4}Sn$; this probably suggests that some of the anode material has become unavailable consistent with the loss of capacity seen in figure 1. At intermediate compositions, the poor crystallinity indicates much disorder from $Li_{2.33}Sn$ to $Li_{2.6}Sn$ and for $Li_{3.5}Sn$. At the current densities used here, 0.1 mA/cm^2 lithium insertion does not proceed to the composition $Li_{4.4}Sn$.

Fig. 4 also shows the XRD patterns for $Li_xSn_{0.5}Bi_{0.5}$. When discharged to $1.5Li/Sn_{0.5}Bi_{0.5}$, only diffraction lines associated with the Li_3Bi phase and pure tin are observed. As additional lithium is added, $Li_{0.4}Sn$ is observed (the peak at 31.5°), then LiSn, followed by the more lithium-rich phases with their broad peaks around 38° 2 theta. As the lithium is removed the reverse is observed with essentially pure tin and pure bismuth being formed. As noted above, the tin formed on lithium removal is essentially bismuth free as expected from the phase diagram shown in figure 3.

The results observed above for the first ten cycles indicate good reversibility for the pure tin foil. The particle size of the tin decreases in size from about 500 nm to 150 nm during these 10 cycles suggesting that particle size at least in this regime does not impact cyclability. Similarly the cell impedance does not change significantly until after 10 cycles when it starts to increase markedly. At the same time, the capacity drops off significantly indicating that resistive products are formed after about 12 to 15 cycles. These results will be discussed in detail elsewhere.

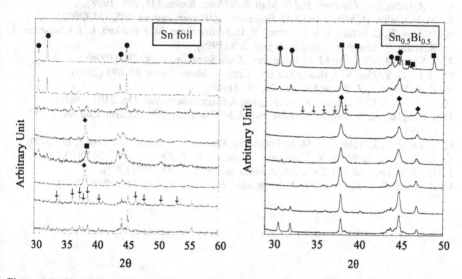

Figure 4. Left): XRD of Li_xSn at different lithiation level, from bottom to the top, x value was 0.83, 1.00, 2.33, 3.54, 2.33, 1.11, 1.00 and 0.4. The last four were obtained during the recharge. ●, ↓, ◆ and ■ represent Sn, LiSn, $Li_{2.33}Sn$ and $Li_{3.5}Sn$, respectively. Right): XRD of $Li_xSn_{0.5}Bi_{0.5}$ at different lithiation level, from bottom to the top, x = 1.50, 2.00, 2.50, 3.40, 2.50, 2.00, 1.51, 0.10. The last four were obtained during the recharge. Symbols: ●, ■, ◆ and ↓ represent Sn, Bi, Li_3Bi and LiSn.

CONCLUSIONS

From the results above, Sn foil shows excellent cycling behavior for several cycles, but the capacity then fades quickly. The Sn-Bi mixtures do not cycle as well as Sn foil. Different Li-Sn and Li-Bi alloys were formed during the cycling, repeated cycling repels Bi from the initial tin rich alloy as expected from the phase diagram. Thus, tin and bismuth become segregated after lithiation and delithiation. The capacity fading mechanism will be discussed in future reports.

ACKNOWLEDGMENTS

We thank the US Department of Energy for supporting this work through the BATT program at Lawrence Berkley National Lab. We also thank Robert Kinyanjui for sample preparation, Henry Eichelberger for SEM studies and Bill Blackburn for SEM-EDS experiments.

REFERENCES

1. K. Ozawa, *Solid State Ionics,* **69,** 212 (1994).
2. C. J. Wen and R. A. Huggins, *J. Solid State Chem.* **35,** 376 (1980).
3. C. J. Wen and R. A. Huggins, *J. Electrochem. Soc.* **128,** 1181 (1981).
4. J. R. Dahn, I. A. Courtney and O. Mao, *Solid State Ionics,* **111,** 289 (1998).
5. M. Wachtler, M. Winter and J. O. Besenhard, *J. Power Sources,* **105,** 151 (2002).
6. O. Mao, R. L. Turner, I. A. Courtney, B. D. Fredericksen, M. I. Buckett, L. J. Krause and J. R. Dahn, *Electrochem. Solid-State Lett.* **2,** 3 (1999).
7. J. Yang, M. Winter and J. O. Besenhard, *Solid State Ionics,* **90,** 281 (1996).
8. H. Li, G. Y. Zhu, X. J. Huang and L.Q. Chen, *J. Mater. Chem.* **10,** 693 (2000).
9. R. A. Huggins, *J. Power Sources,* **22,** 341 (1988).
10. A. Anani, S. C. Baker and R. A. Huggins, *J. Electrochem. Soc.* **135,** 2103 (1988).
11. J. Vaughey, K. Kepler, R. Benedek and M. M.Thackeray, *Electrochem. Commu.* **1,** 517 (1999).
12. J. Vaughey, J. O'Hara and M. M.Thackeray, *Electrochem. Solid-State Lett.* **3,** 13 (2000).
13. L. Shi, H. Li, Z. Wang, X. Huang and L. Chen, *J. Matr. Chem.* **11,** 1502 (2001).
14. D. Guyomard and T. J. Tarascon, *J. Electrochem. Soc.* **139,** 937 (1992).
15. A. Anani, S. C. Baker and R. A. Huggins, *J. Electrochem. Soc.* **134,** 3098 (1987).

Electrochemical Properties of Blockcopolymer Templated Mesoporous Silicates with Heteropolyacids clusters

Hui Suk Yun,[a] Makoto Kuwabara,[a] Hao Shen Zhou,[b] and Itaru Honma*[b]
[a]Dept. of Materials Science, University of Tokyo, Bunkyo-ku, Tokyo 113-8656, Japan
[b]National Institute of Advanced Industrial Science and technology (AIST),
1-1-1 Umezono, Tsukuba, Ibaraki, 305-8568, Japan, *E-mail: i.homma@aist.go.jp

ABSTRACT

In this paper, we have studied a method of loading 12-tungstophosphoric acid ($H_3PW_{12}O_{40}$; PWA) into blockcopolymer templated (designated $EO_{20}PO_{70}EO_{20}$; Pluronic P-123) mesoporous SiO_2 framework without a disordering of mesostructure and without remarkable decrease of specific surface area through one-step condensation process. A precursor solution is initially prepared by TEOS (tetraethoxysilane) hydrolyzed with templating polymer (eg. triblock copolymer), and then mixed directly with PWA to condense mesoporous silicate products that incorporate PWA clusters in the framework. Heteropolyanions are possibly incorporated in the framework silicates. Electrochemical properties of these PWA impregnated mesoporous silicates were studied as electrode materials of Li ion battery as well as solid proton conductors.

INTRODUCTION

Nanostructured mesoporous materials have attracted much attention recently due to their potential applications for catalysts, sensors, size- and shape-selective separation media. The nanostructure can be controlled by a variety of self assembling amphiphiles as template molecules through hydrolysis and polycondensation reaction at the interfaces. Novel structural properties of high surface area and lamellar, hexagonal or cubic arrays of uniform pore sizes within 2-50nm are suitable for advanced catalysts as well as sensors. Since the discovery of M41S families (MCM-41, MCM-48, MCM-50)[1], researches for providing better functional property on the mesoporous materials have been studied. Transition metals (Ti, Ta, V, Sn, Mn, W, Mo) or heteropoly acids (HPAs) were incorporated into mesoporous silica or mesoporous transition metal oxide materials[2-3, 5-10]. In the mean time, Keggin-type HPAs have stimulated considerable research in both heterogeneous and homogeneous catalysis, solid electrolytes, and aqueous solutions in fuel cells due to their excellent thermal, redox stability and special electrocatalytic properties. 12-tungstophosphoric acid ($H_3PW_{12}O_{40}$; PWA) - the strongest acid in the HPAs series- is particularly recommended [3-10]. PWA are normally dispersed on suitable supports such as high surface area SiO_2 since the bulk PWA have very low specific surface area ($<10m^2g^{-1}$). PWA has been reported to be anchored onto MCM-41 mesopores using impregnation method because MCM-41 structure exhibits high thermal stability (up to 900°C) and a large surface area (about $1000m^2g^{-1}$) [5-10]. However, it is difficult to achieve high PWA loading without a loss of mesostructure ordering and surface area through this method because pore size of MCM 41 is too small ($<3nm$) to impregnate the HPAs clusters ($\sim1nm$) into its channel interior [7-9]. It is also a problem that PWA is not uniformly impregnated in the interior surface of MCM-41 since pore sizes of PWA/MCM41 composite are not dramatically changed as increasing amount of PWA while surface area is considerably decreased [8].

In this paper, we have studied a method of loading PWA into mesoporous SiO_2

framework without a disordering of mesostructure and a large decrease of specific surface area by one-step condensation process. A precursor solution is initially prepared by TEOS (tetraethoxysilane) hydrolyzed with templating polymer (eg. triblock copolymer), and then mixed directly with PWA to condense mesoporous silicate products that incorporate PWA clusters in the framework.

RESULTS AND DISCUSSION

Figure 1 shows a synthetic process of the PWA doped mesoporous silicates materials. The template polymers of triblock copolymers $EO_{20}PO_{70}EO_{20}$; Pluronic P-123 (BASF) $(HO(CH_2CH_2O)_{20}(CH_2CH(CH_3)O)_{70}(CH_2CH_2O)_{20}H)$ were dissolved in diluted HCl solution and mixed with tetraethoxysilane $(Si(OC_2H_5)_4)$ for 20h at 45°C. Then, PWA solution (prepared by mixing PWA and distilled H_2O, PWA/TEOS=0 ~50 wt.%) was added into the above solution. After stirring for another 4h, the sol solution was aged at 80°C without stirring for 24h. The precipitated product was filtered, washed by distilled water, and dried at room temperature for 72h, then calcined at 350°C in air, to remove the P123 template. The process can provide silicate mesoporous framework incorporated with PWAs. The obtained PWA containing mesoporous SiO_2 was characterized by X-ray diffraction (XRD; MacScience-M03XHF22 using CuKα irradiation), transmission electron microscopy (TEM; Hitachi-H800, 200kV), energy-dispersive X-ray (EDX), and BET surface area measurements (BELSORP-23SA). The cyclic voltammograms (CV) of the sample was measured by three electrode electrochemical cell using $LiClO_4$/PC liquid electrolytes and Li metal sheets for counter as well as reference electrodes. The proton conductivity of the samples both at non humidified and humidified conditions were characterized by two terminal ac impedance spectroscopy (Solartron 1260) at frequency range 1 Hz – 2Mz.

According the small-angle XRD patterns for the PWA doped mesoporous silicates with different PWA doping concentrations (PWA/TEOS=10 - 50 wt.%), d spacings of 102.63, 59.3, 51.32Å, which can be indexed as 100, 110, 200 reflections were observed, which indicate hexagonal mesostructure. Additionally, no peak for PWA crystalline phases was found on wide-angle XRD patterns of the samples, which indicates that the clusters are highly dispersed in the mesoporous silicates framework [7].

Figure 2 shows TEM photograph of the PWA doped mesoporous silica samples. PWA clusters can be distinguished by the dark image for having higher electron density of W atoms. Apparently, PWA clusters are almost homogeneously dispersed in the products and seem to be aligned with the walls of the hexagonal phase. The dark region of the TEM figure is almost lines with exactly the same sizes and the spacing of approximately equal to the d-value of the hexagonal phase (10nm). The dispersion of PWA clusters were directed by the hexagonal structure of the silicates. The XRD results and TEM observations of the mesoporous structure agree well that the nanostructure of the silicate did not destroyed by doping of PWAs. Moreover, the EDX measurements were made on the samples and it showed the intense W signals were observed at the dark lines while uniform PWA distribution in the hexagonal structures. Additionally, Fourier transformed infrared spectra were measured for the samples and confirmed the Keggin structure of PWA was still existing in the silicate matrix even after the calcination. The typical adsorption peaks of 890, 982, 1080 cm^{-1} corresponding to W-O-W, W-O and P-O were observed and the intensity was increased with the amount of doping of PWA.

Figure 3 shows the impedance measurements of pure mesoporous silicate and PWA doped (30 wt.%) mesoporous silicate, respectively. The powder samples were pressed to form pellets and then subjected to the two terminal ac impedance measurements. The impedance were measured both at non humidified conditions (dry) and humidified conditions (95% relative humidity) at room temperatures, respectively. As the PWA doped samples are electrically insulating, so the obtained impedance indicate the proton conductivity of the samples although the PWA clusters are strongly trapped in the silicate skeletons and reported to have small anion mobility[11]. The pure mesoporous silica does not have a large proton conductivity at both at non (10^{-5} S/cm)and humidified conditions (10^{-4} S/cm). This is because of the small number of mobile proton carrier. Presumably, the remaining protons at the sol-gel synthesis are conducting. However, the PWA doped samples have a large conductivity both at non (10^{-3} S/cm) and humidified conditions (10^{-2} S/cm). Especially, the conductivity of the fully humidified sample (24 hour humidification) possesses a large conductivity exceeding 10^{-2} S/m level, which is an application level for electrochemical devices such as fuel cells. This is simply because that the more mobile proton carriers are provided for PWA doped samples. The protons are dissociated from the PWA bodies and become conductive when the mesopores are incorporated with water. The doped samples are stable proton conducting solid electrolytes with strong acidity.

The cyclic voltammograms (CV) of the PWA doped sample was measured by three electrode electrochemical cell using LiClO4/PC liquid. The potential range of 2.0 – 3.0 V (vs. Li^+/Li) was cycled at various scanning rate from 0.5 mV/sec to 50 mV/sec. Figure 4 shows the CV results for the sample doped with 30 wt.% PWA. In spite of the PWA are incorporated inside the silica frameworks, redox capacity of PWA clusters due to the faradic reaction with lithium ions are clearly observed. At the low scanning rate of 0.5 mV/se, the redox peak at around 2.7 V were remarkable, indicating that the faradic reaction between the PWA clusters and Li ions at this potential. Though the materials are mesoporous with nanostructured pores and large surface area (600 m^2g^{-1} for 30 wt.% PWA doped), the intercalation reaction into the host matrix are not realistic, but psuedocapacitance due to the Li^+ surface reaction with PWA clusters are presumably responsible for the electrochemical activity of the doped materials. As the scan rate increases, the ohmic resistance become significant and at the same time, the redox capacity decreases as in the figure 4. However, the PWA clusters were strongly confined in the silica framework, the doped materials are recognized as a new class of electrodes with faradic reaction of PWA clusters and possible fast ionic mobility in the mesopores.

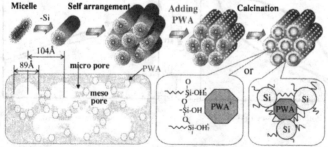

Figure 1   One step synthetic process of PWA doped mesoporous silicates

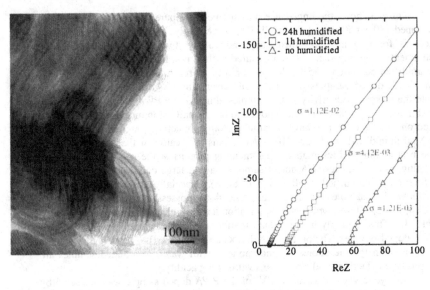

Figure 2   TEM photograph   Figure 3   AC impedance response

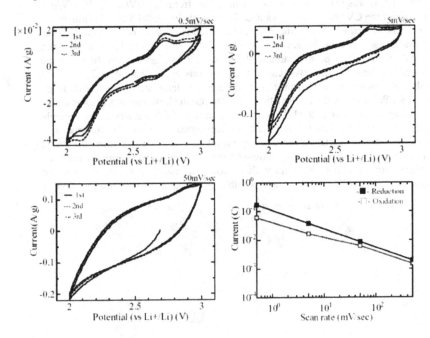

Figure 4   Cyclic Voltammograms of the PWA doped mesoporous silicates

SUMMARY

PWA doped mesoporous SiO_2, with large specific surface area and well ordered mesostructure has been successfully synthesized by one-step condensation process using triblock copolymer (P-123) as a template. The silicate framework have been electrochemically functionalized by the incorporation of PWA in the silicates. The materials become both proton conducting electrolytes with large a conductivity and electrodes for Li with faradic reaction of PWA cluster's surface. The possible fast ionic transport through the mesopores might be suitable for application such as fuel cell inorganic electrolytes at elevated temperatures or high rate electrodes of Li battery.

REFERENCES

1. C. T. Kresge, M. E. Leonowics, W. J. Roth, J. C. Vartuli and J. S. Beck, Nature **359**, 710 (1992)
2. D. M. Antonelli, J. Y. Ying, Chem. Mater., 1996, 8, 874, Z. Tian. W. Tong, J. Wang, N. Duan, V. V. Krishnan and S. L. Suib, Science **276**, 926 (1997), P. Yang, D. Zhao, D. I. Margolese, B. F. Chmelka and G. D. Stucky, Nature **396**, 152 (1998), H. S. Yun, K. Miyazawa, H. S. Zhou, I. Honma and M. Kuwabara, Adv. Mater. **13**, 1377 (2001)
3. I. Honma, Y. Takeda and J. M. Bae, Solid State Ionics **120**, 255 (1999), I. Honma, S. Nomura and H. Nakajima, J. Membrane. Sci. **185**, 83 (2001)
4. L. Marosi, E. E. Platero, J. Cifre and C. O. Arean, J. Mater. Chem. **10**, 1949 (2000),
5. W. Kaleta and K. Nowinska, Chem. Comm. 535 (2001)
6. J. M. Brégeault, J. Y. Piquemal, E. Briot, E. Duprey, F. Launay, L. Salles, M. Vennat and A. P. Legrand, Micropor. Mesopor. Mater. **44-45**, 409 (2001)
7. W. Li, L. Li, Z. Wang, A. Cui, C. Sun and J. Zhao, Mater. Lett. **49**, 228 (2001)
8. A. G. Siahkali, A. Philippou, J. Dwyer and M. W. Anderwon, Appl. Catal. A **192**, 57 (2000)
9. M. J. Verhoef, P. J. Kooyman, J. A. Peters and H. V. Bekkum, Micropor. Mesopor. Mater. **27**, 365 (1999)
10. P. A. Jalil, M. A. A. Daous, A. R. A. A. Arfaj, A. M. A. Amer, J. Beltramini and S. A. I. Barri, Appl. Catal. A **207**, 159 (2001)
11. M.Tatsumisago, H.Honjo, Y.Sakai and T.Minami, Solid State Ionics, 74, 105 (1994)

Mat. Res. Soc. Symp. Proc. Vol. 756 © 2003 Materials Research Society　　　　　EE8.5

Ball-milling : an alternative way for the preparation of anodes for lithium-ion batteries

Daniel Guérard and Raphaël Janot
Laboratoire de Chimie du Solide Minéral, UMR 7555 CNRS,
UHP Nancy, BP 239, 54506 Vandoeuvre lès Nancy Cedex, France

ABSTRACT

The preparation of new anodic materials for lithium-ion batteries is possible by ball-milling within liquid media. Those powders can be as diverse as : highly anisometric graphite particles (geometrical anisotropy around 100), graphite-maghemite ($\gamma\,Fe_2O_3$) composites and graphite-intercalation compounds with high lithium contents (e.g. LiC_3 is obtained by ball-milling of Li + 2C powders). These phases are the first superdense graphite-lithium compounds stable under ambient pressure and lead to a capacity of 1 Ah/g as primary battery and 370 mAh/g as secondary one. The synthesis and characterizations (chemical analysis, XRD, pycnometry, TEM, SEM, ^7Li NMR, magnetic measurements) are presented for each materials as well as the electrochemical behavior.

INTRODUCTION

Due to the important development of portable electronic devices, the needs for lithium-ion batteries is still increasing and the improvement of the electrochemical performances is, therefore, of a large interest. Most of the researches concerning the anodic materials are oriented towards disoriented carbon matrices, since these powders present a larger capacity than graphite. However, this increase is generally linked to a larger irreversible lithium loss, due to the formation of an important passivating layer, and to a large hysteresis of the charge-discharge process. On the other hand, nanosized oxides like CoO are good candidates as anodic materials with very high capacity. Nevertheless, these powders require to be supported on graphite in order to obtain a good electric conduction between the grains and to avoid the agglomeration of the nanoparticles during cycling [1, 2]. Our research on the preparation of new materials by ball-milling covers those two types of anodes for lithium-ion batteries.

EXPERIMENTAL

Graphite is generally used as anodes in commercial batteries. Its capacity, corresponding to the reversible lithium intercalation into graphite leading to LiC_6 (372 mAh/g), is quite high and the charge/discharge process occurs mainly below 0.3 V vs. Li. By using low-temperature carbon materials (as cokes), the capacity is higher (up to 800 mAh/g), but the charge-discharge potentials are around 1.0 V limiting the practical use for such disorganized carbons. On the other hand, the use of small and thin graphite particles can increase the rate of the intercalation /deintercalation process. If one mills graphite, the resulting powder is made of carbon particles less anisotrope than those of starting graphite and presenting a high defects content : as a result, the electrochemical behavior of these ball-milled carbons is quite similar to that of low temperature carbons [3]. In order to limit the formation of defects during the mechanical milling, we have chosen to mill the graphite in the presence of a liquid (water). A planetary mill is used since it favors the friction movements and thus the cleavage of the graphite particles. The experimental conditions are as follows : 5 g of graphite powder (Alfa, mean size 40 μm) are mixed together with 50 cm³ of water and 200 g of stainless steel balls (diameter 5 and 10 mm in the weight ratio 1:1) for 48 h at 200 rpm. Under those conditions, one obtains after drying a black powder whose magnetic properties are such that a simple contamination due to the abrasion of the milling tools is excluded [4]. An XRD pattern of this powder is shown in figure 1.

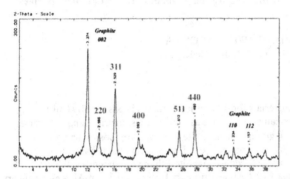

Figure 1. XRD of a graphite powder ball-milled within water. The italic characters are the Miller indices of graphite, the bold characters are those of a cubic phase. ($\lambda_{Mo\,K\alpha}$)

Synthesis of pure maghemite

The magnetic impurity together with graphite is neither metallic iron nor hematite (α Fe$_2$O$_3$), whose diffraction peaks are not observed by X-ray diffraction. It can be either magnetite (Fe$_3$O$_4$) or maghemite (γ Fe$_2$O$_3$). Those compounds present a similar spinel structure and their **a** parameters are respectively 8.39 and 8.35 Å. Anyway, the obtained powder is a composite based on graphite and magnetite (or maghemite).

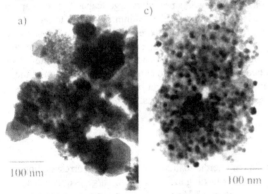

Figure 2. TEM micrographs of maghemite prepared by ball-milling a) after 12 h, b) after 48 h

Since the chemical analysis (Fe^{2+} and Fe^{3+} titration) is made difficult by the presence of graphite, we have performed the milling of iron powder (5 g) within water with the same conditions as before, but without graphite [5]. The milling duration is an important parameter, not only for the grain size, but also for the purity : after 12 hours, the large crystals correspond to a mixture of magnetite and maghemite (as determined by magnetic susceptibility measurements), whereas a 48 hours milling leads to a pure maghemite with a narrow particle size distribution (around 15 nm) (cf. figure 2).

Grinding of graphite in the presence of dodecane

Water is reduced by iron, thus, we have replaced water by a liquid, which does not react neither with iron nor with graphite. The choice of a heavy alcane ($C_{12}H_{26}$) is connected to the boiling point –and low vapor pressure at room temperature- on one hand and its viscosity on the other hand (respectively : 216°C, 16 Pa at 25°C and 1.383 mPa.s at 25°C, this last being close to that of water (0.890) and ethanol (1.074). Its melting point at - 9.6°C makes it a liquid at room temperature. The powder obtained by ball-milling of natural graphite within this alcane is made of flat graphite crystals, typically 1 μm in diameter and 20 nm in thickness, which involves a high anisometry. This powder is called Highly Anisometric Graphite (HAG) [6].

Synthesis of superdense graphite lithium intercalation compounds.

Under high pressure (50 kbars), lithium intercalates into graphite and leads to the so-called superdense LiC_2 phase, characterized by a high lithium in-plane density and an interplanar distance equal to that of the classical LiC_6 compound [7]. The ball-milling allows to obtain conditions close to those used for the high pressure synthesis : due to the high energy of the shocks occurring during ball-milling, the pressure and temperature applied to the grains can reach, momentarily and locally, 40 kbars and 300°C. We have shown that the synthesis of superdense graphite-lithium compounds is therefore possible by ball-milling. The presence of small quantities of dodecane is required (to limit the sticking on the milling tools) and the following conditions are used : 5 g of Li and graphite powders (in the ratio Li/C = 2) mixed with 1 cm^3 of dodecane are placed in the milling container with 200 g of stainless steel balls and milled at 200 rpm for 12 hours. The XRD diagram of the resulting powder is shown in figure 3a.

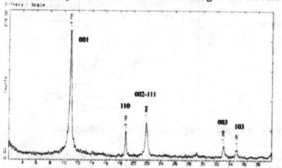
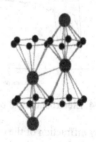

Figure 3. Superdense graphite-lithium compound prepared by ball-milling
a) XRD diagram ($\lambda_{Mo\ K\alpha}$), b) hexagonal unit cell of the LiC_3 phase

The structure of the intercalated phase was determined by the combination of X-ray diffraction and ^{7}Li NMR spectroscopy. Its stoechiometry is LiC_3 : this phase is mixed with free Li, the starting composition of the powder being Li + 2 C. The hexagonal unit cell (a = 4.30 Å, c = 2 Ic = 7.40 Å) belongs to the $P6_3/mmm$ space group with the lithium atoms in position 4f (z = 0.054) and the carbon atoms in 12 j (x= 2/3, y = 0). The splitting of the lithium intercalated layer, along the c axis, leads to a Li-Li distance of 2.60 Å and is responsible of the high stability of these ball-milled compounds. On the contrary, in the high pressure phases, the lithium atoms form Li_7 clusters where the frustration prevents a lengthening of the Li-Li bonds and a stabilization of the compound under ambient pressure [8].

ELECTROCHEMICAL STUDIES

By ball-milling, we have set up the preparation of several materials, which can be used as anodes in lithium-ion batteries : pure maghemite, graphite-maghemite composites, highly anisometric graphite, superdense LiC_3 compound mixed with an excess of free lithium. The electrochemical behavior of those materials was investigated.

Maghemite and graphite-maghemite composites

The behavior of maghemite is classical with the reduction into iron followed by a partial oxidation into wustite (FeO), but it appears that short-circuits occur when the battery is cycled. We have, thus, studied the behavior of graphite-maghemite composites. Two different ways for the synthesis of such composites are investigated according to the starting components : graphite and iron powders milled together for 48 hours within water (see the first part of the chapter "Experimental") or simple mixing of graphite powder and already prepared maghemite. The best results are obtained when Highly Anisometric Graphite (HAG) is mixed with maghemite within water for 2 hours. This synthesis leads to a well-defined material, whose electrochemical properties are shown in figure 4.

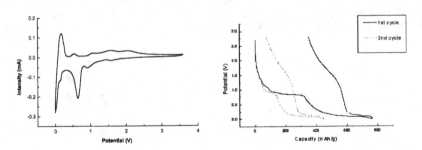

Figure 4. Electrochemical behavior of a graphite-maghemite (22 wt. %) composite

X-ray diffraction of the composite was performed at different potentials when cycling in order to understand the mechanism of the charge-discharge process. During the first reduction, the lithium insertion into the vacant sites of maghemite occurs around 1.1 V. The plateau at 0.8 V on the galvanostatic cycle is the sign of the reduction of maghemite into iron and is followed by the intercalation of lithium into the carbon matrix (below 0.2 V). During the oxidation process, after the deintercalation of lithium from graphite, there is a large hysteresis corresponding to the oxidation of iron into wustite. The capacity (initially higher than 400 mAh/g) decreases while the battery is cycled (down to 320 mAh/g after 15 cycles), mainly due to the coalescence of the iron-wustite nanoparticles leading to a less reversible oxido-reduction process. This last point has to be avoided for potential applications [9].

Highly anisometric graphite

By ball-milling of graphite powder (Alfa graphite, mean diameter 40 μm) within dodecane, the average diameter of graphite particles is not drastically decreased. On the contrary, the thickness is divided by a factor of 200 (from 4 μm to 20 nm). The specific surface area of graphite is, therefore, largely increased but, the irreversible capacity of the ball-milled graphite is lower as shown in figure 5.

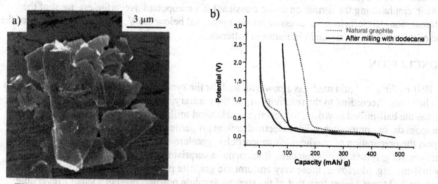

Figure 5. a) TEM micrography of HAG
b) First galvanostatic cycles of natural graphite ($\cdots$) and HAG ($-$)

The explanation is as follows. It is well-known that the formation of the passivating layer does not occur on the graphene sheets, but mainly on the prismatic ones. This would involve a similar value for the irreversible capacity in both cases. In fact, we assume that the milling performed with a planetary mill favors the cleavage, which occurs preferentially at the crystals defects. Decreasing number of defects leads to a smaller irreversible capacity [6].

Superdense graphite-lithium compounds

When the intercalation of lithium into graphite is realized under high pressure, a superdense phase with a LiC_2 stoechiometry is obtained. Unfortunately, this compound is unstable under ambient pressure and decomposes slightly into LiC_6 and free lithium when pressure is released. In electrochemistry, it gives a first discharge capacity around 1 Ah/g, but the superdense phase is not reformed by electrochemical mean and the following cycles are comparable to those given by the reversible intercalation of lithium into graphite to lead to LiC_6 [10].

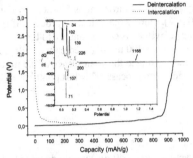
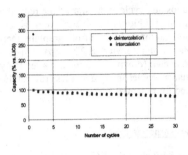

Figure 6. First galvanostatic cycle and cycling of the lithium GIC prepared by ball-milling

The first galvanostatic cycle of the powder obtained by ball-milling of a Li + 2 C mixture is shown in figure 6. During the deintercalation process, the capacity is close to 1 Ah/g, which takes into account for a global LiC_2 composition. On the contrary, the intercalation is limited to 370 mAh/g emphasizing the formation of the classical LiC_6 compound. Nevertheless, most of the intercalation-deintercalation process occurs at a potential below 0.4 V, which is interesting for the use of this material as anode in lithium-ion batteries.

CONCLUSION

Ball-milling in liquid media is a powerful tool for the synthesis of anodic materials for lithium ion batteries. According to the reactivity or, on the contrary, the chemical inertness of the liquid versus the ball-milled powders, the reactive mechanical milling leads to a large variety of compounds, or composites, whose electrochemical properties are promising. We can especially report the preparation of maghemite nanoparticles, graphite-maghemite composites and highly anisometric graphite. In this last case, it looks quite surprising to prepare graphite single crystals by ball-milling. Moreover, those very anisometric graphite particles, whose specific surface area is about 200 times higher than that of the starting graphite powder, present a small irreversible capacity.

REFERENCES

1. P. Poizot, S. Laruelle, S. Grugeon, L. Dupont and J.M. Tarascon, *Nature* **407**, 496 (2000)
2. Y. Lee, R. Zhang and Z. Liu, *Electrochem. & Solid State Let.*, **3**, 167 (2000)
3. T. Nagaura and K. Tozawa, Prog. Batt. *Solar Cells* **9**, 209 (1990)
4. R. Janot, J. Conard and D. Guérard, *Carbon* **40-15**, 2887 (2002)
5. R. Janot and D. Guérard, *J. All. Comp.* **333/1-2**, 302 (2002)
6. R. Janot and D. Guérard, *Carbon* **39-12**, 1931 (2001)
7. V. Nalimova, *Sol. St. Com.* **97**, 583 (1996)
8. R. Janot, J. Conard and D. Guérard, *Proc. Coll. Mat. Tours (Oct. 2002)* CM-14-004
9. R. Janot and D. Guérard, *Proc. Carbon'02, Beijing (Sept. 2002)* H062
10. C. Bindra, V.A. Nalimova, D.E. Sklovsky, Z. Benes and J.E. Fischer, *J. Electroch. Soc.* **145**, 2377 (1998)

Mat. Res. Soc. Symp. Proc. Vol. 756 © 2003 Materials Research Society   EE1.9

Proton conductivity of functionalized zirconia colloids

D. Carrière, M. Moreau, K. Lahlil, P. Barboux and J.-P. Boilot.
Laboratoire de Physique de la Matière Condensée, CNRS UMR 7643C, 1 route de Saclay,
Ecole Polytechnique, 91128 Palaiseau Cedex

ABSTRACT

Colloidal 60 nm ZrO_2 particles have been treated with aqueous solutions of phosphoric
acid and sulphophenyl-phosphonic acid (SPPA). This leads to the covalent bonding of P-OH
and $C_6H_4-SO_3H$ acid groups onto the surface of the particles. The resulting acidity yields a
significant proton conductivity at the surface of the particles.

The proton conductivity of the $ZrO_2-H_3PO_4$ system is quite stable against a change of
relative humidity and temperature. However, it remains limited by the low acidity of the P-
OH group. On the contrary, the proton conductivity of the ZrO_2-SPPA system is almost two
orders of magnitude higher at high relative humidities due to the high acidity of the C_6H_4-
SO_3H group (1.10^{-3} $S.cm^{-1}$ in the same conditions). But, it is much more sensitive to changes
in relative humidity.

A mixed grafting of both H_3PO_4 and SPPA onto the zirconia particles allows to obtain a
conductivity as high as in the SPPA case whereas it remains stable with humidity and
temperature as for the phosphoric acid. This probably indicates that the conductivity arises
from ionization of sulphonic acid but that the weaker phosphoric acid groups contribute to the
conduction mechanism.

INTRODUCTION

Room temperature protonic conductors may find well known applications in the domain
of electrochromic devices, fuel cells or supercapacitors [1]. But they suffer from a weak
stability of the conductivity against changes in temperature or relative humidity. Mineral-
organic hybrid systems may offer a promising compromise between the strong acidity of
some acid polymers and the better stability towards dehydration of the mineral species.
Polymer-phosphoric acid blends such as H_3PO_4 / PBI complexes or H_3PO_4 / N,N
dimethylacetamide / PVDF composites have been proposed [2,3,4]. However, these systems
contain free phosphoric acid that is lost upon water treatment. An alternative is to use acid
and/or basic functions of insoluble polycondensed mineral species. This has been achieved
with some crystallized lamellar zirconium phosphates and phosphonates [5,6,7] or by in-situ
condensation of inorganic phosphates in a proton conducting polymer matrix [8]. Similarly,
our work focuses on the role of the surface conduction on mineral colloidal particles. This
requires an optimization of the synthesis and characterisation of nanometric and organically
modified inorganic species. Apart from higher loadings in the membranes, the nanometric
size of the particles is expected to allow an enhancement of the proton conduction properties
due to a better grain-to-grain transfer and a higher surface groups content.

In a previous work, we have shown that protonic conductors can be prepared by grafting
phosphoric acid at the surface of colloidal zirconia particles [9]. A higher conductivity could
be obtained by grafting of stronger acid such as sulphophenyl-phosphonic acid [10]. The
particles have been treated with aqueous solutions of phosphoric acid and sulphophenyl-
phosphonic acid (SPPA). This leads to the formation of covalent Zr-O-P bonds with a surface

313

acidity of the nanoparticles strongly enhanced by the pendant P-OH and C_6H_4-SO_3H groups. The grafting of H_3PO_4 onto the ZrO_2 particles results in a significant and quite stable proton conductivity against relative humidity whereas the ZrO_2-SPPA system shows a poor stability against relative humidity and temperature changes. For this reason, the mixed grafting of both H_3PO_4 and SPPA onto the particles is presented.

EXPERIMENTAL

Colloidal 60 nm ZrO_2 particles have been synthesized by the thermohydrolysis of aqueous zirconium acetate solutions following a patent by Matchett [11]. After synthesis, the colloidal particles have been purified by centrifugation until all residual acetates and zirconium monomers have been removed. Further experimental details are described elsewhere [9]. The particles, shown in Fig. 1, are 60 nm aggregates of 5 nm crystallites of well crystallized monoclinic zirconia. Before drying, the surface area of the particles is 450 m²/g, as determined by X-ray small angle scattering. Upon drying it decreases down to 150 m²/g as confirmed both by X-ray scattering and nitrogen adsorption isotherms. Therefore, a better reactivity is expected in the colloidal solution than in the dried powder and the graftings were performed on the colloids.

Meta sulphophenyl-phosphonic acid has been obtained by sulphonation of phenylphosphonic acid ($C_6H_5PO(OH)_2$) with SO_3 in CH_2Cl_2 followed by purification with barium chloride and acidification on an ionic exchange resin [12].

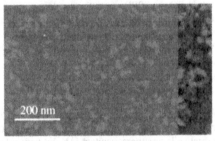

Figure 1. Transmission electronic microscopy of the monoclinic zirconia nanoparticles.

Chemical analysis was performed at the CNRS analysis center in Vernaison (France). Grafting yields were obtained form the elemental analysis using Zr, P and S. ^{31}P MAS-NMR spectra were obtained at 145 MHz on a Bruker MSL360 with a probe rotating at 10 kHz.

Conductivity was measured using a HP 4192A impedancemeter on pressed powders (100 kN/cm²) with gold-evaporated or silver paint electrodes. The relative humidity was controlled at 20°C by equilibration of water vapor with sulfuric acid solutions. Measurements above 100°C were carried out in a furnace under a pressure of 1 atm of water vapor controlled by a water boiler. For a given temperature the relative humidity at temperature T was obtained by the relationship: $RH(T) = \dfrac{P_{H_2O}}{P_{sat}(T)}$ where $P_{H_2O} = 1$ atm and $P_{sat}(T)$ is the equilibrium vapor pressure at temperature T obtained by thermodynamic data or given by the handbooks[13].

RESULTS AND DISCUSSION

Grafting of H₃PO₄ and SPPA:

Phosphoric acid and meta sulphophenyl-phosphonic acid (SPPA) have been grafted onto the particles by heating at 100°C during 4 hours an aqueous mixture of colloidal ZrO_2 and H_3PO_4 (respectively SPPA) with a P/Zr ratio above 0.5. Any excess phosphoric (resp. phosphonic) acid is then washed out by successive centrifugations and redispersions in water, coupled to a test of precipitation with zirconium acetate (resp. to a test by ^{31}P liquid-state NMR). The resulting colloid is then dried at 50°C under vacuum. After grafting, X-ray diffraction patterns are still characteristic of monoclinic zirconia without any change in the nature of the diffraction lines. This shows that no particle dissolution occurs during the grafting. This allows a good conservation of the specific area, as shown by nitrogen adsorption measurements (BET specific area above 100 $m^2.g^{-1}$).

The maximum amount of phosphate groups that can be grafted onto the surface of the zirconia particles leads to a ratio P/Zr = 0.26 which corresponds to a surface concentration of 14 $\mu mol.cm^{-2}$ phosphate groups. The grafting is less efficient for SPPA and a SPPA/Zr = 0.10 ratio is only obtained. This is probably due to the steric hindrance and to the charge repulsion between the ionized sulphonic acid groups during the grafting in aqueous solution.

For the study of mixed grafting, two methods have been used: simultaneous grafting with SPPA and phosphoric acid, and sequential grafting.

1) In the simultaneous grafting, an aqueous solution 0.2 M of ZrO_2 particles is treated 4 hours at 100°C with a mixture H_3PO_4 and SPPA in the ratio H_3PO_4/ASPP/Zr= 0.14/0.13/1. The grafted particles were purified by repeated centrifugations. Chemical analysis yields a composition Zr/P/S = 1/0.20/0.05. The S/Zr ratio is much lower than observed with the single grafting of SPPA onto ZrO_2 (0.05 instead of 0.10). This indicates that the SPPA grafting is inhibited in the presence of H_3PO_4.

2) In the sequential grafting, the zirconia particles are first treated with excess SPPA, purified, and then treated with excess H_3PO_4. Chemical analysis on the final product yields a ratio Zr/P/S = 1/0.24/0.02. This indicates that most of the SPPA molecules grafted in the first reaction has been removed by phosphoric acid. The composition is nearly that of materials grafted with phosphoric acid only.

The ^{31}P MAS NMR spectra of all samples are shown in Fig. 2. By comparison with the literature [14], the resonance peaks for the grafted phosphoric acid (Fig 2a) can be attributed to PO₄ groups connected to one (-6.2 ppm), two (-11.7 ppm) or three zirconium atoms (-20.8 ppm) through Zr-O-P covalent bounds. Thus, most of the phosphates bound to the surface carry an acidic proton and we note the absence of free phosphoric acid. The ^{31}P NMR MAS spectrum of the SPPA-treated particles is shown in Fig. 2b. Three peaks are also observed, with a major one at 7.8 ppm and two minor peaks at 2.9 ppm and – 4.5 ppm. By considering the displacement relative to the position of the free phosphonate (c.a. 15 ppm), the peaks can be also attributed to phosphonates coordinated to one, two and three zirconium atoms respectively. The major species are coordinated to only one zirconium, which is lower than for orthophosphoric acid.

The NMR spectrum of the sample obtained by simultaneous grafting (Fig. 2c) shows a mixture of SPPA and phosphoric acid signals. The average chemical shift indicates a weak coordination of P atoms. This result can be correlated to the low yield of grafting. The NMR signal of the sample obtained by sequential grafting (Fig. 2d) is very similar to that of phosphoric acid only (Fig. 2a) except for the very small signal at 2.9 ppm characteristic of grafted SPPA. Thus, this is a phosphoric acid grafted sample into which a small amount of SPPA has been introduced.

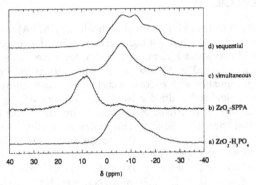

Figure 2. ^{31}P MAS NMR spectra of the zirconia colloid treated during 4 hours at 100°C with aqueous solutions of phosphonates, P/Zr > 0.5 **a)** phosphoric acid **b)** sulphophenyl-phosphonic acid **c)** H₃PO₄ and SPPA, simultaneous treatment **d)** H₃PO₄ and SPPA, sequential treatment. The mixtures were purified before drying.

Conductivity measurements:

The conductivity of non grafted powders is too low to be measured at 20°C ($<10^{-10}$ S.cm^{-1}). For the grafted samples, the conductivity measurements as a function of the relative humidity are shown in Fig. 3. At high relative humidities, the conductivity of the SPPA-grafted zirconia particles is one order of magnitude higher than with the phosphoric acid grafting (Fig. 3a and b): above 70% R.H., it remains around $4\ 10^{-4}$ S.cm^{-1}. This may be attributed to the strong acidity of the sulphonic acid as compared to phosphoric acid. Despite of a lower coverage of the particle surface by the sulphonate groups, a high conductivity is obtained. However, the conduction mechanism probably requires large amounts of adsorbed water forming the conduction pathway or the charge carriers. Indeed, the conductivity is sensitive to relative humidity and rapidly decreases down to $1.5\ 10^{-9}$ S.cm^{-1} at 19% R.H., as compared to $4.3.10^{-7}$ S.cm^{-1} for the H₃PO₄-grafted zirconia.

The conductivity of the phosphate-grafted materials shows a better stability as a function of the vapor pressure. Below RH = 50% it becomes higher than for SPPA. This indicates that, although the phosphate groups are less easily ionized than sulphonic groups, they allow a better coverage of the zirconia particles and offer a better pathway for conduction either by themselves (as charge carriers) of by fixing strongly adsorbed water at their surface.

Although the conductivity of the mixed samples is slightly lower than the SPPA grafted one at high relative humidity, it is more stable towards a change of relative humidity. This effect can be attributed to the presence of phosphate groups. Note that the higher conductivity is obtained with the sample obtained by sequential grafting. It is remarkable that at low relative humidities, the conductivity of the sample is higher than the conductivity of both H₃PO₄ and SPPA graftings. This strongly suggests that the zirconia after sequential grafting behaves like a sulphonate-doped phosphated zirconia. Therefore, a small amount of SPPA (10%) in a phosphoric acid grafted material can allow an increase of conductivity by one order of magnitude.

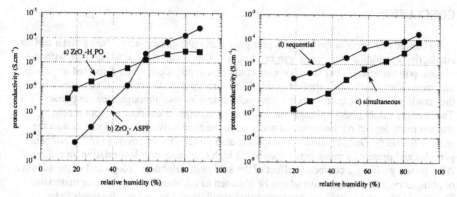

Figure 3. Proton conductivity as a function of relative humidity at 20°C of zirconia nanoparticles grafted with **a)** phosphoric acid **b)** SPPA **c)** H₃PO₄ and SPPA, simultaneous grafting **d)** H₃PO₄ and SPPA, sequential grafting

The conductivity has also been measured at high temperatures between 120°C and 200°C (Fig. 4). These results correspond to equilibrium values since they are nearly the same when obtained with increasing and decreasing temperatures. For the sake of clarity, we only show the data obtained upon increasing temperatures. For all samples, the conductivity decreases with the temperature. At fixed vapor pressure (1 atm of water vapor), higher temperatures correspond to lower relative humidities (partial pressure of water), the values of which are reported in the figures. Therefore, the dehydration effect that occurs upon heating is higher than the activation for the conduction mechanism. Although the less conducting at room temperature and high relative humidity, the phosphoric acid sample is the most stable and becomes the best conducting material at higher temperature. This indicates again, the role played by phosphate groups in the conduction mechanism.

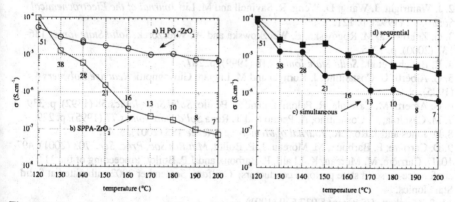

Figure 4. Proton conductivity as a function of temperature under a 1 atm. water pressure of zirconia nanoparticles grafted with **a)** phosphoric acid **b)** sulphophenyl-phosphonic acid **c)** H₃PO₄ and SPPA, simultaneous grafting **d)** H₃PO₄ and SPPA, sequential grafting. The figures on the charts indicate the corresponding relative humidities at a given temperature.

CONCLUSION

The study of zirconia grafted with phosphoric and phosphonic acids shows that a surface conductivity as high as 10^{-3} S.cm^{-1} can be obtained. This is still too low to compete with organic polymeric acids but such powders could find applications when dispersed in polymers. They can be also applied as thin films directly from the colloidal solution, where they could be used as electrolytes for electrochromic windows although they are porous. The stronger sulphophenyl-phosphonic acid yields the highest conductivity but proton transfer must depend on the adsorbed water since the conductivity strongly varies with the relative humidity. On the contrary, phosphoric acid yields a better coverage of the zirconia particles and presents a much better stability at high temperature or low relative humidity. After mixed grafting, a composite effect of the acidity of sulphonic groups and of the stability of phosphoric acid grafted material can be obtained as compared to the separate materials. The role of phosphoric acid on the temperature stability of the conductivity needs to be further explored. This could allow to obtain protonic conductors stable at higher temperatures (100°C-250°C). Rotation of the grafted phosphates species can contribute to the proton transfer mechanism if the surface coverage is high enough. But, phosphate groups may also strongly adsorb a monolayer of water that contributes to the conduction pathway. This will be verified by careful mass change analysis.

Further studies are in progress to optimize the grafting yield and to obtain smaller particles that could allow higher surface and higher surface conductivity. The role of the intergrain and intragrain transfer needs also to be addressed too.

REFERENCES

1. K.D. Kreuer, Solid State Ionics 97 (1997) 1.
2. J. Wainright, J. Wang, D. Weng, R. Savinell and M. Litt *Journal of the Electrochemical Society* 42 (1995), p. 121.
3. G. Zukowska, M. Rogowska, E. Weczkowska and W. Wieczorek, *Solid State Ionics* 136-137 (2000), p. 1205.
4. R. Bouchet et al., *Solid State Ionics* 118 (1999), p. 287.
5. G. Alberti, U. Costantino, J. Kornyei and M. Luciani Giovagnotti, *Reactive Polymers* 4 (1985), p. 1.
6. G. Alberti, M. Casciola, R. Palombari and A. Peraio, *Solid State Ionics* 58 (1992), p. 339.
7. M. Casciola, U. Costantino, A. Peraio and T. Rega *Solid State Ionics* 77 (1995), p. 229.
8. D. Jones and J. Rozière, *Journal of Membrane Science* 185 (2001), p. 41.
9. D. Carrière, P. Barboux, M. Moreau, J.-P. Boilot, *Mat Res Soc. Proc. Ser.* 703 (2001), 497
10. D. Carrière, M. Moreau, K. Lhalil, P. Barboux and J.P. Boilot, Proceeding of the 11th conference on solid state protonic conductors, Guilford, September 2002, submitted at Solid State Ionics.
11. S. Matchett. *US patent* 5,037,579 (1991)
12. E. Montoneri and M. Gallazi *Journal of the Chemical Society, Dalton Transactions* (1989), p. 1819.
13. R.C. Weast, Handbook of chemistry and Physics 56th ed. CRC Press, Cleveland (1976) D-181.
14. N. Clayden *Journal of the Chemical Society, Dalton Transaction* (1987), p. 1877.

Mat. Res. Soc. Symp. Proc. Vol. 756 © 2003 Materials Research Society   EE3.14

Study of bare and functionalized Zirconia Nanoparticles Filled Polymer Electrolytes Based on a Polyurethane

Paulo V. S. da Conceição, Luiz O. Faria, Adelina P. Santos, Clascídia A. Furtado
Centro de Desenvolvimento da Tecnologia Nuclear - CDTN/CNEN, C.P. 941, 30123-970, Belo Horizonte, MG, Brazil.

ABSTRACT

In this work, composite polymer electrolytes based on a thermoplastic polyurethane/LiClO4 amorphous system and on bare and functionalized zirconia nanoparticles as a filler are reported. The ceramic nanoparticles were synthesized via the sol-gel route using zirconium butoxide as the precursor for zirconium oxide nanoclusters and methacrylic acid as an organic modifier group. The salt concentration in the polymer phase was 17 wt% and fillers were added in the range between 2 and 10wt%. Scanning electron microscopy (SEM) was used to characterize the average size and the homogeneity of the nanoparticles in the polymer matrix, while impedance spectroscopy (IS) was used to evaluate the ionic conductivity of the composites. The addition of zirconia fillers results in an increase in ionic conductivity for all filled systems. The results also show that the functionalization of the zirconia nanoparticles promotes a significant increase in conductivity, suggesting that the interaction of the metracrylate-functionalized fillers with the polyurethane matrix was greatly improved. These results raise interest in the study of organically modified ceramic clusters as fillers for electrolyte polymers.

INTRODUCTION

The addition of inorganic fillers has been demonstrated to be an effective method to improve the ionic conductivity, mechanical strength, and thermal stability of polymeric electrolyte systems [1-7]. The changes in their properties are related to both the size and the surface characteristics of the ceramic particles. It has also been reported that the benefits of this approach are greatly increased when the size of the ceramic particles used in the filling process is reduced to the nanoscale. This finding has directed the research of these materials toward the area of nanocomposite polymer electrolytes (NCPE). According to Scrosati and co-workers [3-4], the inclusion of nano-sized fillers into poly(ethylene oxide) (PEO) - Li+ systems not only promotes a stable enhancement in the low-temperature ionic conductivity by an order of magnitude, but also improves both the electrochemical stability and the compatibility between the electrolyte and lithium electrodes. The gain in conductivity is accompanied by an increase in the Li$^+$ transport number, reaching a value as high as 0.6 in the temperature range of 45-90 °C, thus bringing these materials very close for practical requirements to the development of advanced lithium batteries.

Most of the studies on NCPE's are focused on semi-crystalline PEO-based electrolytes due their superior properties as a host polymer matrix for Li$^+$ transport. The effect of the inclusion of inorganic nanoparticles into these systems has been associated with two factors: i) an increase in the amorphicity of the PEO matrix; and ii) the formation of more conductive paths for ionic transport at the polymer-ceramic interfaces. Therefore, the study of the separate contributions of these two factors in NCPE's based on a completely amorphous matrix, is very interesting. In this paper, we report preliminary results obtained by adding nanoscale zirconia nanoclusters synthesized by a sol-gel route into polymer electrolyte systems based in a polyurethane amorphous matrix. In order to investigate the role of ceramic surface states on the zirconia-polyurethane interaction, the zirconia nanoparticles were modified by introducing acid methacrylic during the sol-gel process.

EXPERIMENTAL DETAILS

Three different stable suspensions of zirconia nanoparticles were prepared by controlled hydrolysis and polycondensation of zirconium tert-butoxide (80 wt% in tert-butanol, Fluka) under acidic conditions. Anhydrous tetrahydrofuran, which is a solvent compatible with the polyurethane, was used as the liquid phase for all systems, allowing careful management of the water concentration in the systems in order to avoid an uncontrolled growth of the nanoclusters. Table I summarizes the conditions used to prepare the suspensions. Only in the case of the Zr-Mc-3 suspension, was the water directly added. For the other systems, the hydrolyzing water was provided *in situ* by the esterification of nitric acid and isopropanol, in the case of Zr-Iso, and methacrylic acid and tert-butanol (from alkoxide solution), in the case of Zr-Mc-10, respectively.

The polyurethane was synthesized by mixing poly[(tetramethylene glycol)$_{0.65}$-*co*-(ethylene glycol)$_{0.35}$] (Mn = 880) and hexamethyldiisocyanate (17 wt %) in the presence of dibutyltindilaurate as a catalyst. Details of the synthesis and characterization of the binary system polyurethane - LiClO$_4$ have been previously reported [8]. For each system, a unique batch of polymer was prepared and allowed to react for 24 hours. Stoichiometric aliquots of the nanoparticle suspensions and of LiClO$_4$ were then added when the polymer was still in solution, yielding samples with different ceramic concentrations. Since the size and composition of the nanoclusters cannot be precisely defined, the composite concentration was calculated as a function of the Zr wt% in relation to the total polyurethane + LiClO$_4$ weight. Composites with 0, 1, 2, 3, 5 and 10 wt % filler were prepared for each one of the different zirconia nanoparticles A blank samples for each polymer synthesis batch was also prepared. In this study, all electrolytes were synthesized at the optimized [8] Li salt concentration of 17wt%. Electrolyte membranes were obtained by casting the solution in Petri dishes, removing the solvent slowly at ambient pressure, followed by vacuum evaporation ($\sim 10^{-3}$ torr) for 48 hours at 60°C.

The average size and the homogeneity of the nanoparticles on the polymer matrix were characterized by scanning electron microscopy (SEM) (JEOL JFM 840A). The temperature dependence of the total ionic conductivity was measured with a Hewlett-Packard 4192A electrochemical impedance spectrometer, between 10 Hz and 0.1 MHz with signal amplitude of 50 – 80 mV at temperatures between 20 and 100 °C. For these measurements, the membranes were sandwiched between two stainless steel electrodes.

RESULTS AND DISCUSSION

Figures 1 and 2 show the morphology of some representative composites obtained by introducing bare (Zr-Iso) and functionalized (Zr-Mc-3) zirconia nanoparticles, respectively, into

Table II. Molar concentration ratios used in the preparation of the zirconia nanoparticles sols

Suspension	[H$_2$O]/[ZrBu$_4$]	[HNO$_3$]/[ZrBu$_4$]	[Iso]/[ZrBu$_4$]	[McOH]/[ZrBu$_4$]
Zr-Iso	-	0.1	20	-
Zr-Mc-3	1	-	-	3
Zr-Mc-10	-	-	-	10

where McOH = Methacrylic acid; ZrBu$_4$ = zirconium tert-butoxide; Iso = Isopropanol

the polymer electrolyte polyurethane/LiClO$_4$. The Zr-Mc-10 system (not show) has a similar morphology as the Zr-Mc-3 system. All composites show a rough surface (Fig. 1a and 2a and 2b), as expected for heterogeneous materials, and a uniform distribution of particles throughout the samples. However, while the particles in the Zr-Iso composites, initially of average diameter of 100 nm, aggregated themselves inside the polymer matrix, producing ceramic particles up to 1 μm in diameter (the white particle in the center of the micrograph shown in Fig. 1b), Zr-Mc-3 and Zr-Mc-10 composites show the presence of very small aggregates, with average diameters of approximately 100 nm (see, for example, figure 2c).

Different species of zirconium oxide can be obtained by a sol-gel process via hydrolysis and condensation of zirconium alkoxides. As these precursors are very reactive, the amount of the water plays an important role. At high water concentrations, these reactions take place very fast and an uncontrolled precipitation of colloidal zirconia is observed in the system. At low water conditions, only a few (-OR) groups are hydrolyzed, allowing the preparation of stable sols of oligomeric Zr-oxo species. The size and composition of the oligomers formed depend on the ratio of water, alcohol, alkoxide and catalyst acid used in the synthesis. In this study, the organically modified zirconia particles, prepared *in situ* by introducing methacrylic acid during the sol-gel processing, can be considered as an oligomeric zirconium oxide species with methacrylate groups replacing some alkoxilate groups [9]. Methacrylic acid also serve as a chelating ligand for the zirconium precursor [10]. So, it is expected that the presence of this organic modifier decrease the rate of the hydrolysis reaction, thus leading to the formation of smaller zirconium species. By varying the alkoxide/methacrylic acid ratio, nanoparticles with a different size and a different degree of substitution can be obtained.

Certainly, the presence of organic molecules containing heteroatoms covering the surface of the zirconia nanoparticles benefit from the interaction between the filler and the polyurethane matrix, probably by the formation of hydrogen bonds between the carboxylate groups from the methacrylic acid and the urethanic function [11]. The improvement of the polyurethane-filler interaction and the repulsion force among the carboxylate groups covering the surface of the fillers allow the homogeneous dispersion of the nanosized ceramic to aggregate into the polymeric phase, as showed in Figure 2.

Figure 3 presents the conductivity behavior on the Arrhenius plot for the three different composites with the same composition (5 wt% of filler) and for the pure polymer electrolyte (PUE). A slight enhancement of conductivity in relation to the pure electrolyte was observed with the introduction of the bare zirconia nanoparticles. As the systems studied in this work are

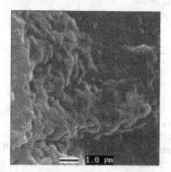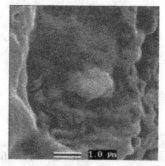

Figure 1. Micrographs for composite materials based on bare ZrO$_2$ (Zr-Iso) nanoparticles.

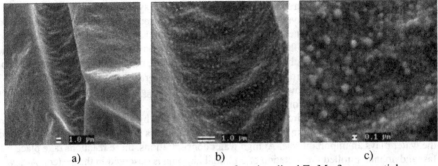

a)   b)   c)

Figure 2. Micrographs of the composites based on functionalized Zr-Mc-3 nanoparticles.

amorphous, the presence of the ceramic nanoparticles in the polymer electrolyte matrix seems to make possible a better conduction pathway, increasing the conductivity values. The previously suggested models describe the formation of alternative channels for the ionic transport in the interface between the polymeric phase and the dispersed filler [2,13]. Furthermore, studies of systems based on TiO_2 and an amorphous polyether matrix suggest that an interfacial region exists, which allows relatively easy hopping from one equivalent site (e.g. ether polyether) to another (TiO_2) [7]. Thus, a better interaction among the constituents of the system results in a higher ionic conduction. This theory is in agreement with our conductivity results for the systems based on the functionalized zirconia nanoparticles. As shown in Figure 3, a pronounced increase in conductivity was observed, mainly with the introduction of the Zr-Mc-3 filler, which showed a good dispersion and interaction with the polymer (Fig. 2). This significant increase in conductivity may be attributed to the presence of a contained-heteroatoms highly flexible phase of methacrylic acid or of low-molecular weight poly(methacrylic acid) chains in the interface ceramic-polymer. A possible explanation for the higher conductivity values of the Zr-Mc-3 composites in comparison to the Zr-Mc-10 composites could be the formation of a thicker and/or less flexible interface layer with the polymerization of the excess of methacrylic acid present in Zr-Mc-10 systems. However, to better explain the conduction behavior and the difference in the conductivity of the two composite systems based on modified zirconia nanoparticles, a careful

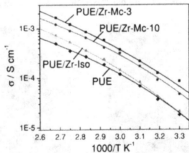

Figure 3. Arrhenius plot for the pure polyurethane electrolyte (PUE) and for the PUE/zirconia composites at 5 wt% of bare (Zr-Iso) and functionalized (Zr-Mc-3 and Zr-Mc-10) nanoparticles. The solid lines refer to the fitting of the experimental curves, log σ vs. 1/T to the VTF equation.

Table II. Fitting parameters for curves of log σ vs. 1/T calculated with the VTF equation for the composite systems at 5wt% of filler.

System	A (S K$^{1/2}$/cm)	B (eV)	T$_0$ °C
Pure electrolyte (PUE)	0.1	0.07	202
PUE/Zr-Iso	0.2	0.08	195
PUE/Zr-Mc-10	0.2	0.08	190
PUE/Zr-Mc-3	0.8	0.11	167

morphologic and spectroscopic characterization of the interface ceramic-polymer electrolyte needs to be performed.

Figure 3 presents also the fitting (solid line) of the temperature dependence of the conductivity for the composite materials to the empirical Vogel-Tamman-Fulcher (VTF) equation [14]: $\sigma_i = A\exp[-B/R(T-T_0)]$. Based on the free-volume model, A is a pre-exponential factor related to the effective number of charge carriers, B is a 'pseudo'-activation energy required for the free volume redistribution, and T_0 is a quasi equilibrium glass transition temperature. The log σ vs. 1/T data were found to follow the VTF model quite well for the three curves, as can be verified by visual inspection in Figure 4, which associates the ionic transport mechanism intimately with the mobility of the polymer chains [14]. The parameters obtained from the VTF curves are shown in Table II. The A values are seen to increase with the addition of the fillers, showing a higher value for the composite based on modified Zr-Mc-3 zirconia. The same behavior is observed for the B parameter, whose values are typical for solid electrolytes of linear polyethers. T_0 shows small values in the composites when compared with the pure electrolytes. The high A value for the Zr-Mc-3 composite system can be associated with a stronger ceramic-polymer interaction in this system, which increases the number of the charge carriers. According to Ribeiro *et al.* [6], the fact that the conduction mechanism becomes more difficult (increase of B) may be a consequence of a higher critical free-volume which permits other pathways for carrier displacement when the ceramic powder is distributed into the electrolyte.

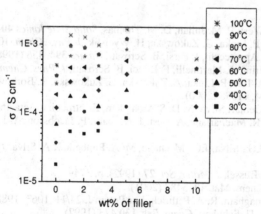

Figure 4. Filler concentration dependence of the conductivity of the total ionic conductivity (σ) for the PU/LiClO$_4$/Zr-Mc-3 system at several temperatures.

The composition of 5 wt% of filler for all composite systems was chosen to illustrate the log σ vs. 1/T behavior in Figure 3, because it corresponds to the concentration value in which a maximum conductivity value is observed. Figure 4 shows the conductivity behavior as a function of concentration for the PU/LiClO$_4$/ Zr-Mc-3 system at several temperatures. A similar behavior was obtained for the other two composite systems. Conductivities at room temperature as high as 1 x 10^{-4} S/cm were measured. Another maximum was also observed at 2 wt% filler. The presence of two maximum conductivity values can be attributed to the presence of two predominant particle sizes, since the position of the maximum as a function of the filler concentration depends of the filler particle size [13]. Following Przyluski et al. [13], the position of the conductivity maximum shifts to a lower volume fraction of filler with a decrease in the grain size.

CONCLUSIONS

Composite polymer electrolytes based on a polyurethane matrix, LiClO$_4$ and bare and functionalized zirconia nanoparticles were prepared. The presence of an organically modified surface containing carboxylate groups in the zirconia nanoparticles improved the ceramic-polymer interface and the dispersion of the fillers in the polymeric electrolyte. Consequently, an enhancement in conductivity was observed for the composite system in comparison with the pure electrolytes. The composite based on the Zr-Mc-3 functionalized zirconia reached conductivities of 1x10^{-4} S/cm at room temperature. The fitting of log σ vs. 1/T for all materials was performed adequately by the VTF equation. Further characterization of the complex interactions among the constituents of the system, the surface groups in the chemically modified zirconia particles and the ceramic-polymeric phase interface are currently being carried out in order to obtain a more complete picture of the systems behavior.

ACKNOWLEDGMENTS

The authors thank the Brazilian agencies CNEN, CNPq and FAPEMIG for financial support.

REFERENCES

1. S.Skaruup, K. West, P. M. Julian, D. M. Thomas, *Solid State Ionics* **40-41**, 1021 (1990).
2. W. Wieczorek, P. Lipka, G. Zukowska, H. Wycislik, *J. Phys. Chem.* 102, 6968 (1998).
3. F. Croce, G. B. Appetecchi, L Persi, B. Scrosati, *Nature* **394**, 456 (1998).
4. F. Croce, R. Curini, A. Martinelli, F. Ronci, B. Scrosati, *J. Phys. Chem. B* **103**, 10632 (1999).
5. M. Marcinek, A. Bac, P. Lipka, A. Zalewska, G. Zukowska, R. Borkowska, W. Wieczorek, *J. Phys. Chem. B* **104**, 11088 (2000)
6. R. Ribeiro, G. Goulart Silva, N. D. S. Mohallem, Electrochim. Acta 46, 1679 (2001).
7. M. Forsyth, D. R. MacFarlane, A. Best, J. Adebahr, P. Jacobsson, A. J. Hill, Solid State Ionics (2002).
8. C.A.Furtado, G.G., Silva, J.C., Machado, M. A. Pimenta, R.A. Silva, *J. Phys. Chem. B* **103**, 7102 (1999)
9. C. Wolf and C. Rüssel, *J. Mater. Sci.* **27** (1992), p. 3749
10. U. Schubert, Chem. Mat. 13, 3487 (2001).
11. A. W. McLennagham, R. A. Pethrick, Eur Polym J. 24/11, 1063 (1988).
12. G. Kickelbick, U. Schubert, *Chem. Ber.* **130** 473 (1997).
13. J. Przyluski, M. Siekierski, W. Wieczorek, *Electrochim. Acta* **40** 2101 (1995).
14. M. A. Ratner, in: J. R. Maccallum, C. A. Vincent (Eds.), *Polymer Electrolyte Reviews*, vol. 1. Amsterdam, Elsevier, 1987 (chap.7).

Mat. Res. Soc. Symp. Proc. Vol. 756 © 2003 Materials Research Society   EE6.1

Development of new ceramic doped ionoconducting membranes for biomedical applications.

Paola Romagnoli[a], Maria Luisa Di Vona[a], Enrico Traversa[a], Livio Narici[b], Walter G. Sannita[c], Simone Carozzo[c], Marcella Trombetta[d] and Silvia Licoccia[a]

[a] Dipartimento di Scienze e Tecnologie Chimiche, and Dipartimento di Fisica [b],Università di Roma "Tor Vergata", Via della Ricerca Scientifica 1, 00133 Rome (Italy)
[c] Centro Farmaci Neuroattivi-Dipartimento di Scienze Motorie, Universita' di Genova, Largo Benzi 10, 16132 Genova, Italy and Department of Psychiatry, State University of New York, Stony Brook, NY,1764-8101 USA.
[d] Interdisciplinary Center for Biomedical Research (CIR), Laboratory of Biomaterials, Università "Campus Bio-Medico", via E. Longoni 83, 00155 Rome (Italy)

ABSTRACT

New ionoconducting composite membranes to be used as an interface between the skin and the actual electrical instrumentation used to produce an electroencephalogram (EEG) have been developed. The gels are based on lithium salts and PMMA (polymethyl methacrylate) and have been doped with nanometric titanium oxide. The samples have been electrochemically characterized by means of impedance spectroscopy and their structure studied by ATR-FTIR and MAS NMR. Spectroscopic studies indicate interactions between the polymer and oxide dopant. The polymeric electrolytes allowed the registration of good electrophysiological cortical signals either spontaneous or stimulus-related.

INTRODUCTION

Conventional EEG recording require accurate preparation of the skin and the use of a semi-fluid electrolyte to establish electrical connection with the scalp. These requirements cannot be met if EEG recording has to be performed in extreme conditions (emergency, experimental conditions "on the field", non-cooperative subjects etc.).

Polymer based materials can represent a solution to these problems since they have the proper electrical and mechanical characteristics. Polymeric gels are extensively studied for a variety of biomedical applications [1-3]. The addition of inert ceramic powders into gel membranes has been reported to enhance ionic conductivity and/or to improve the mechanical stability of these materials [4].

We have then developed new composite polymer electrolyte consisting of nanometric titania dispersed in gels formed by immobilizing liquid solution containing $LiClO_4$ and 1,2-diethoxyethane into a PMMA polymer matrix.

EXPERIMENTAL DETAILS

All chemicals were reagent grade and were used as received. Titania powders were prepared via sol-gel according to a procedure previously described [5]. The electrolyte gels were prepared using the solvent casting technique. $LiClO_4$ was dissolved in 1,2-diethoxyethane (gly). PMMA and TiO_2 were then added along with $EtOH/H_2O$ 9:1. The mixture was heated until complete gelification, then poured on a glass plate and cooled at RT until the volatile casting solvent was completely evaporated favoring cross-linking and solid gel formation. The molar composition (%) was 7.19-64.01-28.80 (Li^+-gly-PMMA).

The ionic conductivity was measured by impedance spectroscopy using a Solartron frequency response analyzer (model 1260) scanning over 1Hz-100 kHz.

ATR-FTIR analysis have been performed using the Golden Gate MK II single reflection diamond ATR systems, Specac mounted on a spectrophotometer FT-IR Nexus 870 E.S.P., Nicolet.

NMR spectra were recorded on a Bruker Avance 400 spectrometer operating at 100.56 MHz. The recycle delay was 4 ms and samples were spun at 4 kHz. All chemical shifts are referenced to TMS.

EEG recordings (subject with eyes closed) were simultaneously obtained from conventional electrodes and with new solid electrolyte gels directly in contact with the scalp.

DISCUSSION

The new membranes developed have been electrochemically characterized by means of impedance spectroscopy and their conductivity values (Table 1) resulted to be highly satisfactory for the required application (ca. 10^{-3} S cm^{-1}).

Label	Membrane	Conductivity (S cm^{-1})
S1	$LiClO_4$–gly–PMMA	3.28×10^{-3}
S2	$LiClO_4$–gly–PMMA +5%TiO_2	2.70×10^{-3}
S3	$LiClO_4$–gly– PMMA +10%TiO_2	1.28×10^{-3}

Table 1 – Composition and conductivity (RT) of gel electrolytes.

The presence of TiO_2 in the polymer electrolyte affects its conductance. Data reported in Table 1 show that the conductivity decreases with increasing concentration of nano-size particles (mean particle size was 11 nm as determined from XRD using Scherrer's equation). This behavior is a direct consequence of the formation of well defined crystalline regions in the presence of high concentration of the ceramic filler as previously demonstrated by SEM observations [6].

Figure 1a shows the impedance plots recorded after different storage time. The linearity of the plots gives convincing evidence of a very good physical integrity of the membranes. In fact, lack of homogeneity, such as that resulting by liquid loss, phase separation phenomena or crystallization would have been reflected by the occurrence of semicircle evolution in the high frequency region [7]. In Figure 1b the conductivity values of samples S1 and S2 (without and with ceramic filler) are compared. It is evident that, although the initial conductivity value of the composite membrane is slightly lower than that of the blank, the presence of TiO_2 greatly improves the stability of the sample which shows almost the same characteristics after over two months. Ceramics are in fact known to improve water retention characteristics, mechanical and electrochemical stability and elasticity of polymeric membranes [8].

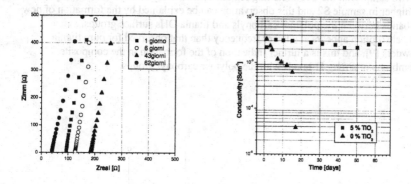

Figure 1. a) Impedance spectra of SS/ $LiClO_4$-gly-PMMA+5%TiO_2/ SS (SS stainless steel) cell at room temperature **b)** Time evolution of the conductivity of composite sample S2 and that of the analogous, ceramic free membrane, sample S1. Data obtained by impedance spectroscopy.

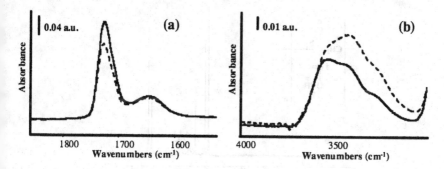

Figure 2. a) ATR/FTIR spectra in the C=O stretching region of the standard (S1, solid line) and sample S2 (5%TiO_2, dotted line) **b)** ATR/FTIR spectra in the OH stretching region of the standard S1 (solid line) and sample S2 (5%TiO_2, dotted line) a.u. = arbitrary units

Figure 2a reports the ATR/FTIR spectra of samples S1 and S2 of in the C=O bond stretching region. The spectra are characterized by two main bands centered at 1728 cm^{-1} and 1645 cm^{-1} due to the νC=O free groups and H-bonded with water respectively.

The first band (1728 cm^{-1}) is almost symmetric and its intensity decreases in the presence of titania indicating that polymer carbonyl groups are involved in interaction with the ceramic phase. At the same time, we observe that the intensity of the second band (1645 cm^{-1}) is quite unperturbed indicating that the amount of water H-bonded with C=O groups of polymer chains is constant. Fig. 2b shows the ATR/FTIR spectra of two samples in the OH stretching region. The intensity of the broad bands present is higher in sample S2 and this observation can be explained by the formation of new H-bands between the polymer carbonyls and titania OHs surface groups.
The absorption appears at a lower frequency than that observed for pure titania powders [9], and this is a further indication of the formation in the composite membrane of strong H-bonds with the polymer carbonyl groups.

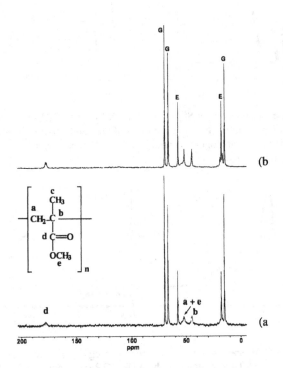

Figure 3: ^{13}C MAS NMR of samples S1(a) and S2 (b). Labels refer to the drawing shown in the Figure. G = gly, E = ethanol.

The solid state ^{13}C NMR spectra of samples S1 and S2 are shown in Figure 3. Assignments are straightforward, in agreement with previously reported data [10], and are reported in the Figure. The most significant difference between the two spectra is observed in the variation of the resonance due to the carbonyl carbon upon addition of titania. In the spectrum of sample S1 there is a broad resonance centered at 177.8 ppm, while in the spectrum of sample S2 a new sharper resonance appears at 178.1 ppm. Furthermore, the resonance due to the quaternary carbon, which in the spectrum of the ceramic free sample appears as a broad peak centered at 44.8 ppm, sharpens and is shifted to 45.1 ppm in the spectrum of the titania-doped membrane. Both the shifts and the variation of line width can be attributed to the increase in the number of H-bonded C=O groups in agreement with FTIR data.

Cerebral bioelectrical signals and visual evoked potential (VEP) were recorded from healthy humans through S2 solid electrolyte and were compared to those obtained via conventional dermal electrodes for human (EEG) recording. Figure 4 shows samples of EEG signals simultaneously recorded by conventional (top trace in plot) and composite with 5% TiO_2 electrode (bottom trace) from adjacent scalp location. The new polymer electrolyte allowed the registration of spontaneous electrophysiological cortical signals morphologically similar to the standard.

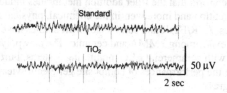

Figure 4. Comparison of EEG traces recording between the new polymer gel S2 (bottom trace) and conventional apparatus (top trace).

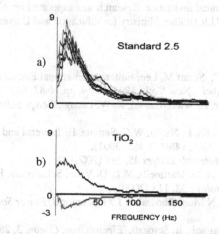

Figure 5. Responses over the occipital following visual stimuli (VEP) presented as detected by a standard electrode (several repetitions) (a) and by the newly designed gel electrode S2 (b). Bottom trace in (b) represents the difference between the registration obtained with S2 and the average of the several registrations obtained with the standard.

One of the drawbacks of conventional EEG recording performed with liquid gels is to avoid hair that the scalp must be accurately prepared to eliminate grease that, being insulating, diminishes the electrical contact between skin and electrode. All the tests carried out with the new membranes were performed on subject with hair and without any special scalp preparation.

Figure 5 compares the visual evoked potential (VEP) responses for the standard and S2 electrolyte. The performance of new membrane is within the spread of measurements carried out with the standard fluid electrolyte and the signal-to-noise ratio is also comparable to that recorded with wet Ag/AgCl electrodes.

CONCLUSIONS

In this work the preparation and characterization of new composite polymer electrolytes obtained by dissolution of a selected ceramic filler TiO_2 in PMMA-based gel are reported. The results confirm that the filler addition membranes induces consistent improvement in liquid retention and moreover in the chemical, physical and electrochemical properties. ATR/FTIR and solid-state NMR studies have clarified the specific interactions that exist among PMMA and ceramic. The new polymer electrolyte membrane is compatible with commercial EEG monitoring systems. Further investigations to improve the performances of composite gel-electrolytes in EEG measurements are in progress.

ACKNOWLEDGMENTS

Tanks are due to Professor F. Babonneau for helpful discussions and to Ms C.D'Ottavi for her valuable technical assistance. Research was supported by A.S.I. (Italian Space Agency) and M.I.U.R (Italian Ministry for Education and University).

REFERENCES

1. "Orthopedic Composites", Stuart M. Lee, Editor. International Encyclopedia of Composites. VCH Publishers, New York, 1991. Vol 4, pp.74-87
2. S. Ramakrishna, J. Mayer, E. Wintermantel, K. W. Leong , *Comp. Science and Tech.* **61**, 1189 (2001).
3. P. Romagnoli, M.L. Di Vona, L. Narici, W.G. Sannita, E. Traversa and S. Licoccia, *J. Electrochem. Soc.*, **148**(12), J63 (2001).
4. S. Rajendran, T. Uma, *Materials Letters* **45**, 191 (2000).
5. M.C. Carotta, M.A. Butturi, G. Martinelli, M.L. Di Vona, S. Licoccia, E. Traversa, *Electron Technology* **33**, 113 (2000).
6. B. Kumar, J.D. Schaffer, N Munichandraiah, L.G. Scanlon *J. Power Sources* **47**, 63 (1994).
7. G.B.Appetecchi, P. Romagnoli, B. Scrosati, *Electrochem. Comm.* **3**, 281 (2001).
8. B. Scrosati, F. Croce and L. Persi , *J. Electrochem. Soc.*, **147** (5), J1718 (2001).
9. G. Martra, Applied Catal., **200**, 275 (2000).
10. J. Straka, P. Schmidt, J. Dybal, B. Schneider and J. Spevacek *Polymer* **36**, 1147 (1995).

Mat. Res. Soc. Symp. Proc. Vol. 756 © 2003 Materials Research Society   EE2.8

POLYMER RELAXATIONAL DYNAMICS ASSOCIATED WITH IONIC CONDUCTION IN CONFINED GEOMETRY

J.-M. Zanotti[1,2], L.J. Smith[3], E. Giannelis[4], P. Levitz[5], D.L. Price[1,6] and M.-L. Saboungi[3,7]
[1] Intense Pulsed Neutron Source, Argonne Nat. Lab., Argonne, IL 60439, USA
[2] Laboratoire Leon Brillouin (CEA-CNRS), CEA Saclay, 91191 Gif/Yvette cedex, France
[3] Material Science Division, Argonne Nat. Lab., Argonne, IL 60439, USA
[4] Department of Materials Science and Engineering, Cornell University, Ithaca, NY, USA
[5] LPMC (CNRS), Ecole Polytechnique, Palaiseau, France
[6] CRMHT (CNRS), Avenue de la Recherche Scientifique, 45071 Orléans, France
[7] CRMD(CNRS), Avenue de la Recherche Scientifique, 45071 Orléans, France

ABSTRACT

Results of a quasi-elastic incoherent neutron scattering study of the influence of confinement on polyethylene oxide (PEO) and $(PEO)_8Li^+[(CF_3SO_2)_2N]^-$ (or $(POE)_8LiTFSI$) dynamics are presented. The confining media is Vycor, a silica based hydrophilic porous glass. We observe a strong slowing down of the bulk polymer dynamics under presence of Li salt. The confinement also affects dramatically the apparent mean-square displacement of the polymer. As supported by DSC measurements, the PEO melting transition at 335 K is strongly attenuated under confinement, suggesting that confinement modifies the global structure of the system, increasing the fraction of amorphous PEO by respect to crystalline phase. Local relaxational PEO dynamics is successfully described by the DLM (Dejean-Laupretre-Monnerie) model usually used to interpret NMR spin-lattice relaxation time data. The scattering vector dependence of the correlation times deduced from inelastic neutron scattering data is found to obey a power-law dependence. DSC and preliminary ionic conduction measurements are also presented.

INTRODUCTION

Conception and industrial production of economically viable fuel cells [1] is a central issue for the developing of non-polluting vehicles. Up to date many studies have shown the interest and the feasibility of hybrid cells where organic and inorganic components are used. One of the most studied systems is PEO (polyethylene oxide) complexed by Li salts. Polymer segmental motions and ionic conductivity are closely related [2]. Bulk PEO is actually a biphasic system where an amorphous and a crystalline state ($T_m \approx 335$ K) coexist. A key to improve ionic conduction in those systems is therefore a significant increase of the amorphous phase fraction where lithium conduction is known to mainly take place [3]. Upon this issue, operation above 80°C is satisfactory but such high temperatures requirements will drastically reduce any broad consumer use. Confinement is known to strongly affect properties of condensed mater and in particular the collective phenomena inducing crystallization [4]. A possible alternative solution to this high temperature mode seems therefore the confinement of the polymer matrice. Polymer-silicate nanocomposites have been proposed and are under extensive study [5,6].

Structure of polymer chains confined in Vycor porous glass [7,8] have already been reported [9] and found in agreement with theoretical predictions [10]. Restricted geometry effects on dynamics of polymers confined in Vycor have been measured by NMR [11]. We have very recently studied by inelastic incoherent neutron scattering the influence of confinement on PEO and $(PEO)_8Li^+[(CF_3SO_2)_2N]^-$ (or $(POE)_8LiTFSI$) dynamics in bulk and confined in Vycor. As measured by small angle neutron scattering, radius of gyration of PEO melt 100,000 g/mol at 371 K is 120 Å [12]. PEO 100,000 g/mol adsorbed into the Vycor 35 Å radius pore [8] is therefore expected to show large confining effects.

EXPERIMENTAL DETAILS

PEO used in the present study is hydrogenated PEO (molecular weight = 100,000 g/mol, Aldrich). $(POE)_8LiTFSI$ film was prepared by casting onto a Teflon coated plate and stepwise evaporating under dry air, an acetonitrile solution (Aldrich) mixed with right proportions of PEO and LiTFSI. A PEO film was prepared by melting at 95°C the needed amount of PEO under dry air into a similar Teflon coated plate. What we refer as the bulk samples of this work are PEO and $(POE)_8LiTFSI$ films of 0.2 ± 0.05 mm thickness. Necessary amount of each of those films was saved and used to prepare the Vycor confined samples.

Vycor 7930 glass [7] is made by heating a homogeneous mixture of boron oxide glass and silica above the melting point and then quenching this mixture to a temperature below the spinodal line where the mixture phase separates into two mutually interpenetrating regions. At a certain stage of preparation, the boron rich region is leached out by acid, leaving behind a silica skeleton with a given distribution of pore sizes. Due to numerous surface pending silanol groups (-SiOH), Vycor is extremely hydrophilic.

In order to avoid multiple scattering effects, a main concerned of a neutron scattering experiment is to ensure a sample neutron transmission of 90%. Assuming a full polymer filling of the Vycor network taking into account the Vycor and polymer densities, total neutron scattering cross sections, the optimal sample thickness was found to be 0.4 mm. Several rectangular plates of that thickness were cut out of a 25 mm diameter Vycor (ANL/APS optics laboratory). Plates were then immersed in 30% hydrogen peroxide and heated to 90°C for a few hours to remove any organic impurities absorbed by the glass. The Vycor was then washed several times in distilled water in order to remove the hydrogen peroxide and stored in distilled water. Samples were then pumped under 10^5 torr for 3 days, defining what we call a "Vycor sample". The average Vycor pore radius is 35 Å. Porosity is 28% of total volume [8], PEO and Vycor density are 1.1 and 1.4 respectively. Total PEO filling of the porous network is then expected for a ratio $m_{PEO}/m_{DryVycor}=0.22$. Confined samples were prepared by melting, under vacuum at 95°C, onto the Vycor plates pieces of the previously prepared PEO and $(POE)_8LiTFSI$ films. After 72 hours, then back to room temperature, mass of Vycor plates had increased of 23% for PEO and 24% for $(POE)_8LiTFSI$, suggesting a full filling of the Vycor porous network.

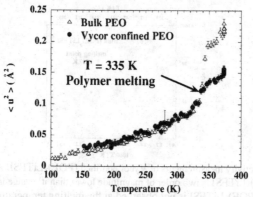

Figure 1. Influence of confinement on PEO mean-square displacement, $<u>^2$, as measured by incoherent inelastic neutron scattering. The melting transition of PEO at 335 K, is strongly affected by confinement (see also Fig. 2). The lower $<u>^2$ of confined PEO compared to bulk at 373 K is to be interpreted as a consequence of strong slowing down of the polymer dynamics.

DISCUSSION

Following of recent MD [13, 14] and experimental [15,16] studies, the ionic conduction of Li^+ in PEO matrices is now better understood. Lithium conduction mainly takes place in the amorphous phase. PEO monomer is $-CH_2CH_2O-$. On average, each lithium ion is complexed by 5 oxygen atoms of neighbouring polymer's ethers bonds [15]. However, this is only a time-averaged structure and due to fast local relaxation modes of the polymer, some transient

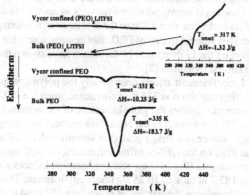

Figure 2. From bottom to top: DSC signal of PEO bulk, confined, (POE)$_8$LiTFSI bulk and confined. When detectable, onset temperatures and enthalpies are indicated.

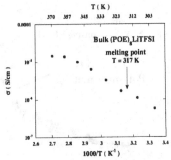

Figure 3. Ionic conductivity measurements of Vycor confined (POE)$_8$LiTFSI. At 273 K, ionic conductivity of (POE)$_8$LiTFSI is two order of magnitude lower than it's value in bulk [17]. The kink observed in the bulk (POE)$_8$LiTFSI is not observed at the melting temperature of 317 K as measured by DSC (see Fig. 2).

Li-O bonds may frequently break. When several of those five bonds are temporary disrupted at the same time, one Li$^+$ ion may be free to diffuse throughout the polymer matrice until new bonds form again. In this study, we extend those results and report new inelastic neutron scattering result of the PEO and (POE)$_8$LiTFSI dynamics in bulk and under confinement. Experiments have been performed on QENS, an "inverse-geometry" time-of-flight inelastic neutron spectrometer installed at the Intense Pulsed Neutron source (Argonne National Laboratory). The sample dynamical structure factor, S(Q,ω) is measured over a Q range (Q is the modulus of the scattering vector) extending from 0.3 up to 2.6 Å^{-1}, with an average energy resolution of 90 μeV. Owing to the large incoherent neutron scattering cross-section (σ_s= 80 barns) of the ^{1}H nucleus and the abundance of this element in polymeric systems, incoherent inelastic neutron scattering measurements are able to give a global view of the fast (ps time-range) local (few angstroems) polymer dynamics as sensed via the averaged individual motions of its hydrogen atoms. Du to the low neutron scattering cross-section of Li atom (σ_s= 1.4 barns), by respect to that of PEO, this neutron scattering experiment did not provide any information on the Li dynamics. NMR results have shown that Li$^+$ and PEO hydrogen atoms have similar dynamics [2].

As sum-up on figure 1, confinement affects dramatically the polymer apparent mean-square displacement. The melting transition at 335 K is attenuated when the PEO is confined, suggesting that confinement modifies the global structure of the system, increasing the fraction of amorphous PEO by respect to the crystalline phase. This interpretation is supported by the disappearance of a Bragg peak (not shown) under confinement observed in the bulk hydrogenated PEO on the QENS diffraction detectors. DSC measurements performed on the same samples (Fig 2) show that confinement induces a strong decrease of melting enthalpy from 183.7 in bulk to 10.25 J/g under confinement. This result suggests that confinement dramatically reduces the fraction of crystalline PEO and is in strong agreement with the neutron scattering picture emerging from figure 1. The cross-linking between polymer chains by Li$^+$ ions being responsible for a strong decrease of the faction of crystalline PEO [15], the melting transition in bulk (POE)$_8$LiTFSI is very weak : -1.8 J/g (Fig. 2). This transition is not detectable any more in the case of confined (POE)$_8$LiTFSI.

This interesting feature is also found in the temperature dependence of the ionic conductivity (Fig. 3). The strong decrease in the ionic conductivity around the melting point temperature usually observed [17] when decreasing the temperature is not observed when (POE)$_8$LiTFSI is confined. Ionic conductivity and polymer segmental motion have been shown to be closely related in bulk [3]. The so-called DLM (Dejean-Laupretre-Monnerie) model [18] successfully describes the bulk polymer dynamics, as measured by ^{13}C spin-lattice relaxation time T$_1$, by the spectral density J(ω) :

$$J(\omega) = K \left(\frac{1-a}{(\alpha+i\beta)^{1/2}} + \frac{a\tau_0}{1+\omega^2\tau_0^2} \right) \tag{1}$$

with $\qquad \alpha = \tau_2^{-2}+2\tau_1^{-1}\tau_2^{-1}-\omega^2$ and $\beta = -2\,\omega(\,\tau_1^{-1}+\tau_2^{-1})$

τ_1 is the correlation time associated with correlated jumps responsible for orientation diffusion along the polymer chain and τ_2 stands for a damping which consists of either of non-propagative specific motions or of distortions of the chain with respect to its most stable local conformation. τ_0 corresponds to the correlation time of libration of C-H vectors inside a cone of half-angle θ and a is a geometrical factor depending of cos(θ). In NMR, the factor K is fixed to 1, but is used as fitting parameter in the present neutron scattering data analysis, to account for the mass and scattering cross section of the different samples. The fitting of the neutron scattering data is performed with the real part of J(ω). The DLM model accounts very well for the neutron quasi-elastic spectra of bulk PEO at 373 K (Fig 4), especially at high Q (above 1.5 Å^{-1}). As compared to NMR, neutron scattering experiments enable to access the Q dependence of the correlation times (Fig. 4). τ_0 is found to smoothly Q dependant as expected for a reorientational and therefore non-dispersive motion. Such a short correlation time in PEO melts has already been reported [19]. τ_1 seems to be consistent with a Q^{-2} power dependence suggesting a diffusive process, while τ_2 is consistent with Q^{-4} dependence, expected to describe the Rouse dynamics [20,21].

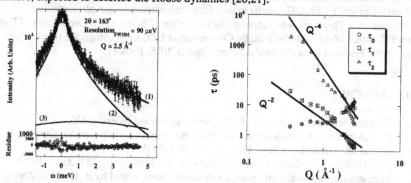

Fig 4. left: Semi-logarithmic plot of the quasi-elastic signal of bulk PEO at 2.5 Å^{-1}, at T= 373 K (Points). Fitting curve (1) using DLM model (Equation 1) shows excellent agreement with the experimental points. Relaxational polymer dynamics contribution (τ_1 = 0.4 ps and τ_2= 2.4 ps, see text) (curve 2) and Lorentzian line standing for the fast librational dynamics (τ_0 = 0.07 ps) (curve3) are shown. Right: Q dependence of the correlation coefficients τ_0, τ_1 and τ_2 expressed in ps. The Q^{-2} and Q^{-4} dependence (full lines) are guides for the eyes.

For bulk $(POE)_8LiTFSI$, at $Q=2.5\mathring{A}^{-1}$ and $T = 373$ K, $\tau_0 = 0.05$ ps and $\tau_1 = 0.4$ ps are found comparable to the value of bulk PEO, while $\tau_2 = 38.5$ ps is more than one order of magnitude larger (see caption of Figure 1). Confinement dramatically slows down the polymer dynamics since in the same conditions as above, but upon Vycor confinement, τ_2 increases to 424 ps for PEO and to 438 ps for $(POE)_8LiTFSI$.

CONCLUSION

Correlation times associated with conformational motions of bulk PEO, extracted from incoherent inelastic neutron scattering data in the frame-work of DLM model, are found to be strongly Q dependant. Confinement induces a spectacular slowing down of the PEO dynamics above 335 K (bulk PEO melting point) and largely prevents PEO crystallization. At T=373 K, $(POE)_8LiTFSI$ dynamics is significantly slower than in bulk PEO as already reported in literature. This slowing down is even more striking under confinement and is correlated with a significantly reduced ionic conductivity compared to bulk $(POE)_8LiTFSI$. Analysis is still in progress, but Vycor seems a good model system to improve our understanding of confinement effects on polymer and polymer electrolytes structure and dynamics. This could be an essential issue for the development of new polymer electrolytes of industrial relevance.

ACKNOWLEGMENTS

Authors would like to acknowledge Rob Connatser (ANL/IPNS) for continual assistance during the neutron scattering experiments, Rafael Herrera Alonso and Kannan Raman (Cornell University/ Prof. Giannelis group) for assistance during the conductivity measurements. This work was supported by the Office of Science, US Department of Energy, under Contract W-31-109-ENG-38, by Commissariat à l'Energie Atomique (CEA, France) and Centre National de la Recherche Scientifique (CNRS, France).

REFERENCES

1.   J. Larminie and A. Dicks, Fuel Cell Systems explained, Wiley, 2000.
2 .   J. P. Donoso, T. J. Bonagamba, H.C. Panepucci and L. N. Oliveira, W. Gorecki, C. Berthier and M. Armand, J. Chem. Phys. 98 10026, (1993)
3.   J.W. Halley , L. A. Curtiss and A. G. Baboul, J. Chem. Phys., 111, 3302 (1999).
4.   International Workshop on dynamics in confinement, edited by B. Frick, R. Zorn and H. Buttner, Journal de Physique, 10, PR7 (2000).
5.   E. P. Giannelis, R. Krishnamoorti and E. Manias, Advances in polymer science 138, 107 (1999).
6.   L.J. Smith, J.-M. Zanotti, M.-L. Saboungi and D.L. Price, this volume.
7 .   Vycor Brand Porous Glass #7930. Corning Glass works.
8.   Levitz, G. Ehret, S. K. Sinha and J. M. Drake, J. Chem. Phys., 95, 6151 (1991).
9.   J. Lal, S.K. Sinha and L. Auvray, J. Phys. II France 7, 1597 (1997).

10. M. Daoud and P.-G. de Gennes, J. de Physique, 38, 8593 (1977).
11. S. Stapf and R. Kimmich, Macromolecule 29, 1638 (1996).
12. G.D. Smith, D.Y. Yoon, R.L. Jaffe, R.H. Colby, R. Krishnamoorti and L.J. Fetters, Macromolecules 29,3462 (1996).
13. F. Muller-Plathe and W. F. van Gusteren, J. Chem. Phys., 103, 4745 (1995).
14. O. Borodin and G.D. Smith, Macromolecules, 33,2273 (2000).
15. G. Mao, M.-L. Saboungi, D. L. Price, M. B. Armand and W. S. Howells, Phys. Rev. Lett., 84:5536 (2000).
16. G. Mao, R. F. Perea, W. S. Howells, D. L. Price and M. L. Saboungi, Nature, 405, 163 (2000).
17. M. Armand, W. Goreki and R. Andreani, in Proceedings of the second international Symposium on Polymer Electrolytes, edited by B. Scrosati (Elsevier Applied Science, New-York, 1990), p91.
18. R. Dejean de la Batie, F. Laupretre and L. Monnerie, Macro-molecules 21, 2045 (1988).
19. A. Triolo, V. Arrighi, R. Triolo, S. Passerini, M. Mastragostino, R.E. Lechner, R. Ferguson, O. Borodin and G.D. Smith, Physica B 301, 163 (2001).
20. B. Mos, P. Verkerk, S. Pouget, A. van Zon, G.-J. Bel, S.W. de Leeuw and C.D. Eisenbach, J. Chem. Phys., 113, 4 (2000).
21. D. Richter, M. Monkenbusch, J. Allgeier, A. Arbe, J. Colmenero, B. Farago, Y. Cheol Bae and R. Faust, J. Chem Phys 111, 6107 (1999).

Mat. Res. Soc. Symp. Proc. Vol. 756 © 2003 Materials Research Society   EE2.9

Characterization of Polymer Clay Nanocomposite Electrolyte Motions via Combined NMR and Neutron Scattering Studies

Luis J. Smith (1), Jean-Marc Zanotti (2), Giselle Sandi (3), Kathleen A. Carrado(3), Patrice Porion(4), Alfred Delville(4), David L. Price(5), and Marie-Louise Saboungi(1,4)

(1) Materials Science, (2) Intense Pulsed Neutron Source, and (3) Chemistry Divisions, Argonne National Laboratory, Argonne, IL, USA
(4) Centre de Recherche sur la Matiere Divisee, CNRS, Orleans, France
(5) Centre de Recherches sur les Materiaux a Haute Température, CNRS, Orleans, France

Abstract

The activation energies for poly(ethylene oxide) motion in a polymer clay composite are reported for the polymer intercalated and external to the clay. PEO intercalated into the clay is found to have a lower activation energy for motion but also a larger Arrhenius prefactor, by almost two orders of magnitude, than for PEO found external to the clay. Neutron scattering measurements confirm the presence of two environments and the effects of confinement on the mean square displacement of the PEO.

Introduction

The use of flexible, long chain polymers has been an approach that has been pursued in the development of rechargeable lithium batteries as a means to replace the liquid electrolytes currently in use. Poly(ethylene oxide) (PEO) is one potential replacement and has shown considerable conductivity at elevated temperatures near or above the melting point of the polymer.[1] Since ionic transport is strongly dependent on the host polymer segmental motion, maintaining the polymer in an amorphous state is important for significant conduction to occur and thus requires high operating temperatures. Plasticizers are often used to increase the percentage of PEO that is amorphous and thus reduce the temperature at which significant conduction can occur. However, many of these plasticizers can also reduce the beneficial mechanical properties that come with the use of flexible, long chain polymers such as PEO. An alternative nanocomposite material that is now being examined is clay, which serves not simply as a filler material but also as an intercalation host for the PEO. Isolation of the polymer segments form the bulk can prevent crystallization of the PEO while confinement of the polymer may also affect its segmental motion thus lowering the temperature at which significant conductivity is achievable. A series of PEO/hectorite clay nanocomposites were studied with variable temperature nuclear magnetic resonance (NMR) and quasielastic neutron scattering (QENS) in order to understand the motion of the intercalated PEO.

Hectorite clays, from the smectite clay family, are composed of two tetrahedral silicate layers that sandwich an octahedral layer. Mg^{2+} is present in the octahedral layer; isomorphous substitution of the Mg^{2+} for Li^+ produces an overall negative charge to the clay layer. The negative charge on the layer is compensated by the presence of cations in the gallery between clay layers. These gallery cations are exchangeable. Polymers can be introduced into the gallery region through exfoliation of the clay in aqueous solution, followed by condensation of the slurry in the presence of PEO. The introduction of PEO into the gallery region is observed through an increase of the lattice spacing of the clay using x-ray powder diffraction.[2]

Deuterium, on the polymer, is an ideal element for the study of motion in these composites. Since ^{2}H has a nuclear spin greater than 1/2, the interaction of the nucleus with the surrounding electric field will cause a perturbation in the observed signal. This perturbation affects not only the lineshape of the spectra but also the spin-lattice relaxation rate of the nucleus. For ^{2}H in deuterated PEO, the electric field gradient has axial symmetry and lies along the C-D bond, almost parallel with the bond vector. Local perturbations in the gradient causes relaxation and can affect the lineshape depending on the time scale of the motion. Thus ^{2}H NMR can serve as a probe of the local motion of the C-D bond vector at the nanosecond time scale through T_1-spin lattice relaxation.

Theory

The temperature dependence of the spin lattice relaxation rate in this case can not be determined using BPP (Bloembergen-Purcell-Pound) relaxation theory since the motion of such long chain polymers like PEO can not be expected to be isotropic. Instead an alternate description of the polymer motions on the nano- and picosecond time scales must be used. For a linear polymer like PEO, the spectral density function for deuterium motion can be described by an autocorrelation function based on the concept of polymer motion as a series of conformational jumps. The propagation of the change in conformation is characterized by a correlation time τ_1 and a second correlation time, τ_2, corresponding to a damping of the propagation due to time dependent barriers. This is the original model described by Hall and Helfand.[3] The inclusion of an additional fast anisotropic reorientation due to librations of the C-D bond about a rest position leads to the addition of a characteristic correlation time τ_0. This modification leads to the Dejean-Laupretre-Monerie expression[4] for the polymer motion (eqn 1) used to analyze the data.

$$G(t) = (1-a)\exp(-t/\tau_2)\exp(-t/\tau_1)I_0(t/\tau_1)$$
$$+a \cdot \exp(-t/\tau_0)\exp(-t/\tau_2)\exp(-t/\tau_1)I_0(t/\tau_1)$$

(1)

In these equations, parameter a is related to the size of the cone in which anisotropic reorientation is occurring and I_0 is a modified Bessel function of order zero. The resulting spectral density function from the Fourier transform of (1) is:

$$J(\omega) = RE\left|\frac{1-a}{\left(\left(\tau_2^{-2} + 2\tau_1^{-1}\tau_2^{-1} - \omega^2\right) - i\left(2\omega\left(\tau_2^{-1} + \tau_1^{-1}\right)\right)\right)^{1/2}}\right| + \frac{a \cdot \tau_0}{1+\omega^2\tau_0^2}$$

(2)

The new spectral density function can now be used in the standard description of quadrupolar relaxation rate for nuclei with a nuclear spin equal to one (eqn 3).

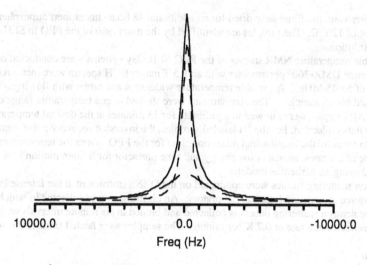

```
10000.0                    0.0                  -10000.0
                        Freq (Hz)
```

Figure 1. ^{2}H spectrum of PEO:SLH =1.0 at T= 90°. Intercalated PEO (long dash), External PEO (short dash)

$$\frac{1}{T_1} = \frac{3}{40}(2\pi C_Q)^2 [J(\omega) + 4J(2\omega)] \tag{3}$$

Since the electric field gradient of the deuterium is due primarily to the covalent bond with the carbon backbone, the field gradient value can be assumed a constant. The quadrupolar coupling value (C_Q) for a C-D bond, 170 kHz, was used with equation 3. Also a standard value of 0.56 for the a parameter was used based on previous PEO studies.[3] Since the libration motions of the C-D bond are much faster than the conformational jumps ($\tau_0 \ll \tau_1$), the minimum value of the T_1 curve in the nanosecond time regime is reduced approximately by (1-a) from the value expected under isotropic motion. Due to the slow nature of the τ_2 related fluctuations, (about 200 times longer than τ_1)[3], only τ_1 is the only parameter on the relevant time scale in the NMR relaxation study.

Experimental Details

The samples were prepared from a synthetic, lithium hectorite (SLH) clay and loaded with either deuterated or hydrogenated PEO. The SLH clay was prepared via hydrothermal crystallization from silica sol, magnesium hydroxide, and lithium fluoride.[5] Colloidal suspensions of the clay were stirred in de-ionized water for one hour. The desired amount of either deuterated PEO (molecular weight = 84000 g/mol) or hydrogenated PEO (molecular weight = 100,000 g/mol) were added to the suspension which was then stirred for an additional 24 hours. The films were produced by casting the slurries onto Teflon coated plated and air-drying. Afterward, the films were dried for an additional 48 hours under inert atmosphere at a temperature of 120 °C. The samples are identified by the mass ratio of the PEO to SLH clay in the initial solution.

Variable temperature NMR studies of the PEO-SLH clay systems were conducted on a Bruker Avance DMX-360 spectrometer with an 8.5 T magnet. ^{2}H spectra were measured with a frequency of 55.45 MHz. A variable temperature wideline static probe with flowing nitrogen gas was used for all samples. The clay samples were studied over a temperature range of 240 K to 363 K. All samples were allowed to equilibrate for 15 minutes at the desired temperature before any data collection. For the ^{2}H labeled samples, the inversion recovery pulse sequence was used to measure the longitudinal relaxation time for the PEO. From the temperature dependence of the rates, an activation energy and time prefactor for lithium motion was determined using an Arrhenius model.

Neutron scattering studies were conducted on the QENS instrument at the Intense Pulsed Neutron Source at Argonne National Laboratory. An elastic scan was collected in which quasielastic neutron scattering data was collected and binned in six minute intervals as the sample was heated at a rate of 0.7 K per minute. The samples were heated from 150 K to 400 K.

Discussion

The ^{2}H spectrum of the PEO could be separated into two components, one corresponding to the intercalated PEO and a second corresponding to the PEO external to the clay (Figure 1). Separate relaxation times were observed for the two components. The temperature dependence of the spin lattice relaxation rate was analyzed using the DLM theory for polymer motion. Based on the ratios of the areas of the two components of the deuterium spectrum, the narrow component of the line was assigned to PEO external to the clay and the broad component to PEO intercalated into the clay. The temperature dependence of the correlation times for the two PEO regions was fit using an Arrhenius-type equation. Separate activation energies for PEO intercalated into the clay and external to the clay were observed and are listed in Table 1.

PEO external to the clay was found to have activation energies in the range of what has been observed for amorphous and crystalline PEO in the bulk.[6] This was not surprising since the external PEO was unconfined. The PEO intercalated in the clay had activation energies that were lower than in bulk PEO. The activation energies varied slightly depending on the PEO content of the nanocomposite film, with the PEO:SLH sample of ratio of 0.8 having the lowest activation energy. The activation energy for the nanocomposites were similar to those observed in other PEO-clay composites[7] and PEO-lithium salt mixtures.[8] For all the nanocomposites, the prefactor value was approximately two orders of magnitude larger than those observed for the external PEO, reflecting the effect of confinement that resulted in slower polymer motion.

The presence of PEO outside of the clay was further confirmed by the neutron scattering results. Measurement of the mean square displacement of the hydrogen in the samples containing hydrogenated PEO showed a significant difference between the PEO:SLH = 0.6 and 1.0 samples.(Figure 2) A larger fraction of the PEO was confined in the 0.6 sample and was reflected by the lower mean square displacement associated with confinement. The 1.0 sample did not have such a small displacement reflecting the greater amount of PEO surrounding the clay.

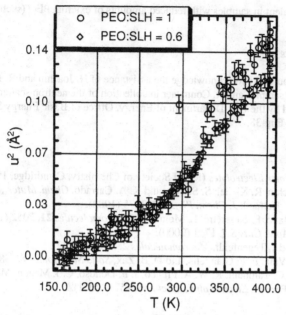

Figure 2. Mean square displacement, u^2, of hydrogenated PEO in PEO/SLH clay nanocomposites at ratio of 0.6 (diamonds) and 1.0 (circles).

Table 1. PEO Activation Energies

Sample Ratio	Intercalated PEO		External PEO	
	E_a (kJ/mol)	τ_{c0} (ps)	E_a (kJ/mol)	τ_{c0} (ps)
0.6	24.7±1.5	0.08±0.04	31.9±3.1	0.001±0.001
0.8	22.6±1.1	0.17±0.07	30.5±1.0	0.002±0.0008
1.0	23.9±2.0	0.10±0.07	30.1±2.9	0.0009±0.0009

Conclusion

Activation energies for the two observed environments of PEO in PEO/SLH clay nanocomposites were determined. PEO confined with the clay gallery had smaller activation energies than PEO external to the clay (24 kJ/mol vs. 30 kJ/mol for 1:1 PEO:SLH, for example). The observed activation energies were similar to those observed for other lithium salt polymer electrolyte systems. The external PEO had activation energies similar to the bulk. A difference by almost two orders of magnitude in the Arrhenius prefactor between the intercalated and external PEO reflected the large effect of confinement on the PEO motion. The effect of confinement was reflected as well in the reduction of the mean square displacement of the PEO and was more prevalent in samples with reduced amounts of external PEO (such as the 1:0.6 PEO:SLH sample).

Acknowledgements

The authors would like to acknowledge the assistance of H. Joachin and R. Kizilel (IIT) in the synthesis of the samples and R. Connatser in collection of the neutron scattering data. This work was supported by the U.S. Department of Energy, Office of Basic Energy Sciences, under contract W-31-109-Eng-38.

References

1. F. M. Gray, *Polymer Electrolytes* (Royal Society of Chemistry, Camdridge, 1997).
2. G. Sandi, H. Joachin, R. Kizilel, S. Seifert, and K. A. Carrado, *Chem. Mater.*, in press (2003).
3. C. K. Hall and E. Helfand, *J. Chem. Phys.* **77**, 3275 (1982).
4 . R. Dejan de la Batie, F. Lauprêtre, L. Monnerie, *Macromolecules* **21**, 2052 (1988).
5 . K. A. Carrado, *Appl. Clay Sci.* **17**, 1 (2000).
6 . A. Johansson and J. Tegenfeldt, *Macromolecules* **25**, 4712 (1992).
7 . S. Wong, R. A. Vaia, E. P. Giannelis, and D. B. Zax, *Solid State Ionics* **86-88**, 547 (1996).
8 . A. C. Bloise, C. C. Tambelli, R. W. A. Franco, J. P. Donoso, C. J. Magon, M. F. Souza, A. V. Rosario, and E. C. Pereira, *Electrochimica Acta* **46**, 1571 (2001).

Mat. Res. Soc. Symp. Proc. Vol. 756 © 2003 Materials Research Society   EE2.10

Increasing the Operating Temperature of Nafion Membranes with Addition of Nanocrystalline Oxides for Direct Methanol Fuel Cells

Vincenzo Baglio[1,3], Alessandra Di Blasi[1], Antonino S. Arico'[1], Vincenzo Antonucci[1], Pier Luigi Antonucci[2], Francesca Serraino Fiory[3], Silvia Licoccia[3], Enrico Traversa[3]
[1]CNR-TAE Institute, via Salita S. Lucia sopra Contesse 98126 Messina, Italy
[2]University of Reggio Calabria, Località Feo Di Vito, 89100 Reggio Calabria, Italy
[3]University "Tor Vergata"of Rome, via della Ricerca Scientifica, 00133 Roma, Italy

ABSTRACT

Composite Nafion membranes containing various amounts of TiO_2 (3%, 5% and 10%) were prepared by using a recast procedure for application in high temperature Direct Methanol Fuel Cells (DMFCs). The electrochemical behaviour was compared to that of a membrane-electrode assembly (MEA) based on a bare recast Nafion membrane. All the MEAs containing the Nafion-titania membranes were able to operate up to 145°C, whereas the assembly equipped with the bare recast Nafion membrane showed the maximum performance at 120°C. A maximum power density of 340 mW cm^{-2} was achieved at 145°C with the composite membrane in the presence of oxygen feed, whereas the maximum power density with air feed was about 210 mW cm^{-2}.

INTRODUCTION

Liquid-fueled Solid-Polymer-Electrolyte Fuel Cells represent a promising alternative to hydrogen based devices as electrochemical power sources for application in portable systems and in electric cars, due to their simplicity of design. However, despite the practical system benefits, the power density and the efficiency of Direct Methanol Fuel Cells (DMFCs) are low compared to Polymer Electrolyte Fuel Cells (PEFCs) operating with hydrogen because of the slow oxidation kinetics of methanol and the methanol cross-over through the membrane [1-5]. In the last years, significant efforts have been addressed to the development of polymer membranes for DMFCs in order to increase their operation temperature. At present, the electrolyte widely used in this field is Nafion, due to its excellent stability, high protonic conductivity at temperatures close to 100°C and good mechanical strength. The main problems associated with this electrolyte are the membrane dehydration at temperatures higher than 100°C, which is a prerequisite for a suitable electro-oxidation of small organic molecules (e. g. CO, methanol, ethanol, etc.), and the methanol cross-over, resulting in a performance decay and a loss of fuel efficiency. An increase in the operation temperature up to 150°C is highly desirable to enhance the kinetics of methanol oxidation. Various approaches have been proposed to solve these problems and allow the operation of liquid-feed solid polymer electrolyte fuel cells at elevated temperatures (higher than 120°C). Recently, the use of Nafion membranes containing finely dispersed nanocrystalline ceramic oxide powders (silica) has been proposed [6-7]. The role of the oxide powders is to improve the water retention, allowing fuel cell operation at 145°C with the oxygen humidifier and fuel conditioner maintained at 85°C.

In the present work, composite titania recast Nafion membranes containing 3%, 5% and 10% wt/wt TiO_2 loadings have been investigated in a DMFC in order to increase the operation temperature of the cell and to investigate the role of titania content into the membrane.

EXPERIMENTAL

TiO_2 nanometric powder was synthesized via a sol-gel procedure, starting from metal alkoxide. The obtained xerogel, after thermal analysis, was calcined at 500°C. X-ray diffraction (XRD) pattern for this oxide was obtained using a Philips X-Pert 3710 X-ray diffractometer using Cu $K\alpha$ source operating at 40 kV and 20 mA. TEM analysis was made by first dispersing the oxide powder in isopropyl alcohol. A few drops of these solutions were deposited on carbon film coated Cu grids and analysed with a Philips CM12 microscope.

Four different types of membranes were prepared having the following composition: 1) Recast Nafion, 2) Recast Nafion – 3% TiO_2, 3) Recast Nafion – 5% TiO_2, 4) Recast Nafion – 10% TiO_2; they are indicated in the text as NAF, NAF3TiO$_2$, NAF5TiO$_2$ and NAF10TiO$_2$, respectively. The membranes were prepared by using a procedure similar to that developed by Stonehart, Watanabe et al. for PEFCs [8]. The thickness of the membranes was about 100 μm.

The catalyst employed for methanol oxidation was 60% Pt-Ru (1:1)/Vulcan purchased from E-TEK, whereas a 30% Pt/Vulcan from E-TEK was used for oxygen reduction. For both anode and cathode, the reaction layer was prepared by directly mixing in an ultrasonic bath a suspension of Nafion ionomer in water with the catalyst powder; the obtained paste was spread on carbon cloth backings. The platinum loading for all the electrodes used in the experiments was 2 ± 0.2 mg cm^{-2}. MEAs were prepared by hot pressing the electrodes onto the membrane at 130°C and 50 atm.

Fuel cell tests were carried out in a 5 cm^2 single cell (GlobeTech, Inc.) connected to a HP 6060B electronic load. Aqueous solutions of methanol of varying concentration (0.5, 1 and 2 M) and cathode oxidant (oxygen or air) were preheated at 85°C and fed to the cell. The methanol solution was fed into the cell at a flow rate of 2.5 ml/min while the cathode feed was passed through a humidification bottle at 500 ml/min. Different operating temperatures for the cell were settled in the range from 90 to 145°C. The anode back-pressure was varied between 1 and 3.5 atm as the temperature was increased from 90 to 145°C; whereas the cathode compartment back-pressure was maintained constant at 3.5 atm.

RESULTS AND DISCUSSION

TiO$_2$ powder characterization

Figure 1 shows the XRD pattern of the of TiO_2 powder. The crystallographic structure of anatase was observed. The peak broadening was used to calculate the particle size of the titanium oxide crystallites by using the Debye-Sherrer formula with Warren correction for instrumental effects. A mean particle size of 11 nm was determined for the TiO_2 particles.

TEM analysis carried out on the nanocrystalline powder evidences a homogeneous distribution of the oxide particles and shows particle dimensions in good agreement with XRD results (Figure 2).

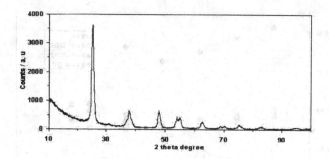

Figure 1. X- ray diffraction pattern of TiO_2 powder calcined at 500°C.

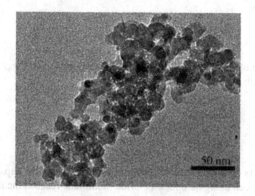

Figure 2. TEM micrograph of TiO_2 powder.

Electrochemical experiments

Figure 3 shows the variation of cell resistance as a function of temperature for the assemblies equipped with composite membranes compared to bare recast Nafion. The cell resistance in the presence of composite membranes containing different contents of titanium oxide is almost unvaried in the range between 90 and 145°C, but it is lower than the cell equipped with bare recast Nafion membrane.

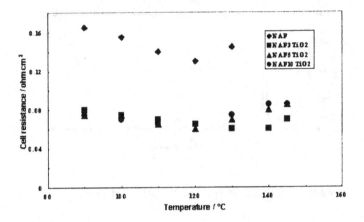

Figure 3. Variation of cell resistance with temperature during DMFC operation for the various MEAs equipped with the different membranes.

All the MEAs equipped with the composite membranes were capable of operation at 145°C, whereas the cell using bare recast Nafion membrane reached the maximum operation temperature of 120°C.

Figure 4 shows the polarization curves obtained for the fuel cell equipped with the different membranes and operating under the same conditions in the presence of oxygen feed at cathode and 2M methanol solution at anode at 120°C. The TiO$_2$-based membranes show better electrochemical behaviour, mainly due to their higher conductivity. The best electrochemical performance is obtained with NAF3TiO$_2$ membrane. Fuel cell tests at 145°C show similar electrochemical characteristics for the cells equipped with the composite membranes with different titanium oxide contents (figure 5). A maximum power density of 340 mW cm^{-2} is reached at current density of about 1.1 A cm^{-2} under oxygen feed operation at 145°C.

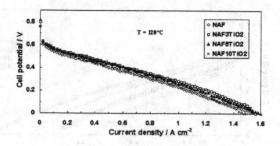

Figure 4. Polarization curves obtained at 120°C for the various MEAs equipped with the different Nafion membranes, in presence of oxygen feed and 2M methanol.

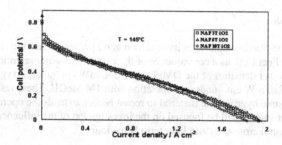

Figure 5. Polarization curves obtained at 145°C for the various MEAs equipped with composite Nafion membranes, in presence of oxygen feed and 2M methanol.

The effect of temperature on electrochemical behaviour of a fuel cell operating with NAF3TiO$_2$ membrane is reported in figure 6. The potential losses decrease progressively as the cell temperature increases in the range from 90 to 145°C, in activation, ohmic and diffusion control regions. Experiments carried out at 150°C gave rise to an unstable electrochemical behaviour. When air is fed to the cathode side, the best electrochemical performance is similarly observed with NAF3TiO$_2$ membrane. The maximum power density is about 210 mW cm^{-2} under air operation and with 1M methanol solution feed to the anode.

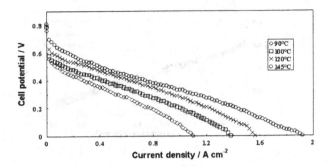

Figure 6. Influence of operating temperature on the DMFC polarization characteristics for the MEA equipped with NAF3TiO2 membrane in presence of oxygen and 2M methanol.

CONCLUSIONS

Composite Nafion-titania membranes have shown good properties for operation at 145°C in a Direct Methanol Fuel Cell, as a consequence of their improved water retention characteristics. The maximum power densities of the DMFC were 340 mW cm^{-2} under oxygen operation with 2M MeOH and 210 mW cm^{-2} under air operation with 1M MeOH. These results indicate that the addition of inorganic hygroscopic material to recast Nafion extends the operation temperature of a DMFC. Further studies will be focused on the investigation of the influence of humidification conditions, methanol cross-over and long term behaviour.

REFERENCES

1. A. Hamnett, *Catalysis Today*, **39**, 445 (1997)
2. S. Surampudi, S. R. Narayanan, E. Vamos, H. Frank, G. Halpert, A. La Conti, J. Kosek, G. K. Surya Prakash, G. A. Olah, *J. Power Sources*, **47**, 377 (1994)
3. A. K. Shukla, P. A. Christensen, A. Hamnett and M. P. Hogarth, *J. Power Sources*, **55**, 87 (1995)
4. X. Ren, M. Wilson and S. Gottesfeld, *J. Electrochem. Soc.*, **143**, L12 (1996)
5. S. Wasmus and A. Kuver, *J. Electroanal. Chemistry*, **461**, 14 (1999)
6. A. S. Aricò, P. Cretì, P. L. Antonucci and V. Antonucci, *Electrochem. Solid-State Lett.*, **1**, 6 (1998)
7. P. L. Antonucci, A. S. Aricò, P. Cretì, E. Ramunni, V. Antonucci, *Solid State Ionics*, **125**, 431 (1999)
8. M. Watanabe, H. Uchida, Y. Seki, M. Emori, P. Stonehart, *J. Electrochem. Soc.*, **143**, 3847 (1996)

Proton Exchange Membrane
Fuel Cells

Ultra CO Tolerant PtMo/PtRu anodes for PEMFCs

Sarah C Ball and David Thompsett
Johnson Matthey Technology Centre
Blounts Court
Sonning Common
Reading, RG4 9NH
UK

ABSTRACT

Progress has been made in designing an anode electrode that can tolerate CO levels of several thousand ppm which are a typical output of a fuel processor consisting of a reformer and water-gas-shift unit only. The combination of a PtMo catalyst that is capable of electrochemically oxidising CO and a PtRu catalyst that can tolerate high levels of CO_2 arranged in a bilayer configuration, has been shown to tolerate 2000 ppm CO in H_2 with a relatively small loss in fuel cell performance. Further improvements in the ability to tolerate higher levels of CO with lower performance losses are expected due to improved PtMo catalyst design.

INTRODUCTION

Difficulties associated with the storage and distribution of hydrogen means that the first generation of commercially available Proton Exchange Membrane Fuel Cells (PEMFCs) for either stationary or transportation use, are likely to operate using a fuel processor that converts hydrocarbons to hydrogen-rich fuel.

A key feature to the successful introduction of PEMFCs will be the effectiveness of the fuel cell stack to tolerate the impurities present in hydrogen derived from reformed hydrocarbons such as methane and gasoline. However, existing present day anode catalyst technology based on PtRu bimetallic particles cannot tolerate even relatively low levels of CO (e.g. > 50 ppm) without significant performance loss at current operating temperatures (c. 80°C). This low tolerance to CO by the fuel cell places a demand on the fuel processor system to produce hydrogen feeds with low CO levels. In general, primary reforming of fuels such as methane produce a hydrogen rich gas containing up to 10% CO. The use of the water-gas-shift reaction can reduce this down to 0.5% CO, as well as increasing the hydrogen concentration. However, the lowering of CO down to ppm levels requires either further selective oxidation or methanation steps. Although effective in lowering CO levels to c. 10 ppm, these extra stages do increase system complexity, system weight and volume, as well as consuming hydrogen. Complementary to this technology is the further use of selective oxidation at the electrocatalyst level within the Membrane Electrode Assembly (MEA) itself. The addition of a low concentration of air into the fuel stream has the effect of further reducing the CO levels to a few ppm. However, this practise is limited to CO levels of less than 100 ppm, due to the exothermic nature of the CO oxidation reaction. It has been shown previously that anode performance can be degraded with air bleed operation over time. This effect can be alleviated by the presence of a separate selective oxidation layer within the anode electrode structure, however higher CO levels cannot be successfully lowered as the catalysts become inhibited with CO at these low temperatures [1].

One alternative approach to dealing with the presence of CO in reformate streams is the development of CO tolerant catalysts which are capable of operating effectively without loss in overpotential on reformate. Many Pt-based catalysts have been investigated for their resistance to poisoning towards CO [2]. Recent work has focussed on the addition of Mo and W to Pt, which can show greater CO tolerance to PtRu [3-6].

However, although much work has been published on tolerance of anode catalysts to H_2 feeds containing 100 ppm CO or less, relatively little has been published on the effect of high CO levels representative of a fuel processor consisting of a reformer and water-gas-shift reactor alone [7]. The ability of the MEA (and hence the fuel cell stack) to tolerate high levels of CO in the fuel mixture without the need for an added air bleed, reduces the requirement on the fuel processing system to produce reformate with ultra low CO levels. The benefits of a high-CO tolerant MEA are two-fold. Firstly, the lack of the need for an air bleed has significant advantages for MEA durability, especially with the use of thin proton conducting membranes. Secondly, a high-CO tolerant MEA will allow the simplification of the fuel processor system, or be used to cope with CO spikes.

This paper presents some recent progress in designing an MEA capable of tolerating high CO containing reformate without significant performance loss.

EXPERIMENTAL DETAILS

The PtMo and PtRu catalysts supported on Vulcan XC72R furnace black (Cabot Corp.) at 20wt% Pt loading were prepared using proprietary methods. The catalysts were characterised for metal content using inductively plasma-emission spectrometry and Pt crystallite size and degree of alloying by powder X-ray Diffraction (XRD). Cyclic voltammetry using CO electro-oxidation was performed on Nafion containing electrode buttons of the catalysts. The electrodes were fabricated using screen printing aqueous Nafion containing inks on poly(tetrafluoroethylene) (PTFE)-impregnated Toray TGP-90 carbon fibre papers [8]. The CO stripping procedure is described in [9]. The electrochemical area value was calculated from the CO stripping charge and assuming that CO adsorbs only onto Pt with a 1:1 stoichiometry.

Similar electrodes were prepared for steady state half-cell experiments. Electrodes (2.5 cm^2) of Pt (20wt%Pt/XC72R), PtRu (40/20wt%PtRu/XC72R) and PtMo (20/2wt%PtMo/XC72R) were fabricated at a Pt loading of $0.1 - 0.2$ mg (Pt) cm^{-2} using the procedure described above. The electrodes were placed in a gas-tight holder and inserted into a glass-filled PTFE half cell together with a Pt foil counter electrode and a Pd/C Reversible Hydrogen Electrode (RHE) [10]. The half-cell was filled with 1M H_2SO_4 and heated to 353 K. A series of steady state polarisation curves were performed firstly with pure H_2 and then with pure CO. The internal resistance of the cell was measured during each polarisation curve using the current interrupt method. At the end of the CO polarisation curves, the electrochemical area was measured using CO stripping voltammetry. The CO oxidation charges was used to correct the measured current for the in-situ Pt surface area and express it as specific activity, $mAcm^{-2}$ (Pt). The applied potentials were corrected for the internal resistance (IR) of the cell and expressed as IR corrected potentials, mV vs RHE.

For single cell evaluations, electrodes (50 cm^2 active area) of the catalysts were prepared as described above. Anode loadings were 0.20 mg (Pt) cm^{-2}. MEAs were fabricated from the anode electrodes and standard cathodes (HiSPEC4000, 40%Pt/XC72R at 0.8 mg (Pt) cm^{-2}) using Nafion 115 membrane (DuPont) by hot pressing. The MEAs were conditioned in the single cell for 2 days on H$_2$/air, before testing on different reformate mixtures proceeded. All the tests were carried out by measuring the cell voltage at a fixed current density of 500 mAcm^{-2}, using fuel and oxidant stoichiometries of 1.5 and 2.0 respectively. Gas inlet temperatures and total pressures were 353 K and 308 kPa, with gases pre-humidified to 100% relative humidity using external membrane humidifiers. Investigation of the reformate tolerance was performed by switching the anode gas stream at $t = 0$ from pure H$_2$ to a reformate mixture, and then monitoring the change in cell voltage. A series of H$_2$ gas mixtures was used containing CO from 10 – 1000 ppm, CO$_2$ up to 25% and N$_2$ up to 50% to mimic possible outputs from a fuel processor system.

For the bilayer experiments, MEAs were tested in Ballard Mk5E hardware (240 cm^2 active area). Similar test conditions were used to the 50 cm^2 tests, using anodes of 0.45 – 0.48 mg Pt cm^{-2} loading (PtMo/PtRu and PtRu respectively). The durability experiment used a reformate that was produced by a HotSpot^{TM} methanol fuel processor which gave a stable H$_2$ feed comprising of 52%H$_2$, 27%N$_2$, 21% CO$_2$ and 5 ppm CO. The reformate was further dosed with CO to bring the CO concentration to 100 ppm.

DISCUSSION

Performance characteristics of PtMo vs PtRu catalysts

It has been well documented that PtMo catalysts can show superior CO tolerance to PtRu catalysts [4,11]. However it has been reported by the present authors that a PtMo catalyst showed greater CO$_2$ intolerance than PtRu, making it less acceptable as a reformate tolerant catalyst [9]. Recently, the effect of PtMo preparation route and Pt:Mo ratio on CO and CO$_2$ tolerance has been investigated. Four PtMo catalysts supported on Vulcan XC72R carbon black were prepared using two different preparation routes (A & B) at a common Pt loading of 20wt%. Two catalysts were prepared at a Pt:Mo ratio of 3:1 and one catalyst each was prepared at a ratio of 5:1 (preparation route A) and 6:1 (route B). Details of the catalysts, together with those of a standard PtRu catalyst are shown in Table 1.

All 4 catalysts show a small contraction of the Pt face-centre-cubic lattice parameter (a) from that of pure Pt (0.3924 nm) suggesting some Mo has been incorporated. However, due to the insensitivity of a to PtMo bulk alloy composition between 10 and 38% it is not possible to determine alloying composition.

The surface area of the 4 PtMo and the PtRu catalysts were assessed using CO stripping voltammetry. The PtRu gave an electrochemical area (eca) of 184 m^{2}g^{-1}, indicating that CO is adsorbing on Ru sites as well as Pt. In contrast, the eca's of the PtMo catalysts are much lower, and somewhat lower than pure Pt catalysts of similar Pt crystallite sizes, suggesting CO does not adsorb onto Mo sites.

All 5 catalysts were assessed with a range of H$_2$ feeds containing varying CO and CO$_2$ contents to determine catalyst tolerance to CO and CO$_2$. Figure 1 shows the effect of PtMo preparation route and Pt:Mo ratio on CO tolerance with CO concentrations up to 1000 ppm CO in H_2, on cell voltage at a current density of 500 mAcm^{-2}, with an anode catalyst loading of 0.2 mg(Pt)cm^{-2}.

Table 1. Physical Properties of PtMo catalysts

Pt:Mo ratio	Preparation Route	Assay		XRD Properties		Electrochemical Metal Area /m^{2}g^{-1} Pt
		Pt (wt%)	Mo (wt%)	Fcc lattice parameter, a/nm	Crystallite size/nm	
3:1	A	19.3	3.6	0.391	1.8	81
3:1	B	20.3	3.5	0.391	3.1	62
5:1	A	20.1	2.1	0.390	2.4	85
6:1	B	19.9	1.8	0.392	3.7	58
PtRu	-	19.2	9.3	0.388	1.9	184

There appears a large effect of catalyst preparation on CO tolerance. The catalysts prepared by route A show similar performance losses despite differences in Mo content with a feed of 100 ppm CO in H_2. In contrast, catalysts prepared by route B do show a large sensitivity to Mo content, The 6:1 PtMo catalyst showed similar performance loss to PtRu at 100 ppm CO, while the 3:1 catalyst showed 100 mV higher performance and somewhat better than the 3:1 catalyst prepared by route A. Increasing the CO concentration to 1000 ppm showed a greater voltage loss for both 3:1 catalysts.

The large CO_2 intolerance shown by PtMo catalysts is illustrated in Figure 2. This shows the effect of switching between pure H_2 and 25% CO_2 in H_2 on cell voltage at a fixed current of 500 mAcm^{-2} for a standard PtRu and the 5:1 PtMo (route A) catalysts. The PtRu catalyst shows only a gradual loss in performance of 25 mV, of which some of this is due to dilution losses (c. 10 mV). In contrast, the PtMo shows a rapid decrease in performance of 80 mV. In addition, a difference between preparation routes is apparent. Figure 3 shows the performance losses at 500 mAcm^{-2} between testing on pure H_2 and 25% CO_2 in H_2. The catalysts prepared by route A show a loss of c. 80 mV, while those of route B show losses of 45 – 55 mV. There appears to be little effect of Pt:Mo ratio.

To probe this effect the CO stripping voltammetry of the four catalysts was studied in 1M H_2SO_4 at 298K (Figure 4). After CO adsorption, all four catalysts show a small oxidation wave at 0.3 – 0.4 V, together with a broad larger peak at 0.70 – 0.75 V on the first cycle, similar to that shown by pure Pt. On the 2nd cycle after CO removal a reversible redox feature is observed at c.0.4 V. It is not clear whether the first oxidation peak on the first cycle is due to CO oxidation or a Mo oxidation process, although the shift between this peak between the first and second cycles suggests that CO is involved. Although all four catalysts show similar features, there are small differences in the intensity and positions of the low potential features. However, no correlations between preparation route and Mo content can be easily seen. It is clear that under the conditions of CO stripping voltammetry, the majority of the CO adsorbed onto the catalyst, is electro-oxidised at potentials similar to pure Pt.

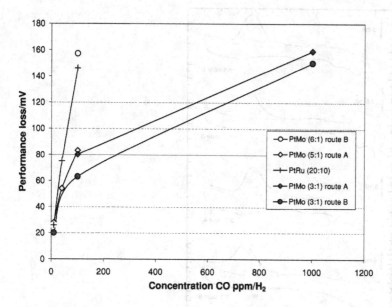

Figure 1. Performance loss vs increasing CO in H_2 concentration for PtMo and PtRu catalysts, $i = 500$ mAcm^{-2}, 308/308 kPa CO-H_2/air, T = 353K.

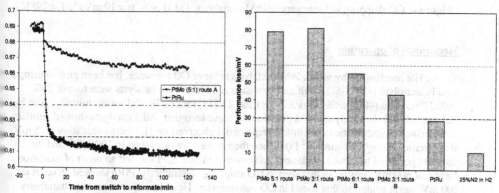

Figure 2. MEA Performance degradation switching between pure H_2 and 25% CO_2/H_2, i = 500 mAcm^{-2}, 308/308 kPa H_2/air, T = 353K

Figure 3. Effect of 25%CO_2 in H_2 on cell performance for PtMo & PtRu catalysts, , i = 500 mAcm^{-2}, 308/308 kPa H_2/air, T = 353K

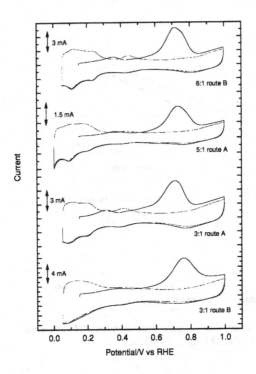

Figure 4. CO stripping voltammetry of PtMo catalysts, 1M H_2SO_4, $\upsilon = 10$ mVs^{-1}, T = 298K

Mechanism of operation

The mechanism by which PtMo catalysts achieve CO tolerance, has been probed using a liquid electrolyte (1M H_2SO_4) half cell approach at 353K. Three catalysts were tested; 20% Pt/XC72R, PtRu (40/20wt% PtRu/XC72R) and 22%PtMo/XC72R (5:1 ratio). Initially, pure H_2 was passed over the catalysts and a polarisation curve recorded. All 3 catalysts showed similar performance characteristics with little overpotential observed for H_2 electro-oxidation (c. 5 mV at a specific activity of 1 mAcm^{-2} Pt)). Then the gas feed was switched to 100% CO and the catalysts polarised to induce oxidation. As shown in Figure 5, Pt showed an onset of oxidation current at 600 mV, similar to that seen in CO stripping voltammetry. PtRu showed an onset of 300 mV, again similar to that found in CO voltammetry. However, in contrast to voltammetry, the PtMo catalyst was found to oxidise CO from an overpotential of 25 mV. The oxidation current increased to a maximum at 250 mV, after which it decreased and only increased at higher potentials similar to Pt and to that found in voltammetry. On a second polarisation curve immediately after the first, no low potential CO oxidation was found, with CO oxidation only occuring at higher potentials similar to Pt. The low potential oxidation was restored if H_2 was reapplied to the catalyst prior to CO polarisation. This strongly indicates that H_2 plays a critical role in achieving low potential CO oxidation.

The role of Mo in promoting CO electro-oxidation on Pt has been discussed by a number of authors including ourselves [10-13]. X-ray Absorption Near Edge studies have shown that Mo at low potentials undergoes reduction from +VI to +IV – V, which corresponds to the redox couple observed in the voltammetry [11,14]. Given that the half-cell studies indicate that H_2 appears important to achieve low potential CO electro-oxidation, this would suggest that the reduction of Mo is assisted by H_2, possibly due to 'spillover' of H from Pt sites.

In contrast, significant CO oxidation does not begin on PtRu until the anode overpotential reaches 400mV. This is consistent with the data observed for CO/H_2 mixtures shown in Figure 8, where anode overpotential rises dramatically as CO levels increase. This suggests that PtRu does not achieve CO tolerance at low CO levels and overpotentials by CO electro-oxidation. It has been proposed by one of the present authors that PtRu achieves CO tolerance by an intrinsic (or ligand) mechanism, where Ru modifies the electronic properties of Pt, reducing the strength of the Pt-CO bond and decreasing CO coverage [5]. This increases the proportion of Pt sites remaining free for H_2 electro-oxidation. This view has gain support recently by a combination of surface science and modelling studies that have demonstrated that Ru does influence the electronic nature of Pt and CO binding strengths [15-19].

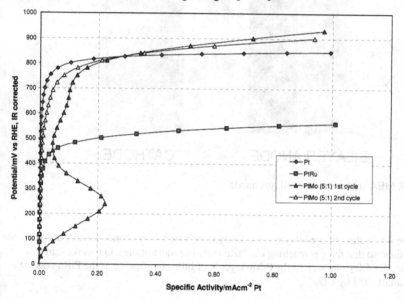

Figure 5. Steady-state 100% CO electro-oxidation on Pt, PtRu & PtMo (5:1) catalysts, 1M H_2SO_4, 101 kPa CO, 353K, H_2 pre-treatment.

<u>**Electrochemical Bilayer**</u>

The ability of PtMo to electro-oxidise CO at low potentials and PtRu to tolerate high concentrations of CO_2, led to the concept of the electrochemical bilayer, in which the PtMo acts as an electrochemical 'selective oxidation' catalyst and the PtRu acts as the H_2 electro-oxidation catalyst.

Figure 6 below shows a schematic representation of the bilayer, where the PtRu layer is adjacent to the membrane and PtMo layer is adjacent to the gas diffusion substrate. The incoming anode gas stream containing CO and CO_2 meets the PtMo catalyst first, where CO is oxidised, whilst CO_2 level is unchanged. Gas progressing to the PtRu layer therefore has a reduced CO content, so PtRu sites are freed for H_2 oxidation, whilst the CO_2 has little effect on the PtRu. It is thought that most of the H_2 oxidation occurs within the PtRu layer as this is closes to the membrane.

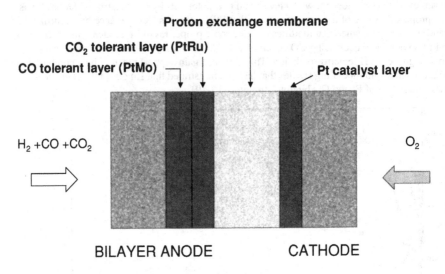

Figure 6. MEA schematic with bilayer anode

The key feature of the bilayer is the abilty to oxidise CO electrochemically at low anode overpotentials, so that the gas reaching the PtRu layer has most of the CO removed. In this context the poor CO_2 tolerance of the PtMo layer is of little significance as the PtRu catalyst is relatively unaffected by CO_2.

<u>**Performance of bilayer to higher CO levels**</u>

Figures 7 and 8 shows the effect of a PtRu single layer and a PtMo/PtRu bilayer on CO_2 and CO containing H_2 feeds. Figure 7 shows the effect of lower CO and CO_2 concentrations on the performance of a PtRu single layer (0.48 mgPtcm^{-2}) and a PtMo (5:1 - route A)/PtRu bilayer

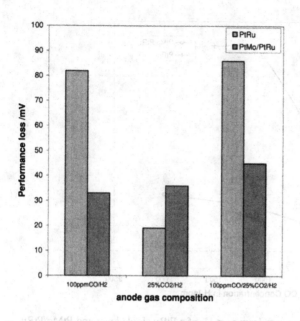

Figure 7. Comparison of effects of CO and CO_2 on a PtRu single layer and PtMo/PtRu bilayer anode, $i = 500$ mAcm^{-2}, 308/308 kPa CO-CO$_2$-H$_2$/air, T = 353K

(0.21/0.24 mgPtcm^{-2}) at a fixed current density of 500 mAcm^{-2}. With a feed of 100 ppm CO/H$_2$, the bilayer shows only a 33 mV loss in performance when compared to the PtRu single layer which shows a loss of 82 mV. With 25% CO$_2$/H$_2$, the PtRu single layer shows a loss of 19 mV (of which 10 mV is due to dilution), while the bilayer shows a loss of 36 mV, a large improvement compared to the performance of the PtMo alone at 0.2 mgPtcm^{-2} (79 mV). On a combination of 100 ppm CO & 25% CO$_2$ in H$_2$, the bilayer shows half the performance loss of the PtRu single layer.

Figure 8 shows the performance of the single and bilayer on H$_2$ feeds containing increasing concentrations of CO. The PtRu single layer shows a large performance loss on increasing the CO from 100 ppm to 1000 ppm, although further increasing to 1500 ppm only results in small additional loss in performance. The performance loss of c. 300 mV corresponds to the onset of CO oxidation in the half-cell experiments and therefore must represent a change in the mechanism between an intrinsic (or ligand) to a promoted (or bifunctional) one. In contrast, the PtMo/PtRu bilayer shows an initial loss in performance at low CO concentrations, but then a relatively small further loss in performance on increasing CO concentration from 100 to 2000 ppm (105 mV). Given that the PtMo is oxidising CO from the H$_2$ feed, the corresponding performance loss is similar to that shown by a PtRu single layer (c. 0.25 mgPtcm^{-2}) on 40 ppm CO in H$_2$. It is expected that given the improvement in PtMo properties for CO tolerance over that shown by the PtMo (5:1) catalyst described above, further improvements in bilayer performance can be expected.

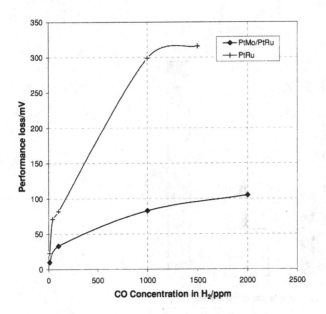

Figure 8. Performance loss from pure H_2 of a PtRu single layer and PtMo/PtRu bilayer anode with increasing CO concentration in H_2, $i = 500$ mAcm^{-2}, 308/308 kPa CO-H_2/air, T = 353K

Durability of bilayer

The durability of the bilayer was investigated over time on MeOH dervived reformate (52% H_2, 27% N_2, 21% CO_2) containing 100ppmCO without air bleed. Figure 9 below shows performance of the bilayer at a constant current density of 500 mAcm^{-2}. The MEA was periodically returned to pure H_2 performance for diagnostic measurements. The performance of the bilayer was found to be durable over at least 700 hours, with no significant decline in performance.

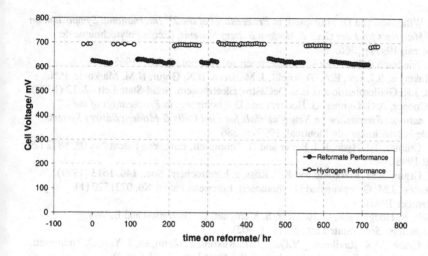

Figure 9. Durability of PtMo/PtRu bilayer anode on reformate containing 100 ppm CO, $i = 500$ mAcm^{-2}, 308/308 kPa H$_2$(reformate)/air, T = 353K

CONCLUSIONS

It has been shown that the higher levels of CO that are typically produced by a fuel processor consisting of a reformer and water-gas-shift reactor alone, can be tolerated by a novel PtMo/PtRu bilayer anode with only a limited performance loss without the use of an air-bleed. The bilayer concept consists of a PtMo 'selective electro-oxidation' layer that is able to lower CO levels to a point where a PtRu 'H$_2$ electro-oxidation' layer can tolerate them without significant performance loss. The intolerance of PtMo catalysts to high CO$_2$ levels is also alleviated by this approach. The bilayer has been found to be durable over 700 hours of continuous reformate testing. Improvements in PtMo catalyst properties have also been shown suggesting that bilayer performance can be further improved.

ACKNOWLEDGEMENTS

We would like to acknowledge the UK Department of Trade and Industry's New and Renewable Energy Programme, for part-funding this work. Also, we would like to thank Brian Theobald, Hazel Davies, Madeline Hurford, Adam Hodgkinson, Philip Birch, Eric Chen and Monica Gonzalez for their contributions to this work.

REFERENCES

1. S. Ball, S. Cooper, K. Dooley, G. Hards and G. Hoogers, ETSU Contract Report F/02/00160/REP (2001).

2. D.P. Wilkinson and D. Thompsett, in *Proceedings of the 2nd International Symposium on New Materials for Fuel Cells & Modern Battery Systems*, (Ecole Polytechnique de Montreal, 1997) p. 266.
3. L.W. Niedrach & I.B. Weinstock, Electrochem. Technol., **3**, 270 (1965).
4. S. Mukerjee, S.J. Lee, E.A. Ticianelli, J. McBreen, B.N. Grgur, N.M. Markovic, P.N. Ross, J.R. Giallombardo and E.S. DeCastro, Electrochem. Solid-State Lett., **2**, 12 (1999).
5. S.J. Cooper, A.G. Gunner, G. Hoogers and D. Thompsett, in *Proceedings of the 2nd International Symposium on New Materials for Fuel Cells & Modern Battery Systems*, (Ecole Polytechnique de Montreal, 1997) p. 286.
6. A.G. Gunner, T.I. Hyde, R.J. Potter and D. Thompsett, European Patent No. 0838872 (29 April 1998).
7. B.N. Grgur, N.M. Markovic and P.N. Ross, J. Electrochem. Soc., **146**, 1613 (1999).
8. J. Denton, J.M. Gascoyne and D. Thompsett, European Patent No. 0731520 (11 September 1996).
9. S. Ball, A. Hodgkinson, G. Hoogers, S. Maniguet, D. Thompsett and B. Wong, Electrochem. Solid-State Lett., **5**, A31 (2002).
10. E.M. Crabb, M.K. Ravikumar, Y.Qian, A.E. Russell, S. Maniguet, J. Yao, D. Thompsett, M. Hurford and S.C. Ball, Electrochem. Solid-State Lett., **5**, A5 (2002).
11. S. Mukerjee and R.C. Urian, Electrochim. Acta, **47**, 3219 (2002).
12. B.N. Grgur, N.M. Markovic and P.N. Ross, in *Proton Conducting Membrane Fuel Cells, PV 98-27*, edited by S. Gottesfeld, T.F. Fuller and G. Halpert (The Electrochemical Society Proceedings Series, Pennington, NJ (1998)) pp. 176-185.
13. B.N. Grgur, N.M. Markovic and P.N. Ross, J. Phys. Chem. B, 102, 2494 (1998).
14. S. Maniguet, PhD. Thesis, University of Southampton, 2002.
15. F. Buatier de Mongeot, M. Scherer, B. Gleich, E. Kopatzki and R.J. Behm, Surf. Sci., **411**, 249 (1998).
16. J.C. Davies, B.E. Hayden and D.J. Pegg, Surf. Sci., **467**, 118 (2000).
17. C. Lu and R.I. Masel, J. Phys. Chem. B, **105**, 9793 (2001).
18. P. Liu and K.K. Nørskov, Fuel Cells, **1**, 192 (2001).
19. M.T.M. Koper, T.E. Shubina and R.A. van Santen, J. Phys. Chem. B, **106**, 686 (2002).

Mat. Res. Soc. Symp. Proc. Vol. 756 © 2003 Materials Research Society

Characterization of Potential Catalysts for Carbon Monoxide Removal from Reformate Fuel for PEM Fuel Cells

Peter A. Adcock, Eric L. Brosha, Fernando H. Garzon, and Francisco A. Uribe
Materials Science and Technology Division, Los Alamos National Laboratory
Los Alamos, NM 87545, USA

ABSTRACT

We are developing a fuel-cell-integrated system for enhancing the effectiveness of an air-bleed for CO-tolerance of hydrogen and reformate PEM fuel cells, with minimal increase in stack cost or specific volume. This is called the reconfigured anode (RCA) system [1]. We report here on properties of several potential catalysts for this system. The materials were characterized by X-ray powder diffraction (XRD), energy dispersive X-ray spectrometry (EDS), and thermal analysis techniques. Surface area was determined using the Brunauer-Emmett-Teller (BET) technique. The XRD results were interpreted using full-profile analysis. In ongoing work, reactivity with CO is being quantified under various conditions, using gas chromatographic (GC) analysis. The results are discussed in terms of effects of the presence of an RCA catalyst on fuel cell performance, using a small air-bleed.

INTRODUCTION

Practical PEM fuel cells for automotive and large stationary applications are expected to operate on reformate, obtained by reforming gasoline, diesel, natural gas, or methanol fuels. A very troublesome impurity in reformate is residual carbon monoxide, CO, which poisons the platinum catalyst surface [2]. CO is still present at a significant level even if water gas shift and preferential oxidation reactors are included. In the on-board reformate production option, complex processing to remove CO would have a high capital cost, volume and weight penalty. To improve the CO-tolerance of Pt-catalyzed anodes, Los Alamos National Laboratory is developing the reconfigured anode (RCA) system for in-cell CO conversion without a significant increase in stack volume or cost. This system relies on adding a small air-bleed to the humidified anode gas stream.

We reported previously that oxides of metals having two stable, low oxidation states showed promise for CO removal from hydrogen and reformate streams at 80 °C [1]. Oxides of Co, Fe, and Cu were studied most extensively, and there was interest also in the manganese based cryptomelane KMn_8O_{16} with potassium exchanged by cobalt [3]. This paper now presents a characterization of some of these catalyst materials as tested in fuel cell experiments.

EXPERIMENTAL

Catalyst Preparation

The acetate decomposition method was reported previously [1] for carbon-supported copper oxide. The mixing ratios and firing temperatures, T_{max}, for each catalyst are given in Table 1. The method for ink preparation was reported previously [1].

The preparation of cryptomelanes was based on a literature method [4] with the following modifications. Sulfuric acid was substituted for nitric acid. Cobalt(II) sulfate was used in place of cobalt(II) nitrate.

Table 1. Details of Preparation of Each Supported Catalyst

Metal	Co	Fe	Cu
Metal Content / g	1.99	4.47	5.09
Carbon Black Used / g	4.00	4.00	4.00
T_{max} / ºC	330 ± 5	260 ± 10	280 ± 5
Time at T_{max} / min	60	80	20
Recovered Material / g	6.69	10.10	10.1

Characterization

The surface areas of the catalyst powders were determined with nitrogen at cryogenic temperatures using the BET method. Prior to the surface area measurement, the sample was dried at 150 °C under vacuum for 12 hours. The sample was re-weighed and the specific surface area was then determined. The surface areas are traceable to NIST standards.

TGA measurements were made using a Perkin-Elmer TGA 7 controlled by a PC. Typical run conditions were from 25 to 800 ºC at 2 ºC per minute under an atmosphere of either forming gas (6% H_2/Ar) or cylinder air (20% O_2/N_2).

EDS quantification and elemental mapping were carried out using an EDAX® instrument coupled to a Philips XL 20 SEM at an excitation voltage of 10.0 kV. Each sample powder was mixed with *iso*-propanol, painted onto a piece of mirror-flat silicon and air-dried prior to the analysis.

X-ray powder diffraction measurements were made with a Siemens D5000 diffractometer fitted with a diffracted beam graphite monochromator, using 40 kV accelerating voltage and 35 mA beam current. However, the cryptomelane samples were measured with a Siemens D5000 fitted with an incident beam Ge single crystal monochromator, using 50 kV accelerating voltage and 40 mA beam current. A zero-background quartz sample holder was used to support the powder samples. The analysis of the XRD data was carried out using the SHADOW® full profile refinement package (Materials Data Corporation). Full profile analysis methods model the entire X-ray pattern. The measured peak profile function was modeled by a convolution of an experimentally-determined instrument function and a Lorentzian peak profile sample function. The average crystallite size of the catalyst powder was determined using a Scherrer crystallite size broadening model after de-convoluting the sample profile from the instrument function.

GC analysis for CO was carried out using a Hewlett Packard 5890 Series II gas chromatograph. The GC was set up in a dual column configuration: the gas to be analyzed was flowed through a VALCO sampling valve (250 µL sampling loop) and was subsequently split into two equal volumes. The carrier gas was UHP argon. One half of the gas sample was sent through a JW Molseive 5A capillary column and onto a thermal conductivity detector (TCD) to measure permanent gases such as N_2 and O_2. The other half of the gas sample was sent through a PoraPlot Q capillary column, which separated CO and CO_2. In order to measure these gas species accurately, the gas sample was passed over a Ni catalyst in the presence of H_2, which converted the CO or CO_2 into CH_4. The gas sample then flowed to a flame ionization detector (FID), which measured the concentration of the analyte gas. All results were averages for 6 injections over a 40 minute period. Both the TCD and FID responses were calibrated using a Scott Specialty Gases calibration gas mix.

RESULTS AND DISCUSSION

The partially cobalt-exchanged cryptomelane and the parent potassium compound were both examined. For the cobalt-containing material, the highest BET surface area recorded was 170 m^2/g, but this was sensitive to crushing conditions. The TGAs in forming gas were practically indistinguishable, whether or not cobalt was present (Figure 1(a)). The traces showed a major transition at about 290 °C, and smaller losses at <180 °C and > 540 °C. In addition to the expected elements, EDS analysis of these materials showed carbon unless finely crushed. This was attributed to surface carbonation (of the potassium ions exposed to air). XRD results for the parent compound and the partially exchanged material are shown in Figure 1(b). The cryptomelane crystal structure (JCPDS 34-168) was verified. Exchanging cobalt produced further line broadening, due to a reduction in grain size from 9.0 to 5.3 nm.

(a)

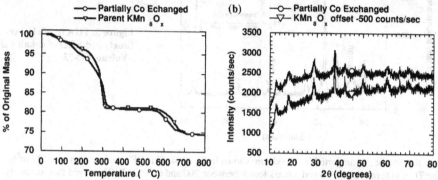

Figure 1. TGAs (in 6%H$_2$/Ar) and XRDs of parent and cobalt-exchanged cryptomelanes

Since the remaining materials were prepared by thermal decomposition of acetates in air, Figure 2 presents the TGA in cylinder air of each of the starting acetates. Cylinder air was found by GC analysis to be a 4:1 blend of nitrogen and oxygen only. On each trace, the firing temperature is indicated by an arrow. The results in Figure 2 will be discussed below, in terms of the XRD and forming gas TGA data for each catalyst, which follows.

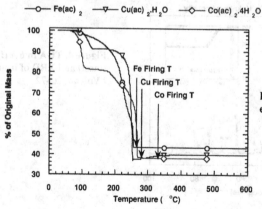

Figure 2. TGAs (in 20%O$_2$/N$_2$) for each acetate salt

Results for characterization of the supported cobalt oxide are shown in Figure 3. The BET surface area was 116 m^2/g compared to 245 m^2/g for Vulcan XC-72 itself. The TGA in forming gas showed a single mass loss of 12.3%, which is on the high side for Co_3O_4 to Co based on 29.7% Co (calculated from Table 1). XRD found mainly Co_3O_4 with about 10% CoO. The main component had a crystallite size of 12.0 nm, and the minor component was of similar size. From EDS elemental maps, there appeared to be some segregation of the cobalt oxide into clusters.

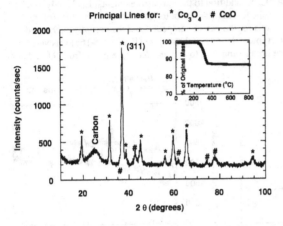

Figure 3. TGA in 6% H_2/Ar (inset) and XRD of cobalt oxide / Vulcan XC-72

Results for the supported iron oxide are shown in Figure 4. The BET surface area was 137 m^2/g. The TGA in forming gas showed 3 mass losses between 200 and 600 °C. XRD showed that the catalyst contains magnetite (Fe_3O_4) and hematite (Fe_2O_3). The crystallite sizes were calculated as 19.5 nm and 8.8 nm, respectively. The distribution was estimated as 52% magnetite and 48% hematite on an iron atom basis. EDS elemental maps showed the iron to be evenly distributed over the carbon. The TGA losses suggest that the material may contain some non-crystalline hydrous oxide or hydroxides, which only convert completely to hematite and magnetite at about 600 °C. Further investigation would require XRD analysis of samples taken to a series of temperatures using the TGA.

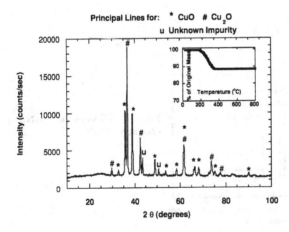

Figure 4. TGA in 6% H_2/Ar (inset) and XRD of iron oxide / Vulcan XC-72

Results for the supported copper oxide are given in Figure 5. The BET surface area was 62 m^2/g, the lowest of all the samples. The TGA in forming gas showed a mass loss of 11.4% between 170 and 370 ºC. XRD showed a mixture of CuO (about 67%) and Cu_2O (about 33%) along with carbon and an as-yet unidentified impurity. The combination of the TGA and the XRD results suggests that the catalyst contained about 54.5% Cu.

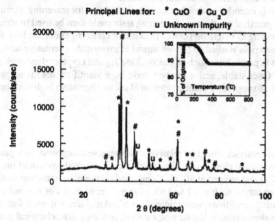

Principal Lines for: * CuO # Cu₂O
u Unknown Impurity

Figure 5. TGA in 6%H₂/Ar (inset) and XRD of copper oxide / Vulcan XC-72

Returning to Figure 2, the cobalt compound showed its major weight loss between 230 and 270 ºC. The end state in air was Co_3O_4. Firing at 330 ºC produced a small amount of CoO as well, which we attribute to a kinetic effect. The iron compound showed weight losses from 60 to 240 ºC. The end state was clearly Fe_2O_3. The supported catalyst fired at 260 ºC contained this in about equimolar mixture with Fe_3O_4. The copper compound showed loss of water around 100 ºC and then a major transition from 170 to 250 ºC. Firing with carbon black (Vulcan XC-72) at 280 ºC gave mainly CuO combined with Cu_2O.

In order to demonstrate how CO, O_2 from the air-bleed, and the components of reformate interact with the catalysts, it is necessary to monitor changes to the gas stream and the catalyst independently of the normal fuel cell anode reactions. We have chosen to use GC analysis on gases and gravimetry on the catalysts for these experiments. The main purpose is to test catalyst activity by non-faradaic mechanisms at temperatures around 80 ºC, where Nafion®-based MEAs operate. Chemical changes to the catalyst could involve modification to a catalytically active form different from the original form, or the uptake of one of the reactants, oxygen or perhaps CO. Initial experiments were carried out on a dry CO/N_2 mixture, passing the gas through the painted backing cloth, like a filter. These experiments failed to show any effect of an air-bleed in the range of 1 to 6%, other than dilution. The initial CO level was 200 ppm. We plan to repeat some of these tests at 50 ppm CO or lower, to make the dilution effect small compared to the expected catalytic oxidation effect. The other reformate components, such as hydrogen, CO_2 and water vapor have yet to be introduced, to quantify the effects of the water gas shift equilibrium, a potential source of different behavior in reformate compared to hydrogen. Other potential "reactor" geometries include (a) using a fuel cell type configuration with the RCA catalyst alone in direct contact with a Nafion membrane which has counter and pseudo-reference electrodes on the opposite side, (b) sampling gases from an operating fuel cell with an RCA layer, and (c) using the TGA isothermally as a microbalance containing catalyst powder. Geometry (a) will allow us to study electrochemical promotion of chemical catalysis [5] in this system, which we now suspect to make a substantial contribution to the activity observed in our fuel cell experiments.

Fuel cell tests have shown excellent results on humidified hydrogen if the supported iron oxide catalyst is used in an RCA layer [1]. However, on humidified, simulated reformate, the supported cobalt oxide catalyst has also performed well [1]. These catalysts can practically eliminate the effect of 250 ppm of CO [1]. Both of the metals can exist in the +3 and +2 valence states. Compounds such as Fe_2O_3 and Co_3O_4 may contain active oxygen sites, which facilitate oxidation of CO and be easily replenished from the oxygen in the air-bleed. The partially cobalt-exchanged cryptomelane showed inconsistent activity, which correlates with the tendency to form surface potassium carbonate.

Based on the GC experiments, it is intended to develop suitable techniques for screening alternative materials for catalytic activity under practical conditions. Screening tests could then be used to identify promising candidates for extensive testing in single fuel cells. Because of concerns about the long-term stability, in a PEM fuel cell, of the materials studied so far, we intend to investigate alternative active materials. Potential candidates include perovskites such as $LaCo_{1-y}Mn_yO_{3-\delta}$ and cryptomelane with all of the potassium substituted by cobalt. Once stable, active catalysts have been identified, the investigation could focus on comparing the effects of supporting them on carbon black or alternative high surface area support materials.

CONCLUSIONS

In the RCA system, CO removal catalysts, based on relatively inexpensive compounds, are painted in a thin layer onto the anode gas distribution cloth on the flow field side. Ideal catalysts would function by facilitating the reaction between CO and oxygen at 80 °C. Compounds containing transition metals, which can be stabilized in two valence states, such as +2 and +3, have proven useful. For maximization of performance under fuel cell operating conditions, with minimal air-bleed requirement, such factors as the equilibrium redox potential, the valence ratio in the starting catalyst, and the electrochemical potential in the catalyst layer may be important. It is possible that the water gas shift reaction could also play a positive role in these systems. This is a promising technology and there remains scope for considerable improvement in the choice of catalyst and its dispersion on a suitable support material.

ACKNOWLEDGMENTS

We gratefully acknowledge funding from the US DOE Office of Hydrogen, Fuel Cells, and Infrastructure Technologies. The assistance of Judith Valerio and Susan Pacheco is also acknowledged.

REFERENCES

1. P.A. Adcock, S. Pacheco, E. Brosha, T.A. Zawodzinski and F.A. Uribe, submitted for publication in *Proton Conducting Membrane Fuel Cells III*, edited by J.W. Van Zee, M. Murthy, T. Fuller and S. Gottesfeld, (Electrochem. Soc. Proc., Pennington, NJ, 2002).

2. S. Gottesfeld and T.A. Zawodzinski, in *Advances in Electrochemical Science and Engineering*, Vol. 5, edited by R.C. Alkire, H. Gerischer, D.M. Kolb and C.W. Tobias, Wiley VCH, Weinheim, Germany (1997), pp. 195-301.

3. G.G. Xia, Y.G. Yin, W.S. Willis, J.Y. Wang and S.L. Suib, *J. Catal.*, **185**, 91-105 (1999).

4. R.N. DeGuzman, Y.-F. Shen, E.J. Neth, S.L. Suib, C.-L. O'Young, S. Levine and J.M. Newsam, *Chem Mater.* **6**, 815-821 (1994).

5. C.G. Vayenas, S. Bebelis, *Catal. Today*, **51**, 581-594 (1999).

Study of the effect of hydrogen on Pt supported Nanoporous Carbon derived from Polyfurfuryl alcohol

Ramakrishnan Rajagopalan, Juan Coronado, Henry C. Foley and Albert Vannice
Department of Chemical Engineering.
Pennsylvania State University.
University Park, Pa 16801,USA

ABSTRACT

Platinum supported on nanoporous carbon (NPC) is promising candidate for using as electrodes in proton exchange membrane fuel cells. Performance of the anode of a fuel cell is markedly influenced by the efficiency of the splitting of hydrogen atoms by platinum and the transference of the produced protons to the carbon support (spillover process). Consequently a better understanding of these elemental processes could prompt the improvement of the materials used. With this aim, a series of Pt/NPC samples varying the metallic content were studied by electron spin resonance (ESR) under controlled gas environment. This spectroscopic tool is especially suitable for the investigation of processes that involve transference of electrons. In the present case, all materials studied, including bare carbon, showed a very narrow signal (H_{pp}=1.5-3 gauss) at g=2.0028±0.0002 after activation in vacuum at 500ºC. In the case of the pure carbon and for the samples with lower platinum content (lower than 0.2 wt%) signals are significantly asymmetric, and their intensity is scarcely affected by the introduction of hydrogen up to 500ºC. In contrast the spin concentration experiences a significant increment when the samples with platinum loading comes in contact with hydrogen at temperatures in the 300-500ºC range. Although the centres originating these signals are located in the carbon matrix, the present results emphasize the importance of platinum for hydrogen activation and electron transference.

INTRODUCTION

Nanoporous carbon (NPC) possess enormous potential in the area of catalysis and gas separations due to their unique size and shape selectivity. The ease of formation and their high temperature stability facilitates the use of these carbons as templates for supporting catalytic materials such as Platinum and alkali metals [1-3]. The ability to form supported catalytic nanoporous membranes with this material expands the horizon of engineering applications that is not limited to gas separation and catalysis. One of the promising applications includes the use of platinum supported nanoporous carbon (Pt/NPC) membranes as electrodes in proton exchange membrane fuel cells (PEM). These materials present graphite-like domains that grant the necessary electrical conductivity along with the porous network, which allows the access of the gases to the catalytically active particles of platinum. Pt/NPC was prepared by pyrolyzing the suspension of platinum precursor dispersed in polyfurfuryl alcohol (PFA) under inert atmosphere.

It has been demonstrated that extremely dispersed and uniform Pt clusters as small as 2 nm can be formed by pyrolysis of chloroplatinic acid suspended in polyfurfuryl alcohol [1,4]. The size of the platinum clusters showed very narrow distribution even with high Pt loading (50 wt%). One of the main drawbacks of this process is the inability to dissolve chloroplatinic acid and polyfurfuryl alcohol in a common solvent. This makes it difficult to form a defect free membrane using chloroplatinic acid as platinum precursor. This difficulty can be overcome by using Pt (II) acetylacetonoate (Pt(acac)2) as the platinum precursor which dissolves in acetone, a cosolvent for PFA [1].

PFA derived nanoporous carbon has a typical pore size of 4 –5 Å. They have been shown to possess excellent "shape selective" molecular transport properties [5]. It has also been demonstrated that uniform and defect free selective membrane, which has O_2/N_2 selectivity of about 8, can be reproducibly formed [6]. This would make it feasible to separate gases such as CO and H_2 very efficiently. This property makes it a very good candidate for the use of the Pt supported nanoporous carbon membranes as fuel cell anodes that operate with direct methanol oxidation.

There has been a lot of interest in the use of carbon supported platinum catalysts as an electrode in fuel cell applications. Some of the materials that are of interest are Graphite Nanofibers and Vulcan carbon [7-9].

In this investigation, we report the characterization and the study of the effect of hydrogen on the Pt/C catalyst formed using Pt(II) acetylacetonoate as platinum precursor. The effect of hydrogen is characterized using Electron Spin Resonance Spectroscopy (ESR), Mass spectrometry and Chemisorption experiments. The size of the platinum particles was calculated using Transmission Electron Microscopy and X-ray Diffraction.

EXPERIMENTAL

Polyfurfuryl alcohol resin (purchased from Monomer Polymer & Dajac Laboratories Inc., Lot A-1-143) was mixed with Pt (II) acetylacetonoate (purchased from Strem Chemicals) and dissolved in acetone. The solution was pyrolyzed in a tube furnace at 600 °C for 8 hours in argon atmosphere. Nanoporous carbon with different platinum loading (0.2 wt%, 0.5wt%, 1 wt%, 2 wt% and 5 wt%) was prepared by using this method.

The as-synthesized Pt/C powders were characterized using ER200D-SRC IBM ESR instrument. ESR measurements were carried out at room temperature systematically after the exposure of the sample to various conditions. The sample was treated in vacuum at 500°C for one hour. This was followed by the introduction of hydrogen into the ESR tube and reducing the catalyst under hydrogen atmosphere at 500°C for half an hour. The intensity of the ESR signals was compared before and after the introduction of hydrogen.

Chemisorption experiments were conducted on the Pt/C catalysts with the introduction of both hydrogen and Carbon monoxide separately. The sample was pretreated under vacuum at 500°C followed by the reduction in hydrogen atmosphere at 500°C.

Thermal physical desorption (TPD) measurements were done using mass spectrometer on the Pt/C catalysts. The reaction conditions were the same as both ESR and chemisorption experiments. The products that evolved were analyzed and correlated with ESR results.

The size of the platinum particles in the carbon formed using different platinum loading was measured using both Philips 420 Transmission Electron Microscopy and using X-ray diffraction measurements.

RESULTS AND DISCUSSION

Transmission Electron Microscopy (TEM)

Figure 1 shows the TEM image of the Pt/C samples formed using 2wt% sample and 0.5wt% carbon sample respectively. Platinum nanoclusters was dispersed in the carbon matrix. Typical size of each platinum particle was about 4-5 nm. The size of the platinum nanoclusters did not change significantly even when the platinum loading was

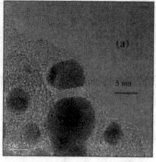
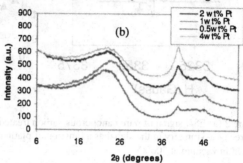

Figure 1 (a) Transmission Electron Micrograph of 5wt% platinum loaded carbon and (b) XRD of the Pt/C as a function of platinum loading

increased to 10 wt%. This is in agreement with the literature. In order to detect the platinum particles, platinum loading of at least 0.5 wt% was required. It was also shown that there was significant rearrangement of carbon around the platinum particles. The amorphous carbon almost shows very short order graphitization.

X-ray diffraction (XRD)

Figure 1 (b) shows the X-ray diffraction pattern of the samples as a function of platinum loading. Nanoporous carbon shows two broad amorphous peaks at 18° and 42° respectively. On the other hand, platinum shows sharp peaks at 40° and 46° respectively. The size of the platinum particle was calculated using Scherrer equation. The average particle size of the platinum was about 5 nm, which agreed well with the TEM data. It was also interesting to note that there was no significant change in the size of the particle with amount of platinum loading.

ESR measurements

Figure 2(a) shows the typical ESR signal obtained from pure nanoporous carbon. The sample shows a strong, narrow asymmetric signal centered at g=2.0028. This signal corresponds to a single electron process and arises due to the localized defects in the carbon such as the presence of dangling bonds in the aromatic carbon. The asymmetry in the carbon can be attributed to the presence of conduction electrons in the sample. It was shown that the asymmetry in the ESR signal decreases with increase in the platinum loading in the carbon as shown in Figure 2(b).

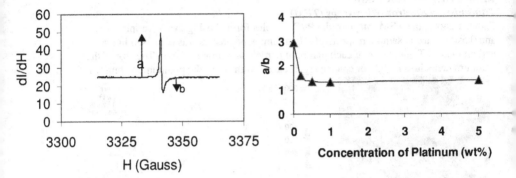

Figure 2 (a) ESR signal of pure nanoporous carbon treated in vacuum at 500°C and (b) Degree of asymmetry in the signal as a function of platinum loading of Pt/C samples treated in vacuum at 500°C

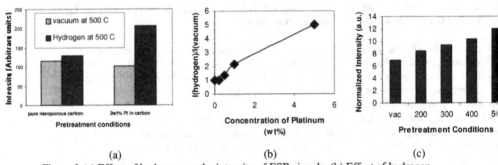

(a)   (b)   (c)

Figure 3 (a) Effect of hydrogen on the intensity of ESR signals, (b) Effect of hydrogen on the ESR signals as a function of platinum loading and (c) Effect of pretreatment temperature on the intensity of ESR signals

Figure 3(a) shows the comparison of the effect of introduction of hydrogen on the intensity of the ESR signal for both nanoporous carbon and 2wt% Pt/C sample. It was shown that there is a significant increase in the intensity of the ESR signal for the Pt/C signal after the introduction of hydrogen at 500°C. On the other hand, pure nanoporous carbon does not show much increase in the intensity of the signal. It was also shown that the increase in the intensity of the ESR signal was proportional to the amount of platinum loading as shown in Figure 3(b).

Figure 3(c) shows the effect of pretreatment temperature on the intensity of the ESR signal. The intensity of the ESR signal increases with increase in the pretreatment temperature. The intensity of ESR signal almost doubled when the sample was pretreated at 500°C under hydrogen atmosphere when compared with pretreatment under vacuum at 500°C (vac).

Chemisorption measurements

Sample	T_{red} (K)	H₂ Uptake		CO Uptake	
		Total	Rev	Total	Irrev
0.5wt% Pt/NPC	473	0.8	0	3	1.2
	773	1.8	0	10.6	0.97
5wt% Pt/NPC	473	5.03	3.3	4.9	1.2
	773	5.4	4.2	22.2	9.7

Figure 4. Amount of H_2 and CO uptake on Pt/C catalysts

H_2 and CO chemisorption was done on Pt/C catalysts. The amount of gas uptake was dependent on the pretreatment temperature. It was also shown that the amount of gas uptake increased with increase in the platinum loading. This is in agreement with ESR results. It was interesting to note that there was some irreversible chemisorption of hydrogen on the carbon catalysts. As the pretreatment temperature increased, the reversible uptake increased. This might be due to the increase in the active sites of platinum with increase in the platinum loading.

TPD measurements

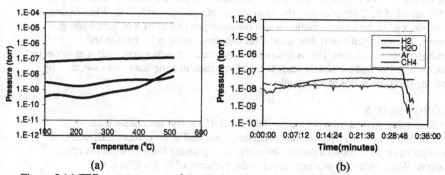

(a)   (b)

Figure 5 (a) TPD measurements of the products evolved in the presence of hydrogen when the temperature is ramped from 100 ^{0}C to 500 ^{0}C and (b) TPD measurements when the temperature is held constant at 500 ^{0}C for 30 minutes in the presence of hydrogen

Figure 5 (a) shows the products evolved as detected using mass spectrometer when the sample is heated under hydrogen atmosphere and the temperature is ramped from 100 ^{0}C to 500 ^{0}C. It was shown that a small amount of water and methane was evolved as the temperature was ramped above 400 ^{0}C. This effect was seen very clearly when the temperature was kept constant at 500 ^{0}C for thirty minutes (Figure 5 (b)). Significant amount of water and methane was released. It was also clearly demonstrated that there was no methane produced as soon as we stop supplying hydrogen. On the other hand, the water was produced for a longer time. The origin of water may be due to the presence of unfragmented polyfurfuryl alcohol still present in the sample.

Mechanism

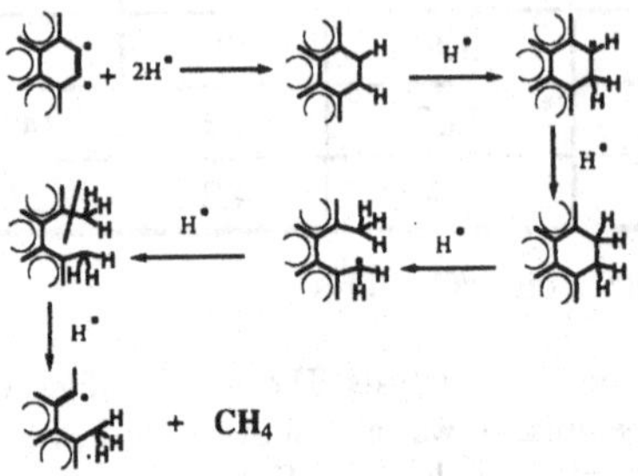

Figure 6. Mechanism for the formation of methane when hydrogen is introduced to the Pt/C catalysts

TPD measurements show that when the Pt/C is exposed to hydrogen atmosphere at 500°C , significant amount of methane was produced. The ESR signal of the exposed sample shows that there is also significant increase in the number of dangling bonds in the carbon. Figure 6 explains the formation of methane and the increase in intensity of ESR signal of Pt/C samples with the introduction of hydrogen at 500°C. When hydrogen is introduced to the sample, platinum adsorbs hydrogen and splits it to form hydrogen atoms. The hydrogen atoms then attacks the dangling bonds in the carbon and hydrogenates the sample. This is followed by the attack of subsequent hydrogen atoms to form sp3 carbon and eventually leads to the elimination of a methane molecule, thereby leaving dangling bonds in the sample.

CONCLUSIONS
Pt/C samples derived from polyfurfuryl alcohol/Pt(II) acetyl acetonoate leads to the formation of 1-2 nm platinum nanoclusters dispersed in the carbon. ESR signal of nanoporous carbon shows a narrow sharp signal originating from the defects in the carbon. When hydrogen was introduced in the presence of Pt, the ESR signal corresponding to the carbon increases in intensity. This is because platinum splits hydrogen to form hydrogen atoms, which reacts with carbon to eliminate methane and form more defects in the carbon.

REFERENCES
1. M.S. Strano and H.C. Foley, *AIChE Journal*, **47**, 66-78 (2001).
2. M.G. Stevens, M.R.Anderson and H.C. Foley, *Chem Commun*, **5**,413-414 (1999)
3. M.G. Stevens and H.C. Foley, *Chem Commun*, **6**,519-520 (1997)
4. S.H. Joo, S.J. Choi, I. Oh, J. Kwak, Z. Liu, O. Terasaki and R. Ryoo, *Nature*,

412, 169-172 (2001).

5. H.C. Foley, *Microporous Mater.*, **4**, 407 (1995).

6. M.B. Shiflett and H.C. Foley, *Carbon*, **39**, 1421-1425 (2001).

7. C.A. Bessel, K. Laubernds, N.M. Rodriguez and R.T.K. Baker, *J. Phys. Chem.B.*, **105**, 1115-1118 (2001).

8. K.E. Swider and D.R. Rolison, *Electrochem.Solid State Lett.*, **3**, 4-6 (2000).

9. E.E. Swider and D.R. Rolison, *Langmuir*, **15**, 3302-3306 (1999).

Mat. Res. Soc. Symp. Proc. Vol. 756 © 2003 Materials Research Society   FF5.8

Fuel Cell Applications of Nanotube-Metal Supported Catalysts

T. Gennett[1], B. J. Landi[1], J. M. Elich[1], K. M. Jones[2], J. L. Alleman[2], P. Lamarre[3], R. S. Morris[3],
R. P. Raffaelle[1], M.J. Heben[2]
[1]NanoPower Research Laboratories, Rochester Institute of Technology
Rochester, NY 14623 U.S.A.
[2]National Renewable Energy Laboratory, Golden, CO 80401 U.S.A.
[3]Viatronix Inc. Waltham, MA 02451 U.S.A.

ABSTRACT

Novel carbon materials with nanometer dimensions are of potentially significant importance for a number of advanced technological applications. Within this report we describe the results for the electrochemical characterization of a series of single walled carbon nanotube (SWNT) metal supported catalysts as cathodes for basic fuel cell systems. Compared to the typical carbon black electrocatalysts, the nanotube supported platinum catalyst resulted in up to a 140% improvement in the efficiency for a proton exchange membrane (PEM) fuel cell.

INTRODUCTION

The successful conversion of chemical energy into electrical energy via fuel cells was realized over 150 years ago. However significant advancement in viable commercial products has been limited because of the inability to compete on a kWh cost basis with other technologies. Recently, as the demand for smaller, more efficient, environmentally friendly power supplies has increased, a significant research effort towards utilizing regenerative fuel-cell systems has developed. The goal of this research is to develop novel solid-state catalytic materials based on carbon single-walled carbon nanotubes (SWNTs) for use as inexpensive electrodes within fuel cells. Of particular interest is the increased surface area of a nanotube supported platinum catalyst as compared to the typical carbon black electrocatalysts, 1000-1500 m^2/g versus 250 m^2/g, and its effect on catalytic activity. Also the direct contact between the nanotubes and the nanostuctured catalysts that results from the synthesis, should dramatically improve the efficiency and performance of fuel cells with electrodes based on these materials.

Utilizing a single wall carbon nanotube laser synthesis reactor with either a traditional furnace or an inductive heating system, significant yields of SWNTs from high refractory catalysts metals are possible. The net result is a composite material with a high dispersion of platinum, ruthenium, iridium, rhodium and/or palladium nanoparticles 2-50 nm in diameter, within the nanotube matrix. The high dispersion of metal catalyst particles via deposition processes has been shown to give rise to electrocatalytic activity with other carbon materials including: carbon black, carbon fibers and ordered nanoporous carbon. Recent reports with ordered nanoporous carbon of a similar surface area showed an increase in the electrocatalytic mass activity. [1-3] The advantages of SWNT-catalyst over processed materials are: the catalyst particles are in intimate contact with the carbon material, free-standing films can be made without a need for a template and the SWNT-catalyst materials can be readily dispersed in a variety polymer matrices. In addition, the SWNTs can also be ultrasonically cut into finite lengths of approximately 1 micron to improve surface area and catalyst distribution.

EXPERIMENTAL

Laser synthesis of SWNT-supported catalysts

A variety of laser vaporization synthetic procedures for the production of single-wall carbon nanotubes (SWNTs) have been described in the literature [4-12]. The resultant materials are a heterogeneous mixture of SWNTs, carbonaceous soots and catalyst metals whose composition is dependent on experimental conditions. The diameter distributions and ratio of the semi-conducting and the metallic SWNTs can be tailored through variation in laser wavelength, catalyst type, raster rate, synthesis temperature and/or peak pulse power.

In previous reports we have described our apparatus that is used for the laser synthesis of single wall carbon nanotubes.[9-14] For the syntheses with high-refractory metals which require higher temperatures, 1400-1800 °C, a previously described apparatus that incorporates a 5 kW self-tuning Radyne Induction heater is utilized.[11] Similar to the previously published procedures, a resultant web-like material was produced, collected and analyzed [9-14]. Typical micrographic images of the as-produced soots are shown in Figure 1.

In our experiments the pressed graphite targets contained a series of metal catalyst dopants at a range of concentration levels and combinations. The catalyst materials include: Co, Ni, Ru, Pt, Pd, Rh and Ir. The actual combinations were varied to maximize the performance of the cathode to oxygen reduction in PEM fuel cells.

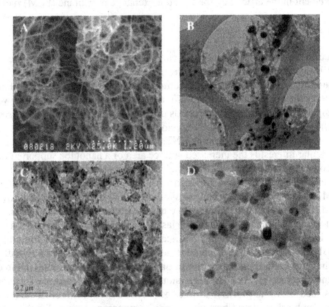

Figure 1. Microscopic images of SWNT-catalyst materials synthesized via laser vaporization. Dark circular areas in B-D are the metal catalyst particles. (A) SEM of Pt/Ni/Ru doped graphite target synthesized at 1200 °C, 100 W/cm^2; (B) TEM of Pt doped target, 1200 °C, 250 W/cm^2; (C) Pt doped target, 1800 °C; (D) HRTEM of (C).

<u>*Electrochemical Analysis of as-prepared SWNT materials.*</u>
Half-cell electrochemical measurements for the as-produced SWNT materials, the non-laser ablated target materials and standard Pt/C fuel cell catalysts (Purchased from DeNora E-Tek Division, Vulcan XC-72, 9.3% Pt) were performed on a CH Instruments Electrochemical analyzer. Linear sweep voltammetry (LSV), cyclic voltammetry (CV) and rotating disk electrode (RDE) electrochemical measurements were performed. Electrodes were prepared as a catalyst material-Nafion composite membrane directly cast onto a 5mm glassy carbon electrode. The electrolyte was a 1.0 M sulfuric acid solution saturated with oxygen. Voltammograms were recorded at room temperature at a scan rate of 0.025 – 0.100 V/sec between 0.7 and -0.25 volts vs. SCE. RDE experiments were recorded at room temperature with a scan rate of 0.010 V/sec and a rotation speed of 1500 rpm for a similar potential window. All electrochemical measurements utilized a 3-electrode cell, with an SCE reference and a platinum auxiliary electrode. In these experiments, the catalytic activity was qualitatively determined by evaluating the potential of the cathodic peak, steady state cathodic current and the current density for the oxygen reduction reaction.

The Nafion composite electrodes, NCE's, were made by blending ~5 mg of the carbon-based catalyst material in a solution that contained 400 μL of a 5% Nafion solution (Aldrich) and 400 uL of water. The solution was subsequently stirred at 500 rpm for 10 minutes to achieve a fairly homogeneous distribution. From this mixture, 10 μL was pipetted onto a pre-cleaned 5 mm diameter glassy carbon electrode. The coated electrode was dried at 75 °C for 10 minutes and then immersed in the oxygen saturated sulfuric acid solution. The LSV, CV and RDE experiments were then performed. This procedure allowed for reproducible composite films to be electrochemically analyzed. Representative preliminary results of the LSV electrochemical analysis of the reduction of oxygen at a selective series of composite materials are shown in Figure 2.

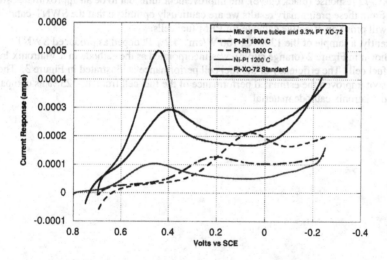

Figure 2: Preliminary voltammetric analysis of oxygen reduction in 1M sulfuric acid saturated with oxygen at the SWNT-catalyst composite electrodes and a Vulcan XC-72, 9.3% w/w Pt standard.

The preliminary results shown in Figure 2 illustrate that both the current response and potential for the onset of oxygen reduction are affected by the as-produced SWNT-catalyst materials as compared to a common standard fuel cell catalyst material. First, the potential at which oxygen reduction occurs for the various samples was found to occasionally shift to less catalytically favorable, more negative potentials. A correlation was found between this potential shift to the amount of encapsulated metals that are observed in the TEM analyses. Therefore, graphitic encapsulation of the catalyst particles appears to partially reduce the catalytic activity of the metals. When ultrasonication is used to disperse the SWNT-catalyst materials into the Nafion solution, the cathodic peak potential shifts up to 300 mV positive to a potential that is comparable to the response for the Pt standard catalyst material. At this time it is not known if this improvement in catalytic reduction of oxygen is caused by a removal of the graphite encapsulation, functionalization of the SWNTs, cutting of the SWNTs and/or a debundling of the SWNTs. We are currently pursuing both microscopic and spectroscopic characterization to fully understand these phenomena.

In order to properly analyze the amperometric response of the oxygen reduction reaction, it is necessary to normalize the current density to the amount of catalyst within the carbon-catalyst matrix. It is important to note that in Figure 2, the voltammetric responses were not normalized. When this calculation is performed, a significantly enhanced response occurs for materials that contained both SWNTs and catalyst particles. For example in Figure 2; the black curve represents the voltammetric response for a standard that contained 9.3% w/w platinum doped Vulcan XC-72. When this material was diluted 50% by weight with pure SWNTs, (>95% w/w), the response is illustrated by the red curve within Figure 2. It is evident that a shift to a more positive potential occurs for the cathodic peak and that the amperometric response approximately doubled for a material that contained 50% less platinum by weight. Furthermore, in the analysis of the normalized SWNT-catalyst amperometric response (orange curve) in Figure 2 to the standard XC-72 response (black curve), the improvement turns out to be an approximate 5-fold increase. From these preliminary results we are cautiously optimistic that the SWNT-catalyst materials will prove an effective cathode for PEM fuel cells.

Recently, a sample of the 1200 °C, 100 W/cm^2, 4.2% Pt doped as produced SWNT material shown in Figure 2 (orange curve), was incorporated as the cathode in a Viatronix Inc. benchtop fuel cell. The efficiency of the fuel cell performance is illustrated in Figure 3. The as-produced soot improved the estimated performance of the fuel cell from 40-125 % as compared to standard Viatronix cathode material.

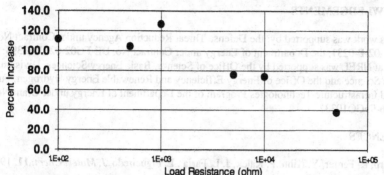

Figure 3: Preliminary results show the percent increase in power performance for as-produced laser-synthesized SWNT-supported 4.2%w/w platinum catalyst as compared to the standard cathode in a Viatronix prototype fuel cell.

CONCLUSIONS

In conclusion we have found that the use of as-produced SWNT-supported catalyst materials can greatly enhance the performance of PEM fuel cells. With the as-synthesized SWNT soots, the nanotubes are in intimate contact with the 2-50 nanometer sized catalyst particles. This apparently results in a synergistic interaction between the nanotubes and the catalyst particles that improves the catalytic behavior of the catalyst towards the reduction of oxygen. In this preliminary study the as-produced SWNT materials increased amperometric response for the reduction of oxygen five fold, while maintaining the oxygen reduction potential at appropriate values. However, it is possible that encapsulation of the catalyst by graphite, which can occur during laser synthesis, may be detrimental to the catalytic nature SWNT-catalyst materials. Therefore, care must be taken during the synthesis of the SWNT-catalyst materials to ensure graphitic encapsulation of the catalyst materials is minimized. This can be achieved experimentally by limiting the power density of the incident laser beam between 75-100 W/cm^2. Preliminary ultrasonication experiments have shown an improvement in the catalytic response of nanotube materials by possible disruption of the metal encapsulation, debundling of the SWNTs and/or shortening the SWNTs. Interestingly, the simple mixing of purified nanotubes with standard catalyst materials increased the amperometric response of the Vulcan XC-72 standard materials two-fold. Finally, independent analysis of the as-produced SWNT supported catalyst composites in a benchtop standard fuel cell showed improved performance of 40-140% over standard catalyst materials.

ACKNOWLEDGEMENTS

This work was supported by: the Defense Threat Reduction Agency under Contract No: DTRA01-02-P-0231; the Department of Energy under Contract No: DE-FG02-02ER63393: and the work at NREL was supported by the Office of Science, Basic Energy Sciences, Division of Materials Science and the Office of Energy Efficiency and Renewable Energy Hydrogen, Fuel Cells, and Infrastructure Technologies Program of the Department of Energy under Grant No. DE-AC36-99GO10337.

REFERENCES

[1] P. Serp, R. Feurer, Y. Kihn, P. Kalck, J. L. Faria . L. Figueiredo, *J. Mater. Chem.*,**11**, 1980, (2001)

[2] E. S. Steigerwalt, G. A. Deluga, D. E. Cliffel, and C. M. Lukehart, *J. Phys. Chem. B*, **105**, 8097, (2001)

[3] S. H. Joo, S. J. Choi, I. Oh, J. Kwak, Z. Liu, O. Terasaki, R. Ryoo, *Nature*, **412**, 169, (2001)

[4] T. Guo, P. Nikolaev, A. Thess, D.T. Colbert and R. E. Smalley, *Chem. Phys. Lett.*, **243**, 49 (1995).

[5] S. Bandow, S. Asaka, Y. Saito, A. M. Rao, L. Grigorian, E. Richter, P.C. Eckland, *Phys. Rev. Lett.* **80**, 3779-82, (1998).

[6] O. Jost, A.A. Gorbunov, W. Pompe, T. Pichker, R. Friedlein, M. Knupfer, M. Reibold, H. D. Bauer, L. Dunsch, M.S. Golden, J. Fink, *Appl. Phys. Lett* **75**, 2217-19, (1999).

[7] H. Katura, y. Kumazawa, Y. Maniwa, Y. Ohtsuka, R. Sen, S. Suzuki, Y. Achiba, *Carbon* **38**, 1691-97, (2000).

[8] E. Munoz, W.K. Maser, A.M. Benito, M.T. Martinez, G.F. de la Fuente, A. Righi, E. Anglare J.L. Sauvajol, *Synthetic Metals* **121**, 1193-94, (2001).

[9] A.C. Dillon, T. Gennett, K. M. Jones, J. L. Alleman, P. A. Parilla, and M. J. Heben *Adv. Mater.*, *11*, No. 16, (1999) 1354-1358 .

[10] A.C. Dillon, P. A. Parilla, J. L. Alleman, J. D. Perkins and M. J. Heben, *Chem. Phys. Lett.* **316**, 13 (2000).

[11] T. Gennett, A.C. Dillon, J.L. Alleman, K.M. Jones, P.A. Parilla, M.J.Heben *MRS Proceedings Fall Meeting, Symposium A*, December 2000.

[12] T. Gennett, A. C. Dillon, J. L. Alleman, K. M. Jones, F. S. Hasoon, and M. J. Heben ; *Chemistry of Materials*; 2000; *12*(3); 599-601.

[13] A. C. Dillon , T. Gennett, J.L. Alleman, K.M. Jones, M.J. *MRS Proceedings Fall Meeting, Symposium A*, December 2000.

[14] P. A. Parilla, A. C. Dillon , T. Gennett, , J. L. Alleman, and, M. J. Heben. *MRS Proceedings Fall Meeting, Symposium A* December 2000.

Mat. Res. Soc. Symp. Proc. Vol. 756 © 2003 Materials Research Society FF5.7

Preparation and Characterization of Metal Sulfide Electro-catalysts for PEM Fuel Cells

Hua Zhang[1], Ysmael Verde-Gómez[3] and Allan J. Jacobson[1]
Alejandra Ramirez[2] and Russell R. Chianelli[2]
[1]Department of Chemistry, University of Houston
Houston, TX 77204
[2]Materials Research and Technology Institute, University of Texas at El Paso,
El Paso, TX 79968
[3]Centro de Investigacion en Materiales Avanzados,
Chihuahua, Chihuahua, 31109, Mexico.

ABSTRACT

Ruthenium sulfide samples were prepared by flowing pure hydrogen sulfide into an aqueous solution of ruthenium chloride followed by further sulfidation in hydrogen sulfide. The final products were characterized by X-ray diffraction and crystallite-sizes were estimated from line broadening. The specific surface areas of catalysts were measured using the multipoint BET method and compositions were determined by thermogravimetric analysis. Ruthenium sulfide loaded gas diffusion electrodes were fabricated by a spraying technique and their electrochemical behavior studied. The electrochemical oxidation of hydrogen was investigated in a three - electrode cell using a ruthenium sulfide loaded gas diffusion electrode as the working electrode with humidified hydrogen containing small amounts of carbon monoxide. Results on the activity and the effects of carbon monoxide with reference to a standard platinum electrode measured at the same conditions show that ruthenium sulfide has a lower activity for hydrogen oxidation but is not susceptible to CO poisoning.

INTRODUCTION

Polymer Electrolyte Membrane Fuel Cells (PEMFCs) are increasingly used for many purposes, from power generation to portable applications and for electric vehicle applications. Platinum or other Pt alloys are used as electro-catalysts, on both anode and cathode sides. For transportation applications, the fuel source is hydrogen produced by reforming or partial oxidation of hydrocarbons. The reformate gases contain impurities (CO and S) which severely poison Pt electro-catalysts and which must, therefore, be reduced to low levels before entering the fuel cell. The development of an alternative CO tolerant electro-catalyst to platinum has been an important aim in order to reduce cost. The carbon monoxide in the reformate fuel must be reduced to 10 ppm by water gas shift for fuel cell operation, which increases the cost of fuel processing and complicates the overall fuel cell system. Several metal alloy fuel cell electro-catalysts with CO tolerance have been developed, such as Pt-Ru [1], Pt-Sn [2], Pt-Bi [3] Pt-Pd [4], and Pt-Mo [5]. These examples all use Pt as hydrogen catalyst and consequently it is of some interest to investigate the possibility of other types of efficient hydrogen and oxygen electro-catalysts with carbon monoxide and sulfur tolerance.

Catalysts based on transition metal sulfides, for example, ReRuS, MoRuS [6] and RuS_x [7] have been characterized and investigated as catalysts for molecular oxygen reduction in acid media, due to their selectivity in the presence of methanol. On the other hand, ruthenium oxide has been found to be a good catalyst not only for hydrogen evolution [8, 9], but also for hydrogen oxidation in acid media [10]. The catalytic activity of ruthenium sulfide for hydrogen dissociation is well known [11], and consequently it would seem to be a good choice to investigate as a fuel cell electro-catalyst. The present work has focused on ruthenium sulfide as a possible anode catalyst for PEMFCs. The catalyst preparation, its physical and chemical properties and crystallographic characterization are presented. Catalyst electrochemical performance using gas diffuse electrodes in presence of pure H_2 and H_2 with CO is also reported.

EXPERIMENTAL

Preparation of Ruthenium Sulfide

Ruthenium sulfide samples were prepared by bubbling H_2S for 3 h into aqueous 0.3~0.4 M ruthenium chloride solution with constant stirring at 70 ºC. The black precipitates were filtered, washed by acetone and dried under vacuum at 80 ºC overnight. The products were then heated in H_2S or H_2-H_2S at 400ºC, and finally heat-treated at different temperatures under argon for 3 h.

Characterization

Powder diffraction patterns were obtained by using a Scintag XDS 2000 powder X-ray diffractometer and CuK$_\alpha$ radiation (λ=1.5406 Å). The intensity data were collected at a scan rate of 2 º/min in the range 70 º $\geq 2\theta \geq 5$ º. The diffraction data show no apparent evidence for either ruthenium metal or ruthenium oxide. The crystallite-size of the ruthenium sulfide particles was estimated from the line broadening of X-ray pattern using the Scherrer equation.

The value of the sulfur content in the ruthenium sulfide samples was determined by thermogravimetric analysis (HI-Res TGA 2950 Thermogravimetric Analyzer, TA Instruments). Approximately 20 mg of sample was placed in a platinum boat and then heated at a rate of 2 °C/min to a maximum temperature of 650 °C. The sample was kept at 650 °C for 6 h until its weight was constant.

The specific surface area of ruthenium sulfide was measured with a NOVA 2000 (Quantachrome Instruments) using the multipoint BET method. The sample was degassed at 280 ºC overnight under vacuum before analysis. Nitrogen was used as the adsorbate. The data was acquired in the P/P_0 range of 0 to 0.1 and the equilibrium time for every P/P_0 point was at least 1 minute.

Electrode Preparation

The three-layer electrodes were prepared by a spraying procedure. A substrate of carbon paper (Toray, TGPH-090, E-TEK) with a diffusion layer was dried at 80°C for 1 h. Catalyst ink was formed by mixing an appropriate amount of ruthenium sulfide, carbon powder (Vulcan XC-72R, Cabot), Nafion solution (5 wt.%, Aldrich) and glycerol with ethanol as a solvent. The suspension was stirred in an ultrasonic bath for 30 min. The homogeneous ink was sprayed onto the weighed composite substrate with an airbrush to produce the catalyst layer. The electrode was then dried for 30 min at 80 °C and cooled down to room temperature in a desiccator. The catalyst loading was determined from the weight of the final electrode.

A Pt/C electrode was made using same procedure described above for comparison with ruthenium sulfide electrodes. The catalyst ink was obtained by mixing 10% Pt/C electro-catalyst

(E-TEK), Nafion solution, glycerol and ethanol. The Pt loading of electrode was 0.17 mg/cm^2. The hydrogen oxidation behavior of the Pt/C electrode made in house was compared with a commercial Pt/C electrode from E-TEK, and showed the same performance.

Electrochemical measurements

The electrochemical measurements were carried out in a conventional three-electrode cell at 25 °C shown as Figure1. The experimental cell was a Teflon holder divided into a gas compartment and a solution compartment by the gas diffusion electrode (working electrode). The area of the gas diffusion electrode exposed to the gas and the solution was 7.55 cm^2. The gas compartment was purged with pure H$_2$ or H$_2$ with 100 ppm CO during the measurements. Gold gauze was used as the current collector for the gas diffusion electrode. A flat platinum electrode was used as the counter electrode. A saturated calomel reference electrode was connected to the cell solution compartment through a Luggin

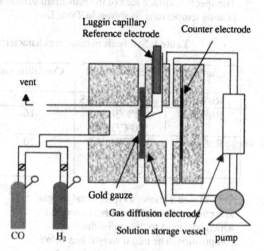

Figure1. Schematic illustration of the experimental setup

capillary whose tip was placed as close as possible to the working electrode surface. The cell at the edge of the gas diffusion electrode and the counter electrode was sealed by O-ring. The electrolyte was 0.1 M H$_2$SO$_4$, which was recycled through a solution storage vessel.

The electrochemical cell was connected to either Solartron SI1287 or Arbin MSTAT potentiostats. Polarization curves for the hydrogen oxidation were recorded at scanning rate of 50 mVs^{-1}.

RESULTS AND DISCUSSION

Ruthenium sulfide was prepared by precipitation from an aqueous solution of ruthenium chloride by flowing pure H$_2$S. The precursor was heated at 400°C under pure H$_2$S or mixture of H$_2$-H$_2$S atmosphere for sulfidation and further heated in argon at 500~700 °C. The powder XRD patterns of the ruthenium sulfide samples obtained in each step are shown in Figure 2. The X-ray data from the reaction product after treatment with H$_2$S at 400°C shows the onset of crystallization. The degree of crystallinity increases with increasing heating

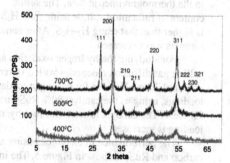

Figure 2. Powder x-ray diffraction patterns for ruthenium sulfide

temperature. The lattice parameter was estimated to be 5.606 A in agreement with the literature [12]. The crystallite-sizes of the ruthenium sulfide prepared at different annealing temperatures are listed in Table I, which indicates the increase in crystallite-size with the heating temperature. The specific surface area of the ruthenium sulfide samples decreases with the increase in the heating temperature as shown in Table I.

Table I. Synthesis parameters characterization of ruthenium sulfide powders.

Sample	Heat treatment	Crystallite size (nm)	Composition (S:Ru)	Specific surface area (m^2/g)
RuS_2-#1	400ºC/2h/H_2/H_2S	14	2.07:1	60
RuS_2-#2	400ºC/3h/H_2S 500ºC/3h/Ar	16	2.31:1	47
RuS_2-#3	400ºC/3h/H_2S 500ºC/3h/Ar 700ºC/3h/Ar	20	2.16:1	24

Figure 3 shows a typical total weight loss curve measured for ruthenium sulfide, which was used to determine the composition. The initial weight loss below 200 °C is most probably due to loss of water and free sulfur. The small weight gain between 300 and 450 °C is due to sulfate formation, and the large weight loss above 450 °C is due to loss of sulfur and the formation of ruthenium oxide. The sulfur to ruthenium mole ratio based on results of the thermogravimetric analysis show in table I. The types of gas used in sulfidation step have an effect on sulfur/ruthenium ratio according to the thermogravimetric data. The sulfur content of ruthenium sulfide using pure H_2S

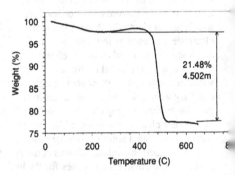

Figure 3. Thermogravimetric curve for ruthenium sulfide heated in argon at 2ºC/min

is higher than that using H_2-H_2S. After annealing under argon, the sulfur/ruthenium ratio of samples was decreased.

For studying the hydrogen oxidation activity of ruthenium sulfide, a series of ruthenium sulfide gas diffusion electrodes with different catalyst loadings and catalyst/C ratio were prepared. The I-V curve for hydrogen oxidation using electrodes with two different catalyst loadings, using as the catalyst 80% RuS_2 and 20% carbon under pure hydrogen are shown in Figure 4. The catalyst is active for hydrogen oxidation and the current is higher when the RuS_2 loading is increased.

The performance of electrodes using a catalyst loading ~0.4 mg/cm^2 with different ratios of carbon and RuS_2 is show in Figure 5. The increase in the amount of carbon in the catalyst layer has a positive effect, increasing the hydrogen oxidation rate. The increase is most likely due to improved electrical conductivity of the porous electrode [10].

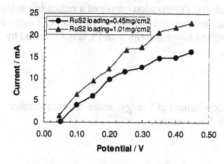

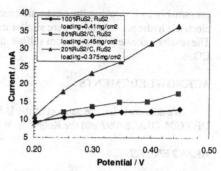

Figure 4. Polarization curves for hydrogen oxidation on RuS_2 electrodes with different loadings in 0.1 M H_2SO_4

Figure 5. Polarization curves for hydrogen oxidation on RuS_2 electrodes with different catalyst/C ratio in 0.1 M H_2SO_4

The polarization properties for electrochemical oxidation of hydrogen on ruthenium sulfide electrode under pure H_2 and H_2 containing 100 ppm CO are shown in Figure 6. For comparison, the performance of a Pt/C electrode is also included in Figure 6. The results show that the performance of Pt/C deteriorates markedly at presence of 100 ppm CO while that of RuS_2 appears almost the same for hydrogen oxidation in the presence and absence of 100 ppm CO in H_2. It is evident that ruthenium sulfide has an improved CO tolerance over the Pt catalyst. The ruthenium sulfide electrodes, prepared to date, however, have much lower activity for hydrogen oxidation in comparison to the Pt/C electrodes. Further modifications to the material aimed at improving the activity are in progress.

CONCLUSIONS

RuS_2 catalysts were synthesized by precipitation from an aqueous solution of ruthenium chloride by flowing pure H_2S followed by a high temperature sulfidation. The synthesis product was shown by XRD to be a single phase of RuS_2. TGA data show that the ruthenium to sulfur ratio is dependent of the type of gas used during the sulfidation heat treatment. The sulfidation temperature affects the specific surface area and particle size of RuS_2.

Electrochemical experiments reveal that RuS_2 is active for H_2 oxidation and that electrodes with higher catalyst loadings are more active. Carbon powder in the catalyst layer has positive effect on electrode performance

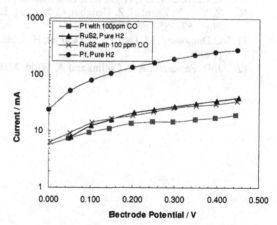

Figure 6. Polarization curves for H_2 oxidation on RuS_2 and Pt electrodes in 0.1M H_2SO_4 with 100 ppm $CO-H_2$. RuS_2 loading: 0.72mg/cm^2

presumably by improving the electronic conductivity. Polarization curves of a ruthenium sulfide electrode hydrogen oxidation were similar in the presence and absence of 100 ppm CO in the H_2. The results indicate that ruthenium sulfide, though of lower activity than Pt, is not affected by CO.

ACKNOWLEDGMENTS

We acknowledge the support of the U.S. Department of Energy, under contract number DE-FC04-O1AL67097 and the Robert A. Welch Foundation.

REFERENCES

1. H.F. Oetjen, V.M. Schmidt, U. Stimming, F. Trila, J. Electrochem. Soc. **143**, 3838-3842 (1996).
2. H.A. Gasteiger, N.M. Markovic, P.N. Ross, Technical Report LBNL-40196, Ernest Orlando Lawrence Berkeley National Laboratory, (1997).
3. B.E. Hayden, Catal. Today **38**, 473-481 (1997).
4. D.C. Papageorgopoulos, M. Keijzer, J.B.J. Veldhuis, F.A. de Bruijn, J. Electrochem. Soc. **149**, A1400-A1404 (2002).
5. H. Igarashi, F. Takeshi, Y. Zhu, H. Uchida, M. Watanabe, Phys. Chem. Chem. Phys. **3**, 306-314 (2001).
6. R.W. Reeve, P.A. Christensen, A. Hamnett, S.A. Haydock and S.C. Roy, J. Electrochem. Soc. **145**, 3463-3471 (1998).
7. N. Alonso-Vante, I.V. Malakhov, S.G. Nikitenko, E.R. Savinova, D.I. Kochubey, Electrochim. Acta **47**, 3807-3814 (2002).
8. S. Ardizzone, G. Fregonara, S. Trasatti, J. Electroanal. Chem. **266**, 191-195 (1989).
9. L. Chen, D. Guay, A. Lasia, J. Electrochem. Soc. **143**, 3576-3584 (1996).
10. Z. Yue. Y. Zhanhui, Z. Hengbin, C. Xuejing, L. Se, L. Shujia, S. Chiachung, Mat. Chem. Phys. **57**, 285-288 (1999).
11. C.Dumonteil, M. Lacroix, C. Geantet, H. Jobic and M. Breysse, J. Catal. **187**, 464-473 (1999).
12. J. D. Passaretti, R.C. Collins and A. Wold, Mat. Res. Bull. **14**, 1167-1171(1979).

Mat. Res. Soc. Symp. Proc. Vol. 756 © 2003 Materials Research Society

Characterizations of Core-Shell Nanoparticle Catalysts for Methanol Electrooxidation

Mathew M. Maye, Jin Luo, Wai-Ben Chan, Li Han, Nancy Kariuki, H. Richard Naslund [a], Mark H. Engelhard [b], Yuehe Lin [b], Randoll Sze, and Chuan-Jian Zhong*
Department of Chemistry, State University of New York at Binghamton, Binghamton, NY 13902. [a] Department of Geological Sciences, State University of New York at Binghamton, Binghamton, NY 13902. [b] Environmental and Molecular Science Laboratory, Pacific Northwest National Laboratory, Richland, WA, 99352.

ABSTRACT

This paper describes the results of an investigation of the structure and composition of core-shell gold and alloy nanoparticles as catalytically active nanomaterials for potential fuel cell catalysis. Centered on the electrocatalytic methanol oxidation, we show three sets of results based on electrochemical, surface, and composition characterizations. First, electrochemical studies have revealed that the nanostructured catalysts are active towards the electrooxidation of methanol and carbon monoxide. Second, X-ray photoelectron spectroscopy (XPS) data have shown that the organic encapsulating shells can be effectively removed electrochemically or thermally, which involves the formation of oxides on the nanocrystals. Thirdly, direct current plasma - atomic emission spectrometry (DCP-AES) has revealed insights for the correlation of the composition of alloy nanoparticles with the catalytic activities. Implications of these results to the design of nanostructured catalysts will also be discussed.

INTRODUCTION

Catalysis plays a vital role in chemical processing, environmental protection and fuel cell technology. The use of gold nanoparticles as catalysts has attracted increasing interest [1-3], since the pioneer work of Haruta [4], which has demonstrated high catalytic activities for CO and hydrocarbon oxidation at gold nanoparticles of less than ~10 nm diameter that are supported on oxides. We have recently shown that the preparation of nanoscale gold catalysts can be achieved using core-shell assembled nanoparticles that consist of metal or alloy nanocrystal cores and organic molecular wiring or linkage to define the interparticle spatial property [5-6]. This approach is important in addressing fundamental issues related to size, shape, aggregation, poisoning, and surface engineering of nanoparticles in fuel cell catalysis.

The focus of this work is to characterize the catalytic structure and composition at gold and alloy nanoparticle assemblies using electrochemical, XPS and DCP-AES techniques. We studied decanethiolate-capped gold and alloy nanoparticles of 2-nm (Au$_{2-nm}$) and 5-nm (Au$_{5-nm}$) core sizes assembled on planar substrates using 1,9-nonanedithiol (NDT) and 11-mercaptoundecanoic acid (MUA) as molecular wiring agents as a model system. XPS was employed to detect the identity of active surface species and to analyze the elemental composition or oxidation states of the nanomaterials, from which we derive structural information about the surface reconstitution of the core-shell nanostructured catalysts. DCP was utilized to analyze the composition of the alloy nanoparticles, which is important in correlating the alloy composition with the catalytic activities.

EXPERIMENTAL

Chemicals. Major chemicals included decanethiol (DT, 96%), 1,9-nonanedithiol (NDT, 95%), 11-mercaptoundecanoic acid (MUA, 98%), hydrogen tetracholoroaurate (HAuCl$_4$, 99%), tetraoctylammonium bromide (TOABr, 99%), sodium borohydride (NaBH$_4$, 99%), toluene (99.8%), hexane (99.9%), and methanol (99.9%). All chemicals were purchased from Aldrich and used as received. Water was purified with a Millipore Milli-Q water system.

Synthesis. The synthesis of DT-capped gold, gold-platinum, gold-nickel, and gold-iron nanoparticles of 2~3 nm core size followed the two-phase protocol [7,8]. DT-capped gold nanoparticles with a core size of ~5-nm (Au$_{5-nm}$) were also studied, which were produced by a thermally-activated processing route [9].

Thin-Film Preparation. Details of the preparation of the molecularly-linked gold nanoparticle films were described in previous reports [10], which involves an exchange-crosslinking-precipitation route. Briefly, NDT or MUA were used as molecular linker agent. A substrate electrode (glassy carbon, HOPG, or graphite) was immersed into a hexane solution of DT-capped nanoparticles (~5 μM) and molecular linker ([NDT] ~ 25 μM or [MUA] ~ 0.3 μM) for 5-10 hours. The thickness of the resulting nanoparticle thin films was controlled by immersion time. The films were thoroughly rinsed with the solvent and dried under argon before characterizations.

Instrumentation. Electrochemical measurements were performed using an EG&G Model 273A potentiostat. A three-electrode cell was employed, with a platinum coil as the auxiliary electrode and a standard calomel electrode (SCE) as the reference electrode.

The XPS measurements were made using a Physical Electronics Quantum 2000 Scanning ESCA Microprobe. This system uses focused monochromatic Al K$_\alpha$ x-rays (1486.7 eV) source for excitation and a spherical section analyzer. The collected data were referenced to an energy scale with binding energies for Cu 2p$_{3/2}$ at 932.67 ± 0.05 eV and Au 4f$_{7/2}$ at 84.0 ± 0.05 eV.

The composition analysis was performed using an ARL Fisons SS-7 Direct Current Plasma - Atomic Emission Spectrometer (DCP-AES) housed in the Department of Geological Sciences on the SUNY-Binghamton campus. Measurements were made on emission peaks at 267.59 nm, 265.95 nm, and 589.59 nm for Au, Pt, and Na, respectively. The nanoparticle samples were dissolved in concentrated aqua regia, and then diluted to concentrations in the range of 1 to 50 ppm for analysis. Na$^+$ was added into the sample solution as an internal standard. Calibration curves were made from dissolved standards with concentrations from 0 to 50 ppm in the same acid matrix as the unknowns. Detection limits, based on three standard deviations of the background intensity, are 0.008 ppm, 0.02 ppm, and 0.003 ppm for Au, Pt, and Na, respectively. Standards and unknowns were analyzed 10 times each for 3 second counts. Instrument reproducibility, for concentrations greater than 100 times the detection limit, results in <± 2% error.

RESULTS AND DISCUSSION

1. Electrocatalytic Characterization

We used a cyclic voltammetric (CV) technique to characterize the electrocatalytic activity towards methanol oxidation at electrodes coated with nanoparticles which were assembled using two types of molecular linkers, NDT and MUA. In the NDT case, the linkage involves Au-thiolate bonding at both ends of NDT [10a]. In the MUA case, the MUA-linking involves an Au-thiolate bonding at the -SH end and a head-to-head hydrogen-bonding at the $-CO_2H$ terminal group [10b]. Upon catalytic activation by either continuous cycling of the potential between -400 and +800 mV or thermal annealing at 200-350 °C, a large oxidation wave ($\sim$+300mV) was observed for both NDT- and MUA-linked gold nanoparticle catalysts, and attributed to methanol oxidation [5,6]. We have also observed electrocatalytic activities using alloy nanoparticles such as Au/Ni, Au/Pt, Au/Fe, etc. Overall, the catalytic activity is dependent on the method of activation and the nanoparticle core composition.

Table 1. Electrocatalytic properties of several activated catalysts on GC electrodes in 0.5 M KOH + 3 M methanol (from CV data at 50 mV/s).

Catalyst	(#eq. layers)	Electrooxidation of Methanol	
		i_{pa} ($\mu A/cm^2$)	E_{pa} (mV vs SCE)
[a] NDT-Au$_{2-nm}$	5	143	+270
[a] MUA-Au$_{2-nm}$	5	357	+250
[a] NDT-AuFe$_{2-nm}$	11	328	+290
[a] NDT-AuPt$_{2-nm}$	8	214	+280
[a] MUA-Au$_{2-nm}$	5	82	+270
[b] MUA-Au$_{2-nm}$	5	166	+250

a) electrochemically activated; b) thermally activated.

Interestingly, the activity was also found to be dependent on the nature of the molecular linkers, which is reflected by differences in E_{pa} and i_{pa} between catalysts assembled using NDT and MUA linkers (first two entries in Table 1). Both films have a similar thickness ($\sim$5 layers) and were activated after continuously cycling between -400 and +800 mV until a constant methanol oxidation wave was observed. The difference in catalytic activity between NDT and MUA-linked nanoparticles is intriguing because it demonstrates the influence of the linker molecular structure on the final catalytic activity of the nanoparticle core. The increased activity is likely due to the presence of a large open framework for easy access of alkaline electrolytes and the higher solubility of MUA than NDT upon potential-induced desorption. To further probe the catalytic properties, especially the catalytic activation and alloy nanoparticles, we next studied the compositional and surface properties of the nanostructured catalysts using XPS and DCP-AES.

2. Structural and Composition Characterizations

The activation-induced structural change was examined using XPS. Figures 1 shows a representative set of XPS spectra for the NDT-Au$_{2\text{-nm}}$ in the S(2p) region. The as-prepared film (*a*), electrochemically-activated film (*b*), and thermally-activated film (*c*), are compared in each spectral region. The result of relative compositional analysis is shown in Table 1.

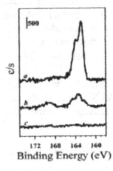

Figure 1. S(2p) XPS spectra in for NDT-Au$_{2\text{-nm}}$ as-prepared (*a*), electrochemically activated (*b*), and thermally activated (*c*).

Binding Energy (eV)

The S(2p) region (Figure 2 A) is characterized by a doublet that arises from spin-orbit coupling (2p$_{3/2}$ and 2p$_{1/2}$). For most neat thiols, this region is generally defined by the more intense 2p$_{3/2}$ band which lies between 163 and 165 eV. In contrast, the binding energy (BE) observed for monolayers derived from thiols in which the sulfur species interact strongly with the surface gold are ~1 eV lower, i.e., ~162 eV for the 2p$_{3/2}$ band [8,11]. After electrochemical activation, the S(2p) band intensity is significantly reduced, but still detectable, indicating a partial removal of the thiolates species. The detection of the small band at 169.2 eV is indicative of the presence of sulfonate species (-SO$_3^-$) [11]. Upon thermal-activation of a similar NDT-Au$_{2\text{-nm}}$ electrode, no sulfur was detected (Table 1).

In addition, a significant increase of the O(1s) band is detected at 532.4 eV after the activation. The peak position is quite close to those observed for surface oxide species [11]. We attribute it to the formation of oxygenated species (oxides) on the surface of gold nanocrystals. Furthermore, the spectral deconvolution of the Au(4f) region band shape and position reveals two sets of Au(4f$_{7/2}$) bands, i.e., 84.1 and 84.9 eV. The lower BE component is attributed to a combination of Au(0) and Au(I) after the partial removal of thiolates, whereas the higher BE component is likely indicative of the presence of both Au(I) and Au(III) oxides (e.g., Au$_2$O$_3$, or Au(OH)$_x$).

In Table 2, the relative changes of the sulfur and the oxygen percentages before and after the electrochemical activation provide quantitative information to assess the surface reconstitution. In the case of MUA-Au$_{2\text{-nm}}$ thin film, the removal of sulfur species upon the electrochemical activation seemed to be more effective. While the net change of the sulfur percentage (from 2.7% to 0.7%), ~74%, appears similar to the NDT case, the detected sulfur species is <1%. In view of the exchange degree of <45% recently found for the MUA-linked Au$_{2\text{-nm}}$ nanoparticles [13], we believe that the thiolates being removed include both DT and MUA molecules on gold nanocrystals. Larger particles (Au$_{5\text{-nm}}$) demonstrated less thiolate removal, and a dramatic increase in oxygenated species. These results may come from the increased surface area of the Au_{5-nm} thin film. The quantitative aspect of this assessment is however yet to be determined. The detected oxygen is largely from the surface oxides of the gold nanocrystals.

Table 2. XPS-determined relative surface composition for nanoparticle films on HOPG.

Catalyst	Treatment	%S	%O
NDT-Au$_{2-nm}$	a	7.7	0.9
	b	2.2	10.4
	c	0.0	1.4
MUA-Au$_{2-nm}$	a	3.0	1.9
	b	0.3	2.6
NDT-Au$_{5-nm}$	a	5.4	2.0
	b	1.9	13.6

a) as-prepared, b) electrochemically activated; c) thermally activated.

We further analyzed several alloy nanoparticle catalysts using DCP-AES for the correlation of the electrocatalytic activity with the alloy composition. Table 3 shows data for Au/Pt nanoparticle systems.

Table 3. DCP-AES analysis results and electrocatalytic activity for several NDT-linked alloy (M_1-M_2) nanoparticle (NP) catalysts upon thermal activation

M_1/M_2 Alloy Nanoparticles			Film Preparation			Electrooxidation of Methanol			
M_1/M_2	M_1/M_2 feed ratio	M_1/M_2 product ratio	NDT/NP ratio	# layers	Scan rate (mV/s)	Wave-A [a]		Wave-B [a]	
						E_{pa} [b] (mV)	i_{pa} (μA/cm^2)	E_{pa} [b] (mV)	i_{pa} (μA/cm^2)
Au	-	-	732 : 1	~1	50	250	92	-	-
					200	350	136	-	-
						220	118		
Au/Pt [c]	5 / 1	99 / 1	200 : 1	~8	50	310	200		
					200	330	360		
Au/Pt	2.5 / 1	88 / 12	8761 : 1	~10	50	300	158	-20	50
					200	330	260	-70	220
						260	224		
Au/Pt	1 / 1	73 / 27	51563 : 1	~10	50	320	148	-50	912
					200	370	238	-50	996
						300	214		

a) Wave-A is the oxidation wave closely-related to that on Au component, whereas wave-B is the wave related to that on Pt component; b) vs. Ag/AgCl, Sat'd KCL; c) Electrochemically-activated

It is evident that the Au vs. Pt ratio in the alloy nanoparticles increases with the feed ratio of $AuCl_4^-$ vs. $PtCl_4^-$ in the synthesis. The alloyed Pt component is likely distributed in both the core and surface, as evidenced by the increased current density for wave-B as the Pt component is decreased. This observation is very important because it demonstrates that the catalytic activity can be tailored by manipulating the composition of the nanocrystal cores. A further understanding of the quantitative correlation of both the peak potentials and current densities for the electrocatalytic oxidation of methanol with the alloy composition is in progress.

CONCLUSION

Two important conclusions have been arrived at for the understanding of the catalytic activity and activation of the molecularly-linked gold and alloy nanoparticle catalysts. First, the electrochemical or thermal activation leads to a partial opening of the capping/linking nanostructure as a result of the removal of thiolates. The thiolate removal is accompanied by the formation of surface oxides on the gold nanocrystals. Second, the catalytic activity of the alloy nanoparticles consists of contributions from both alloy components. Each of the contributions can be systematically manipulated by synthetic feed ratios. The precise assessment of the catalytic mechanism involving each of the individual components or their combination is yet to be determined. In view of the presence of the trace amount of thiolates in the activated thin film of gold nanoparticles, an important question is how it affects the catalytic activity. Recent work using refined electrochemical, thermal, and other catalytic activation routes have demonstrated the viability of a complete removal of the thiolates. Research in this direction is in progress to help unravel mechanisms for the nanostructured gold catalysts.

ACKNOWLEDEMENT

Financial support of this work from the Petroleum Research Fund administered by the American Chemical Society and the 3M Corporation is gratefully acknowledged. M.M.M. thanks the Department of Defense (Army Research Office) for support via the National Defense Science & Engineering Graduate Fellowship. We thank Dr. S. Madan for donating metal compounds. The work was partially performed at the Environmental Molecular Sciences Laboratory, a national scientific user facility sponsored by U.S. DOE's Office of Biological and Environmental Research and located at PNNL. PNNL is operated for the Department of Energy by Battelle.

REFERENCES:

1. G.C. Bond, D. T. Thompson, *Gold Bulletin*, 33, 41 (2000).
2. A. Sanchez, S. Abbet, U. Heiz, W.-D. Schneider, H. Hakkinen, R.N. Barnett, U. Landman, *J. Phys. Chem. A*, 103, 9573 (1999). (b) J.W. Yoo, D. Hathcock, M.A. El-Sayed, J. Phys. Chem. A, 106, 2049 (2002). (c) W.T. Wallace, R.L. Whetten, *J. Am. Chem. Soc.*, 124, 7499 (2002).
3. G. Schmid, S. Emde, V. Maihack, W. Meyer-Zaika, S. Peschel, *J. Mol. Catal. A–Chem.*, 107, 95 (1996)
4. M. Haruta, Catal. Today, 36, 153 (1997). (b) M. Haruta, M. Date, *Appl. Catal. A–Gen.*, 222, 427 (2001). (c) M. Valden, X. Lai, D.W. Goodman, *Science*, 281, 1647 (1998).
5. M.M. Maye, Y.B. Lou, C.J. Zhong, *Langmuir*, 16, 7520 (2000). (b) Y. B. Lou, M.M. Maye, L. Han, J. Luo, C.J. Zhong, *Chem. Commun.*, 473, (2001). (c) J. Luo, Y. Lou, M.M. Maye, C.J. Zhong, M. Hepel, *Electrochem. Commun.*, 3, 172, (2001).
6. C.J. Zhong, M.M. Maye, *Adv. Mater.*, 13, 1507 (2001). (b) J. Luo, M.M. Maye, Y.B. Lou, L. Han, M. Hepel, C.J. Zhong, *Catal. Today*, 77, 127-138 (2002).
7. M. Brust, M. Walker, D. Bethell, D.J. Schiffrin, R. Whyman, *J. Chem. Soc., Chem. Comm.*, 801, (1994).
8. M.J. Hostetler, C.J. Zhong, B.K.H. Yen, J. Anderegg, S.M. Gross, N.D. Evans, M.D. Porter, R.W. Murray, *J. Am. Chem. Soc.*, 120, 9396 (1998).
9. (a) M.M. Maye, W.X. Zheng, F.L. Leibowitz, N.K. Ly, C.J. Zhong, *Langmuir*, 16, 490 (2000). (b) M.M. Maye, C.J. Zhong, *J. Mater. Chem*, 10, 1895 (2000).
10. (a) F.L. Leibowitz, W.X. Zheng, M.M. Maye, C.J. Zhong, *Anal. Chem.*, 71, 5076, (1999). (b) L. Han, M.M. Maye, F.L. Leibowitz, N.K. Ly, C.J. Zhong, *J. Mater. Chem.*, 11, 1258 (2001).
11. M.C. Bourg, A. Badia, R.B. Lennox, *J. Phys. Chem. B*, 104, 6562 (2000).
12. C.D. Wagner, W.M. Riggs, L.E. Davis, J.F. Moulder, G.E. Muilenberg, *Handbook of X-ray Photoelectron Spectroscopy*, Perkin Elmer, Eden Prarie, (1978)
13. N.N. Kariuki, L. Han, N.K. Ly, M.J. Peterson, M.M. Maye, G. Liu, C.J. Zhong, *Langmuir*, 18, 8255-8259. (2002)

Mat. Res. Soc. Symp. Proc. Vol. 756 © 2003 Materials Research Society   FF6.4

The Effect of the Nano and Microstructure of PtRu/C Electrocatalysts Towards Methanol and Carbon Monoxide Oxidation

Wataru Sugimoto, Tomoyuki Kawaguchi, Yasushi Murakami, and Yoshio Takasu
*Department of Fine Materials Engineering, Faculty of Textile Science and Technology,
Shinshu University, 3-15-1 Tokida, Ueda 386-8567, JAPAN*

ABSTRACT

Nanoparticulate $Pt_{50}Ru_{50}$ supported on a carbon black (CB) were prepared by an impregnation method and the effect of the nano and microstructure towards the electrocatalytic oxidation of methanol and carbon monoxide were evaluated.   The in-house prepared PtRu nanoparticles were well-alloyed and finely dispersed on CB, as revealed by XRD and FESEM. FESTEM-EDX analysis of isolated PtRu single nanoparticles indicated that although the majority of the binary particles had a metal composition in accordance to the nominal ratio, some compositional deviation was observed.   The activity towards electrocatalytic oxidation of methanol and preadsorbed carbon monoxide is discussed based on the nano and microstructure.

INTRODUCTION

Carbon supported PtRu alloys are one of the most promising anode electrocatalysts for DMFC (Direct Methanol Fuel Cell) application.   Numerous investigations on the anodic characteristics of PtRu/C electrocatalysts have so far been conducted.[1-6]   In order to increase the utilization of the precious metal, many factors which should effect the electrocatalytic oxidation of methanol must be elucidated.   The influence of the homogeneity, the particle size, the bulk/surface composition of the PtRu alloy nanoparticles towards the electrocatalytic oxidation of methanol are just some of the issues that need to be clarified.   In order to examine these effects, the extent of bimetal alloying, particle size, *etc.* must be controlled, and the evaluation of the composition and crystal structure of PtRu/C electrocatalysts on the nanoscopic level are inevitable.

We have previously reported that well alloyed and highly active PtRu/C electrocatalysts can be prepared by a simple impregnation method using chlorine-free precursors such as $Pt(NO_2)_2(NH_3)_2$ and $Ru(NO)(NO_3)_x$.[7-9]   In this study, the nanoscopic and microscopic characteristics of PtRu/C electrocatalysts were evaluated and their relation to the electrocatalytic oxidation of methanol and preadsorbed carbon monoxide was studied.

EXPERIMENTAL

Carbon supported PtRu electrocatalysts were prepared by an impregnation method using carbon black (Mitsubishi Chemicals CBB; S_{BET}=200 m^2 g^{-1}) as the support and ethanolic solutions of $Pt(NO_2)_2(NH_3)_2$ and $Ru(NO)(NO_3)_x$ as metal precursors.   The alloy composition used was $Pt_{50}Ru_{50}$, (Pt/Ru = 50/50 (mole/mole)) and the metal loading was 30 mass%.

Pyrolysis was conducted under a continuous flow of $H_2(10\%) + N_2(90\%)$ at 150, 200, 300, and 450°C for 2 h.   The bulk crystal structure and microscopic features of the PtRu nanoparticles were characterized using X-ray Diffraction (XRD) and Field Emission-Scanning Electron Microscopy (FE-SEM).   The nanoscopic composition of single PtRu nanoparticles was analyzed by Field Emission-Scanning Transmission Electron Microscope equipped with Energy Dispersive X-Ray Detector (FESTEM-EDX).

Half-cell measurements of the electrocatalytic oxidation of methanol and preadsorbed carbon monoxide were examined in a three-electrode type beaker cell with a platinum gauze counter electrode.   An Ag/AgCl reference electrode was used in the experiments.   The surface area of the PtRu nanoparticles was measured electrochemically by electrocatalytic oxidation of preadsorbed carbon monoxide (CO_{ad} stripping) in 0.5 M H_2SO_4 at 60°C.   The electrocatalytic oxidation of methanol was studied by chronoamperometry in 1 M CH_3OH + 0.5 M H_2SO_4 solution at 60°C.   The current at 500 mV (*vs.* RHE) after 1800 sec was used as the quasi-steady state current.

RESULTS AND DISCUSSION

The average PtRu particle diameter of $Pt_{50}Ru_{50}$/CB was approximately 3-4 nm, and relatively insensitive to the pyrolysis temperature as revealed by HR-FESEM analysis.   Figure 1 shows the XRD patterns of the electrocatalysts pyrolyzed at different temperatures.   The bulk Pt:Ru ratio estimated from the position of the fcc(220) XRD reflection and the particle size estimated from the Scherrer equation (assuming a homogeneous alloying state) are summarized in Table I.   For the 150°C-pyrolyzed electrocatalyst, the Pt:Ru ratio was 63:37, suggesting only partial alloying, probably due to incomplete decomposition of the Ru precursor.   Since no evidence of hcp Ru was provided in the XRD pattern, an unidentified Ru-based phase, either as a nanoparticule or amorphous phase is most likely present in this electrocatalyst. .   The lattice parameter for the other electrocatalysts was in accordance with the nominal Pt:Ru ratio.

The composition of individual particles was studied by STEM-EDX.   A typical bright field STEM image of $Pt_{50}Ru_{50}$/CB pyrolyzed at 450°C is shown in Fig. 2, and the composition of single PtRu nanoparticles (some of which are numbered in Fig. 2) are plotted as a function of the particle size in Fig. 3.   The

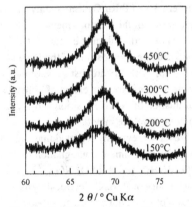

Fig. 1. The XRD patterns of $Pt_{50}Ru_{50}$/CB pyrolyzed in reducing atmosphere at different temperatures. The vertical lines represent literature values for bulk Pt (left) and PtRu (right).

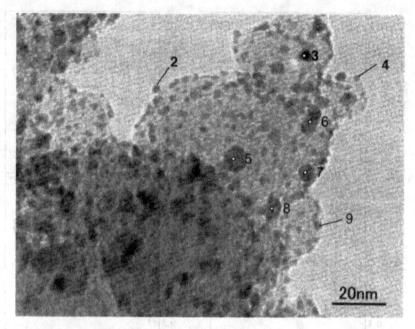

Fig. 3. A typical BF-STEM image of $Pt_{50}Ru_{50}$/CB pyrolyzed in reducing atmosphere at 450°C.

majority of the particles had composition in reasonably good agreement with the nominal composition. However, some Pt or Ru rich particles were also observed; clearly the composition of single particles varies from one another to a considerable degree under the present synthetic consitions, even by increasing the pyrolysis temperature to 450°C. Although it is difficult to correlate the particle size with the composition due to the limited number of analysis points, a tendency for small or large particles to deviate from the nominal composition was observed; some nanoparticles smaller than 2 nm were Ru rich, while those larger than 6 nm were Pt rich.

Figure 4 shows the results for the electrocatalytic oxidation of preadsorbed carbon monoxide (CO_{ad}). The coulombic

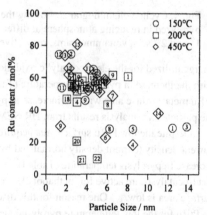

Fig. 3. The Ru content of single PtRu particles pyrolyzed in reducing atmosphere at different temperatures. Numbers represent analysis points, for example in Fig. 3.

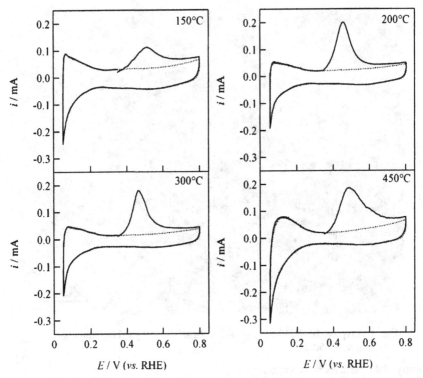

Fig. 4. Cyclic voltammograms showing the electrolytic oxidation of CO_{ad} on $Pt_{50}Ru_{50}$/CB pyrolyzed in reducing atmosphere at different temperatures. Solid and dotted lines show 1st and 2nd scan voltammograms, respectively.

charge utilized for the electrocatalytic oxidation of CO_{ad} assuming a linear adsoption increased with the increase in pyrolysis temperature (Table I).   This result suggests an increase in the PtRu metal surface area with the increase in pyrolysis temperature, which is inconsistent with the particle size analysis results from HR-SEM and XRD.

Despite the low PtRu surface area acquired from CO_{ad} stripping voltammetry, the specific current density (current density normalized by the surface area of PtRu) increased with the increase in pyrolysis temperature (Table I).   This suggests that the low temperature pyrolyzed electrocatalysts are active for methanol electrocatalytic oxidation even though the exposed metal surface area is low.   One reason for this discrepancy may be that CO_{ad} is not linearly adsorbed on PtRu for the low-temperature pyrolyzed electrocatalysts, resulting in a lower estimated PtRu surface area than the actual PtRu surface area.   Another explanation may be that non-metallic (e.g. partially oxidized amorphous ruthenium oxyhydroxides or incompletely decomposed ruthenium) species that cannot adsorb CO but are effective for providing OH groups that can

Table I. Properties of $Pt_{50}Ru_{50}/CB$

Pyrol. Temp / °C	Pt:Ru mole ratio (by XRD)	Particle Size (by SEM) / nm	Particle Size (by XRD) / nm	PtRu Surface Area / cm^2	Current Density / $\mu A \, cm^{-2}_{(PtRu)}$
150	63:37	3±2	4	3.3	230
200	48:52	N.D.	3	4.8	170
300	48:52	N.D.	5	4.5	110
450	45:55	4±1	5	6.8	80

oxidize neighboring "polluting" CO_{ad} exist in the low-temperature pyrolyzed electrocatalysts. It has been reported that partially oxidized ruthenium species may be active for methanol electrocatalytic oxidation.[10,11] Although other possibilites such as differences in surface composition, surface defects, exposed surface planes, *etc.* cannot be ruled out, it may be possible that such a species is responsible for the apparently high specific activity in our electrocatalysts. In fact, the double-layer region in the 2nd scan voltammograms (Fig. 4) shows that the electrocatalysts pyrolyzed at lower temperature tend to have a larger double-layer pseudocapacitance, which is suggestive of non-metallic ruthenium species. XPS studies of these electrocatalysts are presently under investigation.

CONCLUSIONS

The nano and microstructure of $Pt_{50}Ru_{50}/CB$ was studied by FESEM, XRD, and STEM-EDX and the effect on the electrocatalytic oxidation of methanol and carbon monoxide was evaluated. Despite the decrease in PtRu surface area for CO adsorption with the decrease in the pyrolysis temperature, an increase in specific activity towards the electrocatalytic oxidation of methanol (current density normalized by the PtRu surface area) was observed. A possible reason for this discrepancy was discussed based on capability of non-metallic Ru species to electrocatalytically oxidize neighboring CO_{ad}, although such species cannot adsorb CO.

ACKNOWLEDGEMENT

This work was supported in part by a Polymer Electrolyte Fuel Cell Program from the New Energy and Industrial Technology Development Organization (NEDO), in collaboration with Toray Industries, Inc.

REFERENCES

1) J.O'M. Bockris and H. Wroblowa, *J. Electroanal. Chem.*, **7**, 428 (1964).
2) M. Watanabe and M. Motoo, *J. Electroanal. Chem.*, **60**, 267 (1975).
3) B.D. McNicol and R.T. Short, *J. Electroanal. Chem.*, **81**, 249 (1977).
4) J.B. Goodenough, A. Hamnett, B.J. Kennedy, R. Manohara, and S.A. Weeks, *J. Electroanal. Chem.*, **240**, 133 (1988).

5) A. Hamnett, S.A. Weeks, B.J. Kennedy, G. Troughton and P.A. Christensen, *Ber. Bunsenges. Phys. Chem.*, **94**, 1014 (1990).

6) M. Neergat, D. Leveratto, and U. Stimmming, *Fuel Cells*, **2**, 25 (2002).

7) Y. Takasu, T. Fujiwara, Y. Murakami, K. Sasaki, M. Oguri, T. Asaki and W. Sugimoto, *J. Electrochem. Soc.*, **147**, 4421-4427 (2000).

8) Y. Takasu, T. Iwazaki, W. Sugimoto, and Y. Murakami, *Electrochem. Commun.*, **2**, 671-674 (2000).

9) Y. Takasu, H. Itaya, T. Iwazaki, R. Miyoshi, T. Ohnuma, W. Sugimoto and Y. Murakami,*Chem. Commun.*, **2001**, 341-342 (2001).

10) D. R. Rolison, P. L. Hagans, K. E. Swider, and J. W. Long, *Langmuir*, **15**, 774 (1999).

11) J. W. Long, R. M. Stroud, K. E. Swider-Lyons, and D. R. Rolison, *J. Phys. Chem. B*, **104**, 9772 (2000).

Mat. Res. Soc. Symp. Proc. Vol. 756 © 2003 Materials Research Society   FF8.9

Microemulsion-Templated Synthesis of
Highly Active High-Temperature Stable Partial Oxidation Catalysts

Mark Kirchhoff[1], Ullrich Specht[1], and Götz Veser[1,2,*]

[1] Max-Planck-Institut für Kohlenforschung, Mülheim an der Ruhr, Germany
[2] Department of Chemical Engineering, University of Pittsburgh, Pittsburgh PA.

ABSTRACT

Highly active catalysts for high-temperature partial oxidation reactions have been synthesized based on a microemulsion-templated sol-gel synthesis. The catalysts were tested with the direct catalytic oxidation of methane to synthesis gas and showed excellent selectitivites towards syngas combined with very high activity and low ignition temperatures. Furthermore, a surprisingly high long term stability was observed at these high-temperature conditions of T > 900°C. The catalyst therefore seem very promising candidates for high-temperature partial oxidation and hydrogen production from hydrocarbon fuels.

INTRODUCTION

Micelle-templated syntheses have found wide-spread application in materials research during the past ten years, since the self-assembling properties of surfactants offer a unique environment for the engineering of nanostructured materials [1-3]. Beyond allowing the template-directed synthesis of pore morphologies for nano- and mesoporous materials, the nanometer-sized droplets in reverse (water-in-oil) microemulsions also constitute a controlled environment for chemical reactions, in which single micelles act as individual 'nanoreactors' [4]. This has been used in recent years to produce materials with large, high-temperature stable surface areas, either as support or catalyst materials, for example through sol-gel syntheses in microemulsions [5, 6].

In this contribution, we report on the synthesis of novel high-temperature stable catalysts for partial oxidation of hydrocarbons through a microemulsion-templated synthesis route. Starting point were so-called hexa-aluminate catalysts, i.e. a group of high-temperature stabilized aluminas which have already been shown to be rather effective catalysts for high-temperature combustion processes, i.e. for total oxidation of hydrocarbons [7, 8]. We tested the suitability of microemulsion-templated hexaaluminate catalysts for the partial oxidation of hydrocarbons. However, in contrast to their high combustion activity, the synthesised catalysts showed only very poor conversions and selectivities under the fuel-rich conditions of partial oxidation. Therefore, the aim of the study was to combine the known high selectivity of noble metal components towards partial oxidation with the high-temperature stability of the hexaaluminate support and the large surface area attainable through a microemulsion-templated synthesis route.

CATALYST SYNTHESIS

The steps in the catalyst synthesis are summarized in figure 1. Starting with an inverse (i.e. water-in-oil) microemulsion by combining water, iso-octane and one of several different non-ionic surfactants (Neodol, Lutensol, and others), an aqueous solution of the respective metal salt was added to incorporate the catalytically active (noble) metal component. In a next step, a solution of metal-alkoxides in iso-octane was slowly added to the microemulsion. Here, we report results with a barium-hexaaluminate (BHA) based catalyst material, in which case Al- and Ba- isopropoxides were mixed at stoichiometric conditions for this precursor solution.

The microemulsion was aged for typically three days under constant stirring, and then a phase separation was induced through cooling of the mixture. After separating the phases, the gel was washed several times to remove the surfactant, and finally the material was vacuum dried and calcined at temperatures of up to 1300°C.

The obtained materials were composites of very homogeneously distributed metal nanoparticles in a hexa-aluminate matrix, as shown in figure 2. Catalyst particles were agglomerates of nanoparticles with primary particle sizes between 5 and 15 nm and BET surface areas between up to 400 m^2/g at 600°C and up to 20 m^2/g at 1300°C.

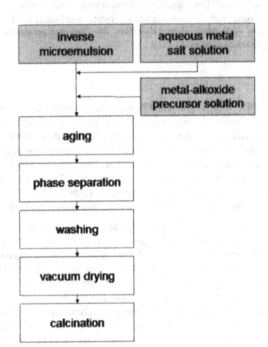

Fig. 1: Steps in the synthesis of the noble metal - hexaaluminate composite catalysts.

Fig. 2: Typical Pt-BHA catalyst after synthesis, calcinations at 600°C and reactor operation for many hours.

CATALYTIC TESTS

These novel catalysts were tested with the high-temperature catalytic partial oxidation of methane. This reaction is a technologically interesting route for the formation of synthesis gas and hydrogen from natural gas, since the reaction proceeds with extremely high reaction rates at temperatures above 900°C, allowing for very high reactor through-puts or very short catalyst contact times (τ < 50 ms) and thus very compact reactor sizes [9, 10]. Furthermore, unlike steam reforming reactions, the reactions proceed autothermally due to the mild exothermicity of the direct oxidation pathway. The combination of these factors renders this reaction particularly interesting for simple and efficient, small-scale, decentralized applications.

The catalysts were tested in a simple tubular flow reactor with 15 mm diameter, 50 – 150 mg catalyst powder and volumetric gas inlet flow rates of about 1 to 5 slpm (standard liters per minute). Methane and air were fed through standard mass-flow controllers and pre-mixed in front of the catalyst bed. Product gases were analysed via a double-oven gas-chromatographic system, which allowed to detect all reactant and product species (including hydrogen and water) with good accuracy (all atom balances closed to better than ±5%).

As shown in figure 3, the Pt-modified BHA-catalysts (Pt-BHA, blue squares) gave excellent syngas selectivities and methane conversions (oxygen conversions were 100% at all conditions reported here), surpassing earlier results over wet-impregnated alumina monoliths (green upward triangles). While optimal CO-yields at autothermal operation of the reactor are very slightly reduced from about 80% to about 78%, H_2-yields are drastically improved from less than 40% to above 80%. Furthermore, both these maxima are shifted from a (molar) methane-to-oxygen ratio of about 1.1 over the conventional catalysts to around 1.5 over the Pt-BHA catalysts, which is an indication that less methane was burned in the reaction and the maximum yield was thus shifted towards the stoichiometric point of 2.0. Interestingly, the novel Pt-BHA catalysts also clearly surpass the results over Rh-impregnated alumina monoliths (black downward triangles), which had previously been shown to be the more the selective partial oxidation catalysts [11]. (While these observed differences are to some degree due to differences between the powderous BHA-catalyst vs the monolithic alumina catalysts, a comparison between the described Pt-BHA catalysts with a catalyst, in which Pt had been post-impregnated onto the pure BHA powder, showed that CO and H_2 yields were still improved significantly – from about 45% H_2 yield and 36% CO yield at the stoichiometric ratio for syngas production over the impregnated catalyst, to 66% and 56%, respectively, over the "co-synthesised" Pt-BHA catalyst [14].)

Since the catalyst is significantly more selective, in particular towards hydrogen production, the autothermal reactor temperatures drop by several hundred Kelvin in comparison to the conventional monolith catalysts (from around 1000°C to about 700°C). To compensate for this effect, which is thermodynamically unfavourable for partial oxidation reactions, the reactor tube was put into an external oven (T= 750°C), minimising heat losses from the reactor and thus increasing reaction temperatures to about 1000°C.

The results from this experiment are also shown in figure 3 (red circles): CO-yields improve to 89%, and hydrogen yields go up to as much as 96% (with H_2 selectivities at this point of close to 100%). At the same time, the maximum in yields also further shifts significantly towards the stoichiometric ratio, indicating a further step to the thermodynamic optimum (which allows for virtually complete methane conversion and 100% selectivity to both CO and H_2 at a stoichiometric ratio of $CH_4/O_2 = 2.0$). Quite clearly, the novel catalysts are far superior partial oxidation catalysts in comparison to conventional supported noble metal catalysts, particularly when hydrogen production is the main interest.

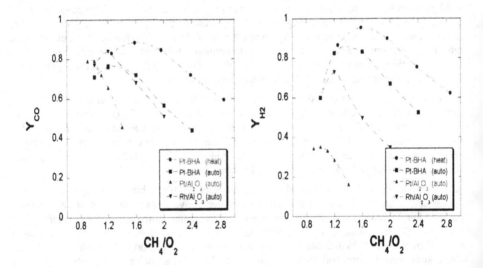

Fig. 3: CO and H_2 yields in autothermal catalytic oxidation of methane over the novel Pt-BHA catalyst (squares) in comparison with a conventional catalyst (wet-impregnation of an alumina monolith with Pt, upward triangles, and Rh, downward triangles). The yields can be further improved through external heating of the reactor (circles, $T_{oven} = 750$°C).

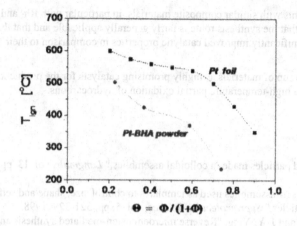

Fig. 4: Ignition temperatures for different CH_4/air mixtures over the Pt-BHA catalyst and a Pt-foil catalyst versus the "equivalence ratio" Θ [12]. Θ runs from 0 for pure oxygen to 1 for pure methane. The stoichiometric ratio for total combustion is at $\Theta = 0.5$ and for syngas formation at $\Theta = 0.8$.

In addition to the selectivity measurements, the ignition temperatures over these catalysts were also measured and compared to previous ignition/extinction experiments over noble metal foil catalysts [12, 13]. The results are shown in figure 4 for the novel Pt-BHA catalyst in comparison with a Pt-foil catalyst.

Clearly, ignition temperatures over the Pt-BHA catalyst are significantly lowered by about 100 K for very fuel lean mixtures, and by about 250 K near the stoichiometric point for syngas formation. While some of this effect might be due to the different catalyst types (i.e. powder vs foil), and while experiments with a conventional Pt-impregnated alumina powder catalysts are therefore necessary for a quantitative comparison, we know from experiments with such catalysts at fuel-rich conditions ($\Theta \approx 0.7$) that ignition temperatures of around 400°C (i.e. close to the ignition temperatures measured over the Pt-foil catalyst) are typically required for light-off of the catalyst. The ignition temperatures thus again confirm the very high activity of the catalyst (in addition to the good selectivity), and indicate that light-off with these novel catalysts should be significantly facilitated in comparison to conventional catalysts.

CONCLUSIONS

We presented results from investigations into the development of novel high-temperature stable catalyst for partial oxidation of hydrocarbons to synthesis gas and hydrogen. The catalyst synthesis is based on a microemulsion-templated sol-gel synthesis route, in which the high-temperature stability of hexa-aluminates and the good selectivity of noble metal components can be combined in a composite catalytic material.

Results with Pt-modified barium-hexa-aluminate catalysts were presented, and it was shown that the novel catalyst are far superior to conventional catalysts, not only in terms of high-temperature stability and hydrogen selectivity, but also with regard to strongly lowered ignition temperatures. Further experiments with similar composite materials, in particular with Rh- and Ni-modified materials indicate that the synthesis route is fairly generally applicable and that the materials all appear to show significantly improved catalytic properties in comparison to their 'conventional' counterparts.

Overall, we therefore see this group of materials as highly promising catalysts for the production of synthesis gas or hydrogen via high-temperature partial oxidation of hydrocarbons.

REFERENCES

[1]   M. P. Pileni, "Nanosized particles made in colloidal assemblies," *Langmuir*, vol. 13, pp. 3266-3276, 1997.
[2]   M. P. Pileni, "Colloidal self-assemblies used as templates to control size, shape and self-organization of nanoparticles," *Supramolecular Science*, vol. 5, pp. 321-329, 1998.
[3]   A. J. Zarur, H. H. Hwu, and J. Y. Ying, "Reverse microemulsion-mediated synthesis and structural evolution of barium hexaaluminate nanoparticles," *Langmuir*, vol. 16, pp. 3042-3049, 2000.
[4]   M. P. Pileni, T. Gulik-Krzywicki, J. Tanori, A. Filankembo, and J. C. Dedieu, "Template design of microreactors with colloidal assemblies: Control the growth of copper metal rods," *Langmuir*, vol. 14, pp. 7359-7363, 1998.
[5]   A. J. Zarur, N. Z. Mehenti, A. T. Heibel, and J. Y. Ying, "Phase behavior, structure, and applications of reverse microemulsions stabilized by nonionic surfactants," *Langmuir*, vol. 16, pp. 9168-9176, 2000.
[6]   A. J. Zarur and J. Y. Ying, "Reverse microemulsion synthesis of nanostructured complex oxides for catalytic combustion," *Nature*, vol. 403, pp. 65-67, 2000.
[7]   H. Arai and H. Fukuzawa, "Research and development on high temperature catalytic combustion," *Catalysis Today*, vol. 26, pp. 217-221, 1995.
[8]   H. Arai and M. Machida, "Thermal stabilization of catalyst supports and their application to high-temperature catalytic combustion," *Applied Catalysis a-General*, vol. 138, pp. 161-176, 1996.
[9]   L. D. Schmidt, M. Huff, and S. Bharadwaj, "Partial Oxidation Reactions and Reactors," *Chem. Eng. Sci.*, vol. 49, pp. 3981-3994, 1995.
[10]   G. Veser, J. Frauhammer, and U. Friedle, "Syngas formation by direct oxidation of methane. Reaction mechanisms and new reactor concepts," *Catalysis Today*, vol. 61, pp. 55-64, 2000.
[11]   D. A. Hickman and L. D. Schmidt, "Production of syngas by direct catalytic oxidation of methane," *Science*, vol. 259, pp. 343-346, 1993.
[12]   G. Veser and L. D. Schmidt, "Ignition and extinction in the catalytic oxidation of hydrocarbons over platinum," *Aiche Journal*, vol. 42, pp. 1077-1087, 1996.
[13]   G. Veser, M. Ziauddin, and L. D. Schmidt, "Ignition in alkane oxidation on noble-metal catalysts," *Catalysis Today*, vol. 47, pp. 219-228, 1999.
[14]   J. Schicks, D. Neumann, U. Specht, and G. Veser, "Nanoengineered catalysts for high-temperature methane partial oxidation", *Catalysis Today* (2003) in print.

Mat. Res. Soc. Symp. Proc. Vol. 756 © 2003 Materials Research Society

Study of Ammonia Formation During The Autothermal Reforming of Hydrocarbon Based Fuels

A.R. Khan, James Zhao, and O. Y. Polevaya
Nuvera Fuel Cells Acorn Park
Cambridge, MA 02140, USA

ABSTRACT

Ammonia formation in autothermal reforming process was studied in Nuvera's Modular Pressurized Flow Reactor facility. Experiments were conducted to study and compare different catalysts for their ammonia formation characteristics. Different hydrocarbon fuels were reformed and effects of fuel structure and operating conditions on ammonia formation were investigated. Reformate generated was analyzed for ammonia contamination by using FTIR spectroscopy.

INTRODUCTION AND BACKGROUND

Recently PEM fuel cell based power systems have gained a lot of focus and attention for their application in stationary and transportation markets. These systems include a reformer that converts hydrocarbon fuel to a hydrogen rich stream of gas. This stream is then fed to a fuel cell that generates electric power. Auto-thermal reforming (ATR) is a widely practiced and accepted process for extracting hydrogen from hydrocarbon based fuels. This type of reforming has been extensively studied in Nuvera Fuel Cells for a variety of fuels including Gasoline [1], Methanol, Ethanol, FT Naphtha , and Bio diesel. Previous studies [2] were mainly focused on the parametric study for syngas production and efficiencies as a function of reforming conditions. For durable fuel cell operation quality of reformate is an important factor, as the reforming process also produces some unwanted trace species (CO, H_2S, NH_3, HC) that are considered to be poisons for PEM fuel cells. The presence of these contaminants in reformate can adversely effect their performance and functioning.

Ammonia is proven to be a fuel cell contaminant and is formed by dissociation of air or fuel based nitrogen in the reforming process. Uribe [3] studied effects of ammonia on PEM fuel cell performance. It was concluded from this study that trace levels of ammonia irreversibly damages the fuel cell over time. These contaminants can be removed from reformate by adding cleanup stages to the process. Depending on the number of contaminants, levels and capacity of clean up material used, these stages can occupy an appreciable volume of the total system and impose pressure drops that penalize efficiencies. Such beds also add another step toward the complexity of the process. Other than the downstream cleanup of such species, catalyst selection could be one way to reduce or eliminate the formation of these contaminants. This study was focused in particular to reduce ammonia formation in auto thermal reformers by choosing the right catalyst and optimizing the process conditions for reforming. The fuels studied included: gasoline, ethanol and natural gas. Ammonia formation characteristic of conventional nickel based catalyst was compared to a precious metal based catalyst.

EXPERIMENTAL SETUP

Figure 1 shows the schematic of the auto-thermal reforming (ATR) process setup used in the current testing. The main focus was on the auto thermal reforming zone, as it is seen to be the site for ammonia formation. Downstream samples did not show any ammonia formation in the later stages of the reforming process. As a result the water gas shift steps were bypassed and all the analysis was done on the pre shift reformate. Nuvera's Modular Pressurized Reactor facility, shown in Figure 2, was used to conduct these studies. This rig is capable of testing a wide variety of fuels and catalyst forms at the full scale. Power input can range from 40-150kW thermal over an operating pressure range of 1-10 atmospheres. Multiple sampling ports are provided for gas and thermal analysis. Analytical equipment used for gas analysis included on line gas analyzers (CO, CO_2), gas chromatograph and FTIR spectrometer. FTIR spectroscopy was mainly used for analyzing contaminants, particularly ammonia, in the reformate stream. The lower detectable limit for the FTIR was under 1ppm.

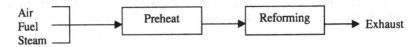

Figure 1. Schematic of the MPR reforming process.

Figure 2. Modular Pressurized Reactor (MPR)

Gasoline, ethanol and natural gas were reformed over different ATR catalysts. The study was divided into two parts. The first part focused on the effect of catalyst on ammonia formation. Fuels tested in this study were denatured ethanol and sulfur free gasoline. The properties of the different fuels are listed in Table1. Ethanol was supplied by Fisher Scientific and sulfur free gasoline from Phillips Chemical Company. The two fuels were reformed over precious metal and nickel based catalysts. A direct comparison was made between reformates generated by two different types of catalyst for the same fuel and two different fuels over the same catalyst. In the second part natural gas was the only fuel and the study was more focused to see the effect of ATR operating parameters, mainly temperature and pressure, on ammonia formation.

Table 1. Fuel properties

Fuels	Hydrogen (WT%)	Carbon (WT%)	Oxygen (WT%)	Sulphur (ppm)	Lower heating value (MJ/Kg)
Gasoline	13.72	86.28	0	<2	42.9
Ethanol	13.04	52.17	34.7	-	26.9
Natural gas	24.72	75.27	0	<5	49.1

STUDY OF AMMONIA FORMATION OVER DIFFERENT ATR CATALYSTS

Gasoline was autothermally reformed over a nickel based ATR catalyst. A preheated fuel, steam and air mixture was sent to the ATR reactor where an exothermic reaction provided the heat for reforming. An equivalence ratio of 3.5 was used in the study where Equivalence ratio = (Fuel/Air)/(Fuel/Air)$_{Stoichiometric}$. Once the steady state was achieved, gas samples were collected along the length of the reactor bed and were analyzed for ammonia using FTIR spectroscopy. The samples taken were kept hot to prevent ammonia from dropping out into water.

As shown in Figure 3 nickel based catalyst test results for gasoline showed an increasing trend for ammonia with higher residence times. Residence times shown in plots are the normalized values. Where residence time is defined as Residence time= Catalyst Volume/Flow rate. When repeated with precious metal based ATR catalyst, reformate showed very low levels (BDL) of ammonia in the gas stream. Ethanol was also reformed under similar conditions over nickel and precious metal based ATR catalysts. A similar increasing trend as a function of higher residence times and effect of catalyst on ammonia formation was exhibited by ethanol.

The results from this testing suggest ammonia formation had a strong dependence on the type of catalyst used for reforming. Compared to precious metal nickel based catalyst promoted the ammonia formation reaction and the levels seen were closer to the equilibrium as shown in Figure 4. This plot shows the calculated equilibrium values for ammonia over a range of ATR temperatures. Precious metal catalyst was less favorable to ammonia formation. As a result reformate generated using precious metal based ATR catalysts was cleaner than conventional nickel. Also from this study it was seen that catalyst behavior to promote ammonia formation exceeded the fuel structure difference as similar ammonia levels and trends were observed for the two very different fuels.

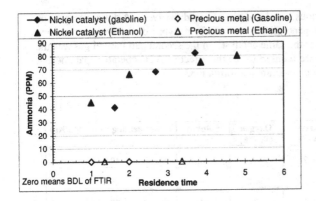

Figure 3. Ammonia formation versus different Catalyst / Fuel

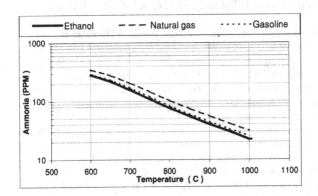

Figure 4. Equilibrium Calculation for ammonia in ATR reformate stream.

STUDY OF ATR OPERATING CONDITIONS ON AMMONIA FORMATION

In this study natural gas was reformed using a precious metal based ATR catalyst. The dependencies of temperature and pressure on ammonia formation were investigated. The reformer was operated at lower temperatures and ammonia levels measured were under detectable limits of FTIR. Once the data was collected at this operating condition ATR bed temperatures were cycled by varying preheats of steam, fuel and air mixture going to the reforming zone. The gradual increase in ammonia levels was observed with higher temperatures ranging from the below detectable levels to a maximum of ~9 ppm. After reaching this point ATR temperatures were again lowered back to starting conditions by reducing preheats to the feeds. This led to the reduced ammonia levels finally going under detectable limits. As shown in Figure 5 continuous ammonia measurements were made and the ammonia profile as a function of ATR temperature was generated. It showed that the ammonia formation was greatly controlled by ATR bed temperatures. Faster kinetics due to higher temperature shifted ammonia levels towards equilibrium. As seen in the graph, optimum conditions made it possible to achieve below detectable levels of ammonia in the natural gas reformate.

In order to observe the effect of pressure on the ammonia formation the reactor was first started under atmospheric conditions. For comparison purposes the conditions were adjusted so as to make ammonia within detectable limits of FTIR. After collecting some steady state data at this condition the pressure was raised during the operation from atmospheric to 3 atms. As shown in Figure 6 little increase was observed in the ammonia formation. The hump seen during the transition was attributed to the pressure variation caused by the reactor in the FTIR cell compartment. This study showed the operating temperature dominated over pressure and turned out to be the key parameter for controlling ammonia formation in the autothermal reforming zone.

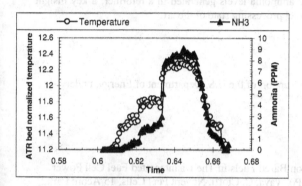

Figure 5. Temperature effect on ammonia formation.

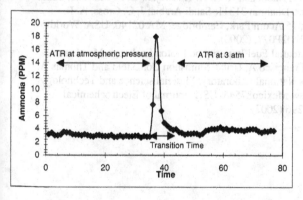

Figure 6. Pressure effect on ammonia formation.

CONCLUSIONS

Gasoline, ethanol, and natural gas were successfully reformed in the modular pressurized flow reactor facility. Nickel based catalyst promoted ammonia formation reaction adding significant levels of contamination in the reformate stream generated through the auto-thermal reforming process. In comparison to nickel, reformate generated using precious metal based ATR catalyst was seen to be much cleaner and suitable for the system. This can greatly reduce the complexity, size, and need of the clean up stages in the reforming system. In comparison to catalyst behavior difference in the fuel structure does not have much of an effect on the ammonia formation. The ammonia formation was greatly governed by the nature of the catalyst. Effects of ATR operating conditions on ammonia formation were investigated and catalyst bed temperature was seen to be the dominant variable over pressure for controlling ammonia formation. The study indicates that the right selection of catalyst and the optimized conditions can greatly reduce the ammonia levels generated in a reformer, a key insight for advancing the transportation fuel processing state of the art.

ACKNOWLEDGEMENTS

This article was prepared with the support of the U.S. Department of Energy, under Award No. DE-FC02-99EE50580 "SFAA".

REFERENCES

1   Evaluation Of Hydrocarbon Based Fuels In The Disintegrated Fuel Cell Power Train. A. R. Khan, O. Y. Polevaya, C. Cioffi. Nuvera Fuel Cells, 15 Acorn Park, Cambridge, MA 02140, USA. American Chemical Society, 47 (2), 564 (2002).
2   Design, Development, and Evaluation of Hydrocarbon Based Fuels for Fuel- Cell On-Board Reformers. B. O. Dabbousi, A. H. Al-Khawajah, G. D. Martinie, Saudi Aramco, Lab R&D Center, Dhahran 31311, Saudi Arabia; J. C. Cross, A. R. Khan, Nuvera Fuel Cells, 15 Acorn Park, Cambridge, MA 02140, USA. World Petroleum Congress, B02F02P027, (2000).
3   Effect of Ammonia as Potential Fuel Impurity on Proton Exchange Membrane Fuel Cell Performance. Francisco A. Uribe, Shimshon Gottesfeld and Thomas A. Zawodzinski.Los Alamos National Laboratory, Material science and Technology Division, Los Alamos, new Mexico87545, USA. Journal of Electrochemical Society, 149 (3) A293-A296 (2002)

Mat. Res. Soc. Symp. Proc. Vol. 756 © 2003 Materials Research Society

Optimization of a Carbon Composite Bipolar Plate for PEM Fuel Cells

Theodore M. Besmann[1], John J. Henry, Jr.[1], Edgar Lara-Curzio[1], James W. Klett[1], David Haack[2] and Ken Butcher[2]
[1]Metals and Ceramics Division, Oak Ridge National Laboratory
Oak Ridge, TN 37831-6063, USA
[2]Porvair Fuel Cell Technology
Hendersonville, NC 28792, USA

ABSTRACT

A carbon composite bipolar plate for PEM fuel cells has been developed that has high electrical conductivity, high strength, light weight, is impermeable, and has the potential for being produced at low cost. The plate is produced by slurry molding short carbon fibers into preform structures, molding features into the green body, and using chemical vapor infiltration to strengthen the material, give it high conductivity, and densify the surface to make it impermeable. Current efforts have focused on optimizing the fabrication process and characterizing prototypical components.

INTRODUCTION

The significant and growing interest in fuel cells for stationary power and transportation applications has been demonstrated by the attention these technologies are receiving from both government and industry, and particularly from the automotive sector [1]. Interest for vehicular applications has focused on the proton exchange membrane fuel cell (PEMFC) because of its low-temperature operation and thus rapid start-up. Currently, challenges for PEMFC technology for automobiles include reducing the cost and weight of the fuel cell stack, the goal being a ~50 kW system of <$40/kW and <133 kg in mass. One of the key components is the bipolar plate, which is the electrode plate that separates individual cells in a stack [2]. The reference design requires the bipolar plate to be high-density graphite with machined flow channels. Both material and machining costs are prohibitive ($100-200/plate), and this has led to substantial development efforts to replace graphite. The bipolar plate requirements include low-cost materials and processing (goal of <$10/kW for the component), light weight, thin (<3mm), sufficient mechanical integrity, high surface and bulk electronic conductivity, low permeability (boundary between fuel and oxidant), and corrosion resistance in the moist atmosphere of the cell (<16 $\mu A/cm^2$) [3].

The bipolar plate approach developed at Oak Ridge National Laboratory (ORNL) uses a low-cost, slurry-molding process to produce a carbon-fiber preform. The molded, carbon-fiber component could have an inherent volume for diffusing fuel or air to the electrolyte surface or impressed, flow-field channels. The bipolar plate is made hermetic through chemical vapor infiltration (CVI) with carbon. The infiltrated carbon also serves to make the component highly conductive. The technique has the potential for low-cost, large-scale production as the material costs are low, the fiber preforms can be produced in continuous processes similar to felt or paper production, and the infiltration can be accomplished in very large-scale batch or possibly continuous processes.

This paper reviews current work in characterizing the carbon composite materials with differing loadings of carbon particulate filler. The filler was added in an attempt to reduce the need for fiber material and make surface sealing easier. Surface roughness was characterized using profilometry and infrared imaging was used to identify flaws in sample coupons. In-plane shear stress was measured to gain an understanding of the materials' behavior under torsion, which may occur during stack assembly. Fracture toughness was measured so that when combined with strength data, previously determined, it could aid in determining the size of cracks that the material can tolerate.

EXPERIMENTAL

Slurry-Molding

Fibrous component preforms for a sub-scale plate (120 x 140 x 2.5 mm) are prepared by a slurry molding technique using 10 μm-diameter, 100-μm-long carbon fibers (Fortafil 3(c) 00 PAN-based, Fortafil Fibers, Inc., Rockwood, TN) suspended in water containing phenolic DUREZ® resin (Occidental Chemical Corp., Dallas, TX) [4]. For some samples, a graphite particulate filler, sieved to ~7 μm, was added to the fiber at 15 and 30 mass % (Asbury M850 graphite, Asbury Carbons, Inc., Asbury, NJ). The fiber-to-phenolic mass ratio is 4:3. A vacuum-molding process produces an ~18 vol % fiber, isotropic preform material containing particles of phenolic. After drying, a set of brass molds are used to impress features, such as gas channels, into the preform material at 150°C and a pressure of 10 kPa. The phenolic binder serves to provide green strength and geometric stability after curing in the mold.

Chemical Vapor Infiltration.

The surface of the preform is sealed using a CVI technique in which carbon is deposited on the near-surface material in sufficient quantity to make it hermetic. The depth of infiltration is governed by the inherent competition between the kinetics of the surface reactions that produce the deposited material and the mass transport mechanism that allows the reactants to diffuse to the internal volume of the material [5]. The result is that the more rapid the kinetics of deposition as compared to mass transport, the more likely material will be deposited near the surface and the reactants will not significantly penetrate into the thickness of the porous preform. In a review of carbon deposition, Delhaes [6] describes depth penetration of CVI carbon as a function of temperature. The higher the temperature the more rapid the surface deposition kinetics, thus the smaller penetration depth allows the surface to be sealed and the bulk volume of the preform to retain a large volume fraction of porosity.

A high surface-to-volume CVI reactor was used to minimize soot formation and allow the efficient infiltration of the fibrous preform. Based on Delhaes [6] and Bammidipati, et al. [7], an infiltration temperature of 1500°C was selected with methane (chemically pure, Air Lquide, Houston, TX) as the precursor. Reduced pressure (8 kPa) also suppresses the formation of soot. Flow rates of 1000 cm^3/min methane in 2500 cm^3/min argon are used. A processing time on the order of 4 h was determined to be necessary to obtain sealed surfaces for typical components without particulate filler. In addition, during the high temperature CVI processing the phenolic present in the preform is pyrolyzed. Further details can be found in Besmann, et al. [8].

Characterization

The surface roughness of the bipolar plates was measured using a Rodenstock RM600 laser profilometer with version 3.27 software (Rodenstock GmbH, München, Germany).

Infrared (IR) images to identify flaws/delamination in bipolar plate material samples were obtained using a high-resolution IR camera (Radiance 1t, Amber/Raytheon, Goleta, CA) with a 25 mm lens. The camera contained a 256 x 256 pixel indium antimonide sensor that was sensitive in the 3-5 μm waveband. Each pixel represented 100μm x 100μm projected area of the bipolar plate and had a temperature resolution of ± 0.1°C throughout the examined temperature range. The camera has a high speed (12 bit) video bus that was used to digitally capture the temperature distribution field. The bipolar plate specimens were imaged while they were on top of a hot plate that induced a temperature gradient through their thickness. Differences in surface temperature, which are recorded with the IR camera, are indicative of inhomogeneities in the thermal conductivity of the material (e.g.- delaminations, defects). Figure 1 illustrates the configuration of the thermal imaging system.

In-plane shear strength tests were carried out per ASTM C1292. Because the test specimens were smaller than those recommended in C1292, stainless-steel end tabs were adhesively bonded to the specimens. Both the specimens and the V-notches were machined using a diamond wheel. The tests were carried out at a constant cross-head displacement rate of 0.05 mm/s using an electromechanical testing machine and an Iosipescu fixture (Fig. 2).

Fracture toughness measurements were determined by a double torsion test. The geometry of the test specimen consists of a plate which is notched to form two beams. Each of these beams is loaded in torsion by four-point bending at the ends, causing the crack to propagate through the center of the specimen (Fig. 3) [9]. In this specimen configuration the crack profile is not straight through the thickness, but extends further along the tensile side of the plate to form a curved crack front. Both the specimens and the notch were machined using a numerically-controlled grinder equipped with a thin diamond wheel. The tests were carried out at ambient conditions using an electromechanical testing machine at a constant cross-head displacement rate of 0.1 mm/s using an in-house developed fixture. To determine the compliance of the load train and specimen, test specimens with different notch lengths were evaluated.

RESULTS

An optical image of a cross-section of a bipolar plate sample can be seen in Fig. 4. The surface is sealed with a layer of carbon, yet significant porosity is visible in the interior of the material. The average density of infiltrated bipolar plates is ~1.2 g/cm^3 (theoretical density of carbon is 2.26 g/cm^3). Figure 4 also contains representative surface profile measurements of a prototypical

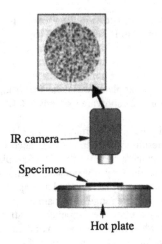

Figure 1. Infrared imaging system used for determining delamination and defects in bipolar plate samples.

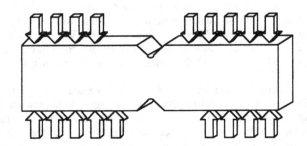

Figure 2. Iosipescu loading configuration for applying asymmetric bending to determine in-plane shear strength.

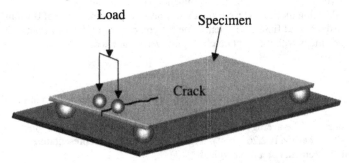

Figure 3. Double-torsion test method for determination of fracture toughness.

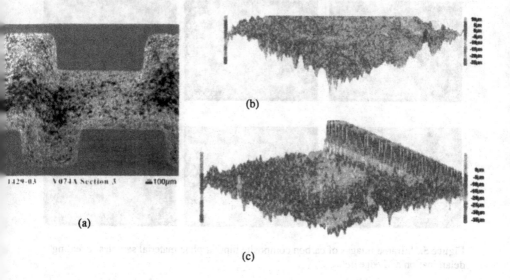

(b)

(a)

1429-03   N074A Section 3   ≈100µm

(c)

Figure 4. Prototypical carbon composite bipolar plate: (a) optical image of a polished cross-section, (b) surface profile of top flat, and (c) surface profile of bottom of a channel and its sidewall.

bipolar plate. Measurements indicate rms surface roughness values that vary between 2.5 and 4.5 µm. It is also apparent that in certain areas, such as in the corner of a sidewall of a channel, there is a higher degree of roughness, with depth variations of ~30 µm.

The integrity of bipolar plate material was assessed using thermal imaging. Figure 5 shows a series of IR images taken of 50x50x2 mm samples. These were cut from a larger plate after slurry-molding and curing, but before CVI. After CVI to seal the surface of the plates they were examined using the thermal imaging system. Gray-scale changes indicate delamination, with all but one sample indicating that they are intact. There does appear to be some damage around the edges of the samples, most likely due to local delamination due to cutting of the larger plate into the smaller samples.

Table I gives the results of the shear stress measurements using the Iosipescu method. An example of a stress-displacement curve obtained for the material is shown in Fig. 6. These values are indicative of a relatively torsion resistant material, particularly given the low density of the carbon composite. There appears to be little effect of filler content within the concentration utilized. The stress-displacement curves indicate little delamination or other failures until ultimate failure. Table II lists the results of the fracture toughness measurements, the values of which are fairly low, again with little effect of filler content.

Figure 5. Infrared images of carbon composite bipolar plate material samples revealing delamination and edge defects.

Table I. Shear stress measurements shows little correlation with percent filler.

Sample (% Filler)	Strength Meas.
PMP10R (0%)	25.9 ± 9.9 MPa
PMP10T (0%)	19.3 MPa (1 test)
PMP09K (15%)	24.4 ± 11.8 MPa
PMP11E (30%)	43.3 ± 2.7 MPa
PMP11G (30%)	17.1 ± 1.1 MPa
PMP11H (30%)	18.7 ± 4.7 MPa

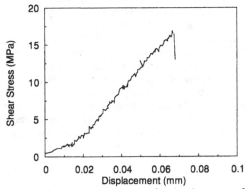

Figure 6. Typical stress-displacement curve for the shear test of the carbon composite bipolar plate material.

Table II. Fracture toughness of carbon composite bipolar plate material for different filler contents (assuming a Poisson's ratio of 0.2).

Sample (% filler)	Fracture Toughness (K_{IC}, MPa·m$^{1/2}$)
PM10 (0)	2.55±0.19
PM09 (15)	3.24±0.11
PM11 (30)	2.84±0.36

DISCUSSION

Surface roughness can be an important issue for bipolar plate materials. The plate must contact the electrolyte membrane, as well as any intervening layers that are often used to support catalyst material. Thus, high surface roughness can affect electrical contact between mating surfaces. The relatively modest surface roughness of much of the bipolar plate areas is within the bounds of acceptability, as indicated by relatively good electrical behavior previously reported [8]. Surface roughness can also affect water management within a PEMFC, causing holdup of water and thus blocking of channels, particularly if wetting is a problem. This issue will need further exploration.

The results of the thermal imaging measurements, which were a first attempt to use this technique in this application, are encouraging. It is expected that in development and production there will be a need for a rapid and definitive technique for non-destructive evaluation of plates in production. Although it is likely that plates with some delamination or other flaws will still meet acceptability criteria, thermal imaging offers at least the opportunity to develop a baseline screening system.

As fuel cell stacks are assembled it is expected that relatively high pressures will be needed to obtain good electrical contact and aid in sealing the edges against leakage. Given that upwards of 100 plates will be needed in a stack, and that it is likely that pressure will be applied via techniques such as posts on each corner running the length of the stack, it is important that cell components be able to withstand significant tension and torsion, as well as compression. The Iosipescu test methodology was therefore viewed as particularly applicable to assess the resistance of the bipolar plates to torsional loading. The relatively low fracture toughness of the material is a concern for handling and assembly of the plates, although the implications are uncertain due to the highly inhomogeneous and anisotropic nature of the components.

CONCLUSIONS

The technique of slurry-molding carbon fibers followed by chemical vapor infiltration with carbon has resulted in relatively strong and light-weight PEMFC bipolar plates with good electrical and corrosion properties, and the potential for low cost manufacturing. Examination of the surface roughness of molded and infiltrated plates revealed relatively smooth surfaces that will facilitate sealing and electrical contact. Thermal imaging has been demonstrated as a potential tool for non-destructive evaluation of bipolar plate components, and has revealed that as-processed materials does not delaminate. Mechanical property measurements of the shear stress resistance of the carbon composite bipolar plate material are encouraging, with relatively good strengths that are likely to withstand assembly and use of the plates. Fracture toughness values are low, but may be misleading due to the inhomogeneity of the material.

ACKNOWLEDGMENTS

The authors wish to thank M. Radovic, who helped with IR imaging and fracture toughness measurements, and S. D. Nunn and D. F. Wilson for their valuable comments. This research was supported by the U. S. Department of Energy, Energy Efficiency and Renewable Energy, Office of Transportation Technology under contract DE-AC05-00OR22725 with UT-Battelle, LLC.

REFERENCES

1. T. F. Fuller, *Interface*, Fall 1997, 26 (1997).
2. T. R. Ralph, *Platinum Metals Rev.*, **41** (3), 102 (1997).
3 R. L. Borup and N. E. Vanderborgh in *Materials for Electrochemical Energy Storage and Conversion – Batteries, Capacitors and Fuel Cells*, D. H. Doughty, B. Vyas, T. Takamura, and J. R. Huff, Editors, PV 393, p. 151, Materials Research Society, Pittsburgh, PA (1995).
4. G. C. Wei and J. M. Robbins, *J. Am. Ceram. Soc.*, **64** (5), 691 (1985).
5. T. M. Besmann, B. W. Sheldon, R. A. Lowden, and D. P. Stinton, *Science*, **253**, 1104 (1991).
6. P. Delhaes, *Chemical Vapor Deposition. Proc. Fourteenth Intl. Conf. and EUROCVD 11*, M. D. Allendorf and C. Bernard, Editors, PV 97-25, p. 486, The Electrochemical Society Proceedings Series, Pennington, NJ (1997), -495.
7. S. Bammidipati, G. D. Stewart, J. R. Elliott, Jr., S. A. Gokoglu, and M. J. Purdy, *AIChE Journal*, **42** (11), 3,123 (1996).
8. T. M. Besmann, J. W. Klett, J. J. Henry, Jr., and E. Lara-Curzio, *J. Electrochem. Soc.*, **147** 4083 (2000).
9. E. R. Fuller, in *Fracture of Mechanics Applied to Brittle Materials*, ASTM STP 678, S. W. Freiman, Editor, (American Society for Testing and Materials, 1979) pp. 3-18.

Corrosion-Resistant Tantalum Coatings for PEM Fuel Cell Bipolar Plates

Leszek Gladczuk*, Chirag Joshi*, Anamika Patel*, Jim Guiheen[#], Zafar Iqbal**, Marek Sosnowski*

* Department of Electrical Engineering and ** Chemistry and Environmental Science, New Jersey Institute of Technology, Newark, New Jersey 07102, USA
[#]Honeywell International Inc, Morristown, New Jersey 07960, USA

Abstract: Tantalum is a tough, corrosion resistant metal, which would be suitable for use as bipolar plates for proton exchange membrane (PEM) fuel cells, if it was not for its high weight and price. Relatively thin tantalum coatings, however, can be deposited on other inexpensive and lighter weight metals, such as aluminum and steel, providing a passive protection layer on these easily formed substrates. We have successfully deposited, high quality α (body-centered-cubic, bcc) and β (tetragonal) phase tantalum coatings that were a few micrometers thick by dc magnetron sputtering on steel and aluminum. The growth of the thermodynamically preferred body-centered-cubic (bcc) tantalum phase was induced by a choice of deposition conditions and substrate surface treatment. The microstructure and corrosion resistance of the α-phase in an environment approximately simulating the electrochemical conditions used in a PEM fuel cell were investigated under potentiodynamic conditions. Preliminary potentiostatic measurements of a β-phase sample are also presented.

INTRODUCTION

Bipolar plates are critical components of a proton exchange membrane (PEM) fuel cell, which enable patterned fuel and oxidant gas flow, electrical charge transport and current collection, cooling and water management. A 10 kW fuel cell stack requires approximately 120 bipolar plates at typical cell efficiencies and footprint levels. For the fuel cell stack to be lightweight and low volume, bipolar plates must combine intrinsic lightness with their ability to be fabricated in thin sheets. Typical metals such as aluminum and steel can easily meet the requirement that they be of the order of 0.05 inches thick when formed for use as regular bipolar plates, and 0.1 inches thick when formed as coolant bipolar plates [1].

Bipolar plates must also have high chemical and corrosion stability, mechanical strength and low electrical resistance. Corrosion resistance is required because the plates are in constant contact with acidic water at pH $\approx$ 5 that is generated under the operating conditions of the stack. Oxides are formed during corrosion, which can migrate and poison the catalyst. Oxides also increase the electrical resistivity of the plates resulting in a drop in fuel cell performance. In order to minimize ohmic losses the specific resistance of the plates need to be maintained at values below 0.1 ohm.cm^2. Hydrogen and oxygen permeation across the bipolar plate also compromises the efficiency of the fuel cell and in addition, results in an unsafe operating condition – hence hydrogen-air permeation rates of not more than about 7.5 x 10^{-10} mol/cm^2-sec must be maintained [1].

Tantalum (Ta) is known to be resistive to corrosion, but is too heavy and expensive to use in pure form as bipolar plates. Tantalum films obtained by sputtering can be produced in two different crystallographic phases, body-center-cubic (bcc) and tetragonal (β-Ta), which have different physical properties: first is tough and highly ductile, second - hard and brittle [2,3]. We have therefore studied the use of thin coatings of two forms of Ta as corrosion-resistant layers on relatively lightweight and formable metals like steel and aluminum.

EXPERIMENTAL DETAILS

Tantalum films were deposited on steel, aluminum and silicon dioxide substrates using a DC magnetron sputtering system. The substrates were highly polished with an abrasive consisting of 0.05 μm alumina particles. They were then cleaned electrochemically and then with alcohol, acetone and alcohol in an ultrasonic bath, for 10 minutes each. The deposition chamber with the substrates was baked under high vacuum at 200°C for 7 h using halogen bulbs inside the chamber. This procedure of pumping and heating allowed a base pressure in the chamber below $8*10^{-8}$ mTorr and an undetectable value of the partial pressure of water, to be obtained. Some substrates were sputter-etched before deposition. Sputter etching was performed in argon at a pressure of 230 mTorr. On some of the steel substrates a thin TaN seed layer was deposited to promote subsequent growth of the bcc Ta phase instead of the less desirable tetragonal β Ta phase. During deposition of the tantalum films the pressure in the chamber was 5 mTorr. The flow of gases was adjusted using mass flow controllers, depending on the process employed. For Ta coatings a flow of 18 sccm of pure (99.998 %) argon was used. For TaN deposition a mixture of Ar and N_2 gases with a 15 % concentration of nitrogen, was used. A 5 cm diameter tantalum target of 99.95 % purity was used as the magnetron sputtering source in both processes.

The structure of deposited films was characterized by x-ray diffraction (XRD). The morphology of the surfaces of the films obtained was imaged by atomic force microscopy (AFM). Corrosion resistance properties of the films under near-ambient conditions were evaluated under potentiodynamic and potentiostatic conditions using a Princeton Applied Research Potentiostat-Galvanostat system.

RESULTS AND DISCUSSION

Figure 1 shows the XRD patterns recorded from Ta films deposited on aluminum substrates after different treatments of the substrate surface. The indexed pattern (Figure 1–top) was obtained from a film deposited on an aluminum substrate after sputter etching its surface in an Ar plasma, prior to Ta film deposition. The diffraction pattern shows that the crystalline structure of the film is that of body-centered-cubic tantalum. A very different pattern (Figure 1-bottom) was obtained from a Ta film deposited on an aluminum substrate for which the surface was not sputter etched. The diffraction pattern for this film is consistent with the tetragonal structure of Ta – the so-called β phase. The differences in crystallographic structure also influence the surface morphology. Figure 2 shows 1 μm x 1 μm AFM scans of the surface of cubic and tetragonal tantalum film deposited on a silicon dioxide substrate. The images show uniform roughness and no obvious indication of porosity. Since tetragonal phase Ta normally grows on silicon dioxide, the growth of the bcc Ta phase was induced by first depositing a thin (10 nm) TaN interlayer. Interlayer TaN was not necessary to grow bcc Ta on a sputter-etched aluminum surface. It appears that sputter etching of aluminum removes its native oxide layer, exposing pure metal, which promotes growth of the bcc Ta phase [4,5].

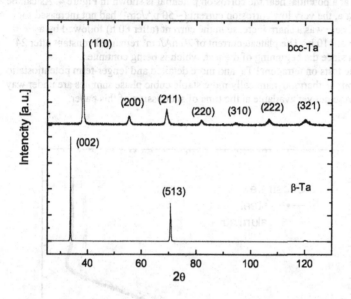

Figure 1. The XRD patterns from tantalum films deposited on aluminum with sputter-etched surface (top) and without sputter-etching (bottom).

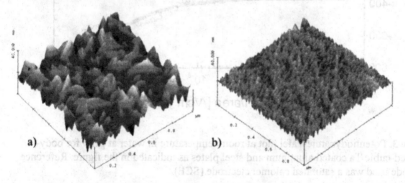

Figure 2. AFM images of a body-centered-cubic phase film surface (left) and a tetragonal phase film surface (right).

Potentiodynamic corrosion testing in pH 5 water at room temperature was performed on samples of cubic Ta films deposited on steel and aluminum plates using a saturated calomel reference electrode. The Tafel plots shown in Figure 3 indicate corrosion current densities that are lower than 10 nA/cm^2. Plateau current densities lower than 10^{-6} A/cm^2 suggest that bcc Ta provides a passive protective layer on both steel and aluminum. Preliminary potentiostatic measurements on a tetragonal film on aluminum over a period of about 6 days at a potential near the corrosion potential is shown in Figure 4. As can be seen from the data, the very low corrosion current (~ 50 pA/cm^2) had not increased for almost 4 days. There was a sharp increase in the current (after 80 h) followed by a plateau, starting at 110 h. The plateau current of 20 nA/cm^2 remained constant after 24 hours and 200 h since the beginning of the test, which is being continued.

Potentiodynamic tests on tetragonal Ta, and more detailed and longer-term potentiostatic measurements on the thermodynamically more stable cubic phase samples are under way, but these data were not yet available at the time of submission of this paper.

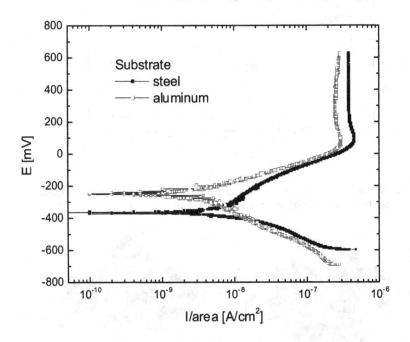

Figure 3. Potentiodynamic Tafel plot at room temperature in water at pH 5 for body-centered-cubic Ta coated aluminum and steel plates as indicated in the figure. Reference electrode used was a saturated calomel electrode (SCE).

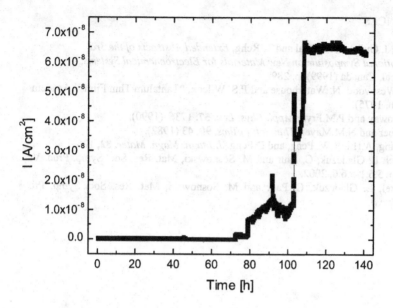

Figure 4. Potentiostatic test plot of corrosion current density in pH 5 water at room temperature for tetragonal Ta coated aluminum.

In conclusion, we have demonstrated the viability of both cubic and tetragonal Ta phases deposited by dc magnetron sputtering on steel and aluminum, as corrosion protection coatings. Potentiodynamic measurements show very low corrosion current densities ($< 10^{-9}$ A/cm^2) for both steel and aluminum surfaces coated with bcc phase Ta. Long term potentiostatic testing showed that the tetragonal phase on aluminum starts to show some corrosive effects after 4 days of continuous testing, but the corrosion current stabilized and remained low after 140 hours. Based on these and other measured properties of Ta films, including morphology and adhesion [6,7], the bcc phase of Ta promises to be an effective corrosion protection coating material for PEM Fuel Cell bipolar plates.

ACKNOWLEDGEMENTS

This research was supported in part by Sustainable Green Manufacturing Program at NJIT, funded by DOD.

REFERENCES

[1] H. Dai, J. Guiheen, Z. Iqbal and T. Rehg, *Extended Abstracts of the 3rd International Symposium on New Materials for Electrochemical Systems*, Montreal, Canada (1999), p. 289.

[2] W.D. Westwood, N. Waterhouse and P.S. Wilcox, "Tantalum Thin Films"(Academ London, 1975).

[3] K Holloway and P.M.Fryer, *Appl. Phys. Lett.* **57**, 1736, (1990).

[4] K. Hieber and N.M.Mayer, *Thin Solid Films*, **90**, 43 (1982).

[5] S.S. Jiang, A.Hu, R.W. Peng, and D.Feng, *J. Magn. Magn. Mater.* **82**, 126

[6] A. Patel, L. Gladczuk, C. Paur and M. Sosnowski, Mat. Res. Soc. Symp. Proc. V(697, pp. 5.6.1-5.6.6, 2002.

[7] A. Patel, L. Gladczuk, C. Paur and M. Sosnowski, Mat. Res. Soc. Symp. Pro *(submitted).*

Mat. Res. Soc. Symp. Proc. Vol. 756 © 2003 Materials Research Society   FF9.2

Miniature Fuel Processors for Portable Fuel Cell Power Supplies

Jamie Holladay, Evan Jones, Daniel R. Palo, Max Phelps, Ya-Huei Chin, Robert Dagle, Jianli Hu, Yong Wang, and Ed Baker
Battelle Pacific Northwest Division, PO Box 999, K8-93,
Richland WA 99352, U.S.A.

ABSTRACT

Miniature and microscale fuel processors that incorporate novel catalysts and microtechnology-based designs are discussed. The novel catalyst allows for methanol reforming at high gas hourly space velocities of 50,000 hr^{-1} or higher while maintaining a carbon monoxide levels at 1% or less. The microtechnology-based designs extremely compact and lightweight devices. The miniature fuel processors, with a volume less than 25 cm^3, a mass less than 200 grams, and thermal efficiencies of up to 83%, nominally provide 25 to 50 watts equivalent of hydrogen, which is ample for the portable power supplies described here. With reasonable assumptions on fuel cell efficiencies, anode gas and water management, parasitic power loss, the energy density was estimated at 1700 Whr/kg. These processors have been demonstrated with a CO cleanup method and a fuel cell stack. The microscale fuel processors, with a volume of less than 0.25 cm^3 and a mass of less than 1 gram, are designed to provide up to 0.3 watt equivalent of power with efficiencies over 20%.

INTRODUCTION

Portable electronic technologies such as PDA's, notebook computers, and microelectrochemical systems (MEMS) have fueled a need for new, high-energy, small-volume power supplies for both military and commercial markets. Wireless electronic devices are currently limited to battery technologies, which, despite recent advances, are insufficient to provide the long-term power these new microelectronic systems require. A solution to this problem would be hybrid systems composed of a microscale fuel processor, proton exchange membrane fuel cells, and secondary batteries. Hybrid systems combine high-energy liquid hydrocarbon fuels (e.g., up to 12.4 kWhr/kg for diesel) with clean fuel cell power, and battery convenience. This paper summarizes work on hydrocarbon selection, fuel reforming catalyst development, and the progress of two fuel processing systems (25 to 100 watts for soldier and personal portable power and 0.025 to 0.500 watts for sub-watt power ranges). In addition to the fuel processor, these systems integrate two fuel vaporizers (one for the fuel processor and one for the catalytic fuel combustor); a catalytic combustor; and two heat exchangers, which preheat reactants and recuperate heat from product streams. This paper reports progress in these projects that has been made since the previous publications [1,2].

FUEL REFORMING

A wide range of hydrocarbon fuels such as methanol, ethanol, propane, butane, gasoline, and diesel are being reformed into hydrogen-rich streams for fuel cells. Diesel has the highest raw energy density; however, the raw energy density of the fuel does not necessarily translate to the ultimate energy density of the device. For example, even though diesel has a raw energy density of 12.4 kWhr/kg, carrying enough water to operate at the stoichiometric steam to carbon ratio of 2:1 reduces its net energy density to 3.5 kWhr/kg. The net energy is also the same for methanol, with a stoichiometric steam to carbon ratio of 1:1.This reduction in net energy demonstrates the need for a water management system if a hydrocarbon-powered device is to reach its full energy density potential. It also shows that if no water management system is employed, methanol's lower reforming temperature makes it more desirable than diesel fuel for microelectronic applications. Methanol additional advantages include: it contains no sulfur; is available from renewable resources; and is completely miscible in water. For these reasons, methanol was selected as the initial fuel for use in developing the portable power supplies.

Battelle's proprietary methanol reforming catalyst is an important part of the portable power supply technology, because it is highly active, has a long life, and has a low selectivity to carbon monoxide. Over 99% methanol conversion has been achieved at relatively low temperatures (<350°C) and high gas hourly space velocities of 24,000-50,000 hr^{-1}. The catalyst was tested for 1000 hours with no detectable deactivation and was shown to be non-pyrophoric at room and low temperatures (<160°C). In addition, it has a low selectivity to carbon monoxide, which provides a product stream (reformate) with a CO content lower than equilibrium.

The ability of this catalyst to achieve lower than equilibrium concentrations of CO can be explained by looking at the steam reforming pathway [3]. Two alternative methanol steam reforming reaction pathways are shown in Figure 1. On the left side is a typical decomposition plus water-gas shift scheme. In this case, methanol first decomposes to CO and H_2, and then the CO reacts with H_2O to form CO_2 and more H_2. As the reaction proceeds, the minimum achievable CO content of the reformate is dictated by equilibrium constraints. At 350°C, this value is 5.4% CO in the dry reformate. On the right side of Figure 1, the pathway converts methanol to CO_2 and H_2 without producing CO. In this case, the minimum CO content of the reformate is dictated by the water-gas shift activity (or lack thereof) of the reforming catalyst. If there is very little shift activity, and the operating temperature is low enough to suppress the thermal decomposition of methanol, the CO levels can be kept well below the equilibrium concentration, which leads to the observed CO concentrations on the order of 1.0% and less (in dry reformate). Separate tests have confirmed that this methanol reforming catalyst has very little water-gas shift activity [4].

Decomposition-Shift Pathway	Alternate Pathway
$CH_3OH \rightarrow CO + 2H_2$	$CH_3OH \rightarrow HCHO + H_2$
$CO + H_2O \leftrightarrow CO_2 + H_2$	$HCHO + H_2O \rightarrow HCOOH + H_2$
	$HCOOH \rightarrow CO_2 + H_2$
Overall Reaction:	Overall Reaction:
$CH_3OH + H_2O \rightarrow CO_2 + 3H_2$	$CH_3OH + H_2O \rightarrow CO_2 + 3H_2$

Figure 1. Methanol reforming pathways.

SOLDIER PORTABLE POWER

Under a contract with the U.S. Army, Battelle is developing a hybrid power system designed to provide 15 to 25 watts of electric power for the dismounted soldier. This methanol-fueled system expected to provide up to 14 days of electric power before refueling. The final system will include an integrated microscale fuel processor, a proton exchange membrane fuel cell, and a rechargeable battery, and is expected to weigh less than 1 kg (excluding fuel). A representative system concept is shown in Figure 2. The device shown is based on hardware components, existing or being developed, capable of supporting a 15 to 25 W_e power supply. Larger power supply (50 W_e and 100 to 300 W_e) systems are also being developed.

The major components of the fuel processor system have been thermally integrated into two different sizes of processors. The smaller device (volume of less than 20 cm^3 and mass less than 150 grams), Figure 3, has been demonstrated at reformate outputs from 8 to 39 watts electrical equivalent. The thermal efficiency was calculated by dividing the lower heating value of the hydrogen in the reformate stream by the total heating value of the methanol fed the reformer plus the heating value of the fuel fed to the combustor.

$$\eta_t = \frac{LHV_{hydrogen}}{LHV_{combustion\_meoh} + LHV_{refor\min g\_meoh}} \tag{1}$$

where LHV is the lower heating value.

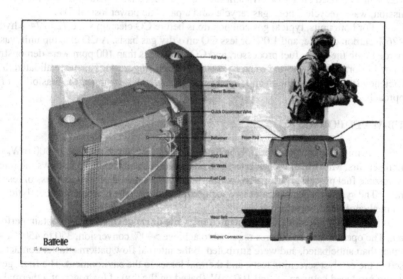

Figure 2. System concept for the soldier-portable power system.

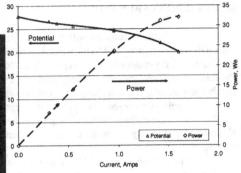

Figure 3. Integrated 15-25 W_e
fuel processor.

Figure 4. Fuel processor + carbon monoxide
cleanup + H-Power fuel cell performance.

These performance numbers result in a thermal efficiency of 62% and a projected specific energy
densities of up to 1500 Whr/kg, based on a 48% efficient fuel cell[1], a 1-kg device hardware
weight, a 14-day mission, water recycle, anode gas recycle, and a parasitic power loss of 3W.
The larger device, less than 25 cm^3 and 200 grams, has produced up to 54 watts electrical
equivalent, with thermal efficiencies up to 83%. This results in higher energy densities of up to
1700 Whr/kg, based on a 48% efficient fuel cell[1], a 3-kg device hardware weight, a 14-day
mission, water recycle, anode gas recycle, and a parasitic power loss of 5W.

For both units, typical gas compositions before CO cleanup were 72 to 74% hydrogen, 2
to 26% carbon dioxide, and 1.0% or less CO on a dry gas basis. A CO cleanup unit was bread-
boarded with the larger fuel processor, and CO levels less than 100 ppm were demonstrated. The
cleaned reformate gas was fed to an H-Power fuel cell (Figure 4). This test validated the 48%
efficiency fuel cell assumption and, more importantly, it demonstrated the feasibility of this
approach.

SUB-WATT POWER

A miniature power supply is being developed that can provide 10 to 500 mW_e power for
microsensors, MEMS, and other autonomous microelectronics. Figure 5 shows the integrated
microscale fuel processor, which has a volume of less than 0.25 cm^3 and a mass of less than 1
gram. The system incorporates two vaporizers/preheaters, a heat exchanger, catalytic combustor
and methanol steam reformer.

The fuel processor was operated over a wide range of conditions to obtain performance
data. The operating temperatures required to achieve >99% conversion, 350 to 450°C, were
higher than anticipated, and were attributed to the internal flow patterns, faster contact times than
used in the catalyst screening tests, and thermal losses to the environment. The first- generation
system produced between 18 and 100 mW_t (based on the lhv) of hydrogen at a thermal efficiency
of 3 to 9%. However, these efficiencies were considerably lower than reported above, because

---

[1] Fuel cell efficiency assumes that 80% of the hydrogen fed to the fuel cell is reacted, and that the fuel cell converts
60% of the chemical energy to electrical energy. This gives a fuel cell efficiency of 80%*60%=48%.

the thermal losses as a percent of the thermal power out increase as with a decrease in thermal power out. The fuel processor was bread-boarded with a mesoscale high temperature (150°C) fuel cell (Figure 6), developed by Case Western University. This micropower system operated on methanol and was able to generate over 20 mW$_e$ of power (Figure 7).

The second-generation fuel processor has the same envelope and packaging as the original, but improves space utilization and thermal management. These improvements resulted in a decrease in operating temperatures from 350°C to 250°C while increasing the output and efficiency to 320 mW$_t$ and 23%, respectively. The CO also decreased from 1.8% to less than 1% and under some operating conditions to less than 0.5%. The decrease in CO was attributed to the decrease in operating temperatures.

This second-generation processor will be further tested with a new mesoscale fuel cell. For the balance-of-plant components, state-of-the art insulation, pumps, and micro-valves are expected to be sufficient; however, for the gas delivery system, alternative designs are being developed.

Figure 5. Microscale fuel processor   **Figure 6.** Mesoscale fuel cell

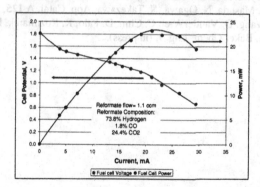

Figure 7. Microscale fuel processor and mesoscale fuel cell operation.

CONCLUSIONS

Lightweight, compact, power supplies are being developed that incorporate a microscale fuel processor, a fuel cell, and a secondary battery and deliver 25 to 300 watts of electric power. Miniature fuel processors, discussed here, have demonstrated 25 to 50 watts equivalent at high thermal efficiencies of up to 83% for an estimated energy density of up to 1700 Whr/kg. These processors have a volume of less than 25 cm^3 and a mass less than 200 grams. Initial tests with CO cleanup show a decrease in CO to less than 100 ppm, which is acceptable for many CO-tolerant fuel cells. A bread-boarded system composed of a Battelle fuel processor, a carbon monoxide cleanup unit, and an H-Power fuel cell has been demonstrated. Balance-of-plant components, such as pumps and blowers, have been identified and will be integrated in the next phase of testing.

A processor that can deliver less than 1 watt equivalent of hydrogen also has been demonstrated. The first-generation system had an efficiency of up to 9%. This system was tested with a mesoscale fuel cell and produced over 0.020 watts of electricity. An improved design (volume of <0.25cm^3 and mass of <1 gram) increased the power to 0.300 watts thermal (an increase of 50% over the original design) and the thermal efficiency to over 20%.

ACKNOWLEDGMENTS

This work has been supported by Jim Stephens of the US Army Communications-Electronics Command under Contract DAAD05-99-D-7014 and by Dr. William Tang and Dr. Clark Nguyen of the Defense Advanced Research Projects Agency (DARPA) under Contract DABT63-99-C-0039, and their support is gratefully acknowledged.

REFERENCES

1. D. Palo, J. Holladay, R. Rozmiarek, C. Guzman-Leong, Y. Wang, J. Hu, Y. Chin, R. Dagle, E. Baker, J. Pwr. Sources **108** 28 (2002).
2. J.D. Holladay, E.O. Jones, M.R. Phelps, and J. Hu, J. Pwr. Sources **108**, 21 (2002).
3. N. Iwasa, S. Masuda, N. Ogawa, N. Takezawa, App. Catal. A **125**, 145 (1995).
4. Hu, J., Y. Wang, D. VanderWiel, C. Chin, D. Palo, R. Rozmiarek, R. Dagle, J. Cao, J. Holladay, E. Baker Chem. Eng. J. (in press).

Mat. Res. Soc. Symp. Proc. Vol. 756 © 2003 Materials Research Society   FF9.6

Effective Hydrogen Separation Using Ion Beam Modified Polymeric Membranes

M. R. Coleman, X. Xu, J. Ilconich, J. Ritchie, and L. Hu,

Univ. of Toledo, Dept of Chem. & Env. Eng., Toledo, Ohio.

Abstract

High purity H_2 gas streams are increasingly important for a variety of applications including feed gases for fuel cells. The potential of hydrogen gas as primary energy source has generated considerable interest in hydrogen separation technologies. We have been investigating ion beam irradiation as a method to modify polymeric membranes to enhance both hydrogen permeability and permselectivities. Combined high permeabilities and permselectivities are required to give high recoveries of high purity hydrogen. Ion irradiation typically results in the formation of numerous crosslinks within the polymer matrix that should enable these materials to maintain selectivities at high temperatures and to resist chemical attack. Helium separations over a range of temperatures of irradiated polyimides were used as a model of the hydrogen system. Finally, the impact of irradiation conditions on gas separations in these materials will be addressed.

Introduction and Background

The purity requirements for H_2 feed streams to fuel cells depend upon the fuel cell membrane. While a PEM fuel cell requires a very high purity stream with no more than 10 ppm CO and a solid oxide fuel cell require lower purity feed streams. The goal of this work is to develop membrane materials to recover H_2 from reforming streams. Based upon economic analysis, a membrane with a H_2/CO selectivity of at least 100 to 150 at the operating temperature would be required to provide a high purity hydrogen stream with 90 % recovery of the H_2[1]. In addition, the material must maintain high selectivities at elevated temperatures. The majority of commercial membrane are polymeric because of their low material costs, ease of processibility and generally attractive gas transport properties (i.e. combined high permeance and permselectivity). While considerable progress has been made in developing membranes for a wide range of gas separations at low temperatures, the primary limitations to existing commercial membranes are fine separations of species at elevated temperatures or from feed streams that contain highly soluble penetrants[2]. The development of advanced membrane materials relies on the ability to manipulate the material microstructure to fulfill the following conditions: (i) high fractional free volume (FFV); (ii) narrow free volume distribution which can precisely sieve the given gas pair; and (iii) a rigid microstructure.

Since synthesis of new polymers is both time consuming and quite expensive, the focus of much recent work has been developing post-synthesis modification techniques. We are investigating ion beam irradiation as a post-synthesis method to modify the structure and transport properties of thin polymer surface layers. With an appropriate choice of irradiation conditions (i.e. ion type, energy, and fluence), energetic ions can modify the surface layer of materials within a well-defined depth. The intensive energy deposition of the incident ion leads to the following chemical processes within the polymer surface layer [3]: (i) degradation of the polymer chains with formation of free radicals and volatile molecules which can leave packing defects in the polymer matrix, (ii) crosslinking between the polymer chains, and (iii) formation of new chemical bonds such as double bonds. These chemical processes typically result in an evolution of the chemical structure of the surface layer from the virgin polymer to a graphite-like material with increasing ion dose [3-4]. There is also a significant evolution in the packing microstructure of polymers with increasing ion dose [5]. Since the transport properties of polymeric materials are dependent upon a combination of the openness of the polymer matrix and the rigidity of the polymer backbone, ion irradiation would be expected to result in a significant evolution in the permeability and permselectivity of polymers [6].

Studies of structural evolution in polyimides demonstrate irradiation results in both formation of crosslinks (increase rigidity) and significant variation in free volume. Two fluorine containing polyimide isomers (6FDA-6FpDA and 6FDA-6FmDA) and the commercially available Matrimid® were used for this study. These polymers have attractive inherent transport properties and have been used as precursors for producing carbon molecular sieving membranes. The 6FDA based polyimide isomers have identical chemical structure with the exception of the bond at the imide linkage [9]. The 6FDA-6FmDA has a meta connect linkage which results in a more rigid and densely packed microstructure. The para connected 6FDA-6FpDA has a more open flexible structure with higher permeabilities and lower selectivities. The impact of ion irradiation on the chemical structure of the three polyimides used in this study has been reported elsewhere [7,8]. While the structural evolution of the polymers following irradiation was similar, the impact of irradiation on the permeation properties was quite different. This paper addresses the impact of irradiation on the He permeance and permselectivities of these polyimides as a function of irradiation conditions.

Experimental

Materials and Membrane Formation

Matrimid® was purchased from Ciba Specialty Chemicals Company and the 6FDA-based polyimides were synthesized using the method established by Husk et.al [10]. In each case, a series of defect free polyimide-composite membranes were irradiated with either H^+ or N^+ at increasing doses. Composite membranes were used to isolate the impact of irradiation on the bulk polymer properties. The thickness of the polymer selective layer was estimated using the bulk O_2 permeability and membrane permeance. All membranes exhibited an O_2/N_2 selectivity that was at least 75 % of the bulk polymer value and were considered to be defect free.

Ion Beam Irradiation

Two set of composite membranes (6FDA-6FpDA and 6FDA-6FmDA) were irradiated with 180 keV H+ ions and one set of Matrimid® was irradiated with 400 keV N+ions. Implantation was performed at the Department of Physics and Astronomy, University of Western Ontario, Canada, using a Tandem Accelerator. All irradiation was performed at room temperature within a vacuum chamber at a pressure less that 1.9×10^{-7} torr.

Permeation Measurements

The permeance of He, O_2, N_2, CO_2, and CH_4 were measured for each virgin and irradiated membrane to isolate impact of irradiation on polymer properties [11]. For most polymers, He has permeabilities that are similar to those of H_2 and CO is typically similar to N_2 and CH_4 permeabilities [12]. Therefore the He/CH_4 systems were used as model probes for H_2/CO. Permeation measurements were made 35 psia and 35 °C in a standard permeation cell [11,12]. The results are presented in terms of relative permeance and permselectivity to eliminate any subtle difference between virgin membrane samples. The relative values are the ratio of irradiated membrane properties to virgin membrane properties.

Results and Discussion

The impact of H^+ irradiation dose on the relative permeances at 35 °C of a series of 6FDA-6FpDA-ceramic composite membranes is shown in Figure 1. Following irradiation at low doses ($<1 \times 10^{14}/cm^2$), there was a general decrease in the permeance relative to the bulk material with a small drop in permselectivity. Irradiation at intermediate doses resulted in general increases in permeance, particularly for large penetrants such as CH_4, and decreases in permselectivity. Following irradiation at doses greater than 2×10^{15} H^+/cm^2, there was a general trend of increase in permeance of small penetrants (He and O_2) with a sharp drop off in the permeance of larger penetrants (i.e. CH_4 and N_2). This results in large increases in permselectivity at higher doses. For example, irradiation at 1×10^{16} H^+/cm^2 resulted in He permeability of 260 Barrer and permselectivities for He/CH_4 of 900, He/N_2 of 220 and He/CO_2 of 5.7. By comparison, the bulk or virgin polymer has a permeability of 140 Barrer with permselectivities for He/CH_4 of 86, He/N_2 of 39 and He/CO_2 of 2.1. In general, helium has slightly lower permeabilities that hydrogen, so similar results are expected for hydrogen separations. Note that for each gas there was a general trend of increase permeance with dose followed by a sharp drop off at higher doses. This would imply a sharp evolution in the microstructure of the material at high doses to a relatively high free volume material with the free volume distributed in the form small packets. Similar results were seen for very preliminary studies of N^+ irradiation of 6FDA-6FpDA with the maximums occurring at much lower doses.

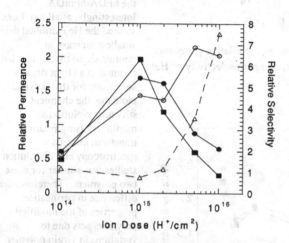

Figure 1: Impact of H^+ ion dose on the relative permeance of He (o), CO_2 (■) and CH_4 (●) and permselectivity of He/ CH_4 (Δ) in 6FDA-6FpDA.

A series of Matrimid® membranes were irradiated with N^+ ions at doses up to 1.5×10^{15} N^+/cm^2. N^+ irradiation provides much greater energy transfer and subsequent modification in polymer structure and properties at lower doses than H^+. Irradiation of Matrimid® composite membranes at low doses ($<6 \times 10^{14}$) resulted in increases in permeance for smaller species (i.e. O_2 and He) combined with sharp decrease in permeance for large molecules (Figure 2). The net results were a simultaneous increase in permeance and selectivity for membranes following low dose irradiation. Irradiation at higher doses resulted in a gradual decrease in He permeance with large improvements in permselectivity. While the bulk polymer has a He/CH_4 selectivity of approximately 80, irradiated samples with selectivities as high as *380* were characterized.

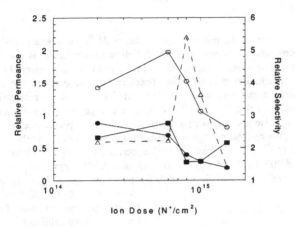

Figure 2: Impact of N^+ ion dose on the relative permeance of He (O), CO_2 (■) and CH_4 (●) and permselectivity of He/CH_4 (Δ) in Matrimid® at 35 °C.

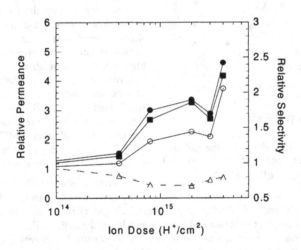

Figure 3: Impact of H^+ ion dose on the relative permeance of He (O), CO_2 (■) and CH_4 (●) and permselectivity of He/CH_4 (Δ) in 6FDA-6FmDA.

A series of 6FDA-6FmDA composite membranes were irradiated with H^+ over a similar dose range and the pure gas permeabilities at 35 °C were measured (Figure 3). In contrast to the results for the para connected 6FDA-6FpDA, there was a larger increase in permeance for each penetrant studied with increasing irradiation dose for the 6FDA-6FmDA. Interestingly, at all ion doses studied the He exhibited the smallest increase in permeance so that irradiation resulted in a slight drop in selectivity for this polymer. Note that the chemical structure evolution as monitored using Fourier transform infrared spectroscopy and dissolution studies was similar for these two polymers. Therefore, the difference in permeation properties of the modified polymers was due to variations in virgin polymer microstructure and chain mobility. The more densely packed polyimide (6FDA-6FmDA) exhibited the largest increases in permeance following irradiation. This is consistent with the increase in the free volume resulting from evolution of gas molecules during irradiation process. Any increase in chain rigidity resulting from crosslinking does not appear to have a significant impact on the selectivity of the more inherently rigid 6FDA-6FmDA [9]. While these results are interesting, the irradiated 6FDA-6FmDA did not exhibit transport properties that were as good as those for the 6FDA-6FpDA. For example, following irradiation at 4×10^{15} H$^+$/cm^2, the 6FDA-6FmDA had a He permeability of 110 Barrer with a selectivity for He/CH$_4$ of 166 and He/N$_2$ of 74. Based upon these results microstructure chain packing of the virgin polymer has a significant effect on the properties of the irradiated sample.

Ion beam irradiation results in significant modification in the chemical structure of polymers including the formation of crosslinking along with modifications in free volume. Crosslinking has been used in several studies as a method to increase rigidity of polymer backbone to decrease losses in selectivity at elevated temperatures. The pure gas permeabilities were determined for a sample of 6FDA-6FpDA irradiated with H$^+$ at 1×10^{16} ion/cm^2 at temperatures up to 80 $^\circ$C. As is typically for polymeric materials, there was an increase in permeability and loss in selectivity with increasing temperature. The irradiated membrane had much higher selecitivities than the virgin polymer at 35 $^\circ$C and maintained those selectivities at temperatures as high as 80 $^\circ$C. Indeed even at higher temperatures the irradiated had improved permeation properties relative to the virgin material at 35 $^\circ$C.

As demonstrated in the trade-off curve for the He/CH$_4$ gas pair in Figure 4, several of the irradiated polymers were well beyond the permeability-permselectivity trade-off curve for traditional polymers (as indicated by solid line on graph). Note that the bulk polymers exhibited permeabilities that were at or near the trade-off curve so that these polymers have inherently attractive transport properties. For example, irradiation of 6FDA-6FpDA at 1×10^{16} H$^+$/cm^2 resulted in He permeabilities of approximately 260 Barrer and He/CH$_4$ of 900, He/N$_2$ of 220 and He/CO$_2$ of 5.7. This is represents a two-fold increase in He permeance and a five-fold increase in He/CH$_4$ selectivity. The H$_2$/CO selectivity is expected to fall between the He/CH$_4$ and He/N$_2$ which puts it in range needed for the hydrogen separations. Similar results were seen for the Matrimid® following N$^+$ irradiation. Note that while irradiation of Matrimid® did not result in a material beyond the trade-off curve at doses studied, the relatively low cost of this polymer combined with an ability to maintain permselectivities at elevated temperatures make it a promising system for further study. Since irradiating ions only penetrant into the near surface layer, all membranes maintained mechanical stability following irradiation even at very high doses.

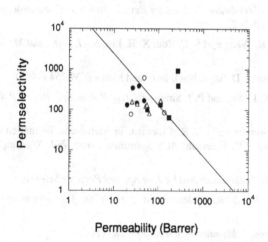

Permeability (Barrer)

Figure 4: He permeability versus He/CH4 selectivity for H$^+$ irradiated 6FDA-6FpDA($\blacksquare$), N$^+$ irradiated Matrimid ($\bullet$) and H$^+$ 6FDA-6FmDA (Δ), and bulk polymers (o). Solid line represents tradeoff curve for traditional membrane polymers.

Conclusions

The impact of H+ irradiation on the permeance and permselectivity of several gases in fluorine containing polyimide isomers (6FDA-6FpDA and 6FDA-6FmDA) and the commercially available Matrimid® were reported. Irradiation at intermediate doses resulted in sharp increases He permeability for the para connected 6FDA-6FpDA and Matrimid® with corresponding large increases in permselectivities. Interestingly, irradiation at similar doses led to increases in permeability with small decreases in permselectivity for the meta connected 6FDA-6FmDA. Therefore, the backbone structure of the virgin material played a vital role in determining properties of gas transport irradiated materials. In addition, the irradiated 6FDA-6FpDA maintained He/CH$_4$ and He/CO$_2$ selectivities at elevated temperatures that would allow for use in membrane systems for production of high purity H$_2$ streams.

Acknowledgements

The support of the National Science Foundation through the Presidential Faculty Fellows (CTS-955367) and CTS 9975452 in funding this project is acknowledged. In addition the authors would like to thank Dr. Peter Simpson of the University of Western Ontario for providing irradiation of the samples.

References:

1. Robeson, L.M., W.F. Burgoyne, Langsam, M., Savoca, A., Tein, C. *Polymer*, **35**, 4970 (1994).

2. Lipscomb, G. G. in *The 1996 Membrane Technology Review*, Business Communications Co., Norwalk, CT, 1996, pp. 23-102.

3. Xu, D., X. L. Xu, G. D. Du, R. Wang and S. C. Zou, X. H. Liu, *Nucl. Instr. and Methods B* **81** 1063 (1993)

4. J. Davenas, X. L. Xu, G. Boiteux, D. Sage, Nucl. Instr. and Meths B **39** 754 (1989)

5. Myler, U., M. R. Coleman, X. L. Xu, and P. J. Simpson, , *J. of Polym. Sci., Part B Polym. Phys.*, **36** 2413 (1998)

6. Xu, X.L., U. Myler, P.J. Simpson, and M.R. Coleman, in Membrane Formation and Modification, ed. Pinnau and B.D. Freeman, *ACS Sympsium Series 744*, Washington, D.C., 205-227 (2000)

7. Hu., L., Xu. X.L., and M.R. Coleman, accepted by *J. of Applied Polymer Science.*

8. Ilconich, J., Hu, L., Xu, X., M.R. Coleman, in preparation for submission to *Macromolecules.*

9. Coleman, M.R. and W.J. Koros, *J. Membrane Science*, 50, 285 (1990)

10. G.R. Husk, P.E. Cassidy, and K.L. Gebert, *Macromolecules*, **21** 1234 (1988)

11. Ilconich, J., Xu, X.L., and M.R. Coleman, *J. Membrane Science*, in press.

12. Koros, W.J. and R.T. Chern in *Handbook of Separation Process Techonology*, ed. R.W. Rousseau, John Wiley and Sons (1987).

Mat. Res. Soc. Symp. Proc. Vol. 756 © 2003 Materials Research Society   FF6.3

In-situ x-ray absorption experiments with a PEM fuel cell in hydrogen and methanol operation mode

Th. Buhrmester, C. Roth, N. Martz, H. Fuess

University of Technology Darmstadt, Department of Materials Science, Structure Research, Petersenstrasse 23, D-64287 Darmstadt, Germany
Fax: +49 6151 16-60 23; E-mail: buhrmester@tu-darmstadt.de

ABSTRACT

In-situ XAS studies have been carried out on PEM fuel cells in methanol and hydrogen operation. This has become possible due to a recently developed PEM fuel cell (single cell) equipped with inherent carbon fibre windows to allow x-ray studies in transmission geometry. The set-up chosen allows the *in-situ* monitoring of the structural changes of the Pt-Ru catalyst, utilised as electrode active material, during cell operation. The analysis of the white line intensities, edge shifts of the Pt-L_{III} edge and the XAFS signal, which was modelled for the first co-ordination shell around Platinum, exhibited no significant changes during operation in terms of the first neighbouring shell. Nevertheless, the white line intensity decreased comparing the *ex-situ* to *in-situ* measurements. From the latter, can be concluded that the catalyst is reduced during operation and further, the redox behaviour of the catalyst does not change the local environment of the Pt-centres in the first co-ordination shell to a measurable extend.

INTRODUCTION

Pt-Ru systems on carbon support are currently regarded as the state of the art anode catalysts in polymer electrolyte membrane fuel cells (PEMFC) and direct methanol fuel cells (DMFC). The performance, the life time and especially the CO tolerance of such cells can somehow be controlled by additional parameters (i.e. particle size distribution, degree of alloying) during manufacturing. Accordingly, this system was under manifold *ex-situ* [1,2] and *in-situ* [3,4,5,6,7] investigation. Catalysts have been characterised by different groups and different methods before and after cell operation but still these so called quenched situations can only be assumed to exist during operation and are not representative for a real *in-situ* measurement. This view is also supported by the work of O'Grady et al. [6], and Russell et al. [7]. To closely relate the behaviour of a PEMFC (*E/i*-curves) to the catalyst performance, the structural features of the material have to be monitored *in-situ* accordingly.

So far, only a few investigations under fuel cell relevant conditions have been published [6,7], and to our knowledge even only one *in-situ* study in reformate-air operation by Viswanathan et al [8,9]. This work is in line with the results of our *in-situ* XAFS experiments in hydrogen and methanol operation, which will be presented in this paper.

EXPERIMENTAL DETAILS

The membrane electrode assemblies (MEA) were produced by utilising a conventional spraying method [10]. Three (MEAs) have been prepared at the ZSW in Ulm by applying a metal loading of approx. 1.2 mg cm^{-2} metal per electrode on Nafion©117 used as proton conducting membrane. This comparatively high metal loading was necessary to improve the signal-to-noise ratio of the spectra. A standard 20 wt. % Pt on Vulcan XC-72 catalyst was used at the cathode, while Pt-Ru (1:1) (20 wt. % on Vulcan XC-72, E-TEK Inc.) served as anode catalyst. After conditioning in hydrogen mode, the MEAs were either dried or stored under distilled water and transferred to the beamline X1 at the synchrotron facility HASYLAB in Hamburg (Germany).

The *in-situ* XAFS experiments were carried out in transmission geometry with the beam perpendicular to the MEA. A severe drawback of this method is that both electrodes are monitored simultaneously and thus contribute to the spectra according to their metal loading. To overcome this disadvantage, a tiny part of the anode (about the size of the incident x-ray beam) was removed although this might lead to a significant change in the current distribution. A commercially-available fuel cell (single cell, Electrochem Inc.) was modified according to fig. 1. Rectangular holes were milled in both end plates, and the graphite blocks were thinned in the same areas such that only a 4 mm distance of weakly-absorbing graphite is interacting with the beam in addition to the catalyst on the MEA. The diminished thickness

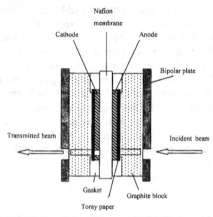

Fig. 1 Schematic cross-section of the fuel cell used for the in-situ XAFS experiments.

has proved to be both, thin enough to give no detectable contribution to the spectra, and thick enough to keep the cell gas-tight. A combined single cell set-up, suitable for both hydrogen as well as direct methanol operation, was constructed and put into operation. The feed gases were supplied by commercially-available mass flow controllers. In hydrogen operation, 150 ml min^{-1} H$_2$ (N 5.0, Linde) were saturated with water vapour at $T = 80$ °C and fed into the fuel cell anode. As cathode feed, high-purity oxygen was supplied (N 4.5, Linde) at 75 ml min^{-1}. High-purity instead of technical purity gases were chosen to guarantee defined conditions and to exclude promoting or inhibiting effects by trace impurities in the feed. In DMFC mode, 1 ml min^{-1} of a 1 M aqueous methanol solution was fed into a vaporizer by a peristaltic pump and evaporated at $T = 150$ °C. The fuel cell supply with the methanol vapour was realized by a supporting nitrogen gas flow (50 ml min^{-1}). A flow of 100 ml min^{-1} oxygen (N 4.5, Linde) served as cathode feed in direct methanol operation. In contrast to a cell temperature of 75 °C for the hydrogen measurements, in methanol mode the cell was heated to $T = 95$ °C and held at this temperature during operation. Subsequent to half an hour of short-circuit operation, E/i curves were measured galvanostatically, beginning at the steady-state current at the short-circuit potential and decreasing it stepwise to zero current. Before recording a single XAFS spectrum, a holding time of 30 minutes at each current value was applied, after which a quasi steady-state potential had been reached.

In fig. 2, one of the measured E/i curves in direct methanol operation at 95 °C is shown. Potentials, at which a spectrum was recorded, are marked by an arrow. The fuel cell performance is slightly inferior to the standard performance, probably due to the partly removed cathode. However, spectra were taken at a current density of 0, 40, 48, 68, 80, 88, 160 and 200 mA cm^{-2}, respectively.

The X-ray spectra were recorded at the beamline X1 at Hasylab, Hamburg (Germany),

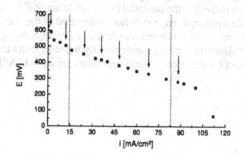

Fig. 2 E/i curve in direct methanol operation; potentials, at which a spectrum was recorded, are marked by an arrow

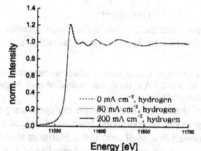

Fig. 3 Comparison of the uncorrected XAS spectra in hydrogen operation.

which offers a large X-ray energy range that also covers the absorption edges of the elements within our interest, at a comparably high relative energy resolution of approx. 0.5 eV at the Pt L_{III}-edge. The storage ring was operated at an energy of 4.45 GeV and an initial positron beam current of 150 mA. All spectra were recorded in transmission geometry in an energy range of 11.550 up to 12.900 keV (Pt L_3-edge at 11.564 keV), using a Pt foil as reference. A Si(111) monochromator was applied in all cases and detuned to 50 % intensity to minimize the presence of higher harmonics. Three gas-filled ion chambers were used in series to measure the intensities of the incident beam, the beam transmitted through the sample and finally by the Pt foil that was used as energy reference. Spectra were recorded in different regions of the E/i curve and processed using the software package Viper© [11,12,13].

The uncorrected spectra of three current densities 0 mA cm^{-2}, 80 mA cm^{-2} and 200 mA cm^{-2} are shown in fig. 3 for hydrogen operation. In fig. 4 the equivalent R-space representation is shown. Fig. 5 shows the uncorrected XAS spectra comparison of the cell in methanol and hydrogen operation as well *ex-situ* with a window to the relevant energy range directly at the Pt-L_{III} edge for comparison with fig. 3. The R-space representation in methanol operation does not differ significantly from fig.4 and is therefore omitted due to limited space.

DISCUSSION

Evidently, no significant changes appear neither in the uncorrected spectra under different conditions with respect to the potential curve (at different E/i values), nor in the r-space representation. No significant edge shift can be detected in fig. 3 as well as no significant difference in the white line intensities in the *in-situ* recorded spectra. In fig. 4 the R-values are shown for similarly corrected and similarly treated spectra using methanol for fuelling. The values taken for the mathematical operations are $k_{min}=3$ [Å^{-1}] and $k_{max}=15.$ [Å^{-1}]

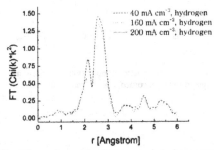

Fig. 4 Comparison of the XAS spectra in hydrogen operation in r-space. d(Pt-O) = 2.1 Å, d(Pt-Pt) = d(Pt-Ru) = 2.7 Å.

The Fourier transforms show two significant distances, which can be attributed to Pt-O (about 2.1 Å) and Pt-Pt and Pt-Ru, respectively, at approx. 2.7 Å in the first co-ordination shell. Higher co-ordination shells could not be fitted to the data sets successfully.

The short-range order of the Pt-Ru anode catalyst in dependence of the potential does not exhibit a significant change as can be concluded from fig. 4. In addition fig. 6 is illustrating a fit of back transformed first shell around the Platinum absorber to the Pt-metal model, which exhibits very good agreement in phase and amplitude. Hereby it can reversibly be concluded, that no significant contribution other than that of Pt-metal is present in the spectra.

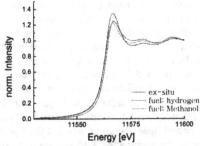

Fig. 5 Comparison of the uncorrected XAS spectra of hydrogen and methanol operation mode in E-space and *ex-situ*.

However, it can not be excluded that the removal of the cathode side catalyst in the beam window region modifies the original current distribution in this part of the electrode to a far greater extent than assumed.

Table I: Results of the Viper fits to the first shell around Pt for different fuels at different current densities. Statistical errors are also supplied.

fuel	current density [mA cm^{-2}]	r [Å]	N	s [Å^2]	E [eV]
hydrogen	40	2.7601 ±4.437E-3	8.3167 ±5.586E-1	6.0800E-3 ±4.772E-4	9.9890 ±6.093E-1
hydrogen	80	2.7616 ±5.014E-3	8.3801 ±6.411E-1	5.8700E-3 ±5.274E-4	10.709 ±6.974E-1
hydrogen	160	2.7590 ±3.957E-3	8.3481 ±4.820E-1	6.4100E-3 ±4.299E-4	9.9330 ±5.197E-1
hydrogen	200	2.7599 ±4.567E-3	8.3314 ±5.590E-1	6.3500E-3 ±4.954E-4	10.023 ±6.023E-1
methanol	0	2.7583 ±5.271E-3	8.3773 ±5.706E-1	7.4600E-3 ±5.852E-4	9.5510 ±6.139E-1
methanol	48	2.7546 ±4.391E-3	8.3821 ±4.990E-1	7.2900E-3 ±5.003E-4	9.4670 ±5.204E-1
methanol	68	2.7534 ±4.801E-3	8.4073 ±5.381E-1	7.3200E-3 ±5.393E-4	9.4320 ±5.679E-1
methanol	88	2.7521 ±4.697E-3	8.3891 ±5.243E-1	7.3800E-3 ±5.311E-4	9.4720 ±5.571E-1

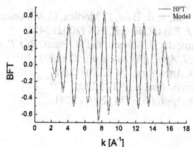

Fig. 6 Comparison of the model (Pt-metal) and a BFT from the first co-ordination shell.

CONCLUSIONS

We performed *in-situ* measurements by applying XAS experiments on a fuel cell in different states of operation (in different regions of the potential curve, E/i curve) using two different fuels (methanol and hydrogen). The uncorrected E-space data sets did not exhibit a significant edge shift or significant changes in the white line intensities within the measured series utilising the two different fuels. However, compared to the *ex-situ* data, the white line intensity decreases. Associated with this behaviour is a reduction of the catalyst from *ex-situ* state to operational conditions.

According to the E-space data sets the corresponding r-space curves do not differ significantly from each other. This leads to the conclusion, that the local environment in the first shell around the Platinum centres does neither vary with fuel nor with different established potential conditions (E/i curve). The values resulting are: $d(Pt-Pt) = 2.758(3)$ Å, co-ordination number $N = 8.3$ and Debye-Waller factor $s = 9.55$ Å^2.

AKNOWLEDGMENTS

Thanks are due to R. Theissmann, and A. Adams for their help during the XAFS measurements. Financial support of the Deutsche Forschungsgemeinschaft (Fu 125/34-1-3) is gratefully acknowledged.

REFERENCES

1 J. McBreen: *Physical Electrochemistry: Principles, Methods and Applications.* I. Rubinstein, Marcel Dekker, Inc. New York Basel Hong Kong (1995) 339-391.

2 J. McBreen, S. Mukerjee: *Interfacial Electrochemistry.* A. Wieckowski (Editor), Marcel Dekker, New York (1998) 895-914.

3 B. Gurau, R. Viswanathan, R. Liu, T. J. Lafrenz, K. L. Ley, E. S. Smotkin, E. Reddington, A. Sapienza, B. C. Chan, T. E. Mallouk, S. Sarangapani, *J. Phys. Chem. B*, **102** (1998) 9997-10003.

4 R. Liu, H. Iddir, Q. Fan, G. Hou, A. Bo, K. L. Ley, E. S. Smotkin, Y.-E. Sung, H. Kim, S. Thomas, A. Wieckowski, *J. Phys. Chem. B*, **104** (2000) 3518-3531.

5 V. Radmilovic, H. A. Gasteiger, P. N. Ross, *J. Catal.*, **154** (1995) 98-106.

6 W. E. O'Grady, P. L. Hagans, K. I. Pandya, D. L. Maricle, *Langmuir*, **17** (2001) 3047.

7 R. A. Lampitt, L. P. L. Carrette, M. P. Hogarth, A. E. Russell, *J. Electroanal. Chem.*, **460** (1999) 80-87.

8 R. Viswanathan, G. Hou, R. Liu, S. R. Bare, F. Modica, G. Mickelson, C. U. Segre, N. Leyarovska, E. S. Smotkin, *J. Phys. Chem. B*, **106** (2002) 3458.

9 C. Roth, N. Martz, Th. Buhrmester, J. Scherer and H. Fuess, *PCCP*, **4** (2002) 3555.

10 M. S. Wilson, S. Gottesfeld, *J. Appl. Electrochem.*, **22** (1992) 1.

11 K. V. Klementev, *J. Phys. D. Appl. Phys.* **34** (2001) 209.

12 K. V. Klementev, *J. Synchrotron Rad.* **8** (2001) 270.

13 K. V. Klementev, *J. Phys. D. Appl. Phys.* **34** (2001) 2241.

Mat. Res. Soc. Symp. Proc. Vol. 756 © 2003 Materials Research Society   FF1.5

Enhanced Graphite Fiber Electrodes for a Microbial Biofuel Cell Employing Marine Sediments

Gregory Konesky
ATH Ventures Inc., 3 Rolling Hill Rd.
Hampton Bays, NY 11946

ABSTRACT

A microbial biofuel cell has been demonstrated utilizing organic material in the sediment as fuel and dissolved oxygen in the overlying seawater as the oxidizer. A graphite electrode placed in the sediment acts as the anode and collects electrons both by mediated and non-mediated processes. Another graphite electrode suspended in the seawater above the sediment acts as the cathode and transfers these electrons to oxygen in the seawater. The sediment serves as a natural permeable membrane that permits hydrogen ions to flow from within the sediment and combine with the oxygen to produce water. Electrons which flow from the anode to the cathode through an interconnecting wire are used to power external circuits. Both fuel and oxidizer are naturally present and self-renewing, and the graphite electrodes are inexpensive and non-toxic. Overall, this is a very "green" fuel cell. A significant improvement in collection efficiency is demonstrated by using graphite fiber electrodes.

INTRODUCTION

Microbial biofuel cells function in a similar manner to traditional hydrogen/oxygen fuel cells. In the latter case, electrons are catalytically removed from hydrogen, producing hydrogen ions (protons). The liberated electrons are collected by an electrode and flow through an external circuit, developing useable power. They then return from the circuit to a second electrode where they are transferred to oxygen. The hydrogen ions migrate across an ion-permeable membrane separating the two electrodes, where these ions combine with the oxygen to create water.

In microbial biofuel cells, nutrients from the surrounding environment are metabolized by respiratory enzymes, liberating electrons. These electrons are ultimately transferred to Terminal Electron Acceptor Processes (TEAP) [1]. In aerobic organisms (such as ourselves), oxygen is the TEAP, and again, water is created. Anaerobic organisms have a much wider range of TEAPs including nitrates, sulfates and a variety of metal ions.

Microorganisms in biofuel cells have their respiratory electrons "stolen" from them, either by mediated or non-mediated processes, and are transferred to a collection electrode. These electrons power an external circuit and then return to a second electrode where they are transferred to a TEAP. A microbial biofuel cell may employ oxygen as the TEAP, even though the microorganism itself is anaerobic, so that ambient air can be used as a source of oxygen. This is possible since the respiratory electron transfer occurs remotely, through a wire, so to speak, rather than by direct exposure to oxygen, which the anaerobic microorganism might find fatal.

Transfer of respiratory electrons to the collection electrode may be significantly enhanced by the use of mediators. These are molecules that penetrate the cell wall of the microorganism and collect electrons from respiratory enzymes. The electron carrying mediators then diffuses back out through the cell wall and eventually transfers these electrons to a collection electrode.

The mediator molecules shuttle between the collection electrode and the microorganism, enhancing electron transfer efficiency. Some examples of mediators are Methylene Blue, Neutral Red and Thionine. Mediators, in general, must be easily oxidized and reduced, be soluble in sufficient concentration to be useful, and not be toxic to the microorganism in question.

Various microorganisms are known to have respiratory enzymes (Cytochromes) on their cell walls [2, 3] which permit them to use, for example, mineral surfaces as a TEAP [4]. Such microorganisms are said to "breathe minerals" [5]. In biofuel cell applications, these microorganisms attach themselves to the collection electrode and transfer their respiratory electrons directly to this electrode without the need for a mediator [6]. These processes are said to be mediator-less.

Microbial metabolic activity in marine sediments produces various chemical horizons with increasing depth in the sediment. This results from the utilization of a succession of TEAPs [1], starting with oxygen at the top of the sediment, and proceeding through nitrate, manganese, iron, and sulfate. As much as 0.75 volt potential difference can exist [7] between an electrode buried a few centimeters into the sediment and an electrode suspended in the overlying oxygenated seawater. Similar electrochemical potential differences are exploited by seawater batteries [8], which use oxygenated seawater and an aluminum anode. Electrical energy developed by electrodes in marine sediments, on the other hand, rely on organic material in the sediment, which is consumed by microorganisms that then maintain the chemical horizons.

Electrical energy collected as a result of these microbe-generated chemical horizons is an example of a mediated microbial fuel cell. However, Bond et al. [9] has demonstrated that non-mediated processes are at work as well. Two identical experiments were performed utilizing buried graphite electrodes in marine sediments and similar graphite electrodes suspended in the seawater above the sediment. One of the setups was left open-circuit, as a control. The other was shorted, permitting a continuous flow of respiratory electrons between the buried electrode and the electrode suspended in the oxygenated seawater. After running both experimental setups for 6 months, the surface of the buried shorted electrode was found to be enriched with Desulfuromonas acetoxidans by a factor of 100 over the control electrode. Similar enrichments were observed when the experiment was intentionally inoculated with Geobacter sulfurreducens and Geobacter metallireducens, all of which can transfer respiratory electrons directly to the electrode surface in a non-mediated fashion.

When "sediment batteries" are first constructed, initial electrical power is derived primarily from mediated processes. As current continues to flow from the buried electrode to the oxygenated electrode, an environment favorable to microbes that can transfer respiratory electrons directly to the electrode surface is established. These microorganisms flourish and eventually contribute a non-mediated component to the overall production of electrical power.

CONSTRUCTION OF MARINE SEDIMENT BIOFUEL CELLS

Microbial biofuel cells utilizing marine sediments are relatively easy to construct and employ low cost materials. Bond et al. [9] utilized solid graphite rods which permitted easy insertion into the marine sediments, but has limited surface area for energy collection. Reimers et al. [7] employed graphite felt, but had compacted it between fiberglass panels to improve its mechanical robustness. This had two unfortunate consequences. First, the available surface area for energy collection is reduced by the compaction process. Second, the fiberglass acts as a filter which reduces the opportunity for growth of non-mediated microbes on the graphite surface, minimizing this electrical power contribution. A factor of 11 improvement in electrical energy collection efficiency has been realized over these previous works on an equivalent geometrical area basis, utilizing 1 cm thick graphite fiber electrodes [10]. The graphite fiber material is sufficiently strong that no additional mechanical support is required. All electrical connection hardware is type 306 stainless steel, as are the interconnecting wires.

The basic microbial biofuel cell is shown in figure 1, below.

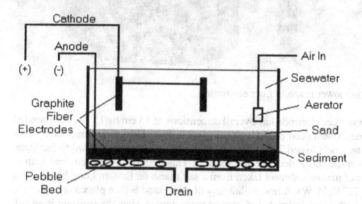

Figure 1a. Microbial biofuel cell outline.

Figure 1b. Two test biofuel cells.

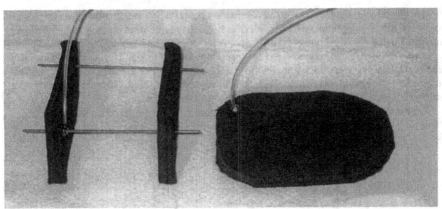

Figure 1c. Upper and lower graphite fiber electrodes.

The lower graphite fiber electrode has overall dimensions of 15 cm by 15 cm, but rounded corners reduce the equivalent geometrical area to approximately 196 square centimeters. This electrode is placed on a pebble bed 2 cm thick (average pebble diameter of 1 cm) to facilitate uniform drainage over the entire electrode surface area. This electrode is then covered with approximately 1 cm of marine sediment taken from a salt marsh on Eastern Long Island, New York (40°50.01'N, 72°30.71'W). A few millimeters of clean sand is then placed on top of the sediment. This acts both to minimize disturbance of the sediment when the seawater is added, and also to reduce consumption of organic matter in the upper layers of the sediment by aerobic microorganisms in the seawater. The latter process reduces the total amount of "fuel" in the sediment and consumes oxygen in the seawater while producing no harvestable electrical energy.

Several centimeters of seawater are carefully added to the cell, into which is then placed the upper electrode (two pieces, 15 cm by 6 cm) and an aerator stone commonly used in an aquarium. An air flow rate of 1 liter/minute is supplied to the aerator to maintain saturated oxygenation in the seawater. About one centimeter of seawater is then drained from the bottom of the cell to help draw sediment into the lower graphite electrode fiber bulk. This reduces the time to stabilize the cell voltage from several days to less than two days.

TEST RESULTS

The variation in performance between the two test cells was generally less than ten percent, and often was identical. Figure 2 illustrates the cell voltage under increasingly heavy loads. The non-linear internal impedance seen in this curve is characteristic of biofuel cells. It has the net effect of lowering the equivalent internal impedance as more current is drawn from the cell.

When a load is first applied, the cell voltage drops from its open circuit value, until the internal impedance matches the external load. It is important to wait a fixed period of time (25 minutes) for the voltage to stabilize before making a measurement, in order to get consistent readings. This effect is seen in figure 3.

The non-linear cell impedance with respect to load is also evident in the output power density versus current density shown in figure 4. Reimers et al [7] attributes this curve shape to

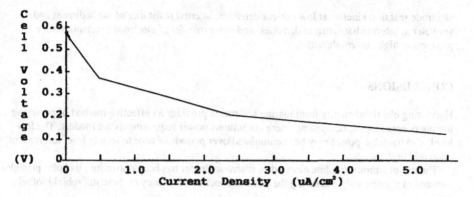

Figure 2. Cell voltage as a function of current density

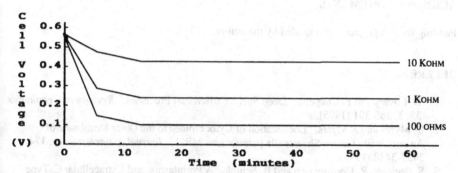

Figure 3. Cell voltage stabilization as a function of time

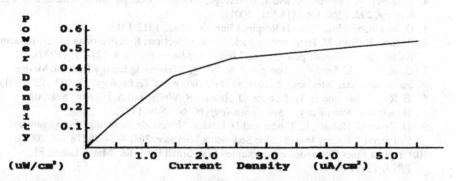

Figure 4. Power density as a function of current density

electrode reaction kinetics at low current densities, internal resistance of the sediment and seawater at intermediate current densities, and mass transfer of electrode reactants and/or products at high current densities.

CONCLUSIONS

Harvesting electrical energy from marine sediments provides an effective method of powering remote oceanographic equipment where continuous power requirements are modest. The low level of extractable power may be accumulated over periods of time to supply brief intervals of relatively high power.

The use of fibrous graphite electrodes, where sediment has been drawn into its bulk, provides a means of significantly enhancing the power collection efficiency of these microbial biofuel cells.

ACKNOWLEDGEMENTS

Funding for this project was provided by the author.

REFERENCES

1. D. Lovley and F. Chapelle, "Deep Surface Microbial Processes," *Reviews of Geophysics,* **33,** 3, 365-381 (1995).
2. C. Myers and J. Myers, "Localization of Cytochromes to the Outer Membrane of Anaerobically Grown Shewanella putrefaciens MR-1," *Journal of Bacteriology,* **174,** 3429-3438 (1992).
3. S. Seeliger, R. Cordruwisch and B. Schink, "A Periplasmic and Extracellular C-Type Cytochrome of Geobacter sulfurreducens," *Journal of Bacteriology* **180,** 3686-3691 (1998).
4. S. Lower, M. Hochella Jr. and T. Beveridge, "Bacterial Recognition of Mineral Surfaces." *Science,* **292,** 1360-1363 (18 May 2001).
5. D. Newman, "How Bacteria Respire Minerals," ibid., 1312-1313.
6. B. Kim, H. Kim, M. Hyun and D. Park, "Direct Electrode Reaction of an Fe(III)-reducing Bacterium Shewanella putrefaciens," *Journal of Bacteriology,* **9,** 127-131 (1999).
7. C. Reimers, L. Tender, S. Fertig and W. Wang, "Harvesting Energy from the Marine Sediment-Water Interface," *Environmental Science and Technology,* **35,** 192-195 (2001).
8. B. Rao, J Giacomo Jr. W. Kobasz, D. Hosom, R. Weller and A. Hinton, "Seawater Battery for Ocean Buoys," *Sea Technology,* 63-66 (Nov. 1992).
9. D. Bond, D. Holmes, L. Tender and D. Lovley, "Electrode-Reducing Microorganisms that Harvest Energy from Marine Sediments," *Science,* **295,** 483-485 (18 Jan. 2002).
10. Graphite Fiber Electrodes were obtained from Small Parts, Inc. Miami Lakes, FL. P/N B-FFHS-0912.

Solid Oxide Fuel Cells

Mat. Res. Soc. Symp. Proc. Vol. 756 © 2003 Materials Research Society EE1.6

New Results for Electron Transport, Chemical Diffusion and Stability of Solid Oxygen Ion Conductors

H.-D. Wiemhoefer, M. Dogan, S. Luebke, V. Ruehrup,
Institute for Inorganic and Analytical Chemistry, University of Muenster,
48149 Muenster, Germany

ABSTRACT

We describe the measurement of electronic conductivity of solid oxide electrolytes by a modified Hebb-Wagner technique based on the use of blocking microelectrodes. Results are presented for a couple of typical solid oxide electrolyte systems mainly derived from ceria and lanthanum gallate. The examples demonstrate a good resolution of the microelectrode technique in particular within the electrolyte domain, i.e. around the minimum of the electronic conductivity. This made possible the detection of deviations from the predicted oxygen partial pressure dependence of simple defect models for the concentrations of electrons and holes. The observed deviations from these defect models, at least partially, reflect the overemphasized ideality of the usually applied semiconductor model.

Whereas the effect of dissolved transition metals with variable valence states such as Fe, and Co on the electronic conduction is well known, it was unexpected to find a strong concentration dependent effect of dopants like Y^{3+} and Zr^{4+} in ceria or Mg^{2+} and Sr^{2+} in the gallates upon the electronic conductivity within the electrolytic domain. Ions like Y^{3+} and Zr^{4+} cause a shift and a partial broadening of electronic states in ceria based materials. Indications have been found for band tailing due to high defect concentrations. In some cases, the dopants cause the appearance of additional localized electron states in the gap which give rise to weak superimposed maxima of the electronic conductivity at a particular oxygen partial pressure within the electrolytic domain.

Accordingly, one cannot expect that electronic conductivities of solid electrolytes are insensitive to a changing concentration of stabilizers such as Y, Ca, etc. For instance, even a moderate doping of ceria by zirconia leads to a considerable electronic excess conductivity in the electrolytic domain.

INTRODUCTION

The onset of electronic conduction above or below certain values of the chemical potentials limits the range of application of solid electrolytes and determines the width of the so-called electrolytic domain. Accordingly, it is of interest to know the material properties and further relevant factors that influence the concentration and mobility of electrons and holes in solid electrolytes.

Various experimental techniques have been used in the past in order to analyze the electronic conductivity of solid oxide electrolytes. The main difficulty is that the electronic conductivity is lower than the ionic one by several orders of magnitude. Therefore, blocking or minimizing of the oxygen ion flow is necessary in the presence of a chemical potential gradient of oxygen, if the electronic conductivity is to be measured with high accuracy around the minimum of the electronic conductivity in the electrolytic domain. The basic ideas with respect to this approach were originally developed by Hebb and C.Wagner [1, 2]. Two main experimental techniques resulted from this with respect to solid oxide electrolytes. The first uses zero current conditions in an oxygen concentration cell with a solid electrolyte membrane [3]. The second technical variant, often called the Hebb-Wagner technique, was first applied to silver and copper halides [4, 5] and later on to oxide electrolytes [6-8]. It uses the electrochemical polarization between a reversible electrode and an electrode that blocks the majority charge carrier flow. In that case, the steady state current corresponds to an exclusive electron flow in a chemical potential gradient. The resulting current-voltage curves normally are highly nonlinear and can be fitted to a theoretical curve, if a defect model is available for the dependence of the electronic conductivity as a function of the chemical potential of the mobile component (i.e. oxygen in the case of solid oxide electrolytes). In this way, electronic conductivities are usually obtained within the range of validity of the underlying defect model. But in principle, the electronic conductivities can be derived as a function of the chemical potential directly from the slope of the steady state I-U curves (differential evaluation) without referring to a known defect model. This was already stated in C.Wagner's original considerations. For a cylindrical or rectangular solid electrolyte sample between two planar electrodes (area A, distance L), the differential conductivity of the electrons as a function of oxygen partial pressure is given by

$$\sigma_e(p_{O_2}) = \frac{L}{A} \cdot \left(\frac{dI}{dU} \right)_{\text{steady state}}$$

(1)

where I and U denote the current and the cell voltage in the steady state. The cell voltage in the steady state is given by

$$U = \frac{RT}{4F} \cdot \ln\left(\frac{p_{O_2}}{p_{O_2}^{\text{ref}}} \right)$$

(2)

with T – absolute temperature, R – gas constant, F – Faraday's constant, and $p^{\text{ref}}(O_2)$– constant oxygen partial pressure at the reference side.

However, the conventional cells with planar electrodes are not easy to handle. Often, the accuracy is limited by drift effects or leakage currents due to non-ideal blocking. Therefore, most investigations in the past relied on fitting the I-U curves with the conventional simple electronic defect model for solid electrolytes in order to circumvent the limited accuracy. For solid oxide electrolytes, the simple defect model assumes a constant oxygen vacancy concentration. Then, the equilibrium with gaseous oxygen in the gas phase yields for the electrons (n-type contribution)

$$\frac{1}{2}O_{2(g)} + V_O^{\cdot\cdot} + 2e' \leftrightarrow O_O^x \qquad \sigma_n \sim [e'] \sim p_{O_2}^{-1/4} \cdot [V_O^{\cdot\cdot}]^{-1}$$

(3)

A similar equilibrium is obtained for the holes. The partial pressure dependence of the total electronic conductivity σ_e is therefore described by the corresponding sum of electron and hole contributions according to

$$\sigma_e = \sigma_n + \sigma_p = \sigma_n^\circ \cdot p_{O_2}^{-1/4} + \sigma_p^\circ \cdot p_{O_2}^{1/4}$$

(4)

456

This gives the usual "V"-like shape in a double logarithmic plot of the electronic conductivity versus the oxygen partial pressure. Depending on the depth of the conductivity minimum described by eq. (4), the electrolytic domain where the electronic conductivity is much lower than the ionic one is more or less extended.

In this context, the electronic states are treated as in crystalline semiconductors assuming that the Fermi level ε_F is far from the band states for all accessible partial pressures of oxygen. Then, in principle, one has a virtual two level model with one level corresponding to the valence band edge ε_V and a corresponding effective density $N_{eff,V}$ of hole states and a second level lying at the conduction band edge ε_C with an effective density $N_{eff,C}$ of electron states. As long as the Fermi level is situated in the gap far from the band edges, the Boltzmann statistics is valid with the Fermi level depending on the hole and electron concentration according to

$$\varepsilon_F = \varepsilon_C + kT \ln \frac{c_e}{N_{eff,C}} = \varepsilon_V - kT \ln \frac{c_h}{N_{eff,V}} = \text{const.} - \frac{kT}{4} \cdot \ln p_{O_2}$$

(5)

But this simple dependence for the electron concentration as a function of oxygen partial pressure cannot remain valid, if additional electronic states are introduced in the band gap or near the band edges. This occurs with certain transition metal dopants such as Ce, V, Mn, Co, Fe, and Ti [9-13]. Analogous observations have been made with solid oxide electrolytes derived from ceramic and nanocrystalline ceria [14, 15], $Gd_2(Ti_{1-x}Mn_x)_2O_7$ [16, 17], thoria [18], and recently for the lanthanum gallates [19-21].

As eq. (5) shows, a changing $p(O_2)$ is equivalent to a sweep of the Fermi level in the band gap. Any additional localized density of electronic states in the band gap will lead to a maximum in the electronic conductivity when the Fermi level crosses the energy of these states. As shown for instance with Mn doped $Gd_2(Ti_{1-x}Mn_x)_2O_7$ [16, 17] and with doped zirconia [13], the excess conductivity due to a dopant with variable valence can be described by

$$\sigma_{excess} \ \square \ c_{total,M} \cdot x_{M^{z+}} \cdot x_{M^{(z+1)+}} = c_{total,M} \cdot x_{M^{z+}} \cdot (1 - x_{M^{z+}})$$

(6)

The electronic transport results from an exchange between M^{z+} and $M^{(z+1)+}$ ions. The molar ratio $x(M^{z+})$ is a function of the distance between the Fermi level and the local electronic levels of the $M^{z+} / Mn^{(z+1)+}$ redox couple. It can easily be calculated with the help of the Fermi-Dirac distribution function and exhibits a maximum, if the two valence states have the same concentration.

Observations from surface spectroscopic investigations on solid electrolytes add another aspect: The band edges of good solid electrolytes usually are not sharp, but show a considerable spread of the electronic state densities above the valence and below the conduction band edges (so called band tails). These states often extend far into the band gap region as demonstrated by the experimental data in figures 1 and 2.

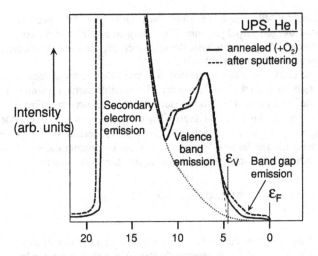

Figure 1. Evidence for additional electronic states in the band gap of Y-stabilized zirconia from UPS measurements. Results for a single crystal of $(ZrO_2)_{0.87}(YO_{1.5})_{0.13}$ [22, 23].

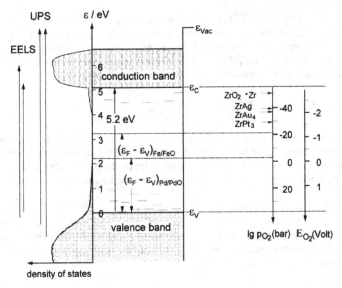

Figure 2. Correspondence between the Fermi level in the band diagram and the electrode potential scale according to experimental results for ZrO_2 $(+Y_2O_3)$ at 800°C [23, 24].

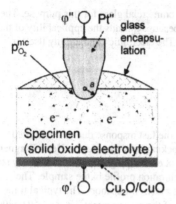

φ'' ⃝ Pt'' glass encapsu- lation

$p_{O_2}^{mc}$

Specimen
(solid oxide electrolyte)

φ' ⃝ Cu$_2$O/CuO

Figure 3. Cell as used for the Hebb-Wagner polarization experiments with microelectrodes (schematically). The cell voltage is given by the potential difference between the two electrodes according to $U = \varphi''-\varphi'$. The reference side with the oxide mixture assures a constant partial pressure of oxygen. The partial pressure at the interface to the microelectrode depends on the voltage and is given by eq. (2).

These conditions are exactly those described by the Anderson model of electron localization in highly disordered semiconductors [25, 26]. It predicts the formation of band tails by a shift of electronic state density from the center to the edge of the band. This is caused by a high disorder which is typical for the high dopant level in solid electrolytes together with the presence of mobile ionic defects. The electronic states within the band tails exhibit quite low electron and hole mobilities which is also typically found in solid electrolytes (strong electron localization, "mobility gap").

The Hebb-Wagner technique is particularly well suited to measure the detailed influences of localized states and band tailing in solid electrolytes, if a sufficient accuracy can be attained. The use of microcontacts has been an important step in order to achieve this aim. The one-to-one correspondence between shifts of the electrode potential at the ion blocking electrode (in a Hebb-Wagner experiment on a solid oxide electrolyte) and shifts of the position of the Fermi level (= electrochemical potential of oxygen) with respect to the band edges of the solid electrolyte opens a straightforward way to characterize local electronic states.

Figure 2 shows experimentally obtained data for the electrochemical electrode potential scale for Y-stabilized ZrO_2 in relation to the band diagram as obtained from electron spectroscopic techniques [23].

EXPERIMENTAL DETAILS

Figure 3 shows the typical cell in our experiments with the sample in the form of a pressed pellet or a single crystal. The contact radius of the microelectrode (usually Pt) ranges between 20 and 100 μm. A metal/metal oxide mixture can be applied as a large area reference electrode. In the temperature range between 500°C and 800°C, the system Cu_2O/CuO was an advantageous choice. Care has to be taken to avoid chemical reactions or in-diffusion of components of the reference.

The thickness of the sample should be at least 10 to 20 times the contact radius of the microelectrode. The region around the microcontact has to be sealed against the outer atmosphere in order to achieve true blocking characteristics. This is much easier as compared to the sealing of a large area planar electrode. We used a commercial glass for this purpose. The conductivity of the glass should be as low as possible, because it limits the applicability of the technique in the range of low electronic conductivities. The electronic conductivity itself is calculated from (a – radius of the microelectrode)

$$\sigma_e(p_{O_2}) = \frac{1}{2\pi a} \cdot \left(\frac{dI}{dU}\right)_{steady\ state}$$

Two of the main advantages of microelectrodes are the fast response due to the limited diffusion length and the easy control of a perfect ion blocking behavior. As the Hebb-Wagner technique measures steady state values, one has to wait at each voltage value until a steady state current is attained corresponding to a steady state concentration profile in the sample. The necessary time is mainly determined by the rate limiting electron transport. The typical time constant τ for this waiting time depends on the radius a of the microcontact, i.e. $\tau \sim a^2/D$ with the chemical diffusion coefficient D. As compared to a sample with planar contacts and typical diffusion lengths around 1 mm, the effective diffusion length of a microelectrode is normally a factor of 10 – 100 lower (the necessary waiting time for measuring a single pair of steady state values of current and voltage will then be reduced by a factor of 100 – 10000). A more detailed description of the experimental considerations can be found in [27].

RESULTS

Doped lanthanum gallates

First results obtained with the microelectrode technique have been already presented for the doped lanthanum gallate system a year ago [28]. Two remarkable results have to be mentioned. First, the n-type range at low oxygen partial pressures did not yield stable current-voltage curves, nor did it show an increasing n-type conductivity. Instead of that, the electronic conductivity was pinned there at a low level. It is not clear up to now whether the formation of a $PtGa_x$ alloy fixes the electronic conductivity due to a two phase equilibrium or whether the reduction of Ga^{3+} already yields volatile Ga_2O which also would lead to a decrease of the electron concentration. Recent unpublished observations, however, showed platinum particles that remained on the lanthanum gallate surface after removing the microelectrode. Their strong adherence to the oxide surface indicated a chemical interaction most probably due to alloy formation. Nevertheless, the composition of the bulk lanthanum gallate did not change even for long experiments. Therefore we think that the quite slow diffusion of gallium limits the reactions to the immediate vicinity of the microcontact.

The second remarkable result is that the dopant concentration clearly influences the electronic conductivity. Figure 4 shows results for three different Mg concentrations on the B sites of $La_{0.9}Sr_{0.1}Ga_{1-x}Mg_xO_{3-\delta}$ in the partial pressure range where p-type conduction prevails. The slope is very near to the expected $p(O_2)^{1/4}$ dependence. Recent results with lower Mg concentrations indicated a further lowering of the p-type conductivity, but also a shift of the conductivity minimum to lower oxygen partial pressures. The latter observation suggests a decreasing tendency of being reduced for low Mg concentration (and accordingly a higher stability).

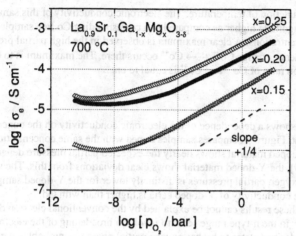

Figure 4. Electronic conductivity in the lanthanum gallates as a function of Mg doping on the B sites.

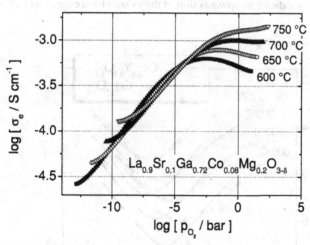

Figure 5. Electronic conductivity of Co doped lanthanum gallate.

We attribute the changes of the p-type conductivity with increasing Mg concentration to a broadening of the valence band due to formation of band tails. This increases the p-type conductivity at medium and high oxygen partial pressures.

Some authors analyzed lanthanum gallates doped with transition metals such as Ni and Co. The idea is to increase the electronic conductivity and to receive a compatible electrode material for the lanthanum gallate based solid electrolytes. Figure 5 shows the electron conductivity for a

Co doped sample as function of temperature. The electronic conductivity of this sample is almost an order of magnitude higher as compared to a corresponding Co free sample with the same Sr and Mg concentrations. A clear maximum is observed in the high partial pressure range indicating that the valence change $Co^{3+} \longleftrightarrow Co^{4+}$ occurs there. The maximum is to be expected when the concentration ratio of Co^{3+} and Co^{4+} is equal to one.

Doped ceria

Doped ceria also shows a dependence of the electronic conductivity on the concentration and the type of dopant. Figure 6 compares samples doped with the same concentration of Y and Gd. Whereas the Gd doped material shows nearly the expected partial pressure dependence as predicted by eq. 4 [29], the Y-doped material shows clear deviations from this. The slope in the n-type range at low oxygen partial pressures is distinctly lower for the Y-doped sample. On the other hand, the p-type conductivity of Y-doped ceria is higher than with Gd.

It is evident that these results cannot be explained by the conventional electron defect model. The deviations in the n-type range seem to involve a broadening of the electronic state density of the Ce(4f) derived states to higher oxygen partial pressures, probably an effect of the dissolved Y^{3+} ions. In general, the change from a sharp density of states towards a distribution of electronic state density near the Fermi level leads to a weaker oxygen partial pressure dependence as simple model calculations showed. The increase of the p-type conductivity of the Y-doped sample may indicate an upwards shift of the valence band edge or an increased band tailing.

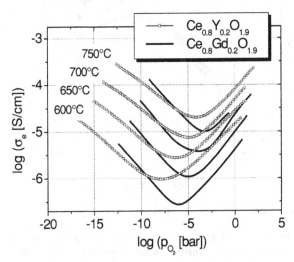

Figure 6. Electronic conductivity of Y- and Gd-doped ceria as a function of temperature and oxygen partial pressure. The data for Gd-doped ceria are derived from [29].

<u>Gd-doped (Ce,Zr)O$_2$</u>

A particularly surprising behavior was found in doped $Ce_{1-x}Zr_xO_{2-x/2}$. These materials have been investigated in many laboratories during the last years because of their interesting catalytic properties and oxygen storage effects [30-38]. It is known from these investigations that the electronic properties are distinctly different from pure doped ceria or zirconia. Some exemplary results of our recent measurements are shown in figures 7 and 8. The presence of Zr in ceria (or vice versa of Ce in zirconia) leads to an additional electronic state density in the center of the electrolytic domain. The corresponding weak maximum of the electronic conductivity is clearly seen in figures 7 and 8.

Figure 9 illustrates a semi-quantitative model for the effects observed in Ce-Zr-O systems. Note that the changes in concentrations can produce a shift and a broadening of the electronic states. In connection to the degree of disorder, the changes may also involve changes in the band tails and the formation of characteristic electronic states due to defect complexes.

CONCLUSIONS

Any detailed theoretical explanation of the strong dopant effects on the electronic conductivity of solid electrolytes must take into account changes of the band structure as well as the influence of local defect complexes. Accordingly, one cannot expect that electronic conductivities of solid electrolytes are insensitive to the concentration of stabilizers such as Y, Ca, etc. Even a moderate doping of ceria by zirconia leads to a considerable electronic excess conductivity in the electrolytic domain which clearly lowers the efficiency of fuel cells driven with corresponding electrolyte materials.

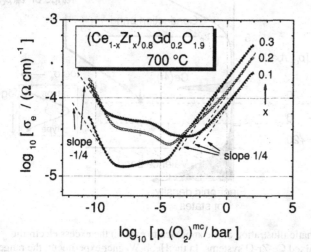

Figure 7. Electronic conductivity of the Gd doped Ce-Zr-O system as a function of Zr concentration and oxygen partial pressure.

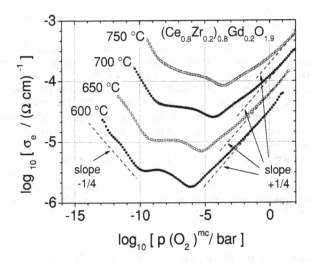

Figure 8. Electronic conductivity of the Gd doped Ce-Zr-O system as a function of temperature and oxygen partial pressure.

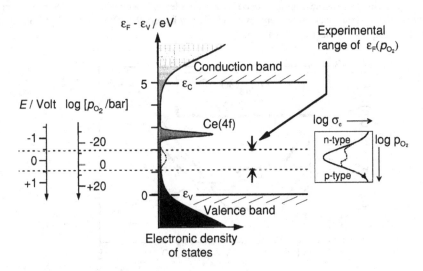

Figure 9. Schematic illustration of the approach to explain the excess electronic conductivity in the mixed Ce-Zr-O systems. In the Hebb-Wagner experiment, the range around the conductivity minima is particularly sensitive to the appearance of additional band gap states.

ACKNOWLEDGEMENTS

The authors would like to thank the Deutsche Forschungsgemeinschaft (DFG) for financial support.

REFERENCES

1. M. H. Hebb, *J. Chem. Phys.* **20**, 1952 (1952).
2. C. Wagner, in *Proc. of the 7th Meeting of the International Committee on Electrochemical Thermodynamics and Kinetics*, Lindau, 1955), p. p. 361ff.
3. H. Schmalzried, *Z. Phys. Chem. N.F.* **38**, 87 (1963).
4. J. B. Wagner and C. Wagner, *J. Chem. Phys.* **26**, 1597 (1957).
5. B. Ilschner, *J. Chem. Phys.* **28**, 1109 (1958).
6. J. W. Patterson, E. C. Bogren, and R. A. Rapp, *J. Electrochem. Soc.* **114**, 752 (1967).
7. L. D. Burke, H. Rickert, and R. Steiner, *Z. phys. Chem. N.F.* **74**, 146 (1971).
8. L. Heyne and N. M. Beekmans, *Proceedings of the British Ceramic Society* **19**, 229 (1971).
9. F. Schilling, U. Vohrer, H.-D. Wiemhoefer, J. Arndt, and W. Goepel, *Sensors and Actuators* **B 4**, 411 (1991).
10. U. Vohrer, H.-D. Wiemhoefer, W. Goepel, B. A. van Hassel, and A. J. Burggraaf, *Solid State Ionics* **59**, 141 (1993).
11. X. Guo and J. Maier, *Solid State Ionics* **130**, 267 (2000).
12. K. Kobayashi, S. Yamaguchi, T. Higuchi, S. Shin, and Y. Iguchi, *Solid State Ionics* **135**, 643 (2000).
13. K. Sasaki and J. Maier, *Solid State Ionics* **134**, 303 (2000).
14. T. S. Stefanik and H. L. Tuller, *Journal of the European Ceramic Society* **21**, 1967 (2001).
15. P. Knauth and H. L. Tuller, *Solid State Ionics* **136**, 1215 (2000).
16. O. Porat, M. A. Spears, C. Heremans, I. Kosacki, and H. L. Tuller, *Solid State Ionics* **86-88**, 285 (1996).
17. H. L. Tuller, *Solid State Ionics* **94**, 63 (1997).
18. H. Nafe, *Solid State Ionics* **59**, 5 (1993).
19. N. J. Long, F. Lecarpentier, and H. L. Tuller, *Journal of Electroceramics* **3**, 399 (1999).
20. N. Trofimenko and H. Ullman, *Solid State Ionics* **118**, 215 (1999).
21. H. Ullmann, N. Trofimenko, A. Naoumidis, and D. Stover, *Journal of the European Ceramic Society* **19**, 791 (1999).
22. K. Schindler, D. Schmeißer, U. Vohrer, H.-D. Wiemhoefer, and W. Goepel, *Sensors and Actuators* **17**, 555 (1989).
23. H.-D. Wiemhoefer and U. Vohrer, *Ber. Bunsenges. Phys. Chem.* **96**, 1646 (1992).
24. H.-D. Wiemhoefer, S. Harke, and U. Vohrer, *Solid State Ionics* **40/41** (1990).
25. P. W. Anderson, *Physical Review* **109**, 1492 (1958).
26. S. N. Mott, M. Pepper, S. Pollitt, R. H. Wallis, and C. J. Adkins, *Proc. Roy.Soc.Lond. A* **345**, 169 (1975).
27. S. Luebke and H. D. Wiemhoefer, *Solid State Ionics* **117**, 229 (1999).
28. J. Weitkamp and H.-D. Wiemhoefer, *Solid State Ionics* **154-155C**, 597 (2002).

29. S. Luebke and H. D. Wiemhoefer, *Berichte Der Bunsen-Gesellschaft-Physical Chemistry Chemical Physics* **102**, 642 (1998).

30. J. H. Lee, S. M. Yoon, B. K. Kim, H. W. Lee, and H. S. Song, *Journal of Materials Science* **37**, 1165 (2002).

31. H. Yokokawa, N. Sakai, T. Horita, K. Yamaji, Y. P. Xiong, T. Otake, H. Yugami, T. Kawada, and J. Mizusaki, *Journal of Phase Equilibria* **22**, 331 (2001).

32. Y. P. Xiong, K. Yamaji, N. Sakai, H. Negishi, T. Horita, and H. Yokokawa, *Journal of the Electrochemical Society* **148**, E489 (2001).

33. N. Sakai, T. Hashimoto, T. Katsube, K. Yamaji, H. Negishi, T. Horita, H. Yokokawa, Y. P. Xiong, M. Nakagawa, and Y. Takahashi, *Solid State Ionics* **143**, 151 (2001).

34. J. H. Lee, S. M. Yoon, B. K. Kim, J. Kim, H. W. Lee, and H. S. Song, *Solid State Ionics* **144**, 175 (2001).

35. K. Kawamura, K. Watanabe, T. Hiramatsu, A. Kaimai, Y. Nigara, T. Kawada, and J. Mizusaki, *Solid State Ionics* **144**, 11 (2001).

36. C. E. Hori, K. Y. S. Ng, A. Brenner, K. M. Rahmoeller, and D. Belton, *Brazilian Journal of Chemical Engineering* **18**, 23 (2001).

37. A. Tsoga, A. Naoumidis, and D. Stover, *Solid State Ionics* **135**, 403 (2000).

38. T. Otake, H. Yugami, H. Naito, K. Kawamura, T. Kawada, and J. Mizusaki, *Solid State Ionics* **135**, 663 (2000).

Mat. Res. Soc. Symp. Proc. Vol. 756 © 2003 Materials Research Society

The Synthesis and Characterisation of Ge Containing Apatite-Type Oxide Ion Conductors

P.R. Slater, J.E.H. Sansom, J.R. Tolchard, M.S. Islam
Department of Chemistry, University of Surrey, Guildford, Surrey. GU2 7XH. UK

ABSTRACT

Apatite-type oxides, $La_{10-x}(Si/Ge)_6O_{26+z}$, have been attracting significant interest recently due to their high oxide ion conductivities. Most of the work so far has focused on the Si based systems, since the Ge based systems suffer from problems attributed to Ge loss. In this paper we show that doping divalent cations on the La site or B on the Ge site helps to stabilise the hexagonal apatite lattice for these Ge based systems. These doped phases show high oxide ion conductivities, although results from extended sintering studies suggest that Ge loss is still a problem. In order to limit Ge loss, we have also examined Bi doping to lower the sintering temperature and preliminary results for the novel Bi containing apatite-type phases, $La_6Bi_2M_2Ge_6O_{26}$ (M=Mg, Sr, Ba) and $La_{8-x}Bi_2Ge_5GaO_{26+y}$, are also reported.

INTRODUCTION

Following the pioneering early work by Nakayama *et al.* on high oxide ion conduction in apatite-type $La_{10-x}Si_6O_{26+z}$ there has been significant interest in the oxide ion conducting properties of these apatite materials [1-14], due to potential technological applications in high temperature devices such as solid oxide fuel cells. Most of the initial work has focused on the Si based materials, since high quality samples can be readily made in this system. Early studies showed the importance of non-stoichiometry, either in terms of cation vacancies or oxygen excess, to maximise the oxide ion conductivity of these phases. Neutron diffraction studies of samples with cation vacancies, e.g. $La_{9.33}Si_6O_{26}$, indicated the presence of disorder within the oxide channels, in the form of Frenkel-type defects [5], while in the oxygen excess samples, disorder in the sense of excess interstitial oxide ions is present. A range of doping studies has also been reported for the Si based samples [6,8,11,13].

In contrast, studies of comparable samples containing Ge in place of Si are limited. Initial work in this area was reported by Arikawa *et al.*, who showed that corresponding Ge containing compositions, $(La/Sr)_{10-x}Ge_6O_{26+z}$, exhibited higher conductivities than the Si based systems at high temperatures [7]. However, the X-ray patterns showed severe peak broadening/extra peaks, and subsequently the same group claimed that the conducting phase was cation deficient La_2GeO_5 [15]. In contrast, studies by Nakayama *et al.* and our previous studies supported the conclusion that the highly conducting phase was indeed apatite-type [9, 12]. In our previous work, we showed that $La_{9.33}Ge_6O_{26}$ with the hexagonal apatite lattice could be prepared at intermediate temperature (1150°C). Heating to higher temperature (required to obtain dense pellets for conductivity measurements) resulted in, firstly peak broadening, and then the occurrence of distinct extra peaks either side of the main apatite peaks (figure 1). The X-ray pattern still resembled that of an apatite-type phase (possibly triclinic), however, rather than La_2GeO_5 [9]. Sintering for an extended time (1500°C for 5 days) was shown to give significant mass loss and the pattern changed to that of La_2GeO_5, while the conductivity dropped substantially (e.g. $\sigma_{800°C}$ changed from 0.01 to 3×10^{-5} Scm^{-1}). The mass loss was attributed to Ge loss, since this is the most volatile component, and is consistent with the change in X-ray pattern to La₂GeO₅ type. These results were subsequently confirmed by Berastegui *et al.*, who reported neutron diffraction studies of $La_{9.33}Ge_6O_{26}$ [14]. They showed that the initial peak broadening could be attributed to a correlated disorder within the apatite channels.

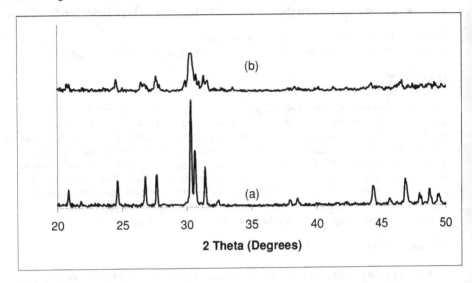

Figure 1. X-ray diffraction patterns for $La_{9.33}Ge_6O_{26}$ prepared at (a) 1150°C and (b) 1500°C

In this paper we report a range of doping studies in this system, aimed at stabilising the hexagonal apatite lattice to higher temperatures, and at limiting Ge loss on sintering.

EXPERIMENTAL

High purity La_2O_3, $BaCO_3$, $SrCO_3$, $CaCO_3$, MgO, MnO_2, GeO_2, H_3BO_3 were used to prepare a range of doped samples. Samples were prepared as follows: the powders were intimately mixed in the correct ratios and heated to 1150°C for 16 hours. The resultant powders were reground and reheated at 1200-1350°C for a further 16 hours. For the B doped samples, prior to the initial heat treatment, the intimately mixed powders were heated first at 300°C for 2 hours and then the temperature was slowly raised over 6 hours to 1150°C. This was to try to limit any B loss before reaction had occurred. For the Bi doped samples, Bi_2O_3 was used, and the intimately mixed powders were first heated at 950-1000°C for 12 hours, the sample was then reground and reheated at 1050-1100°C for 8 hours.

Conductivities were determined from AC impedance measurements (Solartron 1260 Impedance analyser). Pellets (1.6 cm diameter) were pressed under a pressure of $8000kg/cm^2$ and sintered at 1600°C for 2 hours. For the B doped samples a sintering temperature of 1400°C for 10 hours was used, while for the Bi doped samples the sintering temperature was 1050-1100°C for 4 hours. Pellet densities in the range 85-91% theoretical were achieved. Due to general problems, experienced previously with these apatite systems [4, 8, 9], in reliably resolving bulk and grain boundary semicircles, the conductivity data reported represents the total conductivity. Extended sintering studies were performed for some samples, to determine whether Ge loss problems were still observed in the doped samples (Pellets were heated at 1500°C for 5 days, after which the conductivity was measured and an X-ray pattern recorded).

RESULTS AND DISCUSSION
Divalent cation and B doping in $La_{9.33}Ge_6O_{26}$

It was found that doping onto the La site with divalent cations, i.e. $La_8M_2Ge_6O_{26}$ (M=Mg, Mn, Ca, Sr, Ba), and B doping onto the Ge site, i.e. $La_{9.33+x/3}Ge_{6-x}B_xO_{26}$, helped to stabilise the hexagonal apatite lattice at the high temperatures required for sintering to obtain sintered pellets. Particular attention has so far focused on Mg, Ba and B doped samples, and cell parameters and conductivity data for a range of such samples are given in table 1. The X-ray pattern for $La_8Mg_2Ge_6O_{26}$, is shown in figure 2, and conductivity data for selected samples are shown in figure 3. The changes in cell parameters compared to $La_{9.33}Ge_6O_{26}$ are consistent with those expected according to the dopant sizes. As for the Si based systems, the highest conductivities were observed for samples containing oxygen excess. In addition the conductivity of the nominally fully stoichiometric (on both cation and anion sites) sample $La_8Ba_2Ge_6O_{26}$ was low, similar to the situation for the Si based systems [5,11]. Interestingly, the conductivity of the comparable Mg doped sample ($La_8Mg_2Ge_6O_{26}$) was high. This requires further investigation, but it is relevant to note that a similar situation has been observed for the Si based systems: the conductivity of $La_8Mg_2Si_6O_{26}$ being much higher than that of $La_8Ba_2Si_6O_{26}$ [16].

Compared to the undoped parent phase, $La_{9.33}Ge_6O_{26}$, all the doped phases have significantly lower activation energies. Therefore, although the conductivity of $La_{9.33}Ge_6O_{26}$ at high temperature is comparable to that of a number of the doped samples (table 1), the conductivity at lower temperature is significantly lower. This can be clearly seen by comparison of the data for $La_{9.33}Ge_6O_{26}$ and $La_{9.67}Ge_5BO_{26}$ in figure 3.

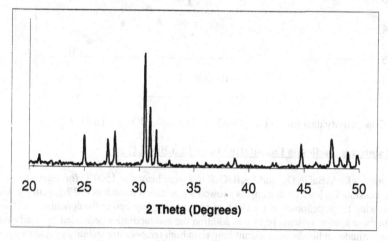

Figure 2. X-ray diffraction pattern for $La_8Mg_2Ge_6O_{26}$

Table 1. Cell parameters (hexagonal) and conductivity data for selected Ge based apatite samples

Sample composition	a=b (Å)	c (Å)	σ/S cm^{-1} (800°C)	E$_a$/eV (low/high temperature)
La$_{9.33}$Ge$_6$O$_{26}$	9.913(3)	7.282(3)	0.01	1.28/1.62
La$_8$Mg$_2$Ge$_6$O$_{26}$	9.853(2)	7.155(2)	0.01	0.89
La$_{8.5}$Mg$_{1.5}$Ge$_6$O$_{26.25}$	9.854(2)	7.164(2)	0.01	0.70
La$_{8.67}$MgGe$_6$O$_{26}$	9.879(3)	7.238(3)	0.01	0.91
La$_8$Ba$_2$Ge$_6$O$_{26}$	10.002(4)	7.422(3)	5.5 x 10^{-5}	1.05/0.62
La$_9$BaGe$_6$O$_{26.5}$	9.940(3)	7.363(3)	0.03	1.06
La$_{8.67}$BaGe$_6$O$_{26}$	9.951(4)	7.355(3)	0.008	0.97
La$_{9.67}$Ge$_5$BO$_{26}$	9.829(4)	7.294(4)	0.01	1.07/0.92

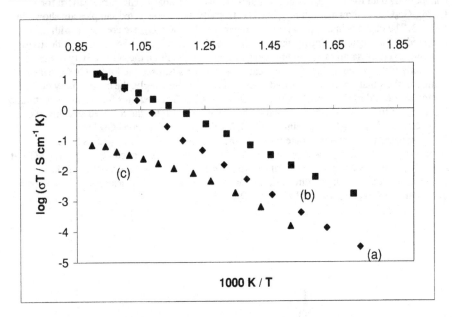

Figure 3. Conductivity data for (a) La$_{9.33}$Ge$_6$O$_{26}$ (b) La$_{9.67}$Ge$_5$BO$_{26}$ (c) La$_8$Ba$_2$Ge$_6$O$_{26}$

Extended sintering studies on La$_{8.67}$BaGe$_6$O$_{26}$ and La$_9$BaGe$_6$O$_{26.5}$

The samples La$_{8.67}$BaGe$_6$O$_{26}$ and La$_9$BaGe$_6$O$_{26.5}$ were heated at 1500°C for 5 days to investigate possible Ge loss. Both samples showed a significant mass loss (>2%), with the X-ray patterns showing the appearance of La$_2$GeO$_5$ as an impurity, supporting the conclusion of Ge volatility in these apatite systems [9,14]. In addition the conductivities decreased by nearly three orders of magnitude, indicating significant long term high temperature stability problems even for the doped phases. Although sintering for this extended time would not necessarily be required for use of these materials as electrolytes in SOFC's, Ge loss could cause problems, as it would mean that while the bulk of the sample may be unchanged, the surface composition would be significantly changed. This could cause interfacial problems with the electrodes, with the formation of La_2GeO_5 at the surface of the electrolyte effectively giving an insulating layer.

This would represent a serious problem at the temperature in SOFC. Investigations into the capability of lowering temperature ($\leq 600 °C$) are planned.

Bi doping in $La_{9.33}Ge_6O_{26}$

The extended sintering studies on the doped samples showed that Ge loss appeared to be still a problem. In order to try to limit this problem, doping studies to lower the sintering temperature have been tried. B doping partially has this effect by lowering the sintering temperature from 1600°C to 1400°C, but it is possible that even at this temperature Ge loss may occur, which needs further investigation.

Bi doping has therefore been tried to significantly lower the sintering temperature, and in addition examine what effect the presence of Bi has on the conductivity. Attempts to prepare the phases $La_{9.33-x}Bi_xGe_6O_{26}$ (x=1, 2, 3) resulted in hexagonal apatite-type phases, with small unidentified impurities. In order to obtain single phase Bi doped samples, codoping on either the La or Ge site was required. In this preliminary work, the Bi content was limited to 20%, although higher contents have been subsequently achieved. Corresponding Si based materials can also be prepared and will be reported elsewhere. Table 2 lists cell parameters and conductivities for selected samples.

Table 2. Cell parameters (hexagonal) and conductivity data for selected Bi doped samples

Sample composition	a=b (Å)	c (Å)	σ/S cm^{-1} (800°C)	E_a/eV
$La_6Bi_2Mg_2Ge_6O_{26}$	9.866(4)	7.211(3)	0.02	1.05
$La_6Bi_2Ba_2Ge_6O_{26}$	10.024(4)	7.399(4)	0.001	1.08
$La_6Bi_2Sr_2Ge_6O_{26}$	9.911(3)	7.310(2)	0.003	1.12
$La_8Bi_2Ge_5GaO_{26.5}$	9.906(3)	7.335(3)	0.02	1.10
$La_{7.67}Bi_2Ge_5GaO_{26}$	9.903(1)	7.333(1)	0.01	1.09
$Nd_6Bi_2Sr_2Ge_6O_{26}$	9.830(3)	7.204(2)	2×10^{-4}	1.31

For these samples, pellets could be sintered at significantly lower temperatures (1050-1100°C), which should limit Ge volatility problems. It should be noted, however, that in these systems, Bi loss could be a problem, and this requires further investigation.

As before, high conductivities were observed for samples containing oxygen excess. The lowest conductivities were obtained for samples nominally stoichiometric on both cation and anion site, with the exception of the corresponding Mg doped sample, for which the conductivity was high (as for $La_8Mg_2Ge_6O_{26}$). Structural studies are planned to try to rationalise this. In changing rare earths from La to Nd, there appeared to be a significant drop in conductivity and corresponding increase in activation energy. Overall the results showed that Bi doping led to high quality samples with high conductivities. The successful synthesis of these phases is also interesting from the viewpoint that very few Bi containing apatite-type phases are known.

The Bi doped samples appeared promising in achieving high conductivities with low sintering temperatures, and so studies of the effect of heating samples in H_2/Ar (5%/95%) at 900°C for 12 hours were made to see if these phases were stable in strongly reducing conditions. This was found not to be the case, with all samples examined showing partial decomposition. This would represent a serious problem at this temperature in an SOFC. Investigations into the stability at lower temperatures (500-700°C) are planned, along with studies into possible Bi loss on synthesis/sintering.

CONCLUSIONS

The results show that doping onto either the La or Ge site can help stabilise the hexagonal apatite lattice for Ge based systems. The conductivities of these phases are high, although the activation energies are higher than for Si based systems. Extended sintering studies at high temperatures resulted in a mass loss, attributed to Ge volatility, which could represent a significant problem for the use of these systems as electrolytes in solid oxide fuel cells.

Bi doping is shown to significantly lower pellet sintering temperatures, but such phases are not stable to decomposition at high temperatures in reducing atmospheres. Further work is ongoing to optimise the conductivities of these Ge based phases, and limit problems associated with Ge loss.

ACKNOWLEDGEMENTS

We would like to thank EPSRC for funding (grant GR/R29239/01), and EPSRC and MERCK Ltd for funding a studentship (JEHS).

REFERENCES

1. S. Nakayama, H. Aono, Y. Sadaoka, Chem. Lett. 431 (1995).
2. S. Nakayama, M. Sakamoto, J. Eur. Ceram. Soc. **18**, 1413 (1998).
3. S. Nakayama, M. Sakamoto, M. Higuchi, K. Kodaira, M. Sato, S. Kakita, T. Suzuki, K. Itoh, J. Eur. Ceram. Soc. **19**, 507 (1999).
4. S. Tao, J.T.S. Irvine, Mater. Res. Bull. **36**, 1245 (2001).
5. J.E.H. Sansom, D. Richings, P.R. Slater, Solid State Ionics **139**, 205 (2001).
6. E.J. Abram, D.C. Sinclair, A.R. West, J. Mater. Chem. **11**, 1978 (2001).
7. H. Arikawa, H. Nishiguchi, T. Ishihara, Y. Takita, Solid State Ionics **136-137**, 31 (2000).
8. J. McFarlane, S. Barth, M. Swaffer, J.E.H. Sansom, P.R. Slater, Ionics **8**, 149 (2002).
9. J.E.H. Sansom, L. Hildebrandt, P.R. Slater, Ionics **8**, 155 (2002).
10. J.E.H. Sansom and P.R. Slater; Solid State Phenomena **90-91**, 189 (2003).
11. P.R. Slater and J.E.H. Sansom; Solid State Phenomena **90-91**, 195 (2003).
12 S. Nakayama and M. Sakamoto; J. Mater. Sci. Lett. **20**, 1627 (2001).
13. J.E.H. Sansom and P.R. Slater; Proc. 5th Euro SOFC forum **2**, 627 (2002).
14. P. Berastegui, S. Hull, F.J. Garcia Garcia and J. Grins; J. Solid State Chem. **168**, 294 (2002).
15. T. Ishihara, H. Arikawa, T. Akbay, H. Nishiguchi, and Y. Takita; J. Am. Chem. Soc. **123**, 203 (2001).
16. J.E.H. Sansom and P.R. Slater; unpublished work.

Mat. Res. Soc. Symp. Proc. Vol. 756 © 2003 Materials Research Society

FF3.3

Synthesis and sintering of $Ce_{1-x}Gd_xO_{2-x/2}$ nanopowders via chemical routes.

Agusti Sin, Antonino S. Aricò[1], Massimo Seregni, Laura Gullo[1], Daniela La Rosa[1], Stefania Siracusano[1], Vincenzo Antonucci[1], Ana Tavares, Yuri Dubitsky, Antonio Zaopo
Pirelli Labs, C.2172, Viale Sarca 222, I-20126 Milan, Italy
[1]CNR-ITAE Via Salita Santa Lucia Sopra Contesse 5, Messina I-98125, Italy

ABSTRACT

Gd-doped ceria ($Ce_{0.80}Gd_{0.20}O_{1.90}$) obtained by several nanopowder synthesis processes are compared structurally, morphologically and electrochemically. The powders have been sintered into pellets and investigated by ac-impedance measurements. The electrochemical properties of the electrolytes have been correlated to the structural morphology of the sintered pellets. When the intergrain region exhibits an elevated interdiffusion, the observation of the grain boundaries becomes difficult while the electrochemical properties are improved. Acrylamide polymerization and oxalic co-precipitation techniques showed the best properties.

INTRODUCTION

In the past years, CeO_2-based materials have been intensively studied as catalysts, structural and electronic promoters for heterogeneous catalytic reactions and oxide ion conducting electrolytes for electrochemical cells. In all these applications, processes for obtaining fine powders to make bulk ceramics, coatings, films and composites are key points.

Our efforts are addressed to develop ceria-based solid electrolytes for solid-state electrochemical devices, including solid oxide fuel cells (SOFCs). The ceria doped electrolytes have higher ion conductivity than conventional YSZ and may operate at lower temperatures (500 - 700°C)[1].

The most commonly used method to produce ceria doped electrolytes is a solid-state reaction, called the "ceramic route". It involves intimate mechanical mixing of oxides, carbonates or nitrates and repeated grinding and heating cycles to achieve complete reaction between all reagents. Synthesis using wet chemistry, often called the "chemical route", can overcome many of the disadvantages present in the ceramic route. The homogeneity of the product is expected to increase because mixing of the reagents occurs at the molecular level in solution. The resulting oxide powders have a high specific surface area and, consequently, a high reactivity, which decreases the final temperature treatment and time of synthesis. Different chemical routes are explored in this work for the preparation of fine ceramic powders, such as several co-precipitation methods and acrylamide gelification method. Their relative ability to achieve high density ceramics ($\rho_{rel} > 95\%$) by sintering is examined by structural and electrical characterization.

EXPERIMENTAL

The starting raw materials were Ce $(NO_3)_3$ xH_2O (99.9%, Aldrich) and $Gd(NO_3)_3$ xH_2O (99.9%, Aldrich), where x is determined by thermogravimetric analysis. Nitrates are easily dissolved in water in the stoichiometric ratio and the solution is treated as function of the selected chemical process.

Oxalic co-precipitation

The process has been carried out according to ref. [2]. Briefly, the cationic nitrate solution is added drop by drop to an oxalic solution at pH = 6.5. The formed precipitate was well washed with water at the same pH as the oxalic solution. Then, the powder has been treated at 700°C for 4h where was crystallized to the desired cubic phase.

Acrylamide polymerisation sol-gel

The acrylamide polymerization gel consists of long polymeric chains, cross-linked to create a tangled network, soaked with an aqueous liquid [3]. Polymerization of the gel proceeds in a first step with the combination of an initiator (e.g. thermo-initiator [4] with the arcylamide monomer which is thereby activated and as the chain of polyacrylamide grows up, the active site shifts to its free end [3]. Bis-acrylamide (i.e., N,N'-Methylenebisacrylamide), which consists in two acrylamide units joined through their-$CONH_2$ groups via a methylene group, can link two growing chains. Hence, bis-acrylamide enables the formation of cross-linked chains resulting in a complex topology with loops, branches and interconnections [3].

The polimerization process started with the chelation of the Ce^{3+} and Gd^{3+} separately, which has been carried out by the addition of the EDTA powder (Flucka, 98%) adjusting the pH with NH_4OH (20%, Carlo Erba) up to complete dissolution (4<pH<5). The EDTA amount corresponds to the molar stoichiometric ratio 1mol EDTA : 1 mol cation. This chelation has been used to prevent the possible reaction of the acrylamide monomers with the cations which inhibit the polymerization reaction [4,5]. However, not all cations react with the acrylamide monomers [6] and more studies are currently done for verifying if it is also the case in the CGO system. Once all the solutions have been chelated they were mixed and subsequently are added the polymerization monomers. Acrylamide (Fluka, 99%), bis-acrylamide (i.e. N,N'-Methylenebiscrylamide, Fluka, 99%) and the initiator AIBN (i.e. 2,2'-Azobis 2-methylbutyronitrile, Fluka 98%) were added in the proportion of 15g Acrylamide : 1.5g Bis-acrylamide in a 200ml of solution. This is a common proportion used taking into account the study done in ref. [4] but could be optimized minimizing the volume solution and presence of polymer as was done in ref. [6]. The solution is vigorously stirred and heated up to the monomers dissolution and the gel has been instantaneously formed when the solution reaches ~80°C. After that, the gel has been dried and calcined at 600°C for 2 hours obtaining the desired phase.

Complex assisted co-precipitation

A complex between Ce-Gd ions and an organic compound has been formed in aqueous solution. This has been subjected to a decomposition step giving rise to a precipitate. The precipitate was filtered and dried at 100°C for 4h and then it is crystallized at 500°C for 1h.

Solid State reaction

The solid state reaction consists in the mixing of commercial oxides and nanoparticles made by the acrylamide polymerization (ratio 70:30). The commercial oxides were CeO_2 (99.999% Aldrich) and Gd_2O_3 (99.999% Aldrich) in the desiderated stoichiometry to form $Ce_{0.80}Gd_{0.20}O_{1.90}$. The commercial powders have been mixed and milled in a ball milling for 2h and then treated at 1000°C for 18h. These powders were mixed with 30% in weight of acrylamide sol-gel powders treated at 600°C. These was done for improving the compaction of the green pellet using a bimodal powder size where the nanophase can act as a kind of binder.

Table 1 Summarizes the sample synthesis processes.

Sample name	Synthesis process
Oxalic co-pre	Oxalic co-precipitation
Acrylamide	Acrylamide sol-gel
Mix solid state	Solid state reaction (70% wt oxides/30% acrylamide nanopowders)
Complex assisted co-precipitation	Complex assisted co-precipitation

All the powders have been pressed uniaxially between 200 and 400 MPa in a cylindrical shape and treated between 1450-1550°C for 4-6h. All samples showed relative densities > 95%. A gold paste (Hereaus) is used to form the current collector. An Autolab electrochemical apparatus equipped with a FRA has been used for the electrochemical analysis.

The structural and compositional analyses have been made by XRD using a Bragg-Brentano configuration with $Cu_{<K\alpha>}$ (Philips Xpert, Netherlands) and XRF (Bruker AXS, Germany). The microstructural analyses were carried out by SEM (Philips XL30, Netherlands) and TEM (Philips CM12, Netherlands).

RESULTS AND DISCUSSION

Structural and Microstructural analysis

The starting powders made by the different procedures showed a CGO pure phase by the XRD analysis. The particle size was calculated from the XRD patterns by use of the Debye-Scherrer formula [7]. It was clearly seen that all the chemical procedures showed small nanometric particles.

Table 2 Particle size and surface area of the different powders synthesized.

Synthesis process	Particle size (nm)	Surface area (m^2g^{-1})
Oxalic co-precipitation	24	34
Complex assisted co-precipitation	6	136
Acrylamide sol-gel	12	68
Solid state reaction	26	31

The TEM imaging (Fig.1) was in agreement with the particle size calculated before.

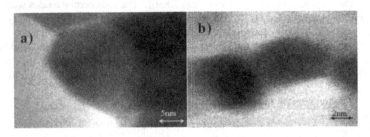

Figure. 1 TEM images a) oxalic co-precipitation powder treated at 700°C b) Acrylamide sol-gel powder treated at 600°C.

The powders have been sintered at high temperature after pellet formation in a cylindrical shape. The acrylamide sol-gel powders are characterized by easy compacting under high uniaxial pressure (200-400MPa); whereas, the solid state powders have been compacted with some difficulty. Moreover, the solid state reaction powders showed a very low sinterability evidencing a lot of cracks and broken zones. When these powders are mixed with a small quantity of Acrylamide sol-gel powders (70% wt. : 30% wt) the compaction and sinterability were improved. The foam morphology, weak agglomeration and small grain size of the acrylamide sol-gel powders might act as a kind of binder for the solid state powders which showed the opposite mophological properties and consequently bad ceramic performing characteristics. The powders made by oxalic and complex co-precipitation showed acceptable powder compaction and good sinterability.

Figure 2. SEM cross-section images of sintered pellets of powders synthesized by a) Acrylamide sol-gel b) Oxalic co-precipitation c) Complex assisted coprec. d) Mix solid state reaction

The SEM cross-section analysis was carried out after high temperature sintering (Fig.2). All samples showed relative densities > 95% of the XRD-derived density. These values were determined by measuring the volume and weighting the sample. The SEM imaging confirmed the high relative density measured. We were able to classify the morphology of the cross-section in two groups which we named: i) *welded grains:* acrylamide sol-gel and oxalic co-precipitation samples and ii) *joined grains:* Complex assisted co-precipitation and mix solid state reaction samples. The welded grains morphology has been characterized by poorly defined grain boundaries between particles, it seems that the material diffused very well making very hard to define the grain boundaries. The joined grain morphology showed a very well defined grain boundary region, in particular the Complex assisted co-precipitation sample exhibits a very marked grain boundary region.

Electrochemical analysis

In Fig.3 is shown the Nyquist plots for all samples measured at 700°C in open circuit voltage in air atmosphere. It is observed that the lowest R_S values (high frequency intercept on the real axis of the Nyquist plot) corresponded to the oxalic co-precipitation and acrylamide sol-gel samples. However, the acrylamide sol-gel sample presents a higher polarization resistance at low frequency than the oxalic co-precipitation. However, this effect is mainly related to the electrode/electrolyte interface and not to the intrinsic material properties.

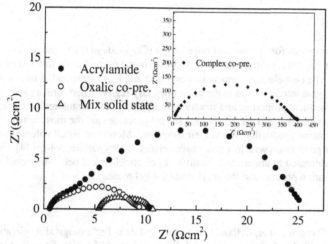

Figure 3. Nyquist diagrams of sintered pellets carried out at OCV and 700°C.

The conductivity values (σ) derived from impedance measurements show the same trend (Fig.4). The best conductivity at 800-700°C corresponds to the so called *welded grains* morphology (oxalic co-precipitated and acrylamide sol-gel samples), but being slightly better for the oxalic co-precipitation. Therefore, it is possible to correlate the grain morphology with the electrochemical properties of the material. From the conductivity measurements the activation energy (E_a) is derived. These values are typically $0.82 < E_a$ (eV) < 0.78 and $0.67 < E_a$ (eV) < 0.57 for the *welded grains* and *joined grains* samples respectively. The welded grains values of activation energy were in agreement with the values reported in the literature [8]. On the other side, the E_a values for the joined grains samples were smaller than the literature reported values

This could be attributed to some structural or compositional heterogeneity in the grains which may affect either the conductivity and the activation energy.

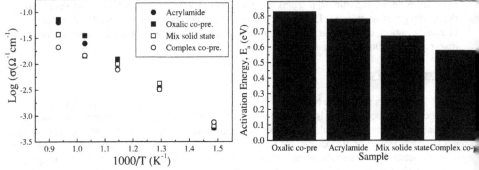

Figure 4. conductivity vs. temperature at OCV   **Figure 5.** Activation energy of all samples

CONCLUSIONS

Several processes for synthesis of doped ceria ($Ce_{0.80}Gd_{0.20}O_{1.90}$) nanopowders have been investigated. The structure, morphology and electrochemical properties of sintered pellets have been compared.. The best electrochemical values (R_S, σ and E_a) correspond to the samples characterized by good interdiffusion in the grain boundary region. Therefore, a clear correlation between electrochemical properties and structural morphology of the sintered pellets exists. The oxalic acid co-precipitation and the acrylamide sol-gel processes are the most suitable synthesis processes for industrial production due to their simplicity. Moreover, small volumes of reactants and large powder production with low cost characterized the acrylamide sol-gel [4]. Further investigation is addressed to understand the different electrochemical behavior recorded for the oxalic co-precipitation process and the acrylamide sol-gel route.

REFERENCES

1. B.C.H Steel, Oxygen Ion and Mixed Conductors and their Technological Applications, Edited by H.L. Tuller, J. Schoonman and I. Riess, NATO ASI Series, Serie E: Applied Sciences Vol.368 (1997) 323-345.
2. J.V.Herle, T.Horita, T.Kawada, N.Sakai, H.Yokokawa, M.Dokiya, Ceramic International 24 (1998) 229-241.
3. T. Tanaka, Sci. Am. 244, (1981) 124-138.
4. A.Sin, P.Odier, Advanced Materials Vol.12, No.9 (2000) 649-652.
5. A.Sin, P.Odier, M.Núñez-Regueiro,Physica C 330 (2000) 9-18.
6. A.Sin, B.El Montaser, P.Odier, F. Weiss, J. Am. Cerm. Soc. Vol.85 No.8 (2002), 1928-33.
7. A. R. West "Solid State Chemistry and its application" Ed. John Wiley & Sons (1996) pag.174.
8. H. Inaba, H. Tagawa, Solid State Ionics 83 (1996) 1-16.

Mat. Res. Soc. Symp. Proc. Vol. 756 © 2003 Materials Research Society FF3.5

Synthesis of Aurivillius Ceramics by the Polymerized Complex Method

Brian S. Luisi and Scott T. Misture[1]
New York State College of Ceramics at Alfred University
2 Pine Street, Alfred, N.Y. 14802 USA
[1] Corresponding author. Electronic mail: misture@alfred.edu

Abstract

Aurivillius phases of the type $Bi_2Sr_2Nb_2TiO_{12}$ and $Bi_{1.6}Pb_{0.4}Sr_2Nb_2Ti_{1-x}Al_xO_{12}$ ($0.0 \leq x \leq 0.8$) were synthesized by the polymerized complex method involving an organometallic precursor. The effect of raising the pH of the solution was investigated through the addition of ammonium hydroxide. Infrared spectroscopy was taken at various points in the reaction to generate a mechanism. The IR data showed the formation of an ester as well as an amide with the addition of ammonium hydroxide. Pure $Bi_2Sr_2Nb_2TiO_{12}$ was formed after heat treatment for 5 hours at 900°C for the unaltered solution and 5 hours at 700°C when the solution pH was raised to 9.00. A solubility limit of aluminum was determined for $Bi_{1.6}Pb_{0.4}Sr_2Nb_2Ti_{1-x}Al_xO_{12}$ when $x \geq 0.4$. Conductivity measurements at 1123 K of $Bi_{1.6}Pb_{0.4}Sr_2Nb_2Ti_{1-x}Al_xO_{12}$ range from 4.76×10^{-3} S•cm^{-1} at $x = 0.8$ to 1.74×10^{-4} S•cm^{-1} at $x = 0.1$.

Introduction

Perovskite structures have been the source of recent interest as potential electrolyte materials in solid oxide fuel cells. This is due to their oxygen conducting behavior. In particular, Aurivillius phases have shown promise as oxygen conducting electrolytes at reduced temperatures [1-5]. Aurivillius phases are composed of perovskite layers with the composition $(A_{n-1}B_nO_{3n+1})^{2-}$ separated by sheets of $(Bi_2O_2)^{2+}$ [1-4]. The general formula for the Aurivillius phase is $Bi_2A_{n-1}B_nO_{3n+3}$ where n denotes the number of perovskite layers, A is a large electropositive cation such as Ba, Sr, Ca, Bi or Pb and B is a small cation such as Ti, Nb, Ta, Mo, W, or Fe [1-4]. Aurivillius structures with n =1 to n=4 layers have been synthesized and tested [1-5]. Doping these phases with aliovalent cations will introduce oxygen vacancies thus theoretically increasing the ionic conductivity [1-5].

Preliminary processing of mixed cation oxides has generally proceeded by solid-state synthesis with some research involved with developing wet chemical methods [6,7]. Solid-state synthesis is a time consuming process that does not easily generate homogenous, phase pure compounds but has enjoyed popularity due to its relative simplicity. Somewhere between two and four days of heat treatment time is usually necessary in order to achieve phase purity in the systems of interest [8]. Nevertheless, some researchers continue to process Aurivillius phases by the solid-state route [9].

Wet chemical methods provide a means of reducing heat treatment time. Wet chemical synthesis naturally lends itself to highly pure, homogenous products. The lengthy processing requirement of the solid-state method is also eliminated with total firing times of two hours having been reported [7]. Potential drawbacks have been the complexity of the initial reactions. Many variables such as temperature, pH, concentration, solvent, and aging time can have dramatic effects on the reaction mechanisms if altered only slightly, and so may greatly affect the final product. If the above-mentioned conditions can be addressed and optimized, the potential benefits of low temperature wet chemical synthesis can be realized.

2. Experimental

Aurivillius phases with target compositions $Bi_2Sr_2Nb_2TiO_{12}$ and $Bi_{1.6}Pb_{0.4}Sr_2Nb_2Ti_{1-x}Al_xO_{12-\delta}$ ($0.0 \leq x \leq 0.8$) were prepared by the polymerized complex method [10,11]. Stoichiometric amounts of Bi_2O_3, $Sr(NO_3)_3$, $NbCl_5$, and Ti(IV) butoxide were weighed out and dissolved in water and a solution of 30% HCl. To this solution, a 30 mole % excess of citric acid was added to chelate the metal cations. Ethylene glycol was then added in a 4:1 mole ratio with citric acid. Ammonium hydroxide was used to raise the pH of the solutions. Solutions with a pH of 3.00, 6.00 and 9.00 were made. Bismuth oxide and niobium(V)chloride were dissolved in 30% hydrochloric acid in water. Strontium nitrate and aluminum nitrate are soluble in water and Ti(IV)butoxide can be dissolved directly into ethylene glycol. Lead oxide was made to dissolve in a heated solution of 30% HCl in water. The solution of mixed cations, ethylene glycol and citric acid was stirred magnetically while heating on a hot plate to 135 °C to cause evaporation of the solvents. Polymerization of the citric acid with ethylene glycol resulted in the formation of a viscous liquid. The polymerization reaction was carried to completion in a drying oven in air at 170 °C for 8 hours. The organics were burned off at 500 °C for 5 hours and the resulting powder fired in air at 900°C for 5 hours with a heating and cooling rate of 5°C per minute.

Infrared spectroscopy was conducted on the polymerized product with a Thermo Nicolet infrared spectrometer. An average of 32 scans with a 0.2 cm^{-1} step was used to generate each spectrum. A Philips12045 X-ray diffractometer was utilized to produce the x-ray diffraction patterns from 10° to 70° 2θ. The measurements were completed at a 0.03 step size and 3 second count time. Impedance spectroscopy data was obtained using a custom system that included an HP 4192A meter and a Centurion Qex furnace at 50-degree intervals from 473 K to 1123 K.

3. Results and Discussion

A solid amorphous gel was produced upon completion of the organic reaction. Figure 1 shows the infrared spectrum of the polymerized product attached to bismuth, niobium, strontium, and titanium ions when the pH was left unaltered and at a pH of 3.00. The presence of the resonant absorption at 1715 cm^{-1} in spectrum (a) in fig. 1 indicates the formation of an ester from the reaction of citric acid with ethylene glycol. This conclusion is supported by the absence of a significant peak at 3300-3400 cm^{-1}, demonstrating the lack of alcohol functional groups left from the ethylene glycol. In the solution where the pH was raised to 3.00, an absorption at 1642 cm^{-1} in addition to the ester peak is observed. The peak at 1642 cm^{-1} is attributed to the formation of an amide resulting from the reaction of the nitrogen in ammonium hydroxide with a carbonyl carbon in citric acid. The peak at 3400 cm^{-1} is due to the presence of water as a solvent.

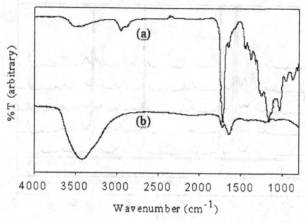

Figure 1: Infrared spectrum of a specimen after the completed organic reaction taken for (a) a solution where the pH has been unaltered, measured using attenuated total reflectance and (b) a solution where the pH was raised to 3.00, measured in transmission after dissolving in water.

After heat treatment at 900°C there was a white powder left in the porcelain crucible. In contrast, $Bi_{1.6}Pb_{0.4}Sr_2Nb_2Ti_{1-x}Al_xO_{12-\delta}$ formed white powders at $x \leq 0.2$ and yellow colored products at $x \geq 0.4$. Figure 2 shows an overlay of the x-ray powder diffraction patterns of specimens of composition $Bi_{1.6}Pb_{0.4}Sr_2Nb_2Ti_{1-x}Al_xO_{12-\delta}$ heat treated at 900°C. Phase pure Aurivillius structures are obtained when $x = 0$ and $x = 0.2$. When 40% aluminum is substituted in for titanium, however, two additional phases begin to develop, indicating a substitutional solubility limit the Aurivillius crystal structure. Data indicating the presence of a solubility limit is supported by the work reported by Modi [5]. The second phase is the result of δ-bismuth oxide evolving out of the structure. As the aluminum concentration increases past 40%, the peaks corresponding to bismuth oxide sharpen and increase in intensity. Several of the peaks associated with the Aurivillius phase are broad at low concentrations of aluminum and begin to sharpen at higher concentrations. Broad peaks are either the result of stacking faults or small particle size. Since the sharpening of the Aurivillius phase peaks occurs at high aluminum concentration, it is likely that the broader peaks at $x \leq 0.4$ are due to stacking faults that have not been annealed out in such a short heating cycle. Figure 3 shows the x-ray powder diffraction patterns for $Bi_2Sr_2Nb_2TiO_{12}$ where the solution pH was raised to 9.00. The pure Aurivillius phase begins to form at firing temperatures as low as 700°C for 5 hours. This marks a significant improvement over both solid-state synthesis where 100 hours of heat treatment is often necessary at temperatures above 1000°C, and wet chemical synthesis where the pH has not been altered. In the latter, firing temperatures of 900°C for 5 hours are typically required. Figure 4 details the conductivity plots of $Bi_{1.6}Pb_{0.4}Sr_2Nb_2TiO_{12-\delta}$ and "$Bi_{1.6}Pb_{0.4}Sr_2Nb_2Ti_{0.2}Al_{0.8}O_{12-\delta}$" from 473 to 1123 K.

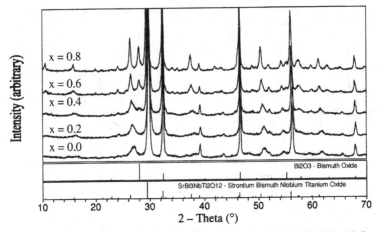

Figure 2. X-ray powder diffraction pattern overlays for $Bi_{1.6}Pb_{0.4}Sr_2Nb_2Ti_{1-x}Al_xO_{12-\delta}$ fired at 900°C in air for 5 hours.

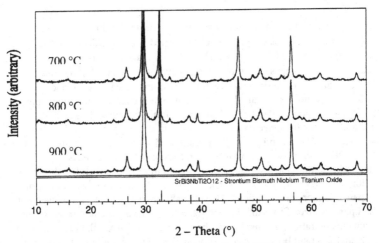

Figure 3. X-ray powder diffraction patterns for $Bi_2Sr_2Nb_2TiO_{12}$ fired from 700°C to 900°C in air for 5 hours. The pH of the solution used in processing this composition was 9.00.

Conductivities at 1123 K ranged from 1.74×10^{-4} S•cm^{-1} when x = 0, to 4.6×10^{-3} S•cm^{-1} when x = 0.8. The higher conductivity of specimen of composition "$Bi_{1.6}Pb_{0.4}Sr_2Nb_2Ti_{0.2}Al_{0.8}O_{12-\delta}$" is due at least in part to the presence of bismuth oxide, which conducts oxygen ions more readily than does the base Aurivillius structure. The number of oxygen vacancies will also increase, further improving conductivity.

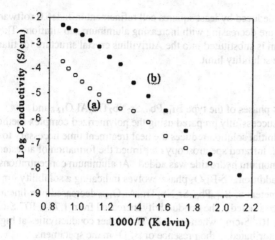

Figure 4: Conductivity of $Bi_{1.6}Pb_{0.4}Sr_2Nb_2Ti_{1-x}Al_xO_{12-\delta}$ at (a) when x = 0.0 and the specimen is a phase-pure Aurivillius ceramic and (b) when x = 0.8, where the specimen contains an Aurivillius phase, Bi_2O_3, and strontium bismuth oxide.

Yttria-stabilized zirconia exhibits a conductivity around 3.2×10^{-2} S•cm^{-1} at 1123K, approximately an order of magnitude higher than "$Bi_{1.6}Pb_{0.4}Sr_2Nb_2Ti_{0.2}Al_{0.8}O_{12-\delta}$" [1]. Figure 5 shows the effect of aluminum substitution on the tetragonal lattice parameters

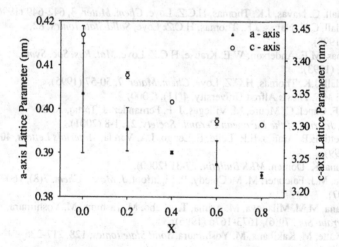

Figure 5. The tetragonal lattice parameter in nm along the a and c axes of $Bi_{1.6}Pb_{0.4}Sr_2Nb_2Ti_{1-x}Al_xO_{12-\delta}$ for increasing values of x. The error bars are plus/minus one standard deviation from the cell refinement.

along the a and c axes deduced by least squares cell refinement in Jade6.0 software [12]. The lattice parameters are decreasing with increasing aluminum substitution. The trend indicates that aluminum is substituted into the Aurivillius crystal structure for titanium even after reaching the solubility limit.

Conclusions

Three layer Aurivillius phases of the type $Bi_{1.6}Pb_{0.4}Sr_2Nb_2Ti_{1-x}Al_xO_{12-\delta}$ and $Bi_2Sr_2Nb_2TiO_{12}$ were successfully prepared using the polymerized complex method. Raising the pH of the initial solution reduces the heat treatment time necessary to form a pure Aurivillius phase. Infrared spectroscopy confirmed the formation of an ester as well as an amide when ammonium hydroxide was added. At aluminum concentrations greater than forty percent, an additional δ-Bi_2O_3 phase evolves indicating a solubility limit. The tetragonal lattice parameters of $Bi_{1.6}Pb_{0.4}Sr_2Nb_2Ti_{1-x}Al_xO_{12-\delta}$ decreased non-linearly over the compositional range studied. The conductivity increased from 1.74×10^{-4} S•cm^{-1} when $x = 0.0$ to 4.76×10^{-3} S•cm^{-1} when $x = 0.8$. The higher conductivities at high aluminum content are attributed to the presence of Bi_2O_3 in the specimens.

Acknowledgements

The researchers would like to thank the NYSCC for support and Mr. Fran Williams for his assistance in collecting the conductivity data.

References

1. K.R. Kendall, C. Navas, J.K. Thomas, H.C.Z. Loye, *Chem. Mater.* **8**, 642-649 (1996).
2. K.R. Kendall, C.N. Navas, J.K. Thomas, H.C.Z. Loye, *Solid State Ionics*, **82**, 215-223 (1995).
3. J.K. Thomas, M.E. Anderson, W.E. Krause, H.C.Z. Loye, *Mat. Res. Soc. Symp. Proc.* **293**, (1993).
4. K.R. Kendall, J.K. Thomas, H.C.Z. Loye, *Chem. Mater.* **7**, 50-57 (1995).
5. V.B. Modi *M.S. Thesis* Alfred University, 4(11), (2000)
6. P. Duran, F. Capel, C. Moure, M. Villegas, J. F. Fernandez, J. Tartaj, A. C. Caballero, *Journal of the European Ceramic Society*, **21**, 1-8 (2001).
7. S. M. Zanetti, E.B. Araujo, E.R. Leite, E. Longo, J.A. Varela, *Materials Letters*, **40**, 33-38 (1999).
8. M. Kakihana, K. Domen, *MRS Bulletin*, 27-31 (2000).
9. S.M. Blake, M.J. Falconer, M. McCreedy, P. Lightfoot, *J. Mater. Chem.* **7**(8), 1609-1613 (1997).
10. M. Kakihana, M.M. Milanova, M. Arima, T. Akubo, M. Tashima, M. Yoshimura, *J. Am. Ceram. Soc.* **79**(6), 1673-1676 (1996).
11. C.P. Udawatte, M. Kakihana, M. Yoshimura, *Solid State Ionics*, **128**, 217-226 (2000)
12. Jade v.6.0.3 Materials Data, Inc., Livermore, CA, 2002

Mat. Res. Soc. Symp. Proc. Vol. 756 © 2003 Materials Research Society   FF4.8

Optimization of the sintering of Aurivillius phases using D-optimal statistical design

M.S. Peterson and S.T. Misture
New York State College of Ceramics at Alfred University, Alfred NY 14802

ABSTRACT

A statistical model has been employed to determine the optimal sintering times and temperatures for $Bi_2Sr_2Nb_2TiO_{12}$. The phase of interest was made in pure form by solid-state synthesis. Bi_2O_3 was used as a sintering aid to decrease sintering temperatures and increase densification. The amount of Bi_2O_3 added and the sintering temperature had a statistical impact on density, while the heat treatment time did not.

INTRODUCTION

In 1949, a series of papers [1-3] were published by Bengt Aurivillius exploring the discovery of mixed-metal oxides having bismuth oxide layers alternating with perovskite structure layers. These materials have the general composition $Bi_2A_{n-1}B_nO_{3n+3}$ (where A=Ca, Sr, Ba, Pb, Bi, Na, K, etc., and B=Ti, Nb, Ta, Mo, W, Fe, etc.). The A cations can be mono-, di-, or trivalent ions; or a mixture. The $Bi_2O_2^{2+}$ sheets are separated by perovskite-type blocks $(A_{n-1}B_nO_{3n+1})^{2-}$ of variable thickness according to the value of n. Because of their ionic structural framework, Aurivillius phases exhibit great flexibility with respect to metal cation substitution; therefore, these phases have high potential for systematic control of their properties [4].

This research explored the sintering behavior of Aurivillius phases in an effort to reduce sintering times and temperatures, while increasing the density. This was accomplished by using a statistical design experiment that compared the sintering temperature, sintering time, a sintering accelerator, and the final density. The $Bi_2Sr_2Nb_2TiO_{12}$ phase was chosen because of its ability to be formed in pure form by solid-state synthesis. One of the inherent problems with the Aurivillius phases is the repeatability of producing single-phase materials. Work done by Peterson and others [5-7] have shown that many of the Aurivillius compositions, in particular the oxygen deficient compositions, are not phase pure. Bi_2O_3 was chosen as a sintering accelerator because of its successful use in other electroceramics [8].

PROCEDURE

The first step was to make $Bi_2Sr_2Nb_2TiO_{12}$ via solid-state synthesis and check for phase purity. Stoichiometric mixtures of precursor powders (Bi_2O_3, Alfa Aesar, 99.975% pure; $SrCO_3$, Alfa Aesar, 97.5% pure; Nb_2O_5, Alfa Aesar, 99.9% pure; TiO_2, Alfa Aesar, 99.9% pure) were intimately mixed in a mortar and pestle. The powders were then calcined at 815° C for 10 hours in air. The powders were reground, pressed into pellets (22 mm in diameter) at 102 MPa, and subsequently put in a powder bed (of the same composition) in a MgO crucible. The pellets were fired twice at 1095° C for twelve hours with regrinding in between. Finally, the pellets were fired at 1095° C for 24 hours. At each regrinding step, the powders were checked for phase purity by XRD. Experimental Design Optimizer [9] was then used to design a full quadratic model that would find the effects of temperature, time, and amount of Bi_2O_3 added on the sintering behavior of $Bi_2Sr_2Nb_2TiO_{12}$. The three variables (temperature, time, and weight percent Bi_2O_3 added) were modeled as shown in Table I. A total of thirteen design experiments were conducted with three confirmation experiments. After sintering, the densities of the sintered pellets were measured using the ASTM standard procedure C 20-97 [10]. Multiple Correlation Analysis [11] was used to model the data.

Table I: Variables in sintering study of $Bi_2Sr_2Nb_2TiO_{12}$.

Temperature (°C)	Time (hours)	Weight % Bi_2O_3 added
950	10	0
1025	40	2.5
1100	70	5.0

RESULTS AND DISCUSSION

The goal of the experimental matrix was to find the lowest temperature and time that would maximize density, and to see if Bi_2O_3 could be used as an effective sintering accelerator. The phase evolution during solid-state synthesis of $Bi_2Sr_2Nb_2TiO_{12}$ can be seen in Figure 1.

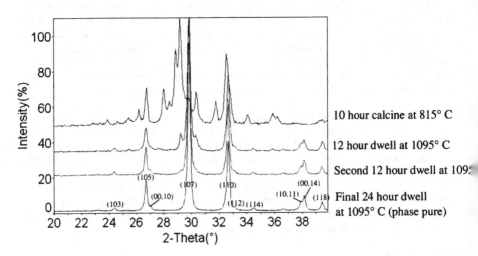

10 hour calcine at 815° C

12 hour dwell at 1095° C

Second 12 hour dwell at 109!

Final 24 hour dwell
at 1095° C (phase pure)

Figure 1. Phase evolution of $Bi_2Sr_2Nb_2TiO_{12}$ after each solid-state synthesis step.

The designed experiments had an R-squared value of 0.96 and a $S_{y.x}$ of 0.24. Three confirmation experiments were done to confirm the model's accuracy. The surface response plot of the significant variables, and the data points are shown in Figure 2. A contour plot is shown in Figure 3. The significant and insignificant variables are shown in Table II with their coefficients and t-values.

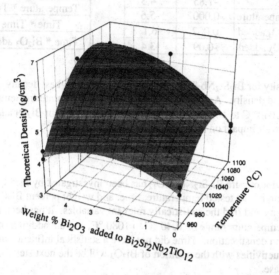

Figure 2. Surface response plot depicting the significant terms in the designed model.

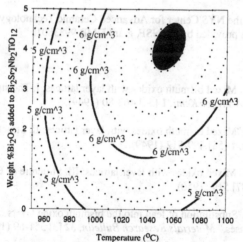

Figure 3. Contour plot showing the optimum temperature and amount of Bi_2O_3 added in relation to density in the designed model (calculated density for $Bi_2Sr_2Nb_2TiO_{12}$ is 6.73 g/cm^3).

Table II: (a) Significant and (b) insignificant variables in the statistical model.

(a) Variables included in model (significant)

Variables	Coefficient	t-variable
Temperature	0.41	5.6
Bi_2O_3 Added	-1.85	-4.3
Temperature * Temperature	-0.0002	-5.6
Temperature * Bi_2O_3 added	0.002	6.1
Bi_2O_3 added * Bi_2O_3 added	-0.09	-3.4

(b) Insignificant variables in model

Variables	t-variable
Time	0.11
Temperature * Time	0.14
Time * Time	0.06
Time * Bi_2O_3 added	0.56

The calculated density for $Bi_2Sr_2Nb_2TiO_{12}$ is 6.73 g/cm^3. The highest density of the model was 96% of the calculated density. According to the model, the ideal sintering temperature range for $Bi_2Sr_2Nb_2TiO_{12}$ is 1040° C to 1080° C with 3.5 to 4.5 weight percent Bi_2O_3 added. Time does not statistically have an impact on the final density.

CONCLUSIONS

The sintering behavior of pure $Bi_2Sr_2Nb_2TiO_{12}$ has been investigated by means of a statistical model. The sintering model tested temperature, time, and weight percent Bi_2O_3 added. The goal was maximize density and find the statistically relevant variables. It has been shown that there is an ideal sintering temperature range of 1040 °C to 1080 °C, and the addition of 3.5 to 4.5 weight percent Bi_2O_3 aids in densification. Time did not have a statistical influence on density. A correlation of conductivities with the addition of Bi_2O_3 will be the next step in the study of the $Bi_2Sr_2Nb_2TiO_{12}$ phase.

ACKNOWLEDGMENTS

Major support was by the NYS Center for Advanced Ceramic Technology at Alfred University and minor support was provided by the NSF, grant DMR 9983801.

REFERENCES

1.   B. Aurivillius, "Mixed bismuth oxides with layer lattices: I. The structure type of $CaNb_2Bi_2O_9$," *Arkiv for Kemi*, **1** [54] 463-80 (1949).

2.   B. Aurivillius, "Mixed bismuth oxides with layer lattices: II. Structure of $Bi_4Ti_3O_{12}$," *Arkiv for Kemi*, **1** [58] 499-512 (1949).

3.   B. Aurivillius, "Mixed oxides with layer lattices: III. Structure of $BaBi_4Ti_4O_{15}$," *Arkiv For Kemi*, **2** [37] 519 (1950).

4.   T. Rentschler, "Substitution of lead into the bismuth oxide layers of the n=2 and n=3 Aurivillius Phases," *Materials Research Bulletin*, **32** [3] 351-69 (1997).

5. M. S. Peterson, C. A. Say, S. A. Speakman, and S. T. Misture, "High Temperature X-ray Diffraction Study of Reaction Rates of Ceramics," *Advances in X-ray Analysis,* in review (2002).

6. A. Snedden, S. M. Blake, and P. Lightfoot, "Oxide ioin conductivity in Ga-doped Aurivillius phases- a reappraisal," *Solid State Ionics,* Article in Press (2002).

7. V. B. Modi, "Electrical and microstructural characterization of N=3 type Aurivillius phases"; Thesis. New York State College of Ceramics at Alfred University, Alfred, 2002.

8. C.-H. Lu and Y.-C. Chen, "Sintering and Decomposition of Ferroelectric Layer Perovskites: Strontium Bismuth Tantalate Ceramics," *Journal of the European Ceramic Society,* **19** 2909-15 (1999).

9. Experimental Design Optimizer, 1.0, Nextbridge Software, 1999.

10. ASTM Standards, "Standard test methods for apparent porosity, water absorbtion, apparent specific gravity, and bulk density of burned refractory brick and shapes by boiling water," *Refractories; Activated Carbon, Advanced Ceramics,* **C 20-97,** Vol. 15.01. (2000).

11. Multiple Correlation Analysis, 1.0, Nextbridge Softwares, 1999.

Mat. Res. Soc. Symp. Proc. Vol. 756 © 2003 Materials Research Society   FF4.5

Structure and Electrochemical Behavior of Plasma-Sprayed LSGM Electrolyte Films

H. Zhang[1], X. Ma[2], J. Dai[1], S. Hui[2], J. Roth[1], T.D. Xiao[1], and D.E. Reisner[1]

1. US Nanocorp®, Inc, 74 Batterson Park Road, Farmington, CT 06032, U.S.A.
2. Inframat® Corporation, 74 Batterson Park Road, Farmington, CT 06032, U.S.A.

ABSTRACT

An intermediate temperature solid oxide fuel cell (SOFC) electrolyte film of $La_{0.8}Sr_{0.2}Ga_{0.8}Mg_{0.2}O_{2.8}$ (LSGM) was fabricated using a plasma spray process. The microstructure and phase were investigated using X-ray diffraction (XRD) and scanning electron microscopy (SEM). The electrochemical behavior of the thermal sprayed LSGM film was investigated using electrochemical impedance spectroscopy (EIS). The study indicates that thermal spray can deposit a dense LSGM layer. It was found that the rapid cooling in the thermal process led to an amorphous or poor crystalline LSGM deposited layer. This amorphous structure has a significant effect on the performance of the cell. Crystallization of the deposited LSGM layer was observed during annealing between 500-600 °C. After annealing at 800 °C, the ionic conductivity of the sprayed LSGM layer can reach the same level as that of the sintered LSGM.

INTRODUCTION

Seeking efficient, economical and advanced fabrication procedures for intermediate temperature SOFC has become one of the key factors to successfully develop integrated SOFCs with high performance, high power density, long service life and low cost. Currently, only a few processing approaches exist in fabricating SOFCs, including electrostatic spray deposition [1], sputter deposition [2], electroless deposition [4], slip casting (SL), and tape casting and calendaring [5]. Disadvantages of these processing techniques include low processing efficiency, complicated processes, and high cost. In this study, a plasma spray approach has been introduced to fabricate electrolyte and electrode layers separately and total-cells consisting of anode/electrolyte/cathode via a sequential plasma spray deposition process in a single operation. In this process, $La_{0.8}Sr_{0.2}MnO_3$ (LSM) was selected as the material for cathode and $Sr_{0.88}Y_{0.08}TiO_3$ (SYT) as the anode material. The electrolyte layer of LSGM was used in plasma spray based on the requirement of a SOFC with operation temperature of 500 – 800°C. The electrolyte layer chemistry and physical properties are very important in SOFC studies. Its ionic conductivity should be high enough at working temperature range, i.e. 0.16 S/cm for LSGM at 800 °C. Also it should be a condensed layer to isolated fuel and air effectively and any cell leakage should be avoided in the cell. This study is to explore the use of an integrated thermal spray technique for the preparation of an LSGM electrolyte layer and explore its electrochemical performance.

EXPERIMENTAL

Thermal spray $La_{0.80}Sr_{0.20}Ga_{0.80}Mg_{0.20}O_{2.80}$ (LSGM) feedstock was obtained from Praxair. A LSGM disk with a diameter of 13 mm and a thickness of 550 um was fabricated using a plasma thermal spray method. In order to obtain high-density LSGM layers, a high plasma power, low feedstock feed rate, and small spray distance were used. For comparison, an LSGM disk with the same dimension as thermal sprayed disk was prepared by sintering at 1450 °C for 24 hours.

The phase and microstructure information of the LSGM powder and sprayed disk were examined using XRD and SEM techniques. The electrochemical properties of the LSGM layer were investigated using electrochemical impedance spectroscopy (EIS). Both sintered and sprayed disks were pasted with Pt-black ink with diameter 9-10mm. The Pt foil was used as the lead to connect with disk on both sides. The EIS was measured at temperatures of 500-800°C under air.

RESULTS AND DISCUSSIONS

The LSGM feedstock exhibits a spherical particle distribution (Fig. 1a). The cross-section microstructure of the sprayed disk is shown in Fig. 1b. It reveals a dense microstructure in plasma sprayed LSGM layers.

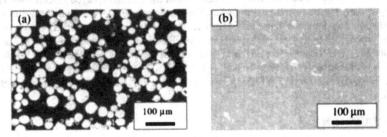

Fig. 1. SEM images for (a) feedstock and (b) cross section of thermal spray layer for LSGM

Typical XRD pattern of the LSGM feedstock is shown in Fig. 2. It indicates that the feedstock is a well-crystallized material. The XRD of the as-sprayed LSGM layer is also shown in Fig. 2. A broad diffraction peak at 2 theta 31.5° indicates that an amorphous phase was formed due to supercooling of the melted feedstock during coating deposition process. This structure may have a significant influence on the electrochemical properties and performance. To investigate the crystallization of the amorphous LSGM layer, another sprayed LSGM sample that has a mixture of amorphous and crystalline structure was post heat- treated at elevated temperatures, ranging from 500 to 900 °C. The XRD patterns for the heat-treated LSGM layer are shown in Fig. 3. When heat-treated at low temperatures, for example at 500 °C, partial amorphous phase is still evident in the XRD pattern. When heat-treated at 700 °C or above, the amorphous peak disappears and at 800 °C, it exhibits a well crystalline phase.

The EIS of sintered LSGM disk are illustrated in Fig. 4a and Fig. 4b. As shown in Fig. 4b, we can see that the absolute value of impedance, $|Z|$, decreases with temperature from 650 to 700°C. It is complicate to analyze the total EIS even for this simple cell. To simplify the analysis, a descriptive equivalent circuit was constructed as shown in Fig. 5. In this circuit, R is resistance of electrolyte that we are most interested in; r_1 is electrochemical polarization resistance at interface between Pt and electrolyte while C_1 is related double layer capacitor. C_2

and r_2 are the same as that of C_1 and r_1, and represent the other side of the electrolyte disk. In this measurement, DC polarization current was not used and the measurement was carried out in an open circuit state. Hence, the r_1, C_1, r_2, C_2 are not very stable. However, when R is lower in value, the r_1 and r_2 become smaller since the higher conductivity of electrolyte makes electrochemical process easier. The R and r_1+r_2 obtained from EIS are outlined in Table 1. The specific conductivities of electrolyte calculated from R and area of disk are also listed in Table 1.

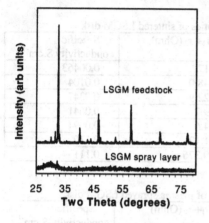

Fig. 2. XRD patterns for the LSGM feedstock and thermal sprayed LSGM layer

Fig. 3. XRD patterns for the sprayed LSGM layer against heat treatment temperature

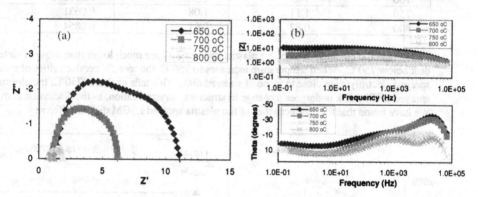

Figs. 4 (a) Complex graph of EIS for sintered disk of LSGM with thickness of 530µm and diameter of 9 mm (effective area) and (b) Bode graph of EIS at different temperatures

The EIS of plasma sprayed LSGM disk is shown in Fig. 6a and 6b. The resistance at temperatures of 650°C and 700°C is much higher than that of sintered disk. When temperature is less than 650°C, the EIS of the plasma sprayed disk is not stable because high resistance makes it like an insulator. The resistance and specific conductivity are listed in Table 2.

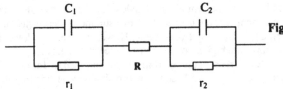

Fig. 5. Schematic of the equivalent circuit of the LSGM electrolyte layer EIS measurement set.

Table 1. Resistances and specific conductivities of sintered LSGM disk

Temperature (°C)	R (Ohm)	$r_1 + r_2$ (Ohm)	Specific conductivity S.cm^{-1}
500	18.0	122	0.00463
550	8.0	48.0	0.0104
600	2.40	22.9	0.0347
650	1.3	9.7	0.0641
700	1.2	4.6	0.0694
750	0.76	1.12	0.110
800	0.75	0.376	0.111

Table 2. Resistance and specific conductivities of plasma sprayed LSGM disk

Temperature (°C)	R (Ohm)	$r_1 + r_2$ (Ohm)	Specific conductivity S.cm^{-1}
650	311	3053	0.000225
700	340	2145	0.000206
750	1.11	1.08	0.0631
800	0.82	0.61	0.0854

The specific conductivities of the plasma sprayed disk are much lower than those of sintere disk below 750°C. When temperature increases to 750°C, the specific conductivities of plasma sprayed electrolyte are close to those of sintered disk. This tells us that at 750°C, the plasma sprayed LSGM disk undergoes a change in structure - crystallization, as the discussion above We have found that the conductivities of the plasma sprayed LSGM disk at different

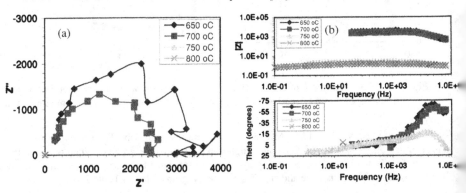

Fig. 6. (a) Complex graph of EIS for sprayed LSGM disk with thickness of 550μm and diameter of 10 mm, and (b) Bode graph of EIS at different temperatures

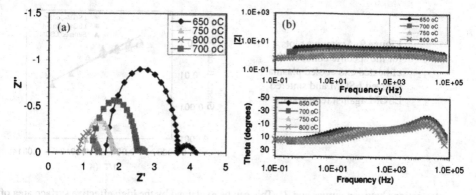

Fig. 7 (a) Complex graph of EIS for sprayed disk of LSGM with thickness of 550μm and diameter of 10 mm (effective area) and (b) Bode graph of EIS at different temperatures during cooling after heated at 800°C

temperatures are similar to those of sintered disk when cooling from 800°C to 500°C. The EIS at different temperatures during cooling for plasma sprayed LSGM disk are illustrated in Figs. 7a and 7b. The resistances and specific conductivities are outlined in Table 3. This is attributed to the well crystallized structure for the sprayed LSGM layer after the heating at 800 °C.

Table 3. Resistances and specific conductivities of plasma sprayed LSGM disk during cooling

Temperature (°C)	R (Ohm)	r_1+r_2 (Ohm)	Specific conductivity S.cm^{-1}
650	1.70	2.36	0.0412
700	1.26	1.50	0.0556
750	1.00	0.97	0.0701
800	0.82	0.61	0.0854

For further comparison of plasma sprayed LSGM disk and sintered disk, EIS at 650°C during cooling are illustrated in Figs. 8. As shown in Fig. 8a, we can see that the left side intersections between curves and Z' are almost the same, which means that their conductivities are nearly the same. However, the r_1+r_2 values are different, which can be seen from the right side

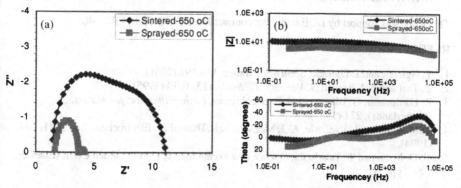

Fig. 8. Comparison of (a) complex of EIS results and (b) Bode graph for the sintered disk and sprayed disk after heated at 800°C and cooled to 650°C.

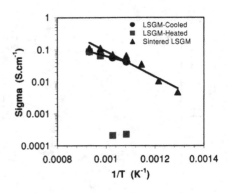

Figure 9. Arrhenius plots for the specific conductivity of heated and cooled plasma sprayed LSGM and sintered LSGM against temperature

intersections between curves and Z'. This can be explained by the high effective surface area of the plasma sprayed disk due to rough surface of the as-sprayed disks. The interface area between electrolyte and Pt is larger for plasma sprayed disk.

The Arrhenius plots of specific conductivity against test temperature for the LSGM samples are shown in figure 9. The sintered LSGM and sprayed LSGM after 800 °C heating show good liner relationship between the $\log(\sigma)$ and $1/T$. While the sprayed LSGM dose not show such relationship due to the existence of amorphous phase below 700 °C.

CONCLUSIONS

An LSGM film has been fabricated by using a plasma spray technique. Plasma spray leads to a dense and amorphous LSGM layer. The amorphous structure results in an LSGM electrolyte layer of low ionic conductivity. The crystallization of the amorphous phase will not take place until over 700°C during post annealing process. The EIS study indicates that the thermal sprayed LSGM layer has much high resistance than that of the sintered LSGM disk before crystallization. This difference in ionic conductivity will disappear when the temperature increases to 800°C. A well-crystallized thermal sprayed LSGM electrolyte layer has the same specific ionic conductivity as that of a sintered LSGM disk.

ACKNOWLEDGMENTS

This work was support by DOE under the contract No: DE-FG-02-01ER83340.

REFERENCES

1. T. Nguyen and E Djurado, *Solid State Ionics*, **138**,191 (2001).
2. T. Tsai and S.A.Barnett, *J. Vac. Sci. Technol.* **A13**, 1073 (1995).
3. R. Forthmann, G. Blass, and H. P. Buchkremer, *Netherlands Soc. for Maters. Sci.* (Netherlands), 271 (1997).
4. M.M. Murphy, J. Vanherle, A. J. McEvoy, K.R.Thampi., J. Electrochem. Soc., **141**, L94 (1994).
5. N. Q. Minh and T. Takahashi, *Science and Technology of Ceramic Fuell Cells*, (Elsevier, 1995) pp. 71.

Mat. Res. Soc. Symp. Proc. Vol. 756 © 2003 Materials Research Society   FF4.12

Synthesis of LaSrXMg-Oxide with X=Ga, Fe, or Cr

Cinar Oncel and Mehmet A. Gulgun,
Sabanci University, FENS, Orhanli Tuzla, Istanbul 34956 Turkiye

ABSTRACT

Strontium and magnesium doped lanthanum gallate (LSGM) is a promising electrolyte material for intermediate temperature range (650-800°C) solid oxide fuel cell (SOFC) applications. Formation of unwanted phases and Ga loss at high temperatures (1100-1500°C) during synthesis and under low oxygen partial pressures during operation are major hurdles that stand in LSGM's way of full utilization. Using a polymeric precursor synthesis method, the feasibility of producing SOFC electrolyte material LSGM is investigated. The method involves complexing each constituent metal ion by the carboxyl and/or hydroxyl group of the citric acid and/or polyvinyl alcohol (PVA) in aqueous solution. The facility of this method compared with the traditional solid state reaction method was shown by synthesis of single phase and pure LSXM (X= Fe, Cr) oxides at reasonable temperatures (800°C). The X-ray diffraction patterns of LSFM and LSCM are also reported here for the first time.

INTRODUCTION

Lanthanum strontium gallium magnesium oxide ($La_{0.8}Sr_{0.2}Ga_{0.8}Mg_{0.2}O_{3-\delta}$), is one of the most promising electrolyte materials for solid oxide fuel cell (SOFC) applications [1,2].   In lanthanum gallate ($LaGaO_3$) structure, Sr-atoms substitute for the La sites and Mg-atoms substitute for the Ga sites. Doping elements introduce oxygen vacancies and thereby increase the oxygen diffusivities through the material at intermediate temperatures [3]. LSGM exhibits oxygen ion diffusivities at intermediate temperature range (650-850°C) that are comparable to the one from ZrO_2 electrolyte at 1000°C [2]. Several hurdles have to be overcome before the electrolyte will be suitable for the fuel cell applications. There are stringent requirements for SOFC electrolyte materials. Besides being a good oxygen ion conductor, the candidate oxide should maintain its stoichiometry, phase character, and crystal structure at the operating temperatures for an extended period of time. The electrolyte must be chemically inert and should have a compatible thermal expansion coefficient with the electrode materials up to the operating temperatures. Traditional production method for multi-cation oxide materials is the solid state reaction technique. In this method repetitive ball-milling and grinding steps are time consuming, and energy intensive. Besides, calcination and sintering with long holding times at high temperatures are costly [4-7]. As one of the alternative methods to produce multi-cation oxide material, the so-called urea method was suggested. The problem with this technique is the high temperatures (1400-1500°C) and long holding times required to obtain a single phase and pure mixed-oxide [8,9]. For another alternative route (citrate synthesis), the problem is the formation of second phases, which influence the stability, reactivity and ionic conductivity of the electrolyte material [10]. One of the major problems with the production and use of LSGM at high temperatures is gallium loss from the material. At high temperatures (>800°C) and low P_{O2} of reducing atmospheres (10^{-18} atm or 10^{-13} Pa) [11], i.e. exactly the operating conditions in the SOFC anode environment, gallium may evaporate from the system as Ga_2O_3. This problem is especially severe in the presence of Pt electrodes. Yet another problem with Pt electrode is the formation of Ga-Pt alloys and formation of unwanted oxide phases [12]. Alternative electrode materials for the anode side of the SOFC are necessary.

In this study, the so-called Steric Entrapment method is employed to produce pure and single phase LSGM powders and alternative electrode materials that will be compatible with LSGM. In the past, highly reactive, highly sinterable, oxide powders were easily synthesized by steric entrapment method [13].

EXPERIMENTAL PROCEDURE

The polymeric pre-ceramic precursors were prepared to obtain LSGM with the following composition : $La_{0.9}Sr_{0.1}Ga_{0.8}Mg_{0.2}O_{2.85}$. Three separate batches were mixed i) in the exact stoichiometry of LSGM, ii) with 10 wt % excess gallium, and iii) with 20 wt. % excess gallium. Calculated amounts of $La(NO_3)_3 \cdot 6H_2O$ (>99%, Sigma Aldrich Chemie GmbH, Taufkirchen, Germany), $Sr(NO_3)_2$ (>99%, Sigma Aldrich Chemie GmbH, Taufkirchen, Germany), and $Mg(NO_3)_2 \cdot 6H_2O$ (>99%, Merck KgaA, Darmstadt, Germany) were dissolved separately in distilled water. $Ga_2(SO_4)_3$ (99.999%, Sigma Aldrich Chemie GmbH, Taufkirchen, Germany) did not dissolve in cold water despite the claim in materials hand books for its high solubility in cold water. It was dissolved in citric acid solution (>99.5%, Sigma Aldrich Chemie GmbH, Taufkirchen, Germany) in such a ratio that there will be one citric acid molecule for each gallium ion. Polyvinyl alcohol (PVA, MW = 72000, >98%, Merck KgaA, Darmstadt, Germany) was dissolved in distilled water. The salt and PVA solutions were mixed in ratios such that for every cation in solution there was one hydroxyl group of PVA in the mixture. The precursor solutions were then heated on a hot plate (< $300°C$) while stirring continuously until the water of solution evaporated, and a crisp light brown gel was obtained. The dried, crisp gel was ground into a powder that was calcined in air at temperatures up to $1200°C$ for 1 hour. The same experimental procedure was used to produce lanthanum strontium iron magnesium oxide (LSFM) and lanthanum strontium chromium magnesium oxide (LSCM). For LSFM, $Fe(NO_3)_3 \cdot 9H_2O$ (>99%, Merck KgaA, Darmstadt, Germany) and for LSCM $Cr(NO_3)_3 \cdot 9H_2O$ (>99%, Merck KgaA, Darmstadt, Germany) were used as iron and chromium sources, respectively. PVA is used as the polymeric carrier for all the cations when producing LSFM and LSCM. The dried powders were heated in air up to $800°C$ and then furnace cooled to room temperature.

Crystal structure and phase distribution of the powders at room temperature was studied by an x-ray powder diffractometer (Bruker AXS-D8, Karlsruhe, Germany). The measurements were performed in the 2θ range of $10°-90°$ at 40 kV and 40 mA, using Cu K_α radiation.

The thermal decomposition profiles of the pre-ceramics precursor and the conversion to the multi-cation ceramics were investigated using a simultaneous differential thermal and thermo gravimetric analysis set-up (Netzsch STA 449C Jupiter, Selb, Germany) up to 1400° C. The analyses were run with the same heating rates as the calcination of pre-ceramic precursors.

RESULTS and DISCUSSION

One of the problems in the production of multiple-cation oxides is to keep the cations homogeneously distributed in the precursor with the desired stoichiometry. Steric entrapment method using a functional polymeric carrier that forms metalorganic complexes was shown to be effective for this purpose. Various, complex multi-cation oxides were produced in the temperature range of 600° C – 900° C as pure single phase powders using this simple solution polymerization technique [13]. The x-ray powder diffraction graphs of LSGM produced from precursors prepared with exact stoichiometry (a), with 10 wt% (b), and 20 wt% (c) excess gallium amounts are shown in figure 1. In all of these experiments, the calcination temperature was 1100°C, heating rate was 10 ^{0}C/min. As the excess gallium in the precursor batch was increased to 10 wt %, strong peaks from LSGM are observed in the x-ray diffraction spectrum. The phase distribution in the powders obtained with 20 wt% excess gallium is very similar to the one from the batch with 10 wt % excess gallium. Thermo gravimetric analysis of the precursors revealed that there is a continuing weight loss above 800° C. It appeared that gallium is leaving the system at the calcination temperatures above 800° C, which is consistent with the previous reported studies [12,14]. At this stage, it is not clear yet how long the gallium loss will continue, and what crystallographic changes will accompany this gallium loss.

When the batch cation composition was prepared for the exact stoichoimetry of the $La_{0.9}Sr_{0.1}Ga_{0.8}Mg_{0.2}O_{2.85}$, no LSGM was obtained under these calcination conditions. X-ray diffraction plots revealed peaks from magnesium gallium oxide ($MgGa_2O_4$) and $(LaO)_2SO_4$. To prove the stability of $(LaO)_2SO_4$ over $LaGaO_3$ up to 1200^{0}C, stoichiometric amounts of gallium sulfate and lanthanum nitrate are dissolved in nitric acid solution. The solution was heated up to different temperatures (150^{0}C, 400^{0}C and 800^{0}C) and the x-ray diffraction plots of the resulting powders are shown (Figure 2).

$(LaO)_2SO_4$ was formed rather than $LaGaO_3$, because former is more stable at the calcination temperatures ($< 1200^{0}$C). The dissociation temperature for lanthanum sulfate is 1150^{0}C. Moreover, in batch I (i.e exact stoichiometry) $MgGa_2O_4$ formed because $(LaO)_2SO_4$ formation leads excess gallium, magnesium and strontium in the environment. The most stable phase among them is supposed to be $MgGa_2O_4$. The stability of $(LaO)_2SO_4$ is larger than lanthanum gallate ($LaGaO_3$) up to 1200^{0}C; and the formation of $(LaO)_2SO_4$ in the LSGM precursor batch causes formation of magnesium gallium oxide ($MgGa_2O_4$).

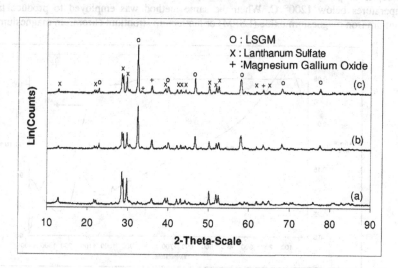

Figure 1. X-ray diffraction peaks of samples with (a) exact Ga content, (b) 10 wt% excess Ga content, (c) 20 wt% excess Ga content

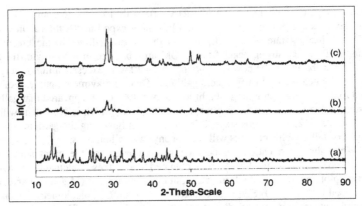

Figure 2. X-ray diffraction peaks of the gallium sulfate-lanthanum nitrate-nitric acid solution heated up to (a) 150° C (b) 400° C, (c) 800° C

Decomposition of lanthanum sulfate to produce lanthanum gallate at about 1150^0C was seen in the simultaneous thermal analysis (TG/DTA) of the sample previously heated up to 800^0C (Figure 3). The x-ray diffraction plots in figure 4 show the formation of $(LaO)_2SO_4$ and $MgGa_2O_4$ rather than LSGM. Only above 1150^0C, the dissociation of $(LaO)_2SO_4$ leads to formation of LSGM. $MgGa_2O_4$ phase appears to be converted to LSGM during the decomposition of $(LaO)_2SO_4$. It is believed that by using gallium nitrate hydrate as the gallium source, it is possible to produce single phase LSGM with the desired stoichiometry at temperatures below 1200° C. When the same method was employed to produce lanthanum strontium iron magnesium oxide (LSFM) or lanthanum strontium chromium magnesium oxide

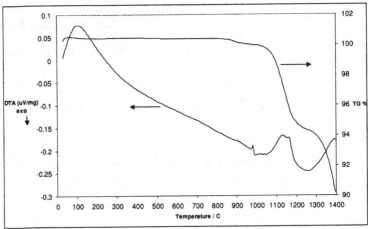

Figure 3. STA of the gallium sulfate-lanthanum nitrate-nitric acid solution heated up to 1400° C (Previously heated up to 800° C)

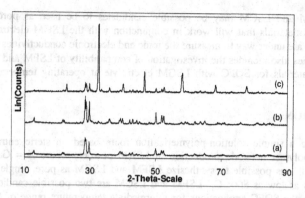

Figure 4. X-ray diffraction peaks of (a) gallium sulfate-lanthanum nitrate-nitric acid solution heated up to 800°C, (b) LSGM precursor, calcined at 1100°C, (c) LSGM precursor, calcined at 1200°C

(LSCM), since no metal sulfate salt was used in the experiments, no such disproportionation of the cations in the precursors was observed. Figure 5 a and b are the x-ray plots of LSFM (a) and LSCM (b). The precursors for LSFM and LSCM were calcined at 800° C with no holding time. Both powders are single phase and their diffraction data is very similar to lanthanum ferrite (LaFeO$_3$) and lanthanum chromite (LaCrO$_3$), respectively. Since the ionic x-ray radii of Ga, Fe and Cr are very similar to each other, it was assumed that the iron and chromium will only introduce slight distortion in the unit cell and will crystallize in the same structure of lanthanum salt.

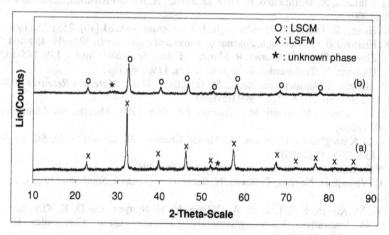

Figure 5. X-Ray diffraction peaks of (a) LSFM and (b) LSCM

LSFM and LSCM may be possible candidates to be used as porous electrode, and interconnect materials that will work in conjunction with the LSGM electrolyte, respectively. Experiments are under way to measure the ionic and electronic conductivities of these materials. Current studies also includes the investigation of compatibility of LSFM and LSCM as possible electrode materials for SOFC with LSGM electrolyte at operating temperatures of 650°C to 850°C.

CONCLUSION

Using a simple solution-polymerization route based on steric entrapment of cations within the polymer network, production of LSXM powders where X = Ga, Fe, or Cr, were attempted. It was possible to synthesize LSFM and LSCM as pure, single phase powders at temperatures as low as 800° C. LSFM and LSCM are two possible candidate oxides for the electrodes in the SOFC applications for intermediate temperature range of 650° C to 850° C. Synthesis of single phase LSGM with the desired chemical composition at temperatures below 1200^0C may be possible by using steric entrapment method with nitrate salts of cations. Another difficulty in the production of LSGM is the continuous loss of gallium from the system at high calcination temperatures (i.e. >800° C). With the proper choice of starting materials and caution in the processing it appears to be possible to circumvent the gallium loss by producing of single phase, pure LSGM with the desired stoichiometry at temperature around 800°C - 900°C.

REFERENCES

1. T. Ishihara, H. Matsuda, T. Takita, *J. Am. Ceram. Soc.*, 116, 3801 (1994)
2. T. Inagaki, K. Miura, H. Yoshida, R. Maric, S. Ohara, X. Zhang, K. Mukai, T. Fukui, *Journal of Power Sources* 86, 347-351 (2000)
3. H. Ullmann, N. Trofimenko, F. Tietz, D. Stöver, A. Ahmad-Khanlou, *Solid State Ionics* 138, 79-90 (2000)
4. K. Huang, R..Tichy, J.B. Goodenough, *J. Am. Ceram. Soc.*, 81 [10], 2581-85 (1998)
5. K. Huang, J.B. Goodenough, *Journal of Alloys and Compounds*, 303-304, 454-464 (2000)
6. X. Zhang, S. Ohara, H. Okawa, R. Maric, T. Fukui, *Solid State Ionics*, 139, 145-152 (2001)
7. H. Ullmann, N. Trofimenko, *Solid State Ionics*, 119, 1-8 (1999)
8. M. Hrovat, A. Ahmad-Khanlou, Z. Samardžija, J. Holc, *Materials Research Bulletin*, Vol. 34, Nos. 12/13, pp. 2027-2034, 1999 (review)
9. T. Mathews, P. Manovari, M.P. Antony, J.R. Sellar, B.C. Muddle, *Solid State Ionics*, 135, 397-402 (2000)
10. G.C. Kostogloudis, C. Ftikos, A. Ahmad-Khanlou, A. Naoumidis, D. Stöver, *Solid State Ionics* 134, 127-138 (2000)
11. N.Q. Minh, *J. Am. Ceram. Soc.*, 76 [3], 563-88 (1993)
12. K. Yamaji, H. Negishi, T. Horita, N. Sakai, H. Yokokawa, *Solid State Ionics*, 135, 389-396 (2000)
13. W. M. Kriven, S. J. Lee, M. A. Gülgün, M. H. Nguyen, and D. K. Kim, invited review paper, in Innovative Processing/Synthesis: Ceramics, Glass, Composites III, Ceramic Transactions, 108, 99-110 (2000).
14. A. Ahmad-Khanlou, F. Tietz, D. Stöver, *Solid State Ionics*, 135, 543-547 (2000)

Mat. Res. Soc. Symp. Proc. Vol. 756 © 2003 Materials Research Society

Structural behavior of zirconia thin films with different level of yttrium content

V. Petrovsky, H.U. Anderson, T. Petrovsky, and E. Bohannan
Electronic Materials Applied Research Center, University Missouri-Rolla, Rolla, MO, USA

ABSTRACT

The preparation of dense, high conductive electrolyte layers is important for the development of intermediate temperature solid oxide fuel cells and other devices based on oxygen-ion conductivity. Thus a number of techniques have been used to produce these structures. This study makes use of one of these methods to produce dense nanocrystalline 0.1 to 1 micron layers of zirconia.

Polymeric precursors were used to prepare zirconia films with different level of yttrium substitution. The films were annealed at a series of temperatures in the range of 400 to 1000°C and were characterized via scanning electron microscopy (SEM) and X-ray diffraction (XRD). It was found that initially (after 400°C annealing) the films had cubic structure and grain size of ~5 nm regardless of Y content. The situation changed when the annealing temperature was increased. Y (16mol %) stabilized zirconia (YSZ) did remain cubic over the entire temperature region investigated (up to 1000°C), but for compositions with lower Y content changes in crystal structure occurred. The samples with 4 and 8mol% Y transformed to the tetragonal phase at about 700°C, and undoped zirconia became monoclinic at the same temperature.

The results were compared with sintered zirconia and it was shown that the behavior of thin films is quite similar to that of the sintered material, if the annealing temperature was high enough (> 700°C). The main differences between polymeric prepared films and sintered material are the existence of the cubic structure at low temperatures (< 600°C) and lower transition temperatures to the high temperature phase, which can be explained by small initial grain size in polymer-derived zirconia.

INTRODUCTION

Yttrium substituted zirconia is a well-known ionic conductor and has been extensively investigated. In particular, the phase transformations have been studied and are well understood [1-8]. Most of the experimental results have been obtained using zirconia (micron grain size) samples prepared by high temperature sintering. The use of high processing temperatures has some limitations in the fact that reactions with the electrode materials can result and grain growth takes place during the sintering procedure, which limits the minimum thickness of the electrolyte layer. As a result alternative techniques for the preparation of zirconia based electrolytes are important.

Polymeric precursor processing is one of the promising techniques which allow for lower processing temperatures.. Previous investigations did show the potential of decreasing the thickness of the electrolyte layer and lowering the processing temperature by use of this type of processing [9-11]. As a result, a number of different nanocrystalline metal oxide coatings have been prepared and investigated including Y stabilized zirconia (YSZ).

The current study focuses on an investigation of Y-ZrO_2 system prepared from polymeric metal oxides precursors. The Y content was varied from undoped zirconia to 16mol%Y. The main goals of this investigation were to study the crystallographic phases present as a function of annealing temperature using SEM and XRD and to compare these results with those observed for sintered microcrystalline zirconia.

EXPERIMENTAL

Polymeric precursors were prepared using zirconium chloride-oxide and yttrium nitrate as a cation sources by the method described elsewhere [9-12]. Yttrium content in the precursors was varied from 0 to 16mol% to overlap the composition region of interest. The precursors were deposited on sapphire substrates by spin coating at 3000 rpm. In order to prevent cracking, the film thickness was maintained at about 20 nm per deposition [10], so multiple depositions were used to obtain 500 nm layers.

The samples were annealed in the temperature range 400 to 1000°C in ambient atmosphere, and SEM and XRD investigations were performed. The SEM investigation of the cross-sections showed after annealing at 400°C, that all of samples had similar nanocrystalline structure with the grain size of about 5 nm. A typical example of this structure is shown in Fig.1.

Figure 1. Cross-section of YSZ (16mol%Y) coating on sapphire substrate annealed at 400°C. Density about 90% of theoretical with the grain size of ~5 nm.

Increase in the annealing temperature leads to the grain growth and the changes in the microstructure, but over the temperature range investigated (up to 1000°C), the grain size remains in the nanocrystalline region (less than 120 nm), with the density becoming higher than 95% of theoretical in the 700 to 800°C range. When the annealing temperature exceeds 700°C, the microstructure does become dependent on Y content. For example, symmetrical grains growth occurs in the films containing 16 mol% Y so that the annealing temperature affects only the size of the grains, but not the shape (see Fig.2 for YSZ film annealed at 1000°C). Whereas, a column structure starts to form in the films with lower Y content. The degree of the out of plain orientation depends on the annealing temperature and Y content which appears to be related to the crystallographic transformations in the material. An example of the well developed column structure is shown in Fig.3 for an undoped zirconia film annealed at 1000°C.

XRD investigation did confirm these results. Initially (after annealing at 400°C), the films have cubic structure independently on Y content. This is connected with the extremely small grain size (~ 5 nm) which provides enough disorder in the lattice to stabilize the cubic phase. This cubic phase is metastable and at higher annealing temperature transforms to either the monoclinic or tetragonal phases depending on Y content because of grains growth.

X-ray diffraction patterns are shown in Fig.4 for YSZ (16mol%Y) film annealed at different temperatures. It can be seen from the figure that YSZ remains cubic in the temperature region investigated. The only differences occurring in the patterns with increasing temperature are decreases in the peak broadening which is connected with the grain growth and very small shifts in the lattice parameters. It is possible to conclude, that Y substitution stabilizes cubic structure by introducing some degree of disorder in the zirconia lattice.

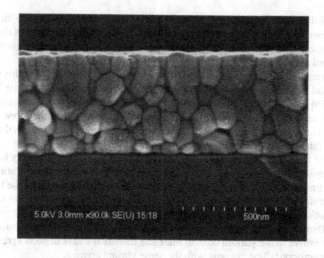

Figure 2. Cross-section of YSZ coating on sapphire substrate annealed at 1000°C. Dense structure with well defined cubic grain structure (grain size of ~100 nm).

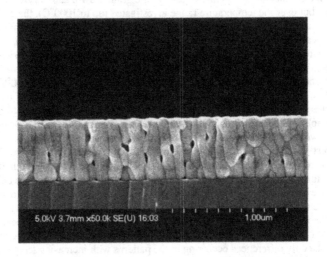

Figure 3. Cross-section of zirconia coating on sapphire substrate annealed at 1000°C. Dense structure with the oriented grain structure (grain size in plane of ~100 nm).

Partially stabilized zirconia (Y content 4 and 8mol %) undergoes transition to the tetragonal phase starting from annealing temperatures of about 600°C. Tetragonal splitting of the peaks and the lattice parameters depend on Y content, which is similar to that observed for microcrystalline partially stabilized zirconia.

Undoped zirconia films undergo transition to the monoclinic phase (Fig.5). This transition starts at 600°C and finishes at 800°C. Two phases (cubic and monoclinic) do coexist in this temperature region (600 to 800°C) and only the monoclinic phase was found after annealing at higher temperatures. The comparison of the XRD peak positions with the microcrystalline zirconia does shows good agreement at both the low annealing (cubic phase) as well as at high annealing temperatures (monoclinic phase) (see Fig.5).

It can be seen from these results, that polymeric precursors give the possibility for the preparation of dense, nanocrystalline zirconia films at very low processing temperature (less than 400°C). These films have initially cubic structure connected with the disorder in the lattice caused by the small grain size. Annealing at higher temperature causes the ordering of the lattice as the grain growth. The transition to the high temperature phase takes place in the grain size range of 20 to 50nm in the temperature region 600 to 800°C. The crystallographic structure of the material is sensitive to the yttrium content after this transition and is in good agreement with the structure of microcrystalline zirconia with the same yttrium content.

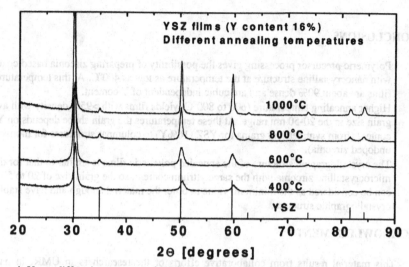

Figure 4. X-ray diffraction patterns of YSZ coating on sapphire substrate annealed at different temperatures. The same cubic YSZ structure can be seen over the entire temperature range.

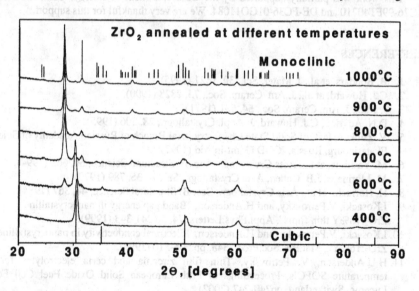

Figure 5. X-ray diffraction patterns of undoped zirconia coating on sapphire substrate annealed at different temperatures. The transition from cubic to monoclinic phase takes place in the temperature region 600 to 800°C.

CONCLUSIONS

- Polymeric precursor processing gives the possibility of preparing zirconia based coatings with nanocrystalline structure at the temperature as low as 400°C. At this temperature, the films are about 90% dense and are cubic independent of Y content.
- Higher annealing temperature (600 to 800°C) yields films with >95% density with average grain size in the 20-50 nm range. At these temperatures the grain shape depends on Y content (from symmetrical grains for YSZ (16%Y) to columnar structures for monoclinic undoped zirconia).
- The crystallographic structure of the annealed polymeric films is similar to that for the microcrystalline zirconia with the same yttrium content, so the grain size of 20 to 50 nm can be considered as the minimum value to order the material enough and have stable crystallographic structure.

ACKNOWLEGEMENTS

This material results from collaborative efforts of the researchers in UMR. In particular authors would like to thank Ms. Clarissa Vierrther for her help with SEM investigation.

This work was partially supported by the US Department of Energy, Contracts No. DE-AC26-99FT40710 and DE-FC36-01GO11084. We are very thankful for this support.

REFFERENCES

1. B. Bondars, et al., J. Mater. Sci., **30**, 1621 (1995).
2. C.J. Howard, at al., J. Am. Ceram. Soc., **73**, 2828 (1990).
3. J. Katz, J. Am. Ceram. Soc., **54**, 531 (1971).
4. D.N. Adyriou, C.J. Howard, J. Appl. Crystallogr., **28**, 206 (1995).
5. V. Zubkov, Inst. of Solid State Chemistry, Ural Branch of Russian Academy of Science, Ekaterinburg, Russia, ICDD Grant-in Aid (1997).
6. M. Yashima, at al., Acta Crystallogr., Sec. B: Structural Science, **50**, 663 (1994).
7. M. Morinaga, J.B. Cohen, Acta Crystallogr., Sec. A, **35**, 789 (1979).
8. H. Horiuchi, et al., Acta Crystallogr., Sec. B: Structural Science, **40**, 367 (1984).
9. I.Kosacki, V.Petrovsky, and H.Anderson, "Band gap energy in nanocrystalline ZrO_2:16%Y thin films", Appl.Phys.Letters, **74**, pp 341-343 (1999).
10. I.Kosacki, V.Petrovsky, and H.Anderson, "Electrical conductivity in nanocrystalline ZrO_2:16%Y", Mat.Res.Soc.Proc., **548**, pp 505-510 (1999).
11. H.U.Anderson, V. Petrovsky, Thin film zirconia and ceria electrolytes for low temperature SOFC's, Proceedings of Fifth European Solid Oxide Fuel Cell Forum, Lucerne, Switzerland, pp240-247 (2002).
12. H.U.Anderson, et al., US patent #5494700, February 27, 1996.

Mat. Res. Soc. Symp. Proc. Vol. 756 © 2003 Materials Research Society   FF3.8

Nucleation And Growth of Yttria-Stabilized Zirconia Thin Films Using Combustion Chemical Vapor Deposition

Zhigang Xu[1], Jag Sankar[1], Sergey Yarmolenko[1], Qiuming Wei[1,2]
[1]NSF Center for Advanced materials and Smart Structures, North Carolina A&T State University, Greensboro, NC 27411
[2]Department of Mechanical Engineering, The Johns Hopkins University, Baltimore, MD 21218

ABSTRACT

Liquid fuel combustion chemical vapor deposition technique was successfully used for YSZ thin film processing. The nucleation rates were obtained for the samples processed at different temperatures and total-metal-concentrations in the liquid fuel. An optimum substrate temperature was found for the highest nucleation rate. The nucleation rate was increased with the total-metal-concentration. Structural evolution of the thin film in the early processing stage was studied with regard to the formation of nuclei, crystallites and final crystals on the films. The films were found to be affected by high temperature annealing. The crystals and the thin films were characterized with scanning electron microscopy.

INTRODUCTION

Yttria stabilized cubic phase zirconia (YSZ) is an oxygen ion conductive material [1]. It is the most widely used electrolyte material available for high-temperature fuel cells. In order to obtain the best performance of the fuel cells, thin film of the YSZ electrolyte is favored to minimize the current path [2]. In the SOFCs, both the air and fuel electrodes are porous materials. The thin film of electrolyte is placed between these electrodes. It has to be gas-tight to avoid any crossover of the oxidant and fuel gases. A fundamental understanding of the nucleation and growth of thin YSZ films is very important to its successful application for SOFCs.

Atmospheric combustion chemical vapor deposition (ACCVD) technique had been employed for YSZ thin film processing, because of its advantages as high growth rate, low setup cost and low run cost.

Nucleation of particles on substrates is a fundamental topic in chemical vapor deposition. Many innovative methods were studied to enhance the nucleation rate and/or to control the quality of the nucleated crystals [3-7]. However, most of the reported studies were required film processing in vacuum reactors. The nucleation and particle growth for ACCVD, conducted in open air, have not been well understood. In this paper, nucleation and growth of YSZ crystals using the ACCVD technique are presented. The emphases of the research were to understand the effects various processing parameters on the nucleation and growth. The methods to enhance the nucleation density were studied. The evolutions of microstructures from nuclei to continuous films were also observed.

EXPERIMENTAL DETAILS

A liquid fuel ACCVD system similar to the one introduced in previous work [8] was used for YSZ thin film processing. The metal-organic reagents were dissolved in toluene that acts as both a solvent and a combustion fuel to supply sufficient thermal energy for chemical reaction in the gas phase and on the substrate. A stable flame of the aerosol of the solution mixed with a high-speed oxygen flow through an atomizer was maintained with a pilot flame. The metal-organic reagents used in all the experiments were zirconium 2-ethylhexanoate (Zr-2EH) and yttrium 2-ethylhexanoate (Y-2EH). Oxygen was used as oxidant. In order to vary the substrate temperature in a relatively larger range, air-blow cooling and oxy-acetylene flame heating under the substrate support were developed. They were used alternatively. The substrate temperature was controlled by front aerosol flame and by the cooling or heating effect. The ranges and values of the parameters employed in the experiments were designed as shown in Table 1.

Si(100) with SiO_2 top layer was used as substrates. The thickness of the SiO_2 layer was about 0.1 μm. The sizes of the silicon substrates were about 10mm×10mm×0.4mm. All the substrates were cleaned in acetone and methanol solutions ultrasonically and rinsed by de-ionized water before the deposition.

For the nucleation experiments, the substrate temperatures from 800 to 1200 °C and total-metal-concentrations from 5×10^{-4} M and 1.05×10^{-2} M were employed. All the nucleation experiments were carries out within the processing time of 130 sec.

To observe the structural evolution of the thin films, a serial of experiments of deposition for different lengths of time had been conducted at the substrate temperature of 1200 °C and the total-metal-concentration of 1.25×10^{-3} M.

The morphologies of the samples were characterized using scanning electron microscopy (SEM, Hitachi S-3000N).

Table 1. Experimental parameter design

Parameters	Ranges and values
Total floe rate of solution (cm³/min)	2.0
Flow rate of oxygen (cm³/min)	1600
Substrate temperature (°C)	800~1200
Total-metal-concentration (M)	$5\times10^{-4} \sim 1.05\times10^{-2}$
Atomizer-to-substrate distance (mm)	76
Ratio of yttria/zirconia (mol %)	10

RESULTS AND DISCUSSION
Nucleation

It is well acknowledged that to produce smooth and dense thin films, high nucleation density is highly demanded. In chemical vapor deposition, the initial step of the nucleation is the impingement of vapor molecules or precursor species on the substrate. After impingement, the vapor species can either absorb and stick permanently to the substrate, or bounce off the substrate. Even the absorbed molecules can re-evaporate in a finite time [9]. Only those molecules that equilibrate rapidly enough with the substrate will become absorbed. By diffusion of the absorbed molecules on the substrate and by incorporation of more impinging molecules from the vapor phase, larger and stable nuclei with certain number of molecules will be formed on the substrate. Substrate temperature is one of the important factors that affect the re-evaporating rate of the absorbed molecules. It is generally found that the nucleation rate decreases with increasing substrate temperature [10]. However, the dependence of the nucleation rate on the substrate temperature is dependent on the deposition system. In our experiments, the nucleation was performed at the substrate temperatures of 800 °C, 900 °C, 1000 °C, 1100 °C and 1200 °C respectively. The as-grown samples were characterized with SEM as show in Fig. 1.

By counting the particle number in the specific area by the application software, it was found that the nucleation rates ranged from the order of magnitude of 10^{10} to 10^{11} cm^{-2} in the tested temperature range. The nucleation rate variation versus the substrate temperature is illustrated in Fig. 2.. It is obviously noticed that the tendency of the nucleation rate versus the substrate temperature can be divided into two sections. With the substrate temperature between 1000 °C and 1200 °C, the nucleation rate decreases with the substrate temperature, while in the substrate temperature range of 800 °C to 1000 °C; the nucleation rate increases with the substrate temperature. The former nucleation rate tendency was well accepted. As far as the author knows, the latter tendency has not been reported. It is proposed that at low substrate temperatures, the material species in the gas phase condense into the solid phase and coagulate into particles. They rebound from the surface instead of becoming adsorbed when they hit the substrate surface. This

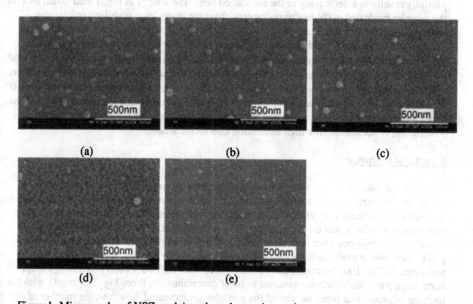

Figure1. Micrographs of YSZ nuclei nucleated at various substrate temperatures, (a) 800 °C, (b) 900 °C, (c) 1000°C, (d) 1100 °C and (e) 1200 °C under conditions of nucleation time of 130 sec and total-metal-concentration of 1.25×10^{-3} M.

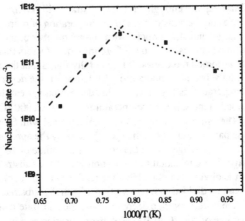

Figure 2. YSZ nucleation rate as a function of the substrate temperature according to the information in Fig. 1.

process also leads to depletion of species in the gas phase that is needed for CVD and then partially results in a decreasing of the nucleation rate. The images in Fig. 1 also reveal that on the samples nucleated at the low temperatures, there are many large nodular particles that support the above statement.

Another method to enhance the nucleation rate was to increase to deposition flux, i.e. total-metal-concentration for our experimental system. At the low concentrations of 5×10^{-4} and 1.25×10^{-3}, only isolated particles presented on the substrate surface after depositions of 130 sec, whereas almost continuous films were obtained when the concentration was increased to 3×10^{-3} M and up. At the high total-metal-concentrations, the particle size reached as large as 30-50 nm. From this set of experiments, it can be concluded that the lowest total-metal-concentration, 5.5×10^{-4} M, is not suitable if a high nucleation rate is demanded.

Structural evolution

The evolution of thin films on substrate consists of several procedures, such as nucleation, coarsening, coalescence and grain growth. Coarsening is one of the way in which the average size of islands increases. This occurs through detachment of atoms from some islands and diffusion of the atoms on the substrate surface to attach to other islands, resulting in the shrinkage and disappearance of some islands and growth of other islands. Coalescence takes place when two or more islands grow to the point of contact. This happens without diffusion processes. In Fig. 3(a), the nuclei can be hardly seen after only 60 second processing except some large particles, which are assumed to be of contamination. From Fig. 3(b) to (d), with the increase of processing time, the size of the particles are increased, however, the number of the particles is reduced. This phenomenon can be interpreted by the mechanisms of coarsening and coalescence of the growing particles. When the processing time reached 720 sec., the film is almost continuous. With increased time of processing, the sizes of the particles increase. Some secondary nucleation and growth on the large particles can be noticed. It is also reasonable to assume that the secondary nucleation should take place on the substrate before all the surface of the substrate is covered with particles.

The film consists of both (111) and (100) oriented crystals. The difference in orientation was originated at the stage when the sizes of particles were very small.

According to the obtained mean particle radii, particle growth rate can be estimated by plotting the mean diameter versus the growing time as shown in Fig. 4. The particle growth rate is approximately linear during the time period studied (130 sec to 430 sec). The intercept of the line on the time axis is about 33.3 sec, which shows the incubation time for nucleation. With prolonged deposition time, Continuous thin films can be obtained. Figure 5 shows a cross-section view of the YSZ/Si(100) system.

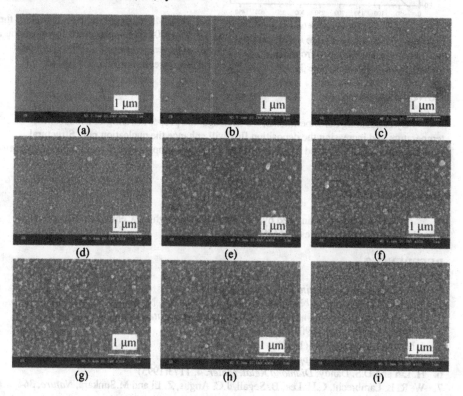

Figure 3. Microstructure of the YSZ particles/crystallites at different processing times (a) 70 sec, (b) 130 sec, (c) 190 sec, (d) 250 sec, (e) 310 sec, (f) 370 sec, (g) 430sec, (h) 490sec and (i) 550 sec on Si(100) substrates, at the substrate temperature of about 1200 °C, total-metal-concentration of 1.25×10^{-3} M.

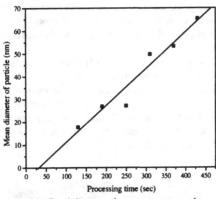

Figure 4. Particle growth rate vs. processing time from the data obtained from the micrographs as shown Figure 3.

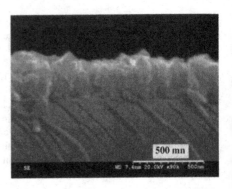

Figure 5. Micrograph of a cross-section of the YSZ/Si(100) system processed for 20 min at a substrate temperature of 1200 °C, and total-metal-concentration of 1.25×10^{-3} M.

CONCLUSIONS

By conducting the experiments, we envisage that proper lowering of the substrate temperature and increasing the deposition flux will enhance the nucleation rate. Structural evolution of the YSZ film and annealing of the nucleated samples give more detailed information on the growing procedures.

ACKNOLEDGEMENT

This research was sponsored by NSF and DOE through the Center For Advanced Materials and Smart Structures (CAMSS).

REFERENCES

1. S.C. Singhal, *MRS Bulletin*, **25**(3), 22(2000)
2. L. J.M.J Blomen and M. N. Mugerwa, *Fuel Cell Systems*, New York, Plenum Press, 1993
3. R.P.H Chang, D. Nelson, and A. Hiraki, *Technology update on diamond films*, MRS Proceedings (EA-19), 1990
4. J. Narayan, *JMR*, **5**, 2414(1990)
5. X. Chen, J. Narayan, *J. Appl. Phys.* **74**, 4168(1992)
6. H. Lin and D.S. Dandy, *Diamond Relat. Mater.* **4**, 1173(1995)
7. W. R. L. Lambrecht, C.H. Lee, B. Segall, J.C. Angus, Z. Li and M.Sunkara, *Nature*, **364**, 607(1993)
8. Z. Xu, Q. Wei and J. Sankar, in *Mechanisms of Surface and Microstructure Evolution in Deposited Films and Film Structures*, MRS proceedings 2001 Spring (O8.29)
9. C.A. Neugebauer, in *Handbook of Thin Film Technology*, ed. LI Maissel, R. Glang Chpt. 8, New York, McGraw-Hill, 1970
10. C.V. Thompson, In *Annu. Rev. Mater. Sci.*, **30**, 159(2000)

Impedance spectroscopy and direct current measurements of YSZ films

T. Petrovsky, H.U. Anderson, and V. Petrovsky
Electronic Materials Research Center, University Missouri-Rolla, MO 65409, USA

ABSTRACT

In this study the electrical properties of thin films of Y substituted zirconia were investigated. The films were prepared using a polymer precursor technique and investigated in the temperature region 250 to 900°C. It was shown, that impedance spectroscopy (IS) and direct current (DC) conductivity measurements results are in good agreement for the films measured in plane for temperatures greater than 400°C. Due to the high resistance resulting from a planar geometry, the DC measurements were found preferable at temperatures <600°C.

Since in planar geometry the films represent a high resistance to the measurement circuit, it is important to minimize sources of electrical leakage, so different sample holders and substrates were investigated. A sapphire substrate and sample holder design using separated alumina single bore tubing for each electrode provided the lowest electrical leakage.

The experimental results showed that electrical behavior of all of the films produced at low annealing temperatures (less than 400°C) was similar regardless of Y content. These films have relatively low conductivity and an activation energy of about 1.5eV. The influence of different Y content started to appear after annealing above 600°C.

The results of the film conductivity measurements were compared with those for the bulk samples of Y stabilized zirconia prepared from 200nm powder by tape casting. These samples were measured by IS in plane and through the tape. It was shown that electrical properties of bulk and thin film material were similar.

INTRODUCTION

The electrical characterization of ionic conductive thin films is more complicated than for bulk material, because thin films cannot be prepared without a supporting structure so that chemical interactions with the substrate can influence measurements, particularly at elevated temperatures. In additions due to the high surface area of the films, substrate surfaces and interfaces can provide electrical leakage paths.

Two different geometrical approaches can be used for the electrical measurements. The first approach is based on measurements through the thickness of the film. In this geometry the influence of the contact resistance can be a serious limitation, and substrates with high electronic and ionic conductivity have to be used. Impedance spectroscopy (IS) is the only way to differentiate the impact of the contact resistance on the overall resistance.

The second approach is based on in plane measurements. In this case substrates with high resistance need to be used and the leakage related to the substrate and sample holder has to be minimized. For this geometry two probe direct current (DC) measurements can be employed because the contact resistance is relatively small compared to that of the film resistance. Typically these measurements can be extended to temperatures as low as 100-200 C. Since the sample holder capacitance is orders higher than film capacitance for this geometry, IS measurements yield the same overall film resistant as DC measurement. The limitations of IS measurements of ionic conductors have been widely discussed in the literature [1-4].

In this study zirconia thin films with different Y content were investigated. This material is known as a good ionic conductor with high potential for practical applications. Bulk ceramic zirconia has been extensively investigated [5-11], so the electrical properties of this material are well known and can be compared with those obtained for thin zirconia films. The main problem with the measurement of thin films conductivity in plane is the high resistance of the sample connected with this geometry, so different leakage sources (connected with the sample holder and substrate) need to be investigated and taken into account. The goals of this investigation were to minimize leakage sources, provide measurements of the electrical conductivity of thin zirconia films and to compare the results with those reported for the bulk material.

EXPERIMENTAL

Polymeric precursors were prepared using zirconium chloride-oxide and yttrium nitrate as cation sources by a method described elsewhere [12]. Yttrium content in the precursors was varied from 0 to 16mol% to overlap the composition region of interest. Two ways were used for the precursor preparation with the different Y content. The first way was based on mixing zirconium chloride-oxide and yttrium nitrate in the proper ratio in the initial solution. The second way was based on the mixing of polymeric solutions containing undoped zirconia and 16mol% Y substituted zirconia in the proper ratio. From the point of view of the electrical properties both techniques are equivalent. The precursors were deposited on dielectric substrates using spin coating at 3000 rpm. The resulting film thickness for each deposition was about 20 nm, so multiple depositions were used to obtain 500nm thick layers. After preannealing at 400°C, the films were measured in plane over the temperature range 250 to 900°C. A Keithley electrometer 6517 was used for the DC measurements and a Solartron impedance analyzer 1260 with a dielectric interface 1296 was used for IS characterization.

Possible leakage sources connected with different elements of measurement system were investigated. The temperature dependences of the leakage for sample holders (SHs) with different design are shown in Fig.1A in comparison with conductance of YSZ (1cm x 1cm x 500 nm thickness). Three SHs were tested (in open circuit regime) as well as alumina cement which was usually used for mounting the sample. Fig.1A shows that the cement layer (1cm x 1 cm x 50μm) has very high conductance and should be excluded from the sample mounting procedure. SH1 and SH2 had a standard design with the both electrode wires in one alumina thermocouple protection tube. It can be seen from Fig.1A that sample holders with this design have relatively high leakage, especially the SH1 which was exposed in the furnace for a long time and was contaminated. It is possible to conclude that separated tubes have to be used to minimize the sample holder leakage. It can be seen that sample holder SH3 in which each wire was contained in a separate single bore alumina tube has more then an order of magnitude less conductance than the other holders. Even at high temperatures when leakage in holder SH3 increases because of the radiation heating, it still is more then two orders lower than the conductance of the YSZ film. This sample holder was used for all films conductivity measurements.

The influence of the substrate on the electrical leakage was also investigated by measuring the conductance of polycrystalline alumina and sapphire substrates without the film present. Figure1B shows the results obtained for two polycrystalline alumina and two sapphire substrates. The conductances of the two polycrystalline substrates are within an order of magnitude of the YSZ film conductance and can influence the sample parameters. The measured conductances of the two polished sapphire substrates were nearly identical and about three orders lower than the

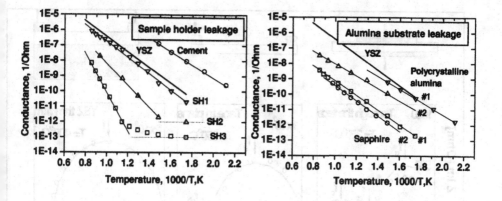

Fig.1. Influence of the sample holders (A) and substrates (B) on conductance measurements.

conductance of the YSZ film (500 nm thick). All the results reported in this investigation were achieved for films deposited on sapphire substrates.

Impedance spectra were investigated for two types of zirconia samples: YSZ tape and thin zirconia films on sapphire substrate. The tape was prepared from YSZ powder (particle size of 200 nm) by tape casting. It was sintered at different temperatures in the range 1000 to 1400°C and measured at different temperatures. The measurement geometry was both through the thickness of the tape and in plane using sputtered Pt as the contacts and probes with capacitance 0.47pF.

Examples of these measurements at 350°C are presented in Fig.2. Fig.2A shows that three distinct semicircles present in IS spectra for measurements through the thickness of the tape (100 µm thickness). These semicircles are connected with grain, grain boundary and contact RC circuits.

Figure 2B illustrates the results when measurements were made using a planar geometry for the same 100 µm thick tape and measurement probes. This figure shows that the contact resistance is relatively small (compare with sample resistance) and due to the measurement probes capacitance (0.47 pF) the grain and grain boundary semicircles merged.

Figure 2C shows that the capacitance (4.5 –5 pF) of the sample holder and measurement cables influenced the IS spectra obtained for thin film measured in the 400-900 °C temperature range. As can be seen only one ideal semicircle occurs for these spectra which yield values for the resistance that are in a good agreement with the DC measurements. The calculated capacitance is equal to the sample holder capacitance instead of the sample. Thus the IS measurements do not give more information about thin film structure than is obtained from DC conductivity measurements.

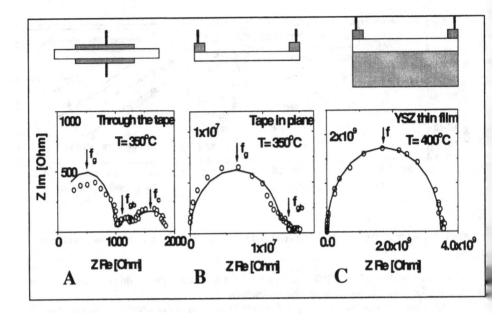

Fig.2. Impedance spectra for three types of YSZ samples.

(A) - YSZ tape measured through the tape (thickness=0.01cm, contact area =0.84cm^2):
f_g =5.18*10^5 Hz, R_g =986 Ohm, C_g=2.9*10^{-10}F; f_{gb} =700Hz, R_{gb}=200 Ohm, C_{gb}=4.8*10^{-7};
f_c=1.5Hz, R_c=810 Ohm, C_c=4.6*10^{-4}F;

(B) - YSZ tape measured in plane (thickness=0.01cm, length=0.6cm, width=0.42cm):
f_g =2.7*10^{4}Hz, R_g =1.25*10^{7}Ohm, C_g=4.7*10^{-13}F; f_{gb} =590Hz, R_{gb} =1.8*10^{6}Ohm,
C_{gb}=1.5*10^{-10};

(C) - YSZ film measured in plane (thickness =450nm, length=1.1cm, width=0.5cm):
f =9.9Hz, R =3.46*10^{9}Ohm, C=4.66*10^{-12}F;

Fig.3 shows a good agreement for IS (for T>400°C) and DC measurements for thin YSZ film over the 250 – 900°C temperature region. The main advantage of DC measurements is the higher accuracy and the possibility of measurements over wider temperature range.

Fig.4. summarizes the results of the measurements for thin films with different Y content. This figure shows temperature dependence of the conductivity for samples annealed at 900°C. It can be seen from the figure that the conductivity and activation energy are close to the bulk YSZ for the samples with 16 and 8mol%Y. The conductivity decreases at lower Y content which is also similar to that observed for bulk zirconia.

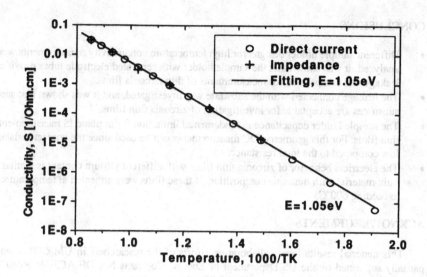

Fig.3. Comparison of IS and DC measurements for YSZ thin film annealed at 900°C.

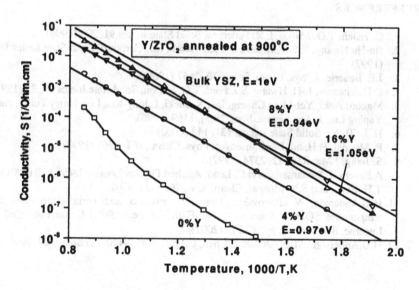

Fig.4. Temperature dependence of the conductivity for zirconia thin films with different Y content annealed at 900°C.

CONCLUSIONS

- Different sample holder designs for high temperature conductivity measurements were analyzed. It was shown that the sample holder with separated electrode tubes provides low leakage and can be used for measurements of thin zirconia films.
- The leakage connected with the substrate was investigated, and it was shown that sapphire substrates are acceptable for investigation of zirconia thin films.
- The sample holder capacitance is fundamental limitation for in plane IS measurements for thin films. For this geometry, DC measurements can be used since the contact resistance is low compared to the total resistance.
- The electrical behavior of zirconia thin films with different yttrium content is similar to the bulk material with the same composition, if these films were annealed at temperatures exceeding 600°C.

ACKNOWLEGEMENTS

This material results from collaborative efforts of the researchers in UMR. This work was partially supported by the US Department of Energy, Contracts No. DE-AC26-99FT40710 and DE-FC36-01GO11084. We are very thankful for this support.

REFFERENCES

1. G. Hsieh, T.O. Mason, L.R. Pederson, Solid State Ionics **91**, 203 (1996).
2. Jin-Ha Hwang, K.S. Kirkpatrick, T.O. Mason, E.J. Garboczi, Solid State Ionics **98**, 93 (1997)
3. J.E. Bauerle, J.Phys. Chem. Solids **30**, 2657 (1969).
4. D.D.Edwards, J.-H. Hwang, S.J. Ford, T.O. Mason, Solid state Ionics **99**, 85 (1997).
5. Makoto Aoki, Yet-Ming Chiang, Igor Kosacki, L.Jong-Ren Lee, Hurry Tuller, and Yaping Liu, J. Am. Ceram. Soc., **79**(5), 1169 (1996).
6. H. L. Tuller, Solid State Ionics, **131**, 143 (2000)
7. P. Mondal, H.Hahn, Ber Bunsenges, Phys. Chem., **101**, 1765 (1997).
8. S. Jiang, J.Mat.Res., **12**, 2374 (1997).
9. A.Rivera, J. Santamaria, and C. Leon, Applied Physics Letters, **78**(5), 610 (2001).
10. T.H. Etsell and S.N. Flengas, Chem. Rev., **70**, 339 (1970).
11. H.U.Anderson, V. Petrovsky, Thin film zirconia and ceria electrolytes for low temperature SOFC's, Proceedings of Fifth European Solid Oxide Fuel Cell Forum, Lucerne, Switzerland, pp240-247 (2002).
12. H.U.Anderson., M. Nasrallah, C.Chen, US patent #5494700, February 27, 1996.

Mat. Res. Soc. Symp. Proc. Vol. 756 © 2003 Materials Research Society

Microstructure, nanochemistry and transport properties
of Y-doped zirconia and Gd-doped ceria

G.Petot-Ervas, C.Petot, J.M.Raulot, J.Kusinski*, I.Sproule**, M.Graham**
CNRS-UMR 8580, SPMS, Ecole Centrale Paris,
Grande voie des Vignes, 92295 Châtenay-Malabry, Cedex (France),
*Academy of Mining and Metallurgy, Krakow, Poland
**CNRC, Ottawa, Canada

Abstract - In this work we have shown the influence of the microstructure and nanochemistry on the transport properties of Y_2O_3-(9mol%) stabilized zirconia and Gd_2O_3 (10 mol%)-doped ceria. Zirconia (YSZ) samples show transport properties (D_O, σ_{gb}) which increase with the grain size, while they decrease in ceria (GSC). This difference was attributed to the presence of glassy grain boundary precipitates in YSZ. On the other hand, it was shown that kinetic demixing processes during cooling, at the end of sintering, play an important role on the grain boundary properties of these oxides.

1-Introduction

The recent works of Badwall [1] and Aoki et al.[2] have related the transport properties of YSZ to the sample microstructure. Badwall commensurated his result to relocation of glassy precipitates to triple points and concludes that the anionic conduction is restricted to grain boundary regions not wetted by amorphous phases. Aoki et al. explain their results by segregation effects occurring during cooling, at the end of sintering.

According to the available results, further progress to improve the electrolyte performance will depend upon a better understanding of the relationships between the microstructure and the associated properties. In the present work, experiments have been performed with samples prepared from different starting powders sintered under different conditions.

2. Experimental

The *Y-doped zirconia* samples have been prepared from powders isostatically pressed at 2000 or 4000 bar [3,4]. The Z_C samples were sintered from commercial submicronic Y_2O_3 (9.9 mol%) doped-zirconia powder, whose main impurity is SiO_2 (~0.42 wt%). The Z_F samples were sintered from nanometric Y_2O_3 (9.0 mol%) doped-zirconia powder prepared by the freeze-drying method [3,4]. Their primary impurity was SiO_2 (~1.0 wt% in samples A, ~1.6 wt% in samples B).

The *Gd-doped ceria samples* have been prepared from two batches of commercial powders (L and H) doped with 10 mole % Gd_2O_3.

The *electrical conductivity* was obtained by complex impedance spectroscopy [3], in the frequency range 10^{-2} - 20×10^6 Hz, with and without a bias voltage.

The *oxygen diffusion coefficient* was determined by the electrochemical method [5]. The principle of the method is to place the opposite sides I and II of the sample, coated with the same electrode material, between reversible electrodes subjected to different P_{O_2} ($P_{O_2}^I$ and $P_{O_2}^{II}$).

In open circuit conditions, the chemical potential of the electrons is the same at the cell terminals $(\mu_e)_2^I = (\mu_e)_2^{II}$ and the cell generates an open circuit voltage, given by :

$$E_{open} = (\eta_e)_2^{II} - (\eta_e)_2^I = \Phi_2^{II} - \Phi_2^I = t_{ion} \frac{RT}{4F} \log \frac{P_{O2}^{II}}{P_{O2}^I} \qquad (1)$$

In short circuit conditions, the terminals of the cell are at the same electrical potential ($\Phi_2^{II} = \Phi_2^I$). The electrons can then flow readily from one electrode to the other, via the external circuit, allowing the interfacial reactions ($O^- => 1/2O_2 + 2e'$) to take place. *The transport processes in the oxide are controlled by the diffusion of the oxygen ions* and the leakage current is zero ($(\eta_e)_1^{II} = (\eta_e)_1^I$). The ionic flux J_O -- through the compound is then proportional to the current density (i_{sc}) in the external circuit ($i_{sc} \approx -2FJ_O$ --). If one neglects the correlation effects, one can show that the oxygen diffusion coefficient (D_O) of the oxygen ions is given by the relation :

$$D_O = g \frac{RT}{4F^2 C_O --} \frac{I_{sc}}{E} \qquad (2)$$

with g=L/s, I_{sc}=s i_{sc} and where "s" is the electrode surface, L the oxide thickness, C_O -- the concentration of oxygen ions per cm^3 and E the open circuit voltage (Eq.1), with P_{O2}^{II}/P_{O2}^I close to one to satisfy the ideal condition to determine D_O.

The short-circuit current (I_{sc}) is obtained by extrapolating the data U=f(I) to U=O, where U is the voltage determined at a variable resistance placed in the external circuit. Silver is used as the electrode material, because it allows to satisfy the reversible electrode condition.

3. Experimental results

3.1 Yttria-stabilized zirconia

3.1.a Grain boundary conductivity

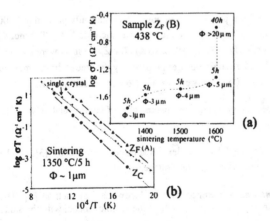

(a)

(b)

Figure 1- Influence of the sintering temperature on the grain boundary conductivity of samples Z_F (a). Comparison of the grain boundary conductivity of samples Z_C and Z_F (b)

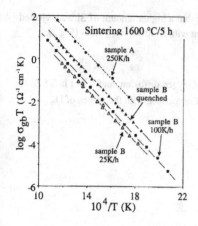

Figure 2- Influence of the cooling rate and silicon contamination on the grain boundary conductivity of samples Z_F

The results reported in Fig.1 show both that the grain boundary conductivity (σ_{gb}) increases with the grain size and that σ_{gb} (sample Z_F)> σ_{gb} (sample Z_C), for materials with close grain sizes, despite the higher amount of Si in the Z_F specimens (Z_F : $SiO_2 \sim 1$ wt%, Z_C : ~0.4 wt%). These results have been explained [3,4] by the poor microstructure of samples Z_C. Indeed, these materials present continuous glassy films at grain boundaries, immediately adjacent to glassy triple points, while samples Z_F contain lenticular amorphous precipitates which do not wet the adjacent grain boundaries. However, nearly the same activation energy of grain boundary conductivity was found for all materials ($E_a = 1.19 \pm 0.02$ eV), in agreement with transport processes restricted to unwetted grain boundary regions, as suggested by Badwal [1].

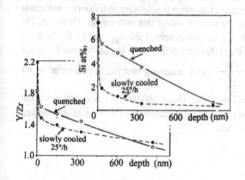

Figure 3 Silicon concentration profile (at%) and change of the atomic ratio Y/Zr near the surface of samples treated at 1600°C and cooled at different cooling rates (air quenched or cooled at 25°/h), as determined by XPS.

Fig.2 shows both that σ_{gb} of samples B ($SiO_2 \sim 1.6$ wt%) increases when the cooling rate at the end of sintering increases and that σ_{gb} (sample B quenched)< σ_{gb} (sample A cooled at 250 °/h), but less contaminated with silicon ($SiO_2 \sim 1.0$ wt%). XPS analysis have allowed us to explain these results by kinetic demixing processes occurring during cooling. The XPS analysis have been performed on the fracture surface of samples B sintered at 1600 °C, for 5 h, and cooled at the end of annealing either at 25 °/h or air quenched. We have found (Fig.3) both a yttrium and silicon enrichment near the surface and a depth profiles of Si and Y/Zr which decreases more rapidly near the surface of the slowly cooled sample. The close amounts of Si in the first monolayers found for the two samples coupled to the Si depth profiles suggests that a higher amount of Si was rejected in the slowly cooled sample grain-boundaries, in agreement with its lower σ_{gb} values (Fig.2).

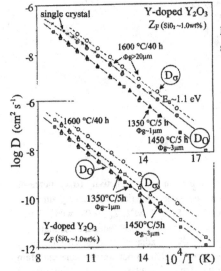

Figure 4- Oxygen diffusion coefficient in YSZ samples. Comparison with the values of D_σ

3.1.b Oxygen diffusion coefficient

Fig.4 shows that the values of Do obtained with both a single crystal and sample Z_F sintered at 1600 °C are in good agreement. The diffusion entities and jump frequencies in these two materials are then very close, in agreement with the clean microstructure of this polycrystal. However, Do (Z_F sintered at 1600 °C > Do (Z_F sintered 1350 °C), as was observed for σ_{gb} (Fig.1). In Fig.4 we have also reported the values of the conductivity diffusion coefficient D_σ, calculated from the generalized Nernst-Einstein relation ($D_\sigma = g\dfrac{\sigma_i kT}{(z_i e)^2 C_i}$) and assuming that the "extra term" g =1 [6,7]. For a given sample we have found that Do $\neq D_\sigma$. These results are consistent with the presence of complex defects whose diffusion entities and jump frequencies depend of the applied driving force (chemical potential gradient or electrical field).

3.2 Gadolinium-doped ceria

3.2.a Bulk and grain boundary conductivities

We have found both that the bulk conductivity (σ_b) is the same for all samples and that the higher σ_{gb} values are shown for the lower grain size materials (Fig5).

3.2.b TEM characterizations and XPS analysis

TEM characterizations show that the grain boundaries are free of detectable glassy phases. XPS analysis performed on the fracture surface of samples P ($0.7<\Phi g<3$ μm) and R ($0.3<\Phi g<1$ μm) show a gadolinium enrichment on the first monolayers, near the surface, but Si was not detected. Furthermore, for an analysis depth ≤ 6 monolayers the Gd enrichment is higher in sample P (Gd/Ce~1.7) in atomic concentration) than in sample R (Gd/Ce~0.9), in agreement with the difference of grain sizes. At a depth analysis ≤ 11 monolayers, the ratio Gd/Ce tends toward the same value (~0.4), confirming that Gd segregates mainly at the periphery of the grains.

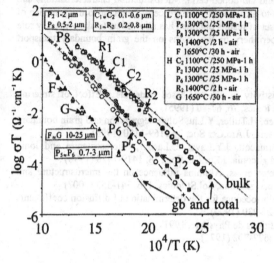

Figure 5 Influence of the grain size on the grain boundary conductivity of Gd- doped ceria.

In the inserts are indicated the grain sizes and the sintering conditions of the samples

3.2.c Oxygen diffusion coefficient

In Fig.6 we have reported the values of D_O (Eq.2) found for samples P_6 and P_8, as well as the values of $D\sigma$ calculated from the simplified Nernst-Einstein relation (g=1). As for YSZ, these results are consistent with the presence of complex defects ($D_O \neq D_\sigma$, except for P_8) in the temperature range investigated. It should be noted that the difference of behavior between samples P_6 and P_8 can be attributed to a change in the near surface structure of the defects, due to kinetic demixing processes occuring during cooling. Indeed, these sample have been sintered in the same conditions, but sample P_8 was cooled more slowly at the end of sintering than sample P_6.

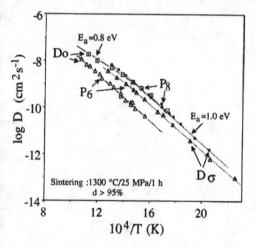

Figure 6 Oxygen diffusion coefficient in P_6 and P_8 Gd-doped ceria samples. Comparison with the values of the conductivity diffusion coefficient (D_σ).

4- Concluding remarks

This work has allowed us to show the opposite influence of the grain size on the grain boundary transport properties (σ_{gb}, Do) of YSZ and Gd-doped ceria. Microstructural characterizations and nanochemical analysis have allowed us to attribute this difference of behavior to the presence of glassy precipitates in YSZ. Finally, the set of results obtained has allowed us to show the key role played by the nanochemistry of the periphery of the grains on the grain boundary transport properties.

References

1- S.P.S.Badwal, Grain boundary resistivity in zirconia-based materials : Effect of sintering temperatures and impurities, Solid State Ionics, 76, 67-80 (1995)

2- M.Aoki,Y.M.Chiang, I.Kosaki, L.J Lee, H.Tuller, Y.Liu, Solute segregation and grain boundary impedance in high-purity stabilized zirconia, J.Am.Cer.Soc, 79,1169-1180 (1996)

3- C.Petot, M.Filal,A.Rizea, K.H.Westmacott, J.Y.Laval, C.Lacour, Microstructure and ionic conductivity of freeze-dried yttria-doped zirconia, J.Eur.Cer.Soc., 18, 1419-1428 (1998)

4- A.Rizea, D.Chirlesan, C.Petot, G.Petot-Ervas, Alumina influence on the microstructure and grain boundary conductivity of yttria-doped zirconia, Sol.St.Ionics, 146, 341-353 (2002)

5- G.Petot-Ervas, C.Petot, Experimental procedure for the determination of diffusion coefficients in ionic compounds, Sol.St.Ionics, 117, 27-239 (1999)

6- J.Philibert, Atom Movements, Les Editions de Physique, 1991

7-H.Sato, R.Kikuchi, J.Chem.Phys.55(2)677-702 (1971)

Microstructural Aspects of the Ionic Transport Properties of Strontium-Substituted Lanthanum Cobaltites

W. Sitte, E. Bucher, W. Preis, I. Papst[1], W. Grogger[1], and F. Hofer[1]
Institute of Physical Chemistry, University of Leoben,
A-8700 Leoben, Austria
[1]Research Institute for Electron Microscopy, Graz University of Technology,
A-8010 Graz, Austria

ABSTRACT

The ionic conductivity and microstructure of selected compositions of the solid solution $La_{1-x}Sr_xCoO_{3-\delta}$ (LSC) were examined with respect to possible vacancy ordering phenomena. Homogeneous samples of LSC were prepared by the glycine nitrate process. The ionic conductivity was obtained as a function of the oxygen partial pressure ($-3.5 \leq \log[p(O_2)/atm] \leq 0.5$) using a recently developed galvanostatic polarization technique. At 825°C the $p(O_2)$-dependence of the ionic conductivity of $La_{1-x}Sr_xCoO_{3-\delta}$ ($x = 0.4$ and 0.6) shows a distinct maximum. Although this behavior has yet to be explained unambiguously it is indicative of decreasing mobility of ionic charge carriers, e. g. due to cooperative vacancy ordering. From the temperature dependence of the ionic conductivity of $La_{1-x}Sr_xCoO_{3-\delta}$ ($x = 0.6$) activation energies at constant nonstoichiometry ($0.20 \leq \delta \leq 0.28$) were obtained. As vacancy association and microstructure are presumed to play a significant role we combined the results of ionic conductivity measurements and electron microscopical investigations. HRTEM images revealed a superstructure within microdomains of about 100 nm in size.

INTRODUCTION

The high ionic conductivity of $(La,Sr)CoO_{3-\delta}$ (LSC) - superior to $La_{1-x}Sr_xMnO_{3-\delta}$ - together with its high surface exchange coefficient make this material attractive for application as intermediate temperature SOFC cathode [1,2]. Compared with the electronic conductivity [3-6] of these mixed-conducting perovskites there are only few investigations regarding the oxygen ion conductivity [7,8]. Whereas the oxygen vacancies are believed to have a tendency to disorder at elevated temperatures, the limits of ideal solution approximation and the necessity of considering interactions between point defects and clusters must be kept in mind at lower temperatures. Further, the ionic conductivity in the temperature range 700°-900°C is heavily influenced by the microstructure. For example, a sharp decrease in the temperature dependence of the ionic conductivity was reported, which has been interpreted in terms of an order-disorder type transition [7].

Vacancy ordering and formation of defect clusters in $La_{1-x}Sr_xCoO_{3-\delta}$ have been reported in a number of studies [9-12] and are frequently assumed to play an important role in controlling the oxygen ion conductivity. If association of defects takes place, an assigned fraction of oxygen vacancies should no longer be free to participate in the conduction process and thus the ionic conductivity would be lowered.

The objective of this study was to obtain data on the ionic transport properties of two different compositions of $La_{1-x}Sr_xCoO_{3-\delta}$ (x = 0.4 and 0.6) and to elucidate the $p(O_2)$-dependence in the intermediate temperature range (775° and 825°C) as well as the temperature dependence at constant oxygen nonstoichiometry. Besides oxygen transport properties, microstructural investigations of the materials with electron microscopy have been performed in an effort to provide information about ordering phenomena.

EXPERIMENTAL DETAILS

For the synthesis of $La_{1-x}Sr_xCoO_{3-\delta}$ (x = 0.4 and 0.6) nitrate solutions of the cations were mixed in appropriate ratios and complexed with glycine. The resulting viscous gels were heated until self-ignition took place. Details on the glycine nitrate method can be found elsewhere [13].

X-ray diffraction analysis of powder samples indicated single phase perovskite structure. Additionally, energy-filtering transmission electron microscopy (EFTEM), energy-dispersive X-ray spectrometry (EDXS), and electron energy loss spectroscopy (EELS) were performed. Results were published in earlier work [14]. To obtain a highly oxygen deficient sample suitable for studies of the microstructure a disk-shaped sample of $La_{0.4}Sr_{0.6}CoO_{3-\delta}$ was annealed at $p(O_2)$= 1×10^{-3} atm at 825°C and subsequently quenched to room temperature. For high resolution transmission electron microscopy (HRTEM) powder particles were mounted on carbon grids. Electron microscopy was performed on a Philips CM20 TEM/STEM equipped with a LaB_6 cathode.

The partial ionic conductivity was measured using the following symmetric electrochemical cell.

$$O_2,Pt \,|\, YSZ \,|\, Pt \,|\, LSC \,|\, Pt \,|\, YSZ \,|\, Pt,O_2. \tag{1}$$

Furthermore, this cell allows variation of the oxygen nonstoichiometry of the oxide by coulometric titration. Details regarding the construction of the cell as well as the polarization method have been described elsewhere [15].

RESULTS AND DISCUSSION

Figure 1 shows the ionic conductivity of $La_{0.4}Sr_{0.6}CoO_{3-\delta}$ and $La_{0.6}Sr_{0.4}CoO_{3-\delta}$ determined from polarization/depolarization experiments as a function of the oxygen partial pressure ($-3.5 \leq \log[p(O_2)/atm] \leq 0.5$) at 775°C. The analogous plot for T = 825°C is given in figure 2. Lines are a guide to the eye. The results of polarization and depolarization experiments at each $p(O_2)$ are in good agreement within the limits of error. At 775°C the ionic conductivity of both sample compositions decreases with decreasing oxygen partial pressure over the whole $p(O_2)$-range. For $La_{0.6}Sr_{0.4}CoO_{3-\delta}$ a plateau is reached at $p(O_2)>10^{-1}$ atm which can be assigned to the saturation of the oxygen sublattice. When δ approaches the minimum equilibrium value at 775°C further increase of $p(O_2)$ does not result in a significant reduction of the vacancy concentration. The same trend is insinuated for $La_{0.4}Sr_{0.6}CoO_{3-\delta}$.

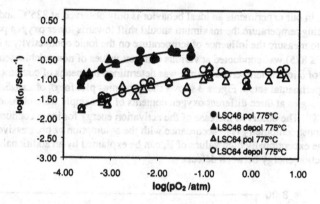

Figure 1. Ionic conductivity of $La_{0.4}Sr_{0.6}CoO_{3-\delta}$ (LSC46) and $La_{0.6}Sr_{0.4}CoO_{3-\delta}$ (LSC64) obtained from polarization and depolarization experiments as a function of oxygen partial pressure at 775°C.

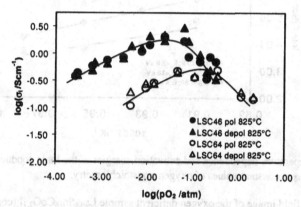

Figure 2. Ionic conductivity of $La_{0.4}Sr_{0.6}CoO_{3-\delta}$ (LSC46) and $La_{0.6}Sr_{0.4}CoO_{3-\delta}$ (LSC64) as a function of oxygen partial pressure at 825°C.

Distinctive maxima in the ionic conductivity of both materials are present at 825°C. The transition from increasing to decreasing ionic conductivity with decreasing $p(O_2)$, i. e. increasing oxygen nonstoichiometry has been interpreted in connection with the formation of vacancy ordered structures which are expected to lower the mobility of oxygen vacancies [15,16].

When the ideal point defect approximation is valid, the ionic conductivity is supposed to increase with decreasing oxygen partial pressure (increasing concentration of ionic charge carriers). Statistically distributed, non-interactive defects can be expected at (i) low vacancy concentrations and (ii) high temperatures. Vacancy ordering is basically a low temperature phenomenon. Therefore, a transition from order to disorder is to be expected with increasing temperature. In our experiments an ideal behavior is only observed at 825°C and $p(O_2)>10^{-1}$ atm. With increasing temperature the maximum should shift towards lower oxygen partial pressures.

In order to measure the influence of temperature on the ionic conductivity a temperature run ($750 \leq T/°C \leq 825$) was conducted at various constant values of nonstoichiometry. The $p(O_2)$-dependence of δ in this temperature range was determined by means of coulometric titration in the same experimental setup. Figure 3 shows the Arrhenius plot for σ_i of $La_{0.4}Sr_{0.6}CoO_{3-\delta}$. The activation energies at three different oxygen contents of the sample were obtained from the slope of $\ln\sigma_iT$ vs. T^{-1}. The observed increase of the activation energy for ionic conduction with increasing nonstoichiometry is in accordance with the assumption of progressive vacancy ordering. The exceptionally high values of E_a can be explained by an additional contribution due to the interaction energy between defects.

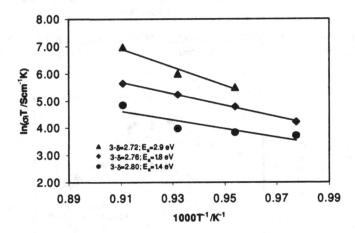

Figure 3. Temperature dependence and activation energies of the ionic conductivity of $La_{0.4}Sr_{0.6}CoO_{3-\delta}$ at constant values of oxygen nonstoichiometry.

The bright field image of the oxygen deficient sample $La_{0.4}Sr_{0.6}CoO_{2.71}$ (corresponding to $\log[p(O_2)/atm] = -2.87$ at $T = 825°C$) in figure 4 shows a single particle with a typical domain structure, i. e. bright and dark regions. The width of the domains ranges from 50 to 200 nm. Figure 5 shows a HRTEM image of a single particle oriented along the [4 2 -1] axis. The Fourier transform (FFT) inset exhibits doubling of the lattice spacing. Superstructure spots which are indicated by arrows correspond to extra spots in the SAED pattern of the same sample [17]. Such a superstructure has previously been attributed to either ordering of the oxygen vacancies within a lattice plane [10] or aperiodically alternate La-rich and Sr-rich (001) lattice planes [11]. Furthermore, it was expected that the ionic conductivity decreases in samples with vacancy ordered structures. This could be confirmed exemplarily for $La_{0.4}Sr_{0.6}CoO_{2.71}$ (see also figure 2 at $\log[p(O_2)/atm] = -2.87$ and $T = 825°C$).

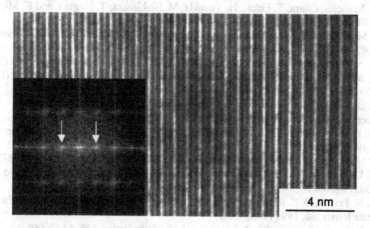

Figure 4. Bright field image of a single particle of $La_{0.4}Sr_{0.6}CoO_{2.71}$ with domain structure.

Figure 5. HRTEM image of $La_{0.4}Sr_{0.6}CoO_{2.71}$. The inset shows the corresponding Fourier transform; spots due to the superstructure are marked.

CONCLUSIONS

The $p(O_2)$-dependence of the ionic conductivity of $La_{1-x}Sr_xCoO_{3-\delta}$ (x = 0.4 and 0.6) was obtained from polarization/depolarization experiments. Regarding the maxima in the $p(O_2)$-dependence of σ_i at 825°C we conclude that defect association has a marked effect on the ionic transport properties of these acceptor-doped mixed conductors. Furthermore, activation energies of ionic conduction at defined δ-values were calculated from the temperature dependence of the ionic conductivity of $La_{0.4}Sr_{0.6}CoO_{3-\delta}$. An increase of the activation energy was found with increasing nonstoichiometry, indicating progressive immobilization of ionic charge carriers.

HRTEM investigations of a sample with high concentration of oxygen vacancies confirmed the existence of a superstructure within microdomains of about 100 nm in size.

In the light of these facts, the approximation of independent point defects is no longer suitable for LSC with high vacancy concentration. Improved defect chemical models for these highly defective perovskite type materials are needed, taking into account interactions between vacancies.

ACKNOWLEDGMENTS

The authors gratefully acknowledge support by the Austrian Science Funds (FWF) within the Special Research Program 'Electroactive Materials' (projects F915 and F923).

REFERENCES

1. H. Uchida, S. Arisaka, and M. Watanabe, Electrochem. and Solid-State Letters, 2, 428 (1999).
2. R. Maric, S. Ohara, T. Fukui, H. Yoshida, M. Nishimura, T. Inagaki, and K. Miura, J. Electrochem. Soc., 146, 2006 (1999).
3. J. Mizusaki, J. Tabuchi, T. Matsuura, S. Yamauchi, and K. Fueki, J. Electrochem. Soc., 136, 2082 (1989).
4. A. N. Petrov and P. Kofstad in *Proc. of the 3rd Int. Symp. on Solid Oxide Fuel Cells*, edited by S. C. Singhal and H. Iwahara, (Electrochem. Soc. Proc. 93-4, Pennington, NJ, 1993) pp. 220-230.
5. E. B. Mitberg, M. V Patrakeev, I. A. Leonidov, V. L. Kozhevnikov, and K. R. Poeppelmeier, Solid State Ionics, 130, 325 (2000).
6. M. H. R. Lankhorst, H. J. M. Bouwmeester, and H. Verweij, J. Solid State Chem., 133, 555 (1997).
7. F. M. Figueiredo, F. M. B. Marques, and J. R. Frade, Solid State Ionics, 111, 273 (1998).
8. H. Ullmann, N. Trofimenko, F. Tietz, D. Stöver, and A. Ahmad-Khanlou, Solid State Ionics, 138, 79 (2000).
9. A. N. Petrov, O. F. Kononchuk, A. V. Andreev, V. A. Cherepanov, and P. Kofstad, Solid State Ionics, 80, 189 (1995).
10. R. H. E. van Doorn and A. J. Burggraaf, Solid State Ionics, 128, 65 (2000).
11. R. Caciuffo, D. Rinaldi, G. Barucca, J. Mira, J. Rivas, M. A. Senaris-Rodriguez, P. G. Radaelli, D. Fiorani, and J. B. Goodenough, Phys. Rev., B59, 1068 (1999).
12. S. Stemmer, A. J. Jacobson, X. Chen, and A. Ignatiev, J. Appl. Phys., 90, 3319 (2001).
13. L. A. Chick, L. R. Pederson, G. D. Maupin, J. L. Bates, L. E. Thomas, and G. J. Exarhos, Materials Lett., 10, 6 (1990).
14. I. Rom, F. Hofer, E. Bucher, W. Sitte, K. Gatterer, H. P. Fritzer, and A. Popitsch, Chem. Mater., 14, 135 (2002).
15. E. Bucher, A. Benisek, and W. Sitte, Solid State Ionics, (2002) (in press).
16. S. Adler, S. Russek, J. Reimer, M. Fendorf, A. Stacy, J. Baltisberger, and U. Werner, Solid State Ionics, 68, 193 (1994).
17. E. Bucher, W. Sitte, I. Rom, I. Papst, W. Grogger, and F. Hofer, Solid State Ionics, (2002) (in press).

Mat. Res. Soc. Symp. Proc. Vol. 756 © 2003 Materials Research Society   EE10.4

Characterization of Multilayer Anodes for SOFC

Axel C. Müller, Albert Krügel, André Weber and Ellen Ivers-Tiffée
Institut für Werkstoffe der Elektrotechnik, Universität Karlsruhe (TH)
D-76131 Karlsruhe, Germany

ABSTRACT

SOFC anodes have to combine various tasks. In anode supported single cells a thick anode substrate is used for current collecting and gas distribution whereas a thin functional layer adjacent to the electrolyte is the electrochemically active part of the anode. This functional anode layer is cofired together with the thin film electrolyte to obtain an enhanced interface with low polarisation losses. This multilayer structure was transferred to an electrolyte supported single cell. The electrochemical active Ni/8YSZ anode layer was screen printed onto a 8YSZ electrolyte green tape and subsequently cofired at 1350 °. Mechanical stresses during cofiring due to shrinkage mismatch of anode and electrolyte were avoided by changing the geometry of the anode layer from a continuous layer to a large number of small sized individual areas. Simulations by finite element modeling indicated that a hexagonal pattern similar to honey-combs is preferable. The second layer which adjoins to the fuel gas channels and which is responsible for current collecting and gas distribution was later on screen printed on top and sintered together with the cathode. Single cells with a multilayer anode and different functional layers were electrochemically characterised under realistic operation conditions. The performance and reduction/oxidation stability of this type of anode was investigated. The electrochemically active layer showed only small degradation during redox cycling and long term operation at high fuel utilisation. In contradiction to single layer anodes Nickel agglomeration was not observed in the functional layer.

INTRODUCTION

The high temperature solid oxide fuel cell (SOFC) is an electrochemical energy converter representing a technology, which allows the economic and environmentally friendly production of electrical energy with high efficiency. To meet the different tasks of a SOFC anode, a multilayer structure is often used in anode supported single cells. A thick anode substrate acts as current collector and gas distributor whereas a thin functional layer adjacent to the electrolyte is the electrochemically active part of the anode. This functional anode layer is usually cofired with the thin film electrolyte to obtain an enhanced interface which exhibits low polarization losses and good mechanical adherence. However, different shrinkage of anode and electrolyte [1] causes bending [2-4] or breaking of the compound. In this work we will show how this multilayer structure was transferred to an electrolyte supported single cell and how the problems of cofiring were encountered. Single cells with a multilayer anode were prepared and electrochemically characterized in respect of performance, redox stability and long term stability.

THEORY

If the length difference Δs caused by different sintering shrinkage of anode and electrolyte, Young modulus E_{an} and Poisson ratio ν of anode are known one can calculate the stress evolving

in the electrolyte [5] according to

$$\sigma_{m,el} = \frac{\Delta s E_{an}}{(1-v)\dfrac{d_{el}}{d_{an}}} \tag{1}$$

if anode thickness d_{an} is much smaller than the thickness d_{el} of the electrolyte.

From eq.1 one can see that the stress in the electrolyte can be reduced by decreasing either the shrinkage difference Δs or the Young modulus E_{an} of the anode. However, shrinkage behavior is difficult to adjust without changing essential properties of anode and electrolyte. An alternative way is the use of an anode pattern which consists of a large number of small sized individual areas instead of a continuous layer. Thus the porosity of the anode is macroscopically decreased and consequently the Young modulus is lowered without changing essential properties of the anode. Some examples of possible pattern types are shown in figure 1. It is expectable that bending radius R and mechanical stress $\sigma_{m,el}$ in the electrolyte depend on the geometry used for the anode pattern. Shape and size of the small electrodes and the width of the channel between them are parameters. Because analytic calculations are not feasible one has to use finite element modeling (FEM) to calculate bending and stress distribution.

A very simple model, which was based on time invariant, isotropic elastic behavior of anode and electrolyte, was applied for the simulations. The shrinkage mismatch was treated analogous to a thermal expansion mismatch and described by linear thermal expansion coefficients i.e. $\Delta s = \Delta T(\alpha_{el} - \alpha_{an})$. Effective constant Young's moduli for anode and electrolyte were used and their ratio was calculated according to [5]

$$\frac{E_{el}}{E_{an}} = d_{an}d_{el}\left[\frac{6(d_{el}+d_{an})\Delta sR}{d_{el}^2}-1\right] \tag{2}$$

Bending radii R of small cofired anode/electrolyte samples were determined by cofiring experiments and shrinkage of electrolyte and anode was measured by dilatometry. Young modulus of anode E_{an} was calculated according to Eshelby [7] for a composite material and corrected according to its porosity resulting in a value of 86 GPa.

Experimental data for patterned anodes with regard to bending were compared with FEM calculations and used as basis for the rating of various pattern types.

It should be emphasized that the applied model was not meant to describe mechanical properties during sintering – which is rather complex – but to help describing bending after cofiring. For some more details on the model see [5].

EXPERIMENTAL

8YSZ electrolyte green tapes with dimensions of 10x10 mm^2 and a thickness of 200 μm

Figure 1. Examples of anode patterns consisting of small individual electrodes instead of a continuous layer. Left: squares; mid: honey-combs; right: hexagons

with a screen printed Ni/YSZ cermet anode (47 vol% Ni, 10 μm thick) were used as model samples. Besides samples with a continuous anode layer, patterned anodes (1.6x1.6 mm^2 squares, 0.5 mm channel width, 66 % of full area) were prepared. Bending radii of these samples were measured after cofiring at 1300 °C and shrinkage of electrolyte and anode were determined by dilatometry under the same conditions as for the cofiring experiments. Based on the parameters evaluated from the experiments FE modeling was conducted with $ANSYS^{TM}$ for the same geometries as used for the cofiring experiments.

SOFC single cells were prepared by screen printing a Ni/8YSZ anode pattern onto a 200 μm thick 8YSZ electrolyte green tape which was 67x67 mm^2 in size. Anode and electrolyte were subsequently cofired at 1350 °C for 5 hours. For current collecting either a Ni/8YSZ slurry (90 vol% Ni) was painted on top or a Ni/8YSZ cermet (47 vol% Ni) was screen printed on top (electrode area 10 cm^2) and sintered together with a $La_{0.75}Sr_{0.2}MnO_3$ cathode at 1250 °C. These single cells were electrochemically characterized under realistic operation conditions in H_2/air at 950 °C. Several reduction/oxidation cycles (redox) were performed by switching off fuel supply and waiting for 30 min so that the anode was oxidized and afterwards fuel was switched on again. Impedance spectra were recorded after each redox cycle in the frequency range from 100 mHz to 1 MHz. Long term stability during 900 hours at constant current load and high fuel utilization was determined. The microstructure of the single cells was analyzed by electron microscopy.

RESULTS

Dilatometric investigations exhibited a total shrinkage of the 8YSZ electrolyte of 24.4 % and 16.1 % for the anode. Cofiring of anode/electrolyte model samples lead to a bending radius of 5.1 mm for a continuous anode layer and 7.1 mm for model samples with a patterned anode. As larger radius means less bending these results demonstrate the validity of our previous considerations that bending is decreased by the use of a patterned anode. The evaluated parameters of the samples with a continuous anode layer were inserted in eq.2 which gave an effective electrolyte Young modulus of 64 MPa. Bending radii of samples with a continuous anode layer and a patterned anode like the one used for cofiring experiments were calculated by FEM and found to be 4.9 mm and 7.1 mm, respectively. These are reasonable approximations for the values measured in actual cofiring experiments and thus the applied model was sufficient to describe bending after cofiring.

Various pattern geometries, which differ in shape and size of single areas and channel width, were rated concerning their bending radius. The used geometries are listed in table I together with their relative area and bending radius calculated by FEM. Obviously, bending radius increased when the covered area was decreased.

Table I. Size and shape of different anode patterns used for simulation. The bending radius was estimated by FEM.

sample id	pattern type	size	channel width	covered area	bending radius R
A	square	8 mm	–	100 %	5.9 mm
B1	square	2 mm	2 mm	25 %	23.3 mm
B2	square	1 mm	0.16 mm	75 %	7.3 mm
C	hexagon	1 mm	0.16 mm	60 %	9.5 mm
D1	honey-combs	1 mm	0.15 mm	71 %	7.9 mm

Furthermore it can be seen that pattern geometry had an influence. A change of the shape type from square (B2) to hexagon lead to a relative decrease of the covered area of 20 %, but on the other hand an increase of the bending radius of 23 % was achieved. Further improvement was accomplished by the use of a honey-comb like pattern (D1, see figure1 mid). Bending radius was increased by 34 % in comparison to the sample with 100 % covered area, whereas effective area was decreased by only 29 %. In addition, mechanical stresses in the electrolyte, when the sample is clamped during cofiring and bending is therefore hindered, were simulated. For a continuous anode layer the stress in the whole electrolyte was uniformly 442 MPa which is definitely too high as the mean strength of 8YSZ is about 258 MPa at 950 °C [6]. Whereas the patterned anode used for cofiring experiments exhibited an average stress of only 179 MPa. Again the advantage of the honey-comb pattern D1 is visible as the average stress is 7 % lower than that of the square pattern B2 (142 MPa in comparison to 158 MPa), although the covered area is only 5 % less.

Based on these results 5x5 cm² single cells with an anode pattern (honey-comb type), which was cofired with an 8YSZ green tape, were prepared. The hexagons had an outer diameter of 1.2 mm before sintering and were separated by channels with a width of 220 μm. The covered area after sintering was 65 %. As it can be seen in figure 2 the single cells showed neither bending nor cracks and good adherence between electrolyte and the first anode layer which was patterned and cofired and between first and top Ni/YSZ cermet layer was seeable (figure 3).

I/V characteristics of single cells with a patterned anode and different top layers were carried out and are shown in figure 4. Without any top layer the performance of the single cell was poor due to a high charge transfer resistance and a high ohmic resistance which was decreased by the use of a current collecting top layer. No difference was visible between performance of a single cell with a Ni slurry and a Ni/8YSZ cermet as top layer. Therefore, it is presumed that electrochemical properties of the anode is determined by the first electrochemical active layer. However, performance was still insufficient due to the low porosity of this layer.

The redox behavior of different anode types was investigated and as one can see in figure 5 all anodes exhibited an increase of the polarization resistance R_{pol} i.e. degradation of the cell performance after several redox cycles. The patterned anode without top layer indicate the largest increase of the polarization resistance. Analysis by electron microscopy revealed partly delamination of the anode and segregation of Ni particles to the surface of the anode. However,

Figure 2. 5x5 cm² cofired single cell with a hexagonal anode pattern.

Figure 3. SEM image of a 2-layer anode with a patterned anode as layer I and a Ni/YSZ cermet on top.

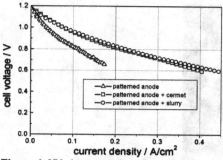
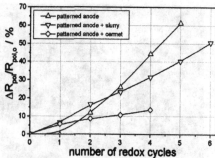

Figure 4. I/V characteristics of single cells with a patterned anode and different top layers.

Figure 5. Increase of polarization resistance of different anode types after several redox cycles.

degradation due to redox cycling was drastically decreased by the use of a Ni/YSZ cermet as top layer whereas the use of a Ni slurry was not so effective.

Long term stability of the multilayer anode with a Ni/8YSZ cermet as top layer was tested for 900 h at a constant current density of 0.3 A/cm² and 20 % fuel utilization, which was simulated by adding appropriate amount of water vapor [8]. As can be seen in figure 6 stable operation without any degradation was possible. Operation at 0.2 A/cm² and 80 % fuel utilization for 900 h exhibited a degradation rate of 3 % that is less than the 5 % degradation rate determined for single layer anodes [8]. Analysis of the anode microstructure by SEM after long term operation (see figure 8) revealed that severe Nickel agglomeration occurred in the current collecting top layer, which was in accordance with former experiments [8]. However, no Nickel agglomeration was observed in the electrochemical active layer I.

CONCLUSIONS

Cofiring experiments of anode and electrolyte showed that shrinkage mismatch caused bending which was decreased by the use of a patterned anode that consisted of a large number of small electrodes. A simple model which described shrinkage analogous to thermal expansion coefficients was used for FE modeling.

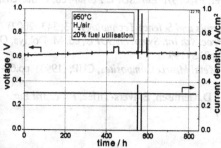
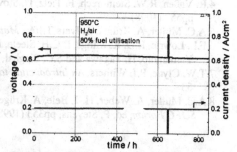

Figure 6. Long term operation at 0.3 A/cm² and 20% fuel utilization. No degradation was visible.

Figure 7. Long term operation at 0.2 A/cm² and 80% fuel utilization. Degradation rate of 3% was determined.

Figure 8. Microstructure after long term operation. No Nickel agglomeration was detectable in layer I.

Comparison of various pattern types indicated that honey-comb like patterns are preferable as they yielded the lowest stresses for a given area coverage. Thus mechanical stresses due to shrinkage mismatch during cofiring can be decreased and also mechanical stress due to TEC mismatch during thermal cycling can be lowered.

Single cells with a honey-comb patterned anode as first layer which was cofired with the electrolyte and a second layer for current collecting, that was screen printed and subsequently sintered, were prepared. Although performance of these kind of single cells was not sufficient, redox stability and long term stability was improved in comparison to single layer anodes. SEM analysis revealed that this was attributed to the fact that no Nickel agglomeration occurred in the electrochemical active layer adjacent to the electrolyte. It is assumed that the top anode layer protected the active layer during redox cycling and long term operation at high fuel utilization.

REFERENCES

1.T. Matsushima, H. Ohrui, T. Hirai, *Solid State Ionics*, **111**, 315 (1998)
2.J.-H. Jean, C.-R. Chang, *J. Am. Ceram. Soc.*, **80** no. 9, 2401 (1997)
3.R.W. Steinbrech, A. Caron, G. Blaß, F. Dias, *Proc. 5th Int. Symp. on SOFC*, ed. U. Stimming, S.C. Singhal, pp727 (1997)
4.R. Vaßen, R.W. Steinbrech, F. Tietz, D. Stöver, *Proc. 3rd Eur. SOFC Forum*, ed. P. Stevens, pp557 (1998)
5.A.C. Müller, A. Krügel, E. Ivers-Tiffée, *Materials and Science Technology*, **33**, 343 (2002)
6.F.L. Lowrie, R.D. Rawlings, B.C.H. Steele, *Proc. 4th Int. Symp. on SOFC*, ed. M. Dokiya, O. Yamamoto, H. Tagawa and S.C. Singhal, pp318 (1995)
7.T.W. Clyne, P.J. Withers, *An Introduction to Metal Matrix Composites*, CUP, (1993) pp.62-63
8.A.C. Müller, A. Weber, H.-J. Beie, A. Krügel, D. Gerthsen, E. Ivers-Tiffée, *Proc. 3rd Eur. SOFC Forum*, ed. P. Stevens, pp353 (1998)

Mat. Res. Soc. Symp. Proc. Vol. 756 © 2003 Materials Research Society

Improvements in Mechanical Behavior of SOFC Anodes

O. Kesler, R. L. Landingham

Lawrence Livermore National Laboratory
Livermore, California, U.S.A.

ABSTRACT

Processing-induced residual stresses in the component layers of Solid Oxide Fuel Cells (SOFC's) can lead to fracture or to cell curvature which impedes stack assembly. Reducing or eliminating residual stresses to improve the mechanical behavior of the cells becomes increasingly important as the area of the cells is increased to increase the power of the fuel cell stack. Residual stresses in SOFC's result primarily from differential thermal expansion and sintering shrinkage between the component layers, such as the electrolyte and the anode support in a planar cell. This work investigates the impact of anode composition on each of these factors, with the ultimate goal of designing a flat, large-area cell. A range of anode compositions is investigated to determine the effect of different additives on the sintering behavior, and on the thermal and mechanical properties. Dilatometry, sintering shrinkage, scanning electron microscopy, and reduction studies are performed to correlate the microstructure, thermomechanical behavior, and composition. The experimental results are used to select an anode composition that leads to low overall cell curvature and improved mechanical behavior with respect to standard SOFC anode cermet materials.

INTRODUCTION

Solid oxide fuel cells (SOFC's) have the potential for the highest efficiency among the different fuel cell types when using pure hydrogen fuel, making them attractive candidates for applications in both transportation [1] and stationary power generation [2]. However, their widespread commercialization remains limited due to materials challenges posed by the high-temperature operating environments (typically 600-1000°C, depending on the materials used). Differential thermal expansion and sintering shrinkage in the fuel cell material layers lead to thermal stresses in production and operation that limit cell lifetime and reliability. This work seeks to improve the mechanical behavior of solid oxide fuel cells and to reduce thermal stresses and curvature in the cells by more closely matching the thermal expansion of the different material layers. We examine anode-supported cells, and study the effect of doping the traditional Nickel-Yttria-Stabilized Zirconia (Ni-YSZ) cermet anode material with Al_2O_3 in order to better match the thermal properties of the anode with that of the electrolyte coating.

EXPERIMENTAL PROCEDURE

Powder specimens were made with a range of compositions by adding Al_2O_3 in varying amounts by volume fraction to NiO and ZrO_2-8mol%Y_2O_3 powders. The volume fraction of NiO in the starting powders was held constant at 51%. Powders were made by wet ball-milling in alcohol, with 1 wt. % of PVB used as a binder and varying amounts of rice starch (0-22 wt. %) used as a pore former. Two compositions were chosen for production of anode tapes prepared by tape casting from NiO, YSZ, and Al_2O_3 powders (Richard E. Mistler, Inc., Morrisville, PA). Sintering shrinkage studies were performed on the tapes. Pressed powder specimens were used for sintering, dilatometry, and reduction studies. Both tape and pressed powder specimens were outgassed by heating 1°C/minute to 330°C, holding at 330°C for one hour, and then heating for 2°C/minute to pre-sintering temperatures ranging from 700-1400°C. Pellets were then sintered at 1430°C for 1-3 hours with 5°C/min heating and cooling rates and used for dilatometry and reduction studies. Spray coatings 10-20 μm thick of YSZ electrolyte were deposited onto pressed discs and cast tapes by colloidal spray deposition, and the resulting bi-layers were co-sintered at 1430°C for 1-3 hours. J. T. Baker NiO powder and Tosoh TZ-8YS YSZ powders were used for all of the specimens. Aldrich <10 micron Al_2O_3 powder was used for the initial specimens used to determine the effect of varying compositions, while ~1 micron A16 Al_2O_3 powder was used for the cast tapes and final powder compositions.

Dilatometry was carried out on a Linseis L75 dilatometer to determine the coefficient of thermal expansion of the specimens as a function of composition and temperature. Some of the specimens were reduced by heating at 10°C/min to 800°C in 4% H_2/96% N_2, holding for 40-120 minutes at 800°C in pure H_2, and then cooling at 10°C/min to room temperature in 4% H_2/96% N_2 to determine the resulting weight loss, volume change, composition, and microstructure. The reduction temperature was chosen to coincide with the temperature used for in-situ reduction of anodes in fuel cell stacks after the assembly of individual cells and interconnects. X-ray diffraction was used to determine the composition of the specimens before and after sintering and reduction, and scanning electron microscopy was performed to correlate the microstructures with the thermo-mechanical properties.

In the anode-supported cells studied, the thickest layer is that of the porous anode, which is approximately 0.5 mm thick. After deposit and densification of a thin (10-20μm) YSZ electrolyte layer, a thin porous cathode layer (40 μm) is then deposited and co-sintered with the two previously deposited layers. Because the cathode layer is porous and very thin, it contributes much less substantially to the curvature of the cell than the anode support and dense electrolyte layer. Therefore, the focus of this work is the minimization of thermal mismatch between the anode and electrolyte materials in order to minimize processing-induced residual stresses and cell curvature prior to stack assembly. To determine the effect of adding alumina in different amounts on the anode thermo-mechanical behavior, powder specimens were prepared with 0, 15, 20, and 25 volume % Al_2O_3, with the NiO content held constant at 51 volume % and the YSZ content forming the balance of the ceramic materials. In addition to the ceramic components, rice starch was added as a pore former to the initial mixtures, holding the volume percentage constant for each of the four compositions, corresponding to weight percentages of 20, 21.1, 21.5, and 21.8, respectively, of the ceramic weight content.

RESULTS

The average thermal expansion coefficients (CTE's) as a function of temperature over the range from 700°C to 1100°C are shown in Figure 1 for anode materials with different alumina compositions, for pure YSZ, and for Al_2O_3 and $NiAl_2O_4$. Addition of Al_2O_3, with a lower CTE than that of YSZ, to the anode results in a lowering of the overall CTE with respect to the standard NiO-YSZ anode material, with increasing amounts of Al_2O_3 resulting in decreasing CTE's in the composition range studied. By replacing approximately 25 volume % of ZrO_2 in NiO-YSZ with Al_2O_3, the CTE of the anode can be matched with that of the YSZ electrolyte.

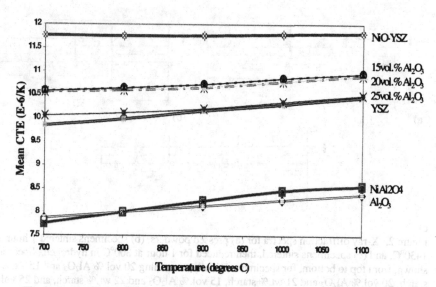

Figure 1. Coefficient of thermal expansion from 20°C to the given temperature for a range of anode and electrolyte compositions.

X-ray diffraction (XRD) studies of the specimens before and after sintering and after reduction indicate that the alumina and nickel oxide in the original specimens react to form NiAl₂O₄, which has a similar CTE to that of alumina. The XRD powder diffraction spectra for the pressed powders, for pellets sintered for 1 hour at 1430°C, and for pellets that were sintered and then reduced for one hour at 800°C in pure hydrogen are shown in Figure 2. The spectra for the pressed powders in Figure 2a show the initial components added to the powders – YSZ, NiO, and Al₂O₃. After one hour of sintering, all of the added Al₂O₃ reacts with available NiO to form NiAl₂O₄, and only NiO, YSZ, and NiAl₂O₄ remain in the structure (Figure 2b). After reduction in pure hydrogen at 800°C, all of the remaining NiO is reduced to Ni, and only peaks for YSZ, Ni, and NiAl₂O₄ remain in the spectrum for the structure (Figure 2c). None of the NiAl₂O₄ reduces to Ni + Al₂O₃, as indicated by the absence of Al₂O₃ peaks in the spectra of the reduced specimens. Therefore, any NiO from the initial composition that reacts with added Al₂O₃ does not reduce to metallic Ni in the final fuel cell anode, and thus does not contribute significantly to the overall electrical conductivity of the anode, since NiAl₂O₄ has a very low conductivity in the fuel cell temperature operating range [3]. NiAl₂O₄ can be partially reduced to Ni and Al₂O₃ at higher temperatures than those typically used for in-situ reduction of fuel cell stacks (>1000°C) [4-5]. However, for stacks that are designed to operate in the 700-800°C temperature range, exposing the metallic interconnect materials to temperatures over 1000°C could cause oxidation and loss of conductivity of the interconnect on the cathode side of the cell. Therefore, reduction of NiAl₂O₄ that forms during the sintering process would not be feasible for SOFC stacks that are designed to operate at 700-800°C. Weight loss measurements after reduction of the specimens in hydrogen at 800°C confirm that the Al₂O₃ which is added to the NiO and YSZ reacts entirely with any available NiO to form NiAl₂O₄, and that the NiAl₂O₄ does not reduce to

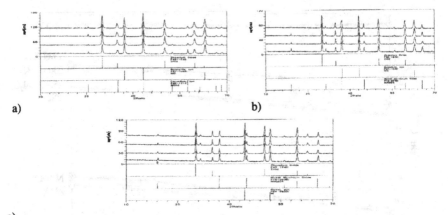

a)

b)

c)

Figure 2. X-ray diffraction spectra for (a) pressed powders, (b) specimens sintered 1 hour at 1430°C, and (c) specimens sintered, then reduced for 1 hour at 800°C in hydrogen. Spectra are shown, from top to bottom, for specimens initially containing 20 vol.% Al_2O_3 and 15 wt.% starch, 20 vol.% Al_2O_3 and 21 wt.% starch, 15 vol.% Al_2O_3 and 22 wt.% starch, and 25 vol.% Al_2O_3 and 22 wt.% starch.

Ni and Al_2O_3 at 800°C in pure hydrogen. The dashed curve shows the weight loss that would be predicted if all of the original NiO was reduced to Ni. The solid curve shows the weight loss that would be predicted if all of the Al_2O_3 added to the original powder reacts with NiO to form $NiAl_2O_4$, and the remaining NiO is reduced to Ni. The data points show the actual weight losses. Since some of the initial NiO reacts with Al_2O_3, the volume fraction of Ni after reduction is decreased in the specimens with higher added Al_2O_3 contents for a constant initial volume fraction of NiO. The volume percent of Ni after reduction decreases from 37.6 for pure NiO-YSZ to 27.5 for the specimens with 25 volume % Al_2O_3 added initially. All of the reduced specimens were electrically conductive, indicating that the Ni was percolated even at the lowest volume fraction.

Based on the results of the dilatometry and reduction studies, two anode compositions were selected for tape casting and further analysis. A composition with 24 volume % added Al_2O_3 and 51 volume % NiO was selected in order to match the CTE of anode and electrolyte to minimize cell curvature and stress. Another composition consisting of 17 volume % added Al_2O_3 and 59 volume % NiO was selected in order to reduce the CTE mismatch between anode and electrolyte with respect to the NiO-YSZ anode, while holding the final volume fraction of Ni after reduction constant at 38%.

In order to make large-area plates, specimens were pre-sintered at a range of temperatures to determine sintering shrinkage. Large plates were cut from the tapes with dimensions calculated to result in plates of at least 10 cm per side. The large plates of the 17 and 24 vol.% Al_2O_3 tapes were pre-sintered at 900, 1150, and 1300°C, and plates of NiO-YSZ tape were pre-sintered at each of those temperatures and also at 1200°C. Spray coatings were deposited on the large plates after pre-sintering, and the resulting bi-layers were co-sintered at 1430°C for one hour. It was found that the minimum overall cell curvature and corresponding residual stresses

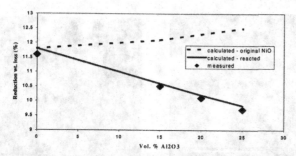

Figure 3. Reduction weight loss of specimens with 51 vol. % NiO as a function of volume % Al₂O₃ (balance YSZ). Dashed curve is calculated based on full reduction of all original NiO. Solid curve is calculated based on reduction of remaining NiO after full reaction of all added Al₂O₃ with NiO to form NiAl₂O₄. Data points are measured weight losses after reduction.

occurred in the plate of 24 vol.% Al₂O₃ that was pre-sintered at 1150°C, while the other plates either cracked or warped substantially. The plate was free of cracks or macroscopic defects, and the center-point deflection of the 10.5 x 10.6 cm plate was 0.36 mm, safely below the target maximum deflection of 0.5 mm desired for minimal fracture during stack assembly. Thus, the plate with the closest thermal expansion match to the electrolyte was found to have the lowest curvature and residual stresses over the full range of pre-sintering conditions and compositions tested.

Reductions were carried out in hydrogen to determine the reduced porosity level and microstructure of the tapes. Tape pieces that had been previously sintered for 1 hour at 1430°C were reduced for 1 hour at 800°C in hydrogen. The resulting microstructures are shown in Figure 4. Figure 4a shows the microstructure of Ni-YSZ, and 4b shows reduced Ni-YSZ-NiAl₂O₄ that started from an initial composition of 24 volume % Al₂O₃. The Ni-YSZ is 58 % dense, while the Ni-YSZ-NiAl₂O₄ is 66 % dense. Although the Ni-YSZ started with 20 weight % of pore former added to the initial ceramic powder while the Ni-YSZ-NiAl₂O₄ had no added starch, the final reduced porosities are closer than expected based on initial starch content alone. Added Al₂O₃ increases the porosity of the final microstructure for a given quantity of initial pore former for similar sintering temperatures and time. Thus, as extra Al₂Oₛ is added to the anode, less pore former is needed in the initial powder formulation.

CONCLUSIONS

By adding Al₂O₃ to the standard NiO-YSZ anode material prior to reduction, the coefficient of thermal expansion mismatch between the anode and electrolyte can be reduced or eliminated. However, reaction of Al₂O₃ with NiO to form NiAl₂O₄ upon high-temperature sintering reduces the overall nickel content of the remaining anode material after reduction, since the NiAl₂O₄ does not reduce to Ni and Al₂O₃ in typical in-situ stack reduction conditions. Additional NiO can be added to the original mixture to retain a constant volume fraction of Ni in the cermet after reduction, creating a trade-off between electrical conductivity of the anode and the stress and curvature resulting from thermal mismatch between the anode and electrolyte layers. It was found that an added Al₂O₃ volume fraction of 24% is sufficient to reduce the

a)   b)

Figure 4. 4a) Microstructure of reduced Ni-YSZ. 4b) Microstructure of reduced Ni-YSZ-NiAl₂O₄.

overall cell curvature to less than 0.5mm deflection for a large-area cell of at least 10 cm on each side, while still retaining a percolated conductive Ni network in the reduced material. Long-term testing is still needed, however, to determine whether the reduced volume fraction of Ni will cause the conductivity to drop over time. However, since microstructure has a strong effect on anode conductivity [6] as well as mechanical behavior, additional adjustments can be made to the final material properties by adjusting factors such as particle size and pore former content to control the overall porosity, and to obtain a more optimized combination of mechanical behavior and electrical conductivity in the anode specimens with added Al_2O_3.

ACKNOWLEDGEMENTS

This work was performed under the auspices of the U. S. Department of Energy by the University of California, Lawrence Livermore National Laboratory under Contract No. W-7405-Eng-48. The authors would like to thank Martin Stratman, Erv See, and Dave Lenz for experimental assistance, and Gregory Woelfel for help with specimen preparation.

REFERENCES

1. G. M. Crosbie, E. P. Murray, D. R. Bauer, H. Kim, S. Park, J. M. Vohs, and R. J. Gorte, J. Soc. Automotive Eng., (2001)-01-2545.
2. N. Q. Minh, Ceramic Fuel Cells, J. Am. Cer. Soc., 76(3), 563-588, (1993).
3. L. Kou and J.R. Selman, J. App. Electrochem., 30, 1433-1437 (2000).
4. E. Ustundag, Z. Zhang, M. L. Stocker, P. Rangaswamy, M.A.M. Bourke, S. Subramanian, K.E.Sickafus, J.A.Roberts, and S.L. Sass, Mat. Sci. & Eng. A238, 50-65 (1997).
5. E. Ustundag, P. Ret, R. Subramanian, R. Dieckmann, S.L. Sass, Mat. Sci. & Eng. A195, 39-50 (1995).
6. D.W. Dees, T.D. Claar, T.E. Easler, D.C. Fee, and F.C. Mrazek, J. Electrochem. Soc. 134(9), 2141-2146 (1987).

Mat. Res. Soc. Symp. Proc. Vol. 756 © 2003 Materials Research Society

Bipolar Plate-Supported Solid Oxide Fuel Cells for Auxiliary Power Units

J. David Carter, Terry A. Cruse, Joong-Myeon Bae, James M. Ralph, Deborah J. Myers,
Romesh Kumar, and Michael Krumpelt
Chemical Technology Division, Argonne National Laboratory
Argonne, IL 60439-4837, U.S.A.

ABSTRACT

This paper presents an advanced design and fabrication concept for a solid oxide fuel cell
(SOFC). The concept is based on a laminate repeat unit comprised of a thin electrolyte, cermet
anode, metallic gas flow fields, and a metallic bipolar plate. The laminate is sintered in a single-
step process in a controlled atmosphere, and the cathode is applied and sintered *in situ* during the
initial heating of the cell (or stack). Observations about the types of cracks that formed in the
electrolyte during the sintering process guided the development of the sintering protocol to yield
the desired product. Cells with power densities exceeding 250 mW/cm^2 have been tested.

INTRODUCTION

Recently, 5 kW SOFCs have been proposed for use in automotive and heavy vehicle
auxiliary power units (APUs) that run on reformed gasoline or diesel [1]. Solid oxide fuel cells
have an advantage over other fuel cell types because of their fuel flexibility, which allows for
simple fuel reforming systems. In APU applications, the SOFC stack must also be able to
withstand mechanical shock and vibration. However, most stack designs are subject to brittle
failure because of their ceramic components. Some researchers have sought to improve the
mechanical properties of the SOFC by supporting cells on metal substrates by using various
fabrication methods [2, 3].

The concept being developed at Argonne, called TuffCell, is shown in Figure 1. It is an
SOFC stack unit consisting of a ceramic electrolyte, cermet anode, metallic flow fields and a
metallic bipolar plate. The brittle ceramic components are bonded to tough metallic layers that
give the laminate improved mechanical strength.

Each functional layer is made by casting oxide or metal slurries into films. The dried films
are laminated and sintered together in a single high-temperature process combining traditional
ceramic and powder metallurgy sintering techniques. The cathode is then applied during the
stack building process and is sintered *in situ* during the initial heating of the stack. Because
layers of the TuffCell are bonded together during sintering, an unbroken electrical current path is
formed between the electrolyte, anode, flow fields and bipolar plate. This approach leaves only a
single electrical contact layer for each repeat unit between the air flow field and the cathode.
Conventional fuel cells have two or more contact areas, which results in a loss of power from
individual cells as they are stacked together.

This fabrication approach also allows flexibility in the composition of each layer. For
example, a multilayered bipolar plate consisting of different alloys in the anode and cathode
environments can easily be incorporated, allowing designers to choose materials best suited to
each environment. In addition, designers can minimize the amount of expensive materials such
as yttria stabilized zirconia (YSZ) and exotic alloy elements by functionally grading the
component layers of the fuel cell, using these materials only where they are needed.

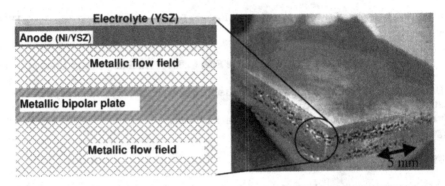

Figure 1. TuffCell stack unit. Essential stack components are integrated into a laminate by sintering in a reducing atmosphere.

The challenge of this fabrication approach was to develop a sintering protocol that allows simultaneous sintering of the ceramic and metal layers, without warping or cracking in the laminate. The sintering protocol was developed by examining cracks that form in the electrolyte during the co-sintering process and adjusting sintering parameters accordingly. In this work, the TuffCells consisted of an YSZ electrolyte, YSZ/nickel (Ni) active anode and anode substrate, 434 stainless steel (SS) gas flow fields, and a 434 SS bipolar plate. Cells that produced dense, flaw-free electrolytes were tested for their electrochemical performance. Power densities greater than 250 mW/cm^2 were obtained.

EXPERIMENTAL PROCEDURE

The desired result of the co-sintering process was to obtain a gas-tight electrolyte and bipolar plate, which are needed for high cell voltage and fuel efficiency. On the other hand, the anode must be porous and possess a good distribution of the nickel and YSZ phase to provide high anodic activity, support the thin electrolyte, and serve as a current collector. The fuel and air flow fields must be open for in-plane gas flow and capable of conducting electricity to the bipolar plate. All of the components must have the same shrinkage rate during processing to prevent curling, warping and cracking. The thermal expansion coefficients of the materials must also be match to prevent internal stresses from forming during thermal cycling.

Fabrication

Tape casting and slurry coating were used to form the component layers and powder metallurgy processing was used to sinter the metallic materials. Slurry formulations were based on an organic solvent-binder system. Polymer spheres, graphite, and coarsened YSZ particles were used to form the pores in the anode, and polymer foam was used to form the gas flow fields. We adjusted the volume fractions of the inorganic solids and the organic components of each layer to match shrinkage during processing.

Slurries were prepared by dispersing powders (YSZ, NiO, or 434 SS) in a xylene/ n-butanol solvent using a compatible dispersant and ball milling for 24 hours. Acrylic binder and a plasticizer were subsequently added to the dispersed slurry and ball milled for at least 4 hours.

The electrolyte slurry was first tape cast on a smooth, flat surface, and subsequent anode layers were cast over it when the previous layer had dried. The gas flow fields were produced by pouring the 434 slurry over reticulated polyurethane foam (60 pores per inch) and squeezing out excess slurry to obtain an open pore structure. The bipolar plate was tape cast, dried, and bonded to the foam flow fields.

Green laminates were cut into 3×3 cm^2 squares and sintered in a controlled-atmosphere furnace. Organic binders were burned out in air at temperatures ranging from 250-400°C. Laminates were then annealed in air to partially oxidize the metallic particles. Afterward, we evacuated the O_2/N_2 atmosphere, introduced a H_2/He atmosphere, and heated the samples to ~1300°C to sinter. Optimized sintering parameters (i.e., flow rates, dwell times, temperatures and ramp rates) are presented in the results and discussion section.

Characterization

When developing the sintering protocol, the electrolyte surfaces were compared from various sintering batches using optical microscopy. Crack patterns on the electrolyte surfaces of unsuccessfully sintered cells were used to characterize the shrinkage mismatch between the layers in the laminate.

Crack-free cells were cemented to a test rig, and a cathode ink was painted onto the electrolyte surface and allowed to dry. Electrochemical cell tests were carried out at 800°C with hydrogen as fuel and air as oxidant. Cell polarization measurements were taken in galvanodynamic mode using a PAR273A potentiostat/galvanostat.

RESULTS AND DISCUSSION

The challenge in processing the laminate was to match the shrinkage of the different layers: electrolyte (YSZ), multi-layer anode (Ni/YSZ), interconnect and flow fields (434 SS). Because nickel oxide was mixed with zirconia in equal amounts, it was assumed that the anode shrinkage behavior roughly matched that of the zirconia. We matched the shrinkage of the 434 SS components with the electrolyte/anode bilayer by partially oxidizing the steel layers in a controlled oxygen flow after the binder burnout process. When heated, the metal began to sinter after its surface oxide was reduced in the hydrogen atmosphere. Under the right conditions, the metals sintering coincided with the zirconia sintering. Optimization of the processing conditions was determined by varying process parameters and examining the sintered laminates with optical microscopy.

The process variables changed during sintering were:

1. Anneal temperature;
2. Anneal time;
3. Oxygen flow during the anneal;
4. Heating rate;
5. Hydrogen flow during heating and sintering;
6. Maximum sintering temperature; and
7. Sintering time at maximum temperature.

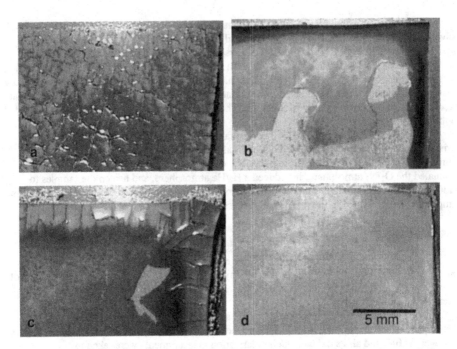

Figure 2. Optical micrographs showing the electrolyte surfaces containing characteristic crack patterns: a) mud cracks, b) delamination, c) combination cracking, and d) dense, crack-free.

The extent of oxidation of the steel particles during the annealing process depended on particle size, particle morphology, porosity, temperature, dwell time, total flow rate, and oxygen partial pressure. Temperature, time, and oxygen partial pressure were chosen as variables.

Under non-optimized conditions, the electrolyte formed cracks characteristic to the type of sintering mismatch. Figure 2 illustrates samples with characteristic crack patterns. The electrolyte in Figure 2a has "mud-cracks". Here, the top layer began to shrink before underlying metal layers. Cracks formed at regular intervals along the edges and moved to the center. The metal was over-oxidized and did not reduce in time to sinter with the YSZ.

Figure 2b shows a case in which the electrolyte sintered after the metal. In this case, the underlying metal layers began to sinter and shrink before the YSZ, causing delamination. In many samples of this type, the pore structure of the foam flow field had over-sintered and collapsed into a dense layer. An intermediate case is observed in Figure 2c. The edges of the electrolyte began to sinter first, forming cracks; then the metal began to sinter, putting the electrolyte surface in compression and causing some delamination. It was necessary to adjust slurry compositions so that the overall shrinkages of the electrolyte and metal were matched. When shrinkages of the electrolyte and metal coincided, the laminates were well-bonded, electrolytes were crack-free and dense (Figure 2d) and gas flow fields had an open structure.

Comparison of the variables of a given sintering run to its corresponding micrograph further validates the explanation of the crack patterns. Figure 3 compares the heating profiles

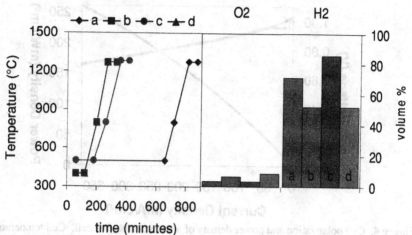

Figure 3. Heating profile (lines) and volume percent (bars) of flowing O_2 (bal. N_2) and H_2 (bal. He) used to sinter the corresponding samples shown in Figure 2.

(lines) and volume percentages (bars) of flowing O_2 and H_2 used in the processing of the corresponding samples in Figure 2.

For example, sample "a" was annealed for 10 h at 500°C in 3 vol. % O_2 (bal. N_2); the atmosphere was then changed to 71% H_2 (bal. He) and the sample was heated to 1275°C at a rate of 300°C/h. The sintering dwell time was 60 minutes. In this case mud cracks formed throughout the surface of the electrolyte because the metal was over-oxidized. Sample "c"had the same profile except for, a different anneal time and more hydrogen during sintering, formed an intermediate crack structure. Samples "b" and "d" had the same sintering profile and hydrogen atmosphere. Yet "d" was slightly more oxidized than "b" because of the increased O_2, allowing the shrinkage of the electrolyte and the metal to better match one another.

Figure 4 shows the polarization and power density of a dense, crack-free cell. The maximum power density of the cell was greater than 250 mA/cm². Because of the cell geometry, it is difficult to pinpoint the exact cause of the lower power density. However, there is some indication from previous work [4] that the anode microstructure needs to be improved to increase the power density of the cell. Unlike the standard approach of sintering SOFCs in air and reducing the NiO of the anode *in situ*, the present anode is reduced during sintering, forming Ni metal. Although the air-sintered anode produces immediate results in higher power density, it is subject to long-term degradation unless a stable microstructure is "built-in" with the anode design. The process discussed here has the potential to form a stable anode microstructure, because the microstructure is fixed at a high temperature in a reducing atmosphere during the sintering process. The immediate challenge is to improve the activity of the anode produced under these sintering conditions.

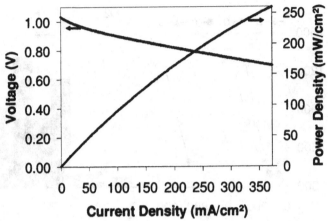

Figure 4. Cell polarization and power density of a dense, crack-free cell. Cell temperature: 800°C, dry hydrogen fuel and air oxidant.

CONCLUSIONS

This paper describes a new design and fabrication process for an SOFC. Argonne's TuffCell, which consists of a thin YSZ electrolyte, Ni/YSZ anode, and 434 SS gas flow fields and bipolar plate. The co-sintering process was developed by examining crack patterns in the electrolyte after sintering and the following variables were used to control the sintering process: temperature, dwell time, heating rate, and O_2 and H_2 volume flow rates. Our work resulted in well bonded laminates with dense, crack-free electrolytes were obtained. Cell measurements in hydrogen/air at 800°C produced a maximum power exceeding 250 mA/cm^2.

ACKNOWLEDGEMENT

This research was supported by the U.S. Department of Energy, Office of Hydrogen, Fuel Cells, and Infrastructure Technology.

REFERENCES

[1] J. Zizelman, S. Shaffer and S. Mukerjee, *SAE Technical Paper Series*, 2002-01-0411, (SAE International, Warrenville, PA, 2002).

[2] T. Kato, S. Wang, N. Iwashita, T. Honda, T. Kaneko, A. Negishi, S. Nagata, and K. Nozaki, in Proc. 7th Int. Symp. Solid Oxide Fuel Cells, edited by H. Yokokawa and S.C. Singhal (Electrochem. Soc., Pennington, NJ, 2001), pp. 1065-1072.

[3] G. Schiller, T. Franco, R. Henne, M. Lang, R. Ruckdäschel, P. Otschik and K. Eichler, in Proc. 7th Int. Symp. Solid Oxide Fuel Cells, edited by H. Yokokawa and S.C. Singhal (Electrochem. Soc., Pennington, NJ, 2001), pp. 885-894.

[4] J.D. Carter, J.M. Ralph, J.-M. Bae, T.A. Cruse, C.C. Rossignol, M. Krumpelt, and R. Kumar in *Program and Abstracts of the 2002 Fuel Cell Seminar*, (Courtesy Associates, Washington, D.C., 2002) pp 874-877.

Mat. Res. Soc. Symp. Proc. Vol. 756 © 2003 Materials Research Society

The Effect of Composition on the Wetting Behavior and Joint Strength of the Ag-CuO Reactive Air Braze

K. Scott Weil, Chris A. Coyle, Jin Yong Kim, and John S. Hardy
Department of Materials Science, Pacific Northwest National Laboratory,
Richland, WA 99352, U.S.A.

ABSTRACT

As interest in high temperature electrochemical membrane devices for energy and gas generation has intensified, it has become apparent that developing an appropriate method of hermetically joining the ceramic and metallic components in these devices will be critical to their success. A recently developed technique referred to as reactive air brazing (RAB) has shown promise in the joining of components for planar solid oxide fuel cells (pSOFC) and oxygen generators. In the study described below, the relationship between braze composition, substrate wetting, and joint strenth was investigated to gain further understanding of the RAB process. It was found that braze wettability and joint strength are inversely related for the simple binary Ag-CuO braze system.

INTRODUCTION

A solid state electrochemical device such as a pSOFC functions because of an oxygen ion gradient that develops across the yttria stabilized zirconia (YSZ) electrolyte membrane under ionic transport. To maintain this gradient, and thereby maximize the performance of the device, the electrolyte and the joint that seals this membrane to the device chassis must be hermetic. That is, the YSZ layer must be dense, must not contain interconnected porosity, and must be connected to the rest of the device structure with a high temperature, gas-tight seal. Recent efforts on the tape casting of thin, anode-supported ceramic bilayers have successfully addressed the first two issues [1], the remaining challenge is in joining the ~10μm thick electrochemically active YSZ electrolyte to the metallic structural component such that the resulting seal is hermetic, rugged, and stable under both thermal cycling and continuous high temperature operation.

The RAB technique was recently developed as method of joining complex oxides to heat resistant metals for use in fabricating high temperature electrochemical devices such as pSOFCs and oxygen generators [2,3]. RAB differs from traditional ceramic-to-metal joining techniques such as active metal brazing and the Mo-Mn process in two important ways: 1) RAB utilizes a liquid-phase oxide-metal melt as the basis for joining and 2) the process is conducted in an air muffle furnace. The latter difference appears to be critical in electrochemical device fabrication. Exposure of the device to a reducing atmosphere at a temperature greater than ~800°C, typical processing conditions in active metal brazing, is too demanding for many of the complex oxide materials used in these devices. For example, many of the mixed ionic/electronic conducting perovskites that are employed as electrodes will reduce under these heat treatment conditions and may undergo irreversible deterioration via phase separation, causing severe degradation in device performance. An additional advantage of air brazing is that the capital expenses and operating costs associated with heat treatment in air are a fraction of the costs for comparable vacuum or reducing gas heat treatments.

Since joining is conducted in an oxidizing atmosphere, bonding occurs between the functional ceramic component and an oxide scale that grows on the structural metallic component during brazing. The goal in RAB is to reactively modify one or both oxide faying surfaces with an oxide compound dissolved in a molten noble metal alloy such that the newly formed surface is readily wetted by the remaining liquid filler material. Potentially, there are a number of metal oxide-noble metal systems that can be considered for RAB, including Ag-CuO, Ag-V_2O_5, and Pt-Nb_2O_5 [4]. In our research, we have focused on the Ag-CuO system, investigating its potential use in brazing high temperature ceramics such as alumina and lanthanum strontium cobalt ferrite [5,6] and in joining heat resistant metals to electro-chemically active ceramics [7]. Our interest in the present study is in examining the relationship between RAB braze composition, substrate wetting, and joint strength with respect to eventual application in a pSOFC stack.

EXPERIMENTAL

Anode-supported bilayers, consisting of NiO-5YSZ as the anode and 5YSZ as the electrolyte, and thin gauge FeCrAlY (Fe, 22% Cr, 5% Al, 0.2% Y) were employed as the model SOFC electrolyte membrane/structural metal system in this study. Bilayer coupons were fabricated by traditional tape casting and co-sintering techniques [1] and measured nominally 600μm in thickness, with an average electrolyte thickness of ~8μm. Prior to their use in the wetting and joining experiments, the samples were cleaned with acetone and ethanol and dried in air at 300°C for 1hr. As-received 12mil thick FeCrAlY sheet was sheared into 2cm square pieces for the wetting experiments, polished lightly on both sides with 1200 grit SiC paper, and ultrasonically cleaned in acetone for 10 minutes. Flat washer-shaped specimens measuring 4.4cm in diameter with a concentric 1.5cm diameter hole were punched from the 12mil sheet and the surfaces of these were polished and cleaned in the same manner prior to use in the joining experiments.

The braze pellets employed in the wetting experiments were fabricated by mixing and cold pressing copper (10μm average particle diameter; Alfa Aesar) and silver (5.5μm average particle diameter; Alfa Aesar) powders in the appropriate ratios to yield the target compositions given in Table I. The copper powder was allowed to oxidize in-situ during heating in air to form CuO. The wetting experiments were conducted in a static air box furnace fitted with a quartz door through which the heated specimen could be observed. A high speed video camera equipped with a zoom lens was used to record the melting and wetting behavior of the braze pellet on a given substrate. The experiments were performed by heating the samples at 30°C/min to 900°C, where the temperature remained for fifteen minutes, followed by heating at 10°C/min to a series of set points and fifteen minute holds. In this way, the contact angle between the braze and substrate was allowed to stabilize for measurement at several different soak temperatures during one heating cycle: 900°C, 950°C, 1000°C, 1050°C, and 1100°C. Select frames from the videotape were converted to computer images, from which the wetting angle between the braze and substrate could be measured and correlated with the temperature log for the heating run.

Table I. Target compositions of the brazes investigated in this study

Braze ID	Ag Content (in mole%)	CuO Content (in mole%)
Ag-69Cu	31	69
Ag-34Cu	66	34
Ag-8Cu	98	2
Ag-4Cu	99	1

Joining samples were prepared by placing a piece of previously fabricated Ag-Cu braze foil between a FeCrAlY washer specimen and the YSZ side of a 2.5cm diameter bilayer coupon, heating the combination in air to 1050°C, and holding at this temperature for 30min. Again, the copper oxidizes in-situ to form CuO. 10°C/min heating and cooling rates were employed during brazing. The braze foil was synthesized by diffusion bonding copper and silver foils of the same areal dimension, but with thicknesses appropriate to achieve the target braze compositions listed in Table I. Diffusion bonding was conducted in an Ar/4% H_2 cover gas at 720°C for 10hrs under a static load of ~½ psi after which the foil was rolled to a thickness of 0.07mm and cut into 2mm wide strips for use in joining.

A rupture strength test was developed to facilitate quantitative comparison of the RAB seal joint strengths. Figure 1 shows a photo of the test specimens, prior to and after joining, and a schematic of the rupture test equipment. The joining sample is clamped into the test fixture and the air pressure behind the joined bi-layer disk is slowly increased until the seal or the disk breaks. A series of six specimens was tested for each joining condition. Microstructural analysis of the wetting and joining specimens was performed on polished cross-sectioned samples using a JEOL JSM-5900LV scanning electron microscope (SEM) equipped with an Oxford windowless energy dispersive X-ray analysis (EDX) system.

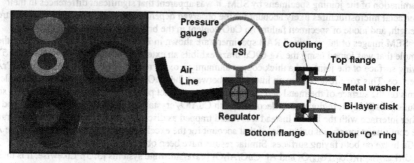

Figure 1. Far left: Components of rupture strength test specimen; from bottom, FeCrAlY washer, braze foil ring, bi-layer disk. Middle: Assembled test specimen. Right: Rupture strength test schematic.

RESULTS AND DISCUSSION

Plotted in Figure 2 are the contact angles that each braze forms on 5YSZ and fecralloy in air at 1050°C and the rupture strength of the corresponding joint. The results show that rupture strength

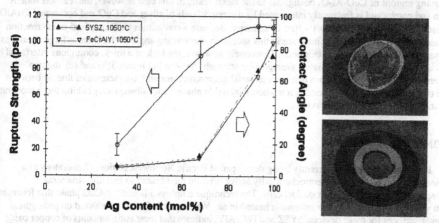

Figure 2. Left: Room temperature rupture strength and contact angle as a function of Ag content. Right: The Ag-69Cu (top) and Ag-2Cu (bottom) joining samples after rupture testing.

improves significantly with increasing Ag content, whereas wettability, which is the inverse of contact angle, moves in the opposite direction. Thus, the specimen joined using the braze containing the highest level of CuO (Ag-69Cu) displays excellent wetting on 5YSZ and fecralloy, but has an average rupture strength of only 23 psi while the specimen brazed with the lowest CuO-containing alloy (Ag-1Cu) exhibits moderate-to-poor wetting on the two substrates but displays greater than a four-fold increase in rupture strength of 109 psi. In fact, the Ag-1Cu seal does not fail during testing but instead the bilayer disc ruptures leaving behind a ring of ceramic still joined to the metal washer as shown in Figure 2. The Ag-69Cu braze specimen on the other hand fails along the interface between the FeCrAlY and the braze. This inverse relationship between braze wettability and joint strength is not commonly observed in ceramic-to-metal brazing. Typically, improvements in wetting lead to increased strength. Upon examination of the joining specimens by SEM, it was apparent that significant differences in their interfacial microstructures likely account for the observed dependences of wetting behavior, rupture strength, and mode of specimen failure on CuO content in the braze.

SEM images of the as-joined RAB specimens are shown in Figure 3. As seen in Figure 3(a), the sample that was joined using the Ag-69Cu braze exhibits an extensive CuO phase that covers the 5YSZ faying surface of the joint and a thick copper alumunium oxide reaction zone along the braze/FeCrAlY interface. This zone results from an interaction between the CuO in the braze and the alumina scale that forms on the surface of the metal and gives rise to two product phases: a continuous mixed-oxide solid solution $CuO-Al_2O_3$ region that is decorated with $CuAlO_2$ crystallites. Very little silver is observed along either interface with the braze. Instead it is found trapped as discrete particles in the bulk of the braze. It is the continuous interfacial oxide phases that account for the excellent wetting characteristics of the Ag-69Cu braze on both faying surfaces. Similar results have been observed in the Ag-$CuO/(La_{0.6}Sr_{0.4})(Co_{0.2}Fe_{0.8})O_3$ and Ag-CuO/Al_2O_3 braze/substrate systems [3, 5]. Likewise, it is these brittle phases, which exhibit poor coefficient of thermal expansion matching with the 5YSZ and FeCrAlY, that contribute to the low rupture strengths observed in the corresponding joining specimens.

At the other extreme, the braze/5YSZ interface in the Ag-1Cu specimen shown in Figure 3(d) is covered by a nearly continuous layer of pure silver, with discrete micron-size CuO particles occasionally found on the 5YSZ faying surface. The silver acts essentially as a matrix for the CuO particulate found in the bulk of the braze as well. Along the FeCrAlY interface in this joining specimen is a continuous ½ - 1μm thick alumina scale, as was observed in the Ag-69Cu specimen, that includes a small amount of iron and chromium, ~5 mol% and 3 mol% respectively. Again, an apparent alloying reaction takes place forming regions of $CuO-Al_2O_3$ contiguous to the metal scale. In this case however, the reaction zone is thin and patchy and is frequently interrupted by discrete islands of silver and CuO and occasional $CuAlO_2$ crystallites measuring roughly 1 - 3μm in diameter. Because a continuous CuO or mixed-oxide reaction phase does not form on either interface in this specimen, the wetting angle of the braze is larger than was observed in the high CuO content braze. Conversely however, the lack of a thick, continuous interfacial oxide phase leads to a much higher average joint strength. As seen in Figures 3(b) and (c), the two intermediate braze compositions display interfacial microstructures with characteristics that lie between the two extremes, i.e. thinner or discrete interfacial oxide phases, and subsequently exhibit more moderate wetting and joint strength behaviors.

CONCLUSIONS

Reaction air brazing is currently being developed at Pacific Northwest National Laboratory as a potential method of joining electrochemically active ceramic membranes to metallic structural components for use in solid oxide fuel cells. The technique employs a liquid Ag-CuO phase that forms in-situ when a solid silver-copper mixture is heated in air. Wetting experiments conducted on protoypical faying surfaces for these devices, 5YSZ and FeCrAlY, indicate that increasing amounts of copper oxide in the braze significantly improve its wetting behavior on both substrates. In general, the observed decrease in contact angle is related to increasing coverage of the faying surface by CuO, in the case of

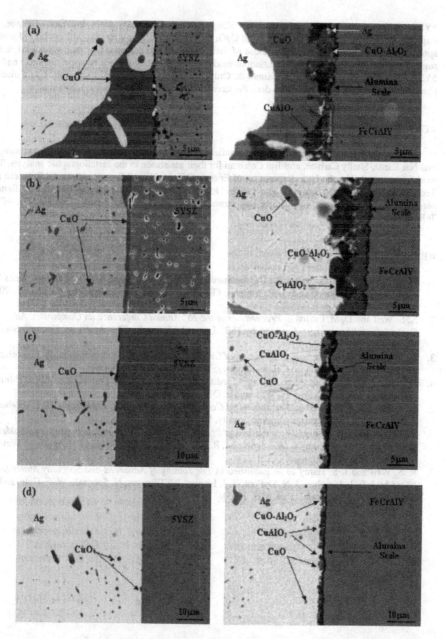

Figure 3. Cross-sectional SEM micrographs of braze/5YSZ interfaces (on the left) and braze/FeCrAlY interface (on the right) of: (a) Ag-69Cu, (b) Ag-31Cu, (c) Ag-2Cu, and (d) Ag-1Cu.

5YSZ, or by a CuO-Al$_2$O$_3$ reaction zone along the braze/FeCrAlY scale interface. Despite the improvement in wetting due to the addition of CuO, the resulting rupture strengths of these brazed joining specimens is significantly poorer than those of their low-CuO counterparts. Our microstructural results suggest that the presence of a thick, continuous oxide phase along the interface between the braze and the 5YSZ or the FeCrAlY degrades joint strength. Only when the CuO or mixed-oxide forms as a discrete phase along these respective interfaces does the strength of the RAB joint increase substantially.

ACKNOWLEDGMENTS

The authors would like to thank Nathan Canfield for his preparation of the ceramic bilayer samples and Nat Saenz, Shelly Carlson, and Jim Coleman for their assistance in the metallographic analysis. This work was supported by the U. S. Department of Energy, Office of Fossil Energy, Advanced Research and Technology Development Program. The Pacific Northwest National Laboratory is operated by Battelle Memorial Institute for the United States Department of Energy (U.S. DOE) under Contract DE-AC06-76RLO 1830.

REFERENCES

1. S. P. Simner, J. W. Stevenson, K. D. Meinhardt, and N. L. Canfield in *Solid Oxide Fuel Cells VII*, edited by H. Yokokawa and S. C. Singhal, (The Electrochemical Society, Pennington, NJ, 2001) pp.1051.
2. K. S. Weil and D. M. Paxton in *Proceedings of the 26th Annual Conference on Composites, Advanced Ceramics, and Structures: A*, edited by H-T. Lin and M. Singh, (The American Ceramic Society, Westerville, OH, 2002).
3. K. S. Weil and J. S. Hardy in *The Joining of Ceramics*, edited by C. A. Lewinsohn, R. E. Loehman, and M. Singh, (The American Ceramic Society, Westerville, OH, 2002).
4. *Phase Diagrams for Ceramists, Volume VI*, edited by R. S. Roth, J. R. Dennis, and H.F. McMurdie, (The American Ceramic Society, Westerville, OH, 1987).
5. J. Y. Kim and K. S. Weil in *The Joining of Ceramics*, edited by C. A. Lewinsohn, R. E. Loehman, and M. Singh, (The American Ceramic Society, Westerville, OH, 2002).
6. K. S. Weil and J. S. Hardy to be published in *Proceedings of the International Brazing and Soldering Conference 2003*, edited by C. A. Lewinsohn, R. E. Loehman, and M. Singh, (The American Welding Society, Miami, FL, 2003).
7. K. S. Weil and J. S. Hardy to be published in *The Joining of Advanced and Specialty Materials V*, edited by J. E. Indacochea, J. N. DuPont, T. J. Lienert, W. Tillmann, N. Sobczak, W. F. Gale, and M. Singh, (The ASM, Materials Park, OH, 2003).

Metal Dusting Problem with Metallic Interconnects for Solid Oxide Fuel Cell

Z. ZENG and K. NATESAN
Argonne National Laboratory
Energy Technology Division
Argonne, IL 60439, USA

ABSTRACT

Metallic interconnects in the solid oxide fuel cell (SOFC) are oxidized on the cathode side by air and carburized on the anode side by natural gas. Metallic alloys can be attacked by metal dusting corrosion in carbonaceous gases of high carbon activity in the temperature range of 350-1000°C. Under these conditions, pits form on the alloy surface and can become large holes through the alloy plate, with subsequent disintegration into a powdery mixture composed of carbon, fine particles of metal, and carbide. Fe and Ni-base alloys were tested in carbonaceous gases around the SOFC operating temperature. It was found that the oxide scales on the alloy surface prevent metal dusting corrosion. If the major phase in the oxide scale is chromic oxide, the alloys have good resistance to metal dusting corrosion. However, the alloys are easily attacked if the major phase is spinel.

INTRODUCTION

Recently, significant progress on reducing the operating temperature of an SOFC has been achieved by reducing the thickness of the electrolyte [1]. The lower operating temperature (550~800°C vs. 1000°C) makes it possible to consider high-temperature alloys as interconnect candidates. The advantage is that a metallic alloy has lower cost, higher electrical conductivity, and higher heat conductivity and is easier to fabricate than the ceramic interconnect presently used. Therefore, research on the metallic interconnect has attracted much attention recently [2,3].

Metallic interconnect acts as a gas separator and distributor; therefore, it works in a dual-gas environment: it faces air on one side, and natural gas on the other. On the side with air, metallic interconnect needs to be protected from oxidation. On the other side with natural gas, the interconnect needs to be protected from the carburizing gas. The metallic interconnect could be carburized at the working temperature of 550~800°C. Moreover, another catastrophic carburization phenomenon, "metal dusting" could occur in an atmosphere containing carbon [4,5]. Pits form on alloy surfaces during the metal dusting corrosion. Metal dusting occurs in environments with carbon activity (a_c)>1. Gaseous reactions that lead to or cause metal dusting are:

$$CO+H_2=C+H_2O \qquad a_c=K_1 \bullet P_{CO} \bullet P_{H2}/P_{H2O} \qquad (1)$$
$$2CO=C+CO_2 \qquad a_c=K_2 \bullet P^2_{CO}/P_{CO2} \qquad (2)$$
$$CH_4=C+2H_2 \qquad a_c=K_3 \bullet P_{CH4}/P^2_{H2} \qquad (3)$$

Inlet gases in the SOFC could be natural gas (mainly methane) or reformed gases containing H_2, CO, CO_2, and H_2O. The composition of the fuel flow gradually changes from high concentration methane or H_2 at the gas inlet to its oxidation products H_2+H_2O and $CO+CO_2$. The humidity is low at the fuel cell inlet and high at the outlet. The reverse is true for carbon activity

For the SOFC-gas turbine system, the SOFC would work at high pressure. Carbon activities increase with increasing pressure for reactions 1 and 2. However, a_c decreases with increasing pressure for reaction 3. To further investigate the corrosion of metallic interconnects, we exposed 20 alloys in different gas compositions and pressure.

EXPERIMENTAL

Thirteen Fe-base and 7 Ni-base alloys were tested in carburizating atmosphere. Their compositions are given in Table 1. Metal dusting tests were conducted in a horizontal furnace with a quartz tube (38 mm in diameter). Specimens were hung in an alumina boat and exposed to a flowing carburizing atmosphere at 593°C. Three gas compositions were used for this study. Gas 1 is a simulation of a reformer outlet gas consisting of 52% H_2-5.6% CO_2-18% CO-1.1%CH_4-23% H_2O. The carbon activity of the gas is ~7.9 at 593°C. Gas 2 consisted of 66.2% H_2-7.1% CO_2-23% CO-1.4%CH_4-2.3% H_2O. Its steam content is only 1/10 of the water concentration of Gas 1, but the other constituent gases have similar relative compositions as in Gas 1. Gas 3 consisted of 79.83% H_2-0.018%CO_2-18.3%CO-1.9%H_2O. In addition to the test at 1 atm, the alloys were exposed to Gas 1 at 593°C and 15 atm. After each exposure period, specimen weight change was determined after removal of adhering coke in an ultrasonic bath with acetone. The microstructure of each sample was examined with a JSM-6400 scanning electron microscope. To study the metallographic cross section, the samples were electrolytically etched with 10% acetic acid at 10 V for 30 sec. Raman spectra were excited with 60 mW of 476-nm radiation from a Kr-ion laser. The incident beam impinged on the sample at an angle close to 45° from the normal. Scattered light was collected along the surface normal with an f/1.4 lens. The scattered light was analyzed with a triple Jobin-Yvon grating spectrometer and detected with a CCD detector from Princeton Instruments. All of our spectra were acquired in 300 sec at room temperature.

RESULTS AND DISCUSSION

Effect of alloy composition on metal dusting rate

The metal loss rates of the alloys are given in Table 2. Alloys with Al and Si are not attacked by metal dusting in 1000 h. Low Cr alloys such as T22 and T91 lose weight rapidly. Increasing Cr content in alloys decreases the metal dusting rate. Ni-base alloys perform better than Fe-base alloys. The Ni-base alloys are not attacked by metal dusting after exposure at 593°C for 1000 h in carburizing gas 2. Figure 1 shows metal dusting pits that formed in Alloy 321 and 800 after 1000 h in gas 2. A pit with 380 µm depth was grown in Alloy 321 after 1000 h. Oxide layers on the surface of Alloy 800 were observed by SEM. However, the oxide layer was not present in the pit area. Raman spectra also showed Cr_2O_3 and $Fe_{1+x}Cr_{2-x}O_4$ spinel phases in the area without pits (Fig. 2), and no oxides in the pit area. This result indicates that the oxide layer protects alloys from metal dusting corrosion. When the oxide layer is removed, metals are directly exposed to carburizing gas and are attacked by metal dusting. Therefore, a dense, adhesive oxide layer is important to prevent an alloy from undergoing metal dusting.

Figure 3 shows that Raman peak intensities of Cr_2O_3 increase with increasing Cr content in alloys. Meanwhile, metal dusting rates decrease with increasing Cr content in alloys. Therefore, Cr_2O_3 may be a better phase than spinel to protect alloys from metal dusting.

Table 1. Composition (in wt%) of alloys selected for metal dusting experiments

Alloys	Fe	Ni	Cr	C	Mn	Si	Mo	Al	Others
T22	95.4		2.3	0.2	0.6	0.5	1	---	---
T91	89	0.1	8.6	0.08	0.5	0.4	1	---	N 0.05, Nb 0.07, V 0.2
153MA	69.7	9.5	18.4	0.05	0.6	1.4	0.2	---	N 0.15, Ce 0.04
253MA	65.4	10.9	20.9	0.09	0.6	1.6	0.3	---	N 0.19, Ce0.04
321L	70.7	9.3	17.4	0.02	1.8	0.5	---	---	N 0.02, Ti 0.3
321	70.4	10.3	17.3	0.04	1.2	0.4	---	---	N 0.01, Ti 0.4
310	52.6	19.5	25.5	0.03	1.7	0.7	---	---	---
800	45.9	31.7	20.1	0.08	1.0	0.2	0.3	0.4	Ti 0.31
803	34.8	36.6	25.6	0.08	0.9	0.7	0.2	0.5	Ti 0.6
38815	63.2	15.3	13.9	0.01	0.6	5.8	1.0	0.2	---
MA956	74.4	---.	20.0	---	---	---	---	4.5	Ti 0.5, Y_2O_3 0.6
APMT	69.9	---	21.7	0.04	0.1	0.6	2.8	4.9	---
4C54	71.4	0.3	26.7	0.17	0.7	0.5	---	---	N 0.19
600	9.7	74.6	15.4	0.04	0.2	0.1	---	---	
601	14.5	61.8	21.9	0.03	0.2	0.2	0.1	1.4	Ti 0.3, Nb 0.1
690	10.2	61.4	27.2	0.01	0.2	0.1	0.1	0.2	Ti 0.3
617	0.9	0.9	53.6	21.6	0.1	0.1	9.5	1.2	Co 12.5, Ti 0.3
214	2.5	77	15.9	0.04	0.2	0.1	0.5	3.7	Zr 0.01, Y 0.006
602CA	9.3	62.6	25.1	0.19	0.1	0.1		2.3	Ti 0.13, Zr 0.19, T 0.09
230	1.2	60.4	21.7	0.11	0.5	0.4	1.4	0.3	W 14, La 0.015

Fig. 1. SEM micrograph of metallographic cross section of alloys 321 (left) and 800 (right) after exposure in Gas 2 (66.2% H_2-7.1% CO_2-23% CO-1.4%CH_4-2.3% H_2O) at 593°C for 1000 h.

Table 2. Weight loss data for alloys after exposure at 593°C for 1000 h in a carburizing gas mixture of Gas 2 (66.2% H_2-7.1% CO_2-23% CO-1.4% CH_4-2.3% H_2O)

Alloys	Mass loss (mg/cm²h)	Visual examination
T22	0.5	Heavy carbon deposition
T91	0.066	Pits
153MA	0	Clean surface
253MA	0	Clean surface
321L	2.2×10^{-4}	Pits
310	0	Clean surface
800	0.045	Pits
803	0	Pits
38815	0	Clean surface
MA956	0	Clean surface
321	3.8×10^{-3}	Pits
APMT	0	Clean surface
4C54	0	Clean surface
600	0	Clean surface
601	0	Clean surface
690	0	Clean surface
617	0	Clean surface
214	0	Clean surface
602CA	0	Clean surface
230	0	Clean surface

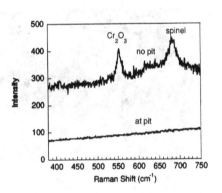

Fig. 2. Raman spectra of Alloy 800 at pit and no pit areas.

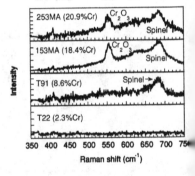

Fig. 3. Raman spectra of Alloys T22, T91, 153MA, and 253MA after exposure in Gas 2 (66.2% H_2-7.1% CO_2-23% CO-1.4%CH_4-2.3% H_2O) for 1000 h.

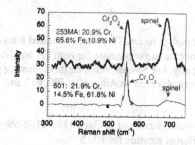

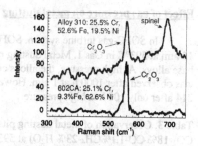

Fig. 4. Alloy 253MA and 601 exposed in
Gas 3 for 1000h at 593°C

Fig. 5. Alloy 310 and 602CA exposed in
Gas 3 for 1000h at 593°C

Figures 4 and 5 show the difference of Raman spectra for two pairs of alloys: Alloy 253MA and 601, and Alloy 310 and 602CA. These alloys were exposed to Gas 3 at 593°C for 1000h. The Cr contents in Alloy 253MA and 601 are close. However, Fe-base alloy 253MA has a much stronger spinel peak than that of Ni-base alloy 601CA. Pits were observed on Alloy 253MA, but not on Alloy 601CA at the same experimental condition. The Cr contents in Alloy 310 and 602CA are also close. Figure 5 shows the strong spinel peak for the Fe-base Alloy 310, but almost no suck peak for the Ni-base Alloy 602CA. Pits were again observed only on Alloy 310, but not on Alloy 602CA. Less spinel in the oxide scale of Ni-base alloys may be a possible reason that the performance of Fe-base alloys is inferior to Ni-base alloys.

Effect of gas composition on metal dusting

Since gas composition in SOFC changes from low to high humidity, we compare the metal dusting corrosion of alloys in two gases with different humidity. The steam content of Gas 2 is only 1/10 of the water concentration of Gas 1, but the other constituents are similar for the two gases. Also, the carbon activities of both gases are larger than 1 at 593°C. Figure 6 shows that pits were observed in Alloy T91, 321, 800, 321L and 803 when exposed in low water concentration Gas 2 for 1000h. In contrast, no metal dusting pits were observed when these alloys were exposed to Gas 1 (high water concentration) for the same period. Therefore, high humidity retards metal dusting.

Fig. 6. Alloys exposed in Gas 1
(52% H_2-5.6% CO_2-18% CO
-1.1%CH_4-23% H_2O) and Gas 2
(66.2% H_2-7.1% CO_2-23% CO
-1.4%CH_4-2.3% H_2O) at 593°C
for 1000h.

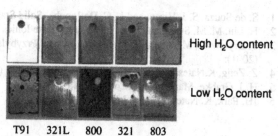

High H_2O content

Low H_2O content

T91   321L   800   321   803

Effect of gas pressure on metal dusting

In SOFC-gas turbine system, SOFC works at high pressure. Therefore, we tested alloys at 15 atm at 593°C in Gas 1. Metal dusting rates are sensitive to gas pressure for both Fe- and Ni-base alloys. At high pressure, the alloys corroded rapidly. For Ni-base alloy, no metal dusting pit was observed after 1000 h at 1 atm. However, pits were observed on Alloy 601, 690, 617 and 214 after only 100 h at high pressure.

Table 3. Observation of metal dusting pits on Ni-base alloys exposed in Gas 1 (52% H_2-5.6% CO_2-18% CO-1.1%CH_4-23% H_2O) at 593°C and different pressure for 100 h

Gas pressure	600	601	690	617	602CA	214	230
1 atm	Clean	Clean	Clean	Clean	Clean	Clean	Clean
15 atm	Clean	pits	pits	pits	Clean	pits	Clean

CONCLUSION

Metallic interconnects could be attacked by metal dusting in SOFC environment. Al and Si addition in alloys are helpful to prevent alloys from metal dusting. Al_2O_3 and SiO_2 may act as protective oxide layer to prevent alloys from metal dusting. However, the electric resistances of those oxide layers were reported to be too high for SOFC application. Electric resistance of Cr_2O_3 is low at high temperature [6]. Oxide scale of Cr_2O_3 is also good to protect alloys from metal dusting corrosion. Whereas, spinel in oxide scale is not as good as Cr_2O_3. Ni-base alloys have less spinel and perform better than Fe-base alloys. Metal dusting corrosion becomes severe in low humidity and high pressure environment.

AKNOWLEDGEMENTS

This work is supported by the U.S. Department of Energy, Office of Industrial Technologies, under Contract W-31-109-Eng-38. The authors thank D. L. Rink for his assistance in conducting the metal dusting experiments.

REFERENCES

1. S. de Souza, S. J. Visco, and L. C. De Jonghe, Solid State Ionics **98**, 57 (1997).
2. H. Liu, M. M. Stack, and S. B. Lyon, Solid State Ionics **109**, 247 (1998).
3. T. Brylewski, M. Nanko, T. Maruyama, K. and Przybylski, Solid State Ionics **143**, 131 (2001).
4. Z. Zeng, K.Natesan, V. A. Maroni, Oxid. Met. 58, 147 (2002).
5. H. J. Grabke, Materials and Corrosion **49**, 303 (1998).
6. JH. Park, K. Natesan, Oxid. Met. **33** 31 (1990)

Mat. Res. Soc. Symp. Proc. Vol. 756 © 2003 Materials Research Society FF4.10

A Performance Based, Multi-Process Cost Model for Solid Oxide Fuel Cells

Heather Benson-Woodward, Mark Koslowske, Randolph Kirchain[1], and Isa Bar-On.
Mechanical Engineering Department, Worcester Polytechnic Institute
Worcester, MA 01609, USA.
[1]Department of Materials Science and Engineering and Engineering Systems Division,
Massachusetts Institute of Technology, Cambridge, MA 02139, USA.

ABSTRACT

Cost effective high volume manufacture (HVM) is a major challenge to the success of solid oxide fuel cells (SOFCs). More than fifteen processing methods have been reported in the literature many of which could be used in various combinations to create the desired product characteristics. Modeling tools are needed to aid in the selection of the appropriate process combination prior to making expensive investment decisions.

This paper describes the development of a multi-process cost model that permits the comparison of manufacturing cost for different processing combinations and various materials. Two specific processing methods are discussed, tape casting and screen printing.

The results are compared with data and experience from the fuel cell and electronic packaging industries. Initial comparisons show good agreement with this experience base. Sensitivity of manufacturing costs to SOFC performance requirements such as maximum power density and operation temperature is investigated.

INTRODUCTION

The success of SOFC technology depends on producing a cost competitive product within performance specifications. Several materials/process technologies have been proposed for the electrolyte in SOFCs [1,2]. The designer will need an analytical tool to select between technology routes to meet the performance and cost goals for these devices.

This project describes a multi-process cost model that takes required performance data (maximum power density, P_{max}, and operating temperature) and maps it to cost regimes for HVM conditions. The model consists of three parts: i) a device performance model that calculates the required electrolyte thickness based on materials properties; ii) process tolerance models deriving from processing experience, and iii) a process based cost model that uses results from i and ii as inputs

In the absence of HVM data for SOFCs the cost model is validated in two ways. First, the model predictions are compared with SOFC cost model data from the literature. Second, the cost model is used to estimate the cost of commercially available alumina layers. Finally, the sensitivity of the cost results to variations in some of the model parameters is presented. This paper reports only a preliminary effort in this direction. It is expected that the model will be refined with the availability of further data.

BACKGROUND

Two planar SOFC architectures are currently being investigated: anode supported and electrolyte supported stack geometries. In the anode supported architecture the 0.5 to 1 mm thick anode is tape-cast from a Nickel Cermet material. The electrolyte layer made of Yttria Stabilized Zirconia (YSZ) is either tape-cast or screen printed onto the anode. The electrolyte thickness can vary from 10 to 150 μm. The cathode consists of an approximately 50 μm thick Lanthanum Strontium Manganese (LSM) oxide layer. For the electrolyte supported architecture the electrolyte is more than 150 μm thick and the anode is correspondingly thinner. This paper focuses on anode supported architectures.

Recently, interest has focused on optimization of SOFC cell performance at reduced (<800 °C) operation temperature. This will allow use of less-costly materials for cell interconnect and system components [3,4]. One approach to accomplish this lower temperature operation uses reduced electrolyte layer thicknesses of 5-10 μm [5].

Of the many processes suggested for HVM of SOFC's [6], tape casting, screen printing, electrochemical vapor deposition (EVD), thermal spraying and RF sputtering are the most widely employed albeit in small scale settings. In the absence of HVM expertise the challenge becomes predicting economic viability of a process in a cost-challenged high volume manufacturing conditions.

METHODOLOGY

The unit cell analyzed within this cost model consists of a 1 mm thick tape-cast anode. The electrolyte thickness is varied according to the performance requirements and can be either tape cast or screen printed. The 50 μm thick cathode uses the same process as the corresponding electrolyte layer. The cells are co-fired in a batch process. The area of such a cell is assumed to be 10 cm by 10 cm. Interconnect material and processing was not incorporated into the model at this early stage.

The modeling effort consists of three steps: 1) the development of a device performance model to calculate the required film thickness for a given operating temperature, maximum power density and performance tolerances for each of these parameters, 2) the calculation of the process yield at each layer for a given process at the required film thicknesses tolerances and 3) synthesis of these into the overall cost to produce a cell stack.

Device Performance Model

The dependence of the film thickness on operating temperature is derived from a general polarization model of the cell voltage as a function of the current density. This model includes corrections for ohmic losses as well as anode activation and concentration losses [7]. Each of these corrections is discussed separately in the literature [7-9] and is integrated here into the expression for the cell voltage. Cathode effects are neglected due to cell geometry in anode and electrolyte supported devices as supported by refs [7-9]. This results in an equation for power density with respect to current density and temperature as shown in equation 1 below. This performance model is compared to experimentally determined maximum power density results from the literature for YSZ in Figure 1 [7-13]. Correlation to the literature results is very good throughout the range of power densities.

$$P(i) = i*V(i) = i*[E_o - iR_i - a - b\ln(i) + \frac{RT}{2F}\ln(1 - \frac{i}{i_{as}}) - \frac{RT}{2F}\ln(1 + \frac{p_{H2}^0 i}{p_{H2O}^0 i_{as}})] \tag{1}$$

where P = power density (W/cm^2), i= current density (A/cm^2),
   i_o= effective exchange current density(A/cm^2), V= Voltage (Volts),
   E_o= open circuit voltage (Volts), R= gas constant (J/mol deg),
   T=Temperature (K), F= Faraday constant (C/mol),
   a = $-RT/4\alpha F * \ln i_o$ b = $-RT/4\alpha F$
   p_{H2}^o = partial pressure of hydrogen at the anode/electrolyte interface (atm)
   p_{H2O}^o = partial pressure of water vapor in the fuel (atm)
   p_{O2}^o = partial pressure of oxygen in the oxidant (atm)

$$R_i = \text{area specific resistance of the electrolyte (Ohm cm^2)} = R_i = R_{el} + R_{ct}^{eff} = \frac{l_e}{\sigma_e} + R_{ct}^{eff}$$

where $R_{ct}^{eff} = \sqrt{\dfrac{BR_{ct}}{\sigma_e(1-V_v)}}$

i_{as}=anode limiting current density (A/cm^2) = $\dfrac{2Fp_{H2}^o D_{eff.a}}{RTl_a}$

   $D_{eff.a}$ = effective diffusion coefficient on the anode side cm^2/s
   l_a = anode thickness, cm l_e = electrolyte thickness, μm
   R_{ct} = intrinsic (area specific) charge transfer resistance, $(Ohm\ cm^2)$
   σ_e = ionic conductivity of the electrolyte (S/cm) V_v=layer porosity
   B=microstructural dimension (grain size of material) (um)

Device operation temperature, power density and a nominal anode thickness are entered into the performance model. Layer thickness tolerances are calculated using this model. The process yield is then calculated for each layer based on this variation using the process tolerance models as outlined in the next section.

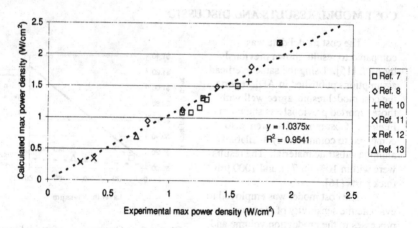

Figure 1. Comparison of calculated maximum power density to published experimental power density results.

Process Tolerance Models

For a given film deposition process, the film deposition rate will vary across the deposition surface. This variation results in thickness variation which can be measured by external measurement of film thickness across the surface. From these measurements, a standard deviation, or film thickness tolerance, at the target film thickness can be determined.

The cell performance models use nominal, minimum and maximum layer thicknesses to determine process yield for each layer. The standard deviation is calculated at the nominal thickness assuming a normal distribution. The probabilities for the minimum and maximum film thicknesses are calculated based on the nominal thickness and process standard deviation values. These probabilities are converted to a percentage upper and lower yield loss for each layer.

Process Based Cost Model

A process-based cost model (PBCM) maps a process and its operating conditions to cost [14]. Materials and process information are used to build up the manufacturing cost for anode supported SOFC's processed by either tape-casting for all three layers or tape-casting of the anode and screen printing of the other two layers. The parameters of the model are based on information from the literature augmented by the experience of the authors and specific data obtained from suppliers.

The model considers slurry preparation, film deposition, and co-firing in a batch process. The material choices are kept constant, as is the price of the materials. The expected variation of yield is estimated based on the process described above and adjusted to values based on experience with tape-casting of conventional materials. Materials cost and yield will greatly affect the final cost and further refinement is needed in this area. At this initial stage disposal costs have been ignored. This should be incorporated in the future.

COST MODEL RESULTS AND DISCUSSION

The cost model data was compared to results shown previously by ADL [15]. Using the same overhead assumptions published in ADL reports, the cost model results agree well with those reported previously as shown in Figure 2. Cost model data were also compared to commercially available alumina substrate material. The results were within 10% for 711 and 1000 μm thick parts [16].

The cost model was employed to evaluate the sensitivity of the two processes to the production volume and to the performance parameters of the cell. For these studies the assumptions were changed from those employed by

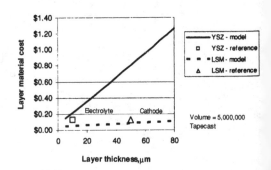

Figure 2. Comparison of model results with data from the literature [15].

ADL. An example is that the number of shifts worked was reduced from 21 to 10 per week. Sensitivity to production volume is shown in Fig. 3. Both graphs show a rapid decrease in cost as the production volume increases from 50,000 to 150,000 units per year. The slight increase in cost at about 180,000 units/year is due to additional equipment. Materials cost dominates the total cost for high production volumes. It is expected that this cost will decrease as high volume materials pricing takes effect.

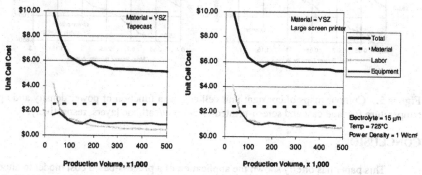

Fig. 3. Total cost and cost elements versus production volume for tape-casting or screen printing of electrolyte/cathode layers

Figure 4 shows total cost and cost elements for the electrolyte layer as a function of thickness of the layer. The added material dominates the cost increase. Higher costs for thin layers are due to reduced yields. Labor cost for screen printing is significantly larger, but the equipment cost is less.

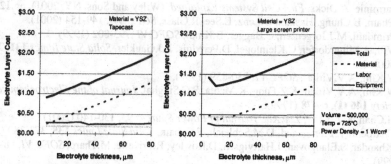

Figure 4. Total cost and cost elements for electrolyte layers of varying thickness.

Figure 5 shows constant cost contours per unit cell as a function of operating temperature and power density for tape casting and screen printing. A decrease in operating temperature requires a decrease in electrolyte thickness in order to maintain the same power density. This decrease in thickness results in a decrease in unit cell cost as shown in these figures. Over the performance ranges of interest, manufacturing costs range from $4.54 to $16.27 per unit cell with tape casting and $4.71 to $ 15.35 per unit cell for screen-printing.

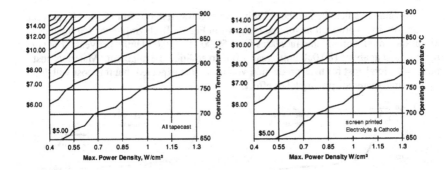

Figure 5. Contour maps of constant unit cell cost as a function of power density and operating temperature for tape-cast and screen-printed electrolyte/cathode layers, respectively.

CONCLUSIONS

This paper has briefly shown the application of a process-based cost model to anode-supported SOFC architectures. By incorporating engineering models of cell performance, it has been possible map out expected production cost and preferred technology for specific cell characteristics.

REFERENCES

1. *SOFC VI*, 127-640 (1999).
2. *SOFC VII*, 275-769, 875-1089 (2001).
3. J.Larminie, A.Dicks, *Fuel Cell Systems Explained*, (Wiley and Sons, NY, 2001), pp.124-180.
4. A.Pham, B.Chung, Haslam, D.Lenz, E.See, R.Glass, *SOFC VII*, 149-154 (2001).
5. S.Primdahl, M.J.Jorgensen, C.Bagger, B.Kindl, *SOFC VI* ,793-802 (1999)
6. J.Will, A.Mitterdorfer, C.Kleinlogel, D.Perednis, L.J.Gaukler, *Solid State Ionics* **131**, 79-96 (2000).
7. J.W.Kim, A.V.Virkar, *SOFC VI*, 830-839 (1999).
8. J.W.Kim, A.V.Virkar, K.Z.Fung, K.Mehta, S.C.Singhal, *Journal of the Electrochemical Society* **146 (1)**, 69-78 (1999).
9. S.H.Chan, K.A.Khor, Z.T.Xia, *Journal of Power Sources* **93**, 130-140 (2001).
10. P.Charpentier, P.Fragnaud, D.M.Schleich, E.Gehain, *Solid State Ionics* **135**, 373-380 (2000).
11. A.Khandar, S.Elangovan, J.Hartvigsen, D.Rowley, R.Privette, M.Tharp, *SOFC VI*, 88-94 (1999).
12. A.V.Virkar, J.Chen, C.W.Tanner, J.W.Kim, *Solid State Ionics* **131**, 189-198 (2000).
13. K.Okamura, Y.Aihara, S.Ito, S.Kawasaki, *Journal of Thermal Spray Technology* **9(3)**, 354-359 (2000).
14. J.P.Clark, R.Roth, F.R.Field, *Techno-economic issues in materials selection*, in: Materials Selection and Design, 1997, PP. 256-265.
15. http://www.netl.doe.gov/publications/proceedings/01/seca/adlstack.pdf
16. Coors Tek, Personal communication.

AUTHOR INDEX

SUBJECT INDEX

Printed in the United States
By Bookmasters